Outskirts of Galaxies

Astrophysics and Space Science Library

More information about this series at http://www.springer.com/series/5664

Johan H. Knapen • Janice C. Lee •
Armando Gil de Paz
Editors

Outskirts of Galaxies

Springer

Editors
Johan H. Knapen
Inst de Astrofísica de Canarias
San Cristobal de la Laguna, Spain

Janice C. Lee
Space Telescope Science Institute
Baltimore, USA

Armando Gil de Paz
Dept. Astrofisica
Universidad Complutense de Madrid
Madrid, Spain

ISSN 0067-0057 ISSN 2214-7985 (electronic)
Astrophysics and Space Science Library
ISBN 978-3-319-56569-9 ISBN 978-3-319-56570-5 (eBook)
DOI 10.1007/978-3-319-56570-5

Library of Congress Control Number: 2017942746

Cover illustration: This image of the spectacular galaxy NGC 4725 shows evidence of dust lanes in the area surrounding its active nucleus, a bright ring-like region of star formation, and outer spiral arm structure. Credit: DAGAL, Nik Szymanek, SDSS, and S^4G, www.dagalnetwork.eu.

Printed on acid-free paper

This Springer imprint is published by Springer Nature
The registered company is Springer International Publishing AG
The registered company address is: Gewerbestrasse 11, 6330 Cham, Switzerland

Preface

The outskirts of galaxies are mostly unexplored territory. Great advances have been made in particular in studying the star formation (through UV imaging) and gas (HI radio emission), but exploration of the stellar component, observed through optical and infrared imaging, can be considered to be still in its infancy. Yet the outskirts are key to understanding how galaxies form and evolve and how they have diversified into a class of objects with the wide range of morphologies of properties that we observe today. Their importance stems from two facts: the timescales in the outer regions are long, and stellar and gas densities are low. Both lead to slower evolution in the outskirts, implying that we are observing conditions at an earlier state there relative to the denser inner regions of galaxies which have been observed traditionally. In addition, accretion of pristine gas happens in the outskirts, stars are thought to migrate outwards and the material in the outer regions, when seen projected against the emission from background quasars, yields important information about the properties of the interstellar and intergalactic medium.

This volume brings together the views and insights of a number of world-renowned experts in this field, who have written a total of ten chapters summarizing our current knowledge of the outer regions of galaxies, as well as their views on how this field is likely to evolve in the near future. The topics described in detail range from studies of the structure and star formation history of our own Milky Way and the nearest external galaxies on the basis of individual star counts, via the deepest possible imaging of the integrated light of galaxies, to the study of the outskirts of galaxies at cosmological distances through the study of the light of background quasars passing through their outer regions. Other reviews discuss recent observations of molecular and atomic gas in the outskirts of galaxies and what we can learn from those about topics as varied as the current and past star formation and the shape of the dark matter haloes. Observed metallicities and their radial gradients are discussed in the context of chemical composition and star formation in the outskirts, touching on mechanisms such as metal-rich infall and metal mixing in disks. Stellar migration in galaxies is discussed in detail, paying particular attention to how observations and theoretical insights are improving our understanding of galaxy evolution, as is star formation in the outskirts of galaxies, which shines a new light not just on the properties of the outer regions but also on the process of star formation itself.

Our knowledge of the outer regions of galaxies is rapidly improving because new data are now enabling detailed study at a variety of wavelengths and with a variety of techniques. As the authors of this volume discuss, the next generation of telescopes and instruments will accelerate the exploration of galaxy outskirts, which will without any doubt lead to breakthroughs in our understanding of how galaxies have formed and evolved. We hope that this collection of reviews will provide a resource for a full range of workers in the field—expert investigators in theory and observation, those intrigued by recent discoveries who wish to learn more and students and other researchers who are interested in entering this exciting field.

Acknowledgements

The current volume owes its existence to the research programme developed in the context of the Initial Training Network Detailed Anatomy of Galaxies (DAGAL), funded through the People Programme (Marie Curie Actions) of the European Union's Seventh Framework Programme FP7/2007–2013/ under REA grant agreement number PITN-GA-2011-289313, and to the partnership with Springer developed within that network. Editors and authors of this book met and developed ideas during the International Astronomical Union Symposium number 321, held in March 2016 in Toledo, Spain, and thank the organizers of that meeting for bringing them together in such a beautiful and stimulating environment.

San Cristobal de La Laguna, Spain — Johan H. Knapen
Baltimore, MD, USA — Janice C. Lee
Madrid, Spain — Armando Gil de Paz
December 2016

Contents

Chapter 1
Outer Regions of the Milky Way

Francesca Figueras

Abstract With the start of the *Gaia* era, the time has come to address the major challenge of deriving the star formation history and evolution of the disk of our Milky Way. Here we review our present knowledge of the outer regions of the Milky Way disk population. Its stellar content, its structure and its dynamical and chemical evolution are summarized, focussing on our lack of understanding both from an observational and a theoretical viewpoint. We describe the unprecedented data that *Gaia* and the upcoming ground-based spectroscopic surveys will provide in the next decade. More in detail, we quantify the expected accuracy in position, velocity and astrophysical parameters of some of the key tracers of the stellar populations in the outer Galactic disk. Some insights on the future capability of these surveys to answer crucial and fundamental issues are discussed, such as the mechanisms driving the spiral arms and the warp formation. Our Galaxy, the Milky Way, is our cosmological laboratory for understanding the process of formation and evolution of disk galaxies. What we learn in the next decades will be naturally transferred to the extragalactic domain.

1.1 Introduction

The kinematical and chemical characterization of the stellar populations in the outer regions of the galactic disks is a crucial and key element to understand the process of disk formation and evolution. These outer regions are areas in the low-density regime and thus hard to observe in external galaxies. It is in this context that our Galaxy, the Milky Way, can truly be a laboratory and an exceptional environment to undertake such studies. The disk formation in our Milky Way was an extended process which started about 10 Gyr ago and continues to the present. Throughout this evolution, time-dependent dynamical agents such as radial migration or resonant scattering by transient or long-lived structures have been driving the orbital motion of the stars. Concerning star formation and evolution,

F. Figueras (✉)
Institute of Cosmos Science (IEEC-UB), University of Barcelona, Barcelona, Spain
e-mail: cesca@fqa.ub.edu

J.H. Knapen et al. (eds.), *Outskirts of Galaxies*, Astrophysics and Space Science Library 434, DOI 10.1007/978-3-319-56570-5_1

key factors such as the initial mass function and the star formation history are fundamental ingredients to describe the growth of disks. Furthermore, it has to be kept in mind that the process of disk formation also involves agents which are not yet well understood, from the dynamical influence of the puzzling three-dimensional structure of the dark matter halo to the characterization of the infalling gas which gradually builds up the disk and forms stars quiescently. The scenario becomes even more complex when other decisive factors such as mergers or gravitational interactions with satellites come into play.

Two approaches have usually been considered. From the extragalactic point of view, disks at different redshifts can be studied. This approach is limited to global information integrated over the disk stellar populations but has the advantage of tracing the evolution of disk properties with time. In a second approach, normally referred to as galactic archaeology (the approach used for the Milky Way), the disk evolution is reconstructed by resolving the stellar populations into individual stars. Disk evolution is fossilized in the orbital distribution of stars, their chemical composition and their ages. A drawback of this approach is that this information may be diluted through dynamical evolution and radial mixing. Tracers can be stars, open clusters and/or gas. In this Chapter we will concentrate on the stellar component. Molecular gas is described by Watson and Koda (2017). Nonetheless, when studying the Milky Way, we should not forget the perspective that in several aspects, our Galaxy is not a typical late-type spiral galaxy. The unusually quiescent merger history of the Milky Way has been discussed in detail by van der Kruit and Freeman (2011), who remark

> ...the unique possibility to make very detailed chemical studies of stars in the Milky Way provides an independent opportunity to evaluate the merger history of our large disk galaxy...

In Sect. 1.2 we describe the current knowledge of the stellar content in the outer regions of the Galactic disk. Section 1.3 deals with the non-axisymmetric structures that most contribute to the dynamics of these regions. Later on, in Sect. 1.4, we introduce some of the essential pieces towards a future chemo-dynamical model of the Milky Way. Finally, in Sect. 1.5, we present a brief overview of the upcoming new astrometric and spectroscopic data, both from the *Gaia* space astrometry mission and from ongoing and future ground-based spectroscopic surveys.

1.2 The Outer Disk of the Milky Way: Stellar Content

In this Section we first describe some of the best tracers of the structure and kinematics of the outer regions of the Milky Way. Very useful tools such as the Besançon stellar population synthesis model (Robin et al. 2014; Czekaj et al. 2014) allow us to quantify the number of field stars belonging to each tracer population (thin and thick disk or halo spheroid in the case of the outer regions). The key ingredients used in the strategy to generate simulated samples are shown in Fig. 1.1

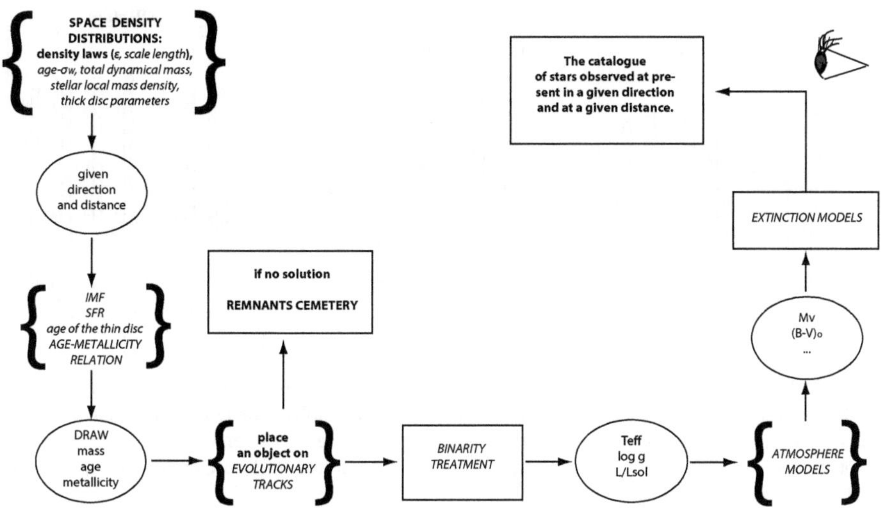

Fig. 1.1 General scheme describing the Besançon Galaxy model ingredients (Czekaj et al. 2014). Credit: M. Czekaj, et al. A&A, 564, 3, 2014, reproduced with permission © ESO

(from Czekaj et al. 2014). Although the model is continuously being updated, caution has to be taken when analysing the properties of the generated samples. The model includes several critical assumptions such us the radial scale length or the disk cut-off, which can induce significant discrepancies between model and observations. A separate paragraph is devoted to Cepheid variables, whose high intrinsic brightness makes them excellent tracers of the radial metallicity gradient (Lemasle et al. 2013). Properties of other very good tracers such as open clusters are not treated here. The reader can find an updated characterization of this population in recent papers such as the one by Netopil et al. (2016). A summarizing discussion on the present knowledge of the outer reaches is presented in Sect. 1.2.2.

1.2.1 Resolved Stellar Populations

Stellar Tracers As an example, in Fig. 1.2 we show the radial galactocentric distribution of three different stellar populations: Red Clump K-giant stars and main-sequence A- and OB-type stars. The estimated number of stars that *Gaia* will observe in the outer Milky Way region, i.e. at galactocentric radii between 9 and 16 kpc, is shown there. These estimates come from the work of Abedi et al. (2014) where test particles were generated fitting the stellar density at the position of the Sun, as provided by the Besançon galaxy model (Czekaj et al. 2014). In this case, the three-dimensional Galactic extinction model by Drimmel et al. (2003) has been used. From this model we estimate that the visual extinction, A_V, in the mean does not exceed $\sim$2 magnitudes when looking towards the Galactic anticentre

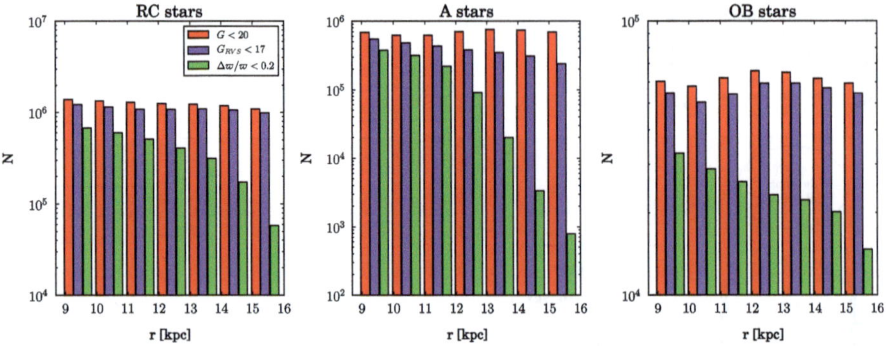

Fig. 1.2 Histograms of the numbers of stars in galactocentric radius bins of 1 kpc to be observed by the *Gaia* satellite (see Abedi et al. 2014 for details). The samples with *Gaia* magnitudes $G \leq 20$, $G_{RVS} \leq 17$ and those with expected parallax accuracy less than 20% (end of mission) are shown, respectively, in *red*, *purple* and *green*. The histograms are plotted for Red Clump stars (*left panel*), A stars (*middle*) and OB stars (*right*)

at low galactic latitudes. These low values for the interstellar extinction and the large number of stars to be fully characterized by *Gaia* and the future spectroscopic surveys (see Sect. 1.5) encourage the studies of resolved stellar populations towards the Galactic anticentre and the outer regions of the Milky Way.

Classical and Type II Cepheids These pulsating intrinsically bright variable stars are excellent tracers of the extent of the thin/thick disk surface densities and also of the abundance distribution of numerous chemical elements. Whereas classical Cepheids are young stars (<200 Myr; Bono et al. 2005) associated with young stellar clusters and OB associations, Type II Cepheids, with some characteristics similar to the classical ones (e.g. period and light curve), are fainter and much older, although their evolutionary status is still not firmly established. Both are interesting targets for the outer disk. Classical and Type II Cepheids can probably be associated with the thin and thick disk populations, respectively. Cepheids have been observed at radial galactocentric distances up to $R = 18$ kpc. Furthermore, as reported by Feast et al. (2014), a few classical Cepheid stars have been found at approximately 1–2 kpc above the plane in the direction of the Galactic bulge, at distances 13–22 kpc from the Galactic centre. Their presence, far from the plane, suggests that they are in the flattened outer disk and thus are excellent tracers of this at present very unknown structure.

1.2.2 The Outer Reaches

The Disk Cut-Off An apparent and sudden drop was reported by van der Kruit and Searle (1981) in the surface brightness of several edge-on galaxies at a radius of about four disk scale lengths. This has been a long-standing question both in external galaxies and in our Milky Way. Nowadays, when looking at external galaxies, one plausible explanation is that the reported edges are, in fact, inflections in the stellar density, i.e. breaks in the exponential density profiles (Bland-Hawthorn and Gerhard 2016). Does our Milky Way have such a truncation? A first analysis using deep optical star counts at a low-extinction window in the Galactic anticentre direction showed a clear signature of a sharp cut-off in the star density at about 5.5–6 kpc from the Sun (Robin et al. 1992). Later on, Momany et al. (2006), using 2MASS data, inferred robust evidence that there is no radial disk truncation at $R = 14$ kpc. More recently Minniti et al. (2011), using the UKIDSS-GPS and VVV surveys, pointed out that there is an edge of the stellar disk at about $R_{GC} = 13.9 \pm 0.5$ kpc along various lines of sight across the galaxy. When analysing these data, it has to be kept in mind that changes in the star counts induced by the warp and flare may not be negligible. New data seem to disagree with a sharply truncated nature, and proof of this is the presence of stars and even star formation regions beyond the break radius. López-Corredoira and Molgó (2014), using also star count techniques and Sloan Digitized Sky Survey (SDSS) data, quantified the change of the vertical scale height with galactocentric radius of the Galactic thin and thick disks. The presence of the Galactic flare, quite prominent at large R, can explain the apparent depletion of in-plane stars that is often confused with a cut-off at $R \sim 14$–15 kpc. Furthermore, in a recent paper, Carraro et al. (2016) studied the spatial distribution of early-type field stars and open clusters in the third Galactic quadrant of the Milky Way. Their Fig. 12 summarizes the spatial distribution of the young population. According to these authors, the field star sample extends up to 20 kpc from the Galactic centre, with no indication of the disk cut-off truncation at 14 kpc from the Galactic centre previously postulated by Robin et al. (1992).

The Monoceros Ring and Beyond The Monoceros Ring, a coherent ring-like structure at low Galactic latitude spanning about 100° and discovered by Newberg et al. (2002), thanks to the SDSS, deserves special attention. This structure was first identified as an overdensity of stars at ~10 kpc, with a metallicity in the range $-1 <$ [Fe/H] < 0 (with a substantial scatter). It is a poorly understood phenomenon with no clear association to other structures such as the Canis Major or the Triangulum-Andromeda overdensity. Recent observations from PAN-STARSS1 (Slater et al. 2014) indicate a larger extent of the stellar overdensity, up to $|b| \sim 25$–35° and about 130° in Galactic longitude. Its origin and gravitational interaction with the Milky Way are unclear (see Sect. 1.3.3). More recently, Xu et al. (2015), using SDSS data, have reported the existence of an oscillating asymmetry in the main-sequence star counts on either side of the Galactic plane in the anticentre region. These stellar overdensities, identified in a large Galactic longitude range, $[110°, 229°]$, are oscillating above and below the plane and have been observed up to distances of about 12–16 kpc from the Sun. As shown in Fig. 1.3, the three more distant asymmetries seem to be roughly concentric rings, open in the direction of the Milky Way spiral arms. The Monoceros Ring is identified as the northernmost of these structures, and the others are the so-called Triangulum-Andromeda overdensities (first detected by Rocha-Pinto et al. 2003), which could extend up to at least 25 kpc from the Galactic centre.

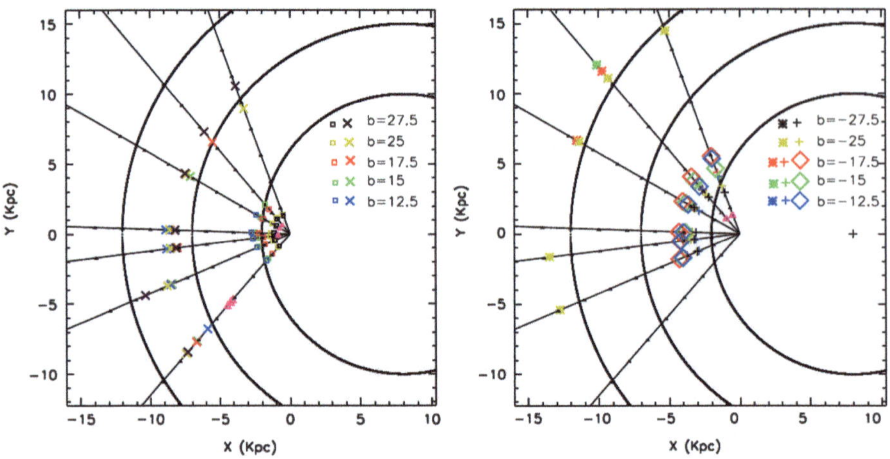

Fig. 1.3 Stellar overdensities above (*left*) and below (*right*) the plane reported by Xu et al. (2015), using SDSS data. The (X, Y) coordinates are centred at the position of the Sun

1.3 The Milky Way Outer Disk: Structure and Dynamics

Here we will focus on those non-asymmetric structures that most contribute to the dynamics of the outer regions of the Milky Way. For an exhaustive and updated description of other components such as the Galactic bar or inner spirals, the reader is referred to the recent review by Bland-Hawthorn and Gerhard (2016).

1.3.1 Spiral Arm Impact on Disk Dynamics and Structure

As is well known, spiral arms have a great impact on the evolution of galactic disks, from driving the formation of massive clouds to the perturbation of the stellar orbits of the old populations. Phenomena such as resonant trapping at corotation and Lindblad resonances, streaming motion and radial migration (Sellwood and Binney 2002) can be understood in this context. Furthermore, the impact of the evolution of these non-axisymmetric structures on the current radial chemical gradient of the Galactic disk or on the age-metallicity relation cannot be understood without considering spiral arms (see Sect. 1.4).

Accurate kinematic data are, without doubt, a first requirement to distinguish between spiral structure theories and thus the mechanisms of formation and evolution of spiral arms. Some of the current theories under investigation are the density wave theory (Lin and Shu 1964), with a rigid spiral density wave travelling through the disk; swing amplification (Toomre 1981; Masset and Tagger 1997), with perturbations on the disk that can be swing amplified; invariant manifolds (Romero-Gómez et al. 2006; Athanassoula 2012), with manifolds originated in the periodic orbits around the equilibrium points; external interactions (Sellwood and Carlberg 1984), for a certain fraction of mass accretion; or chaotic orbits (Voglis et al. 2006; Patsis 2006), important for a large perturbation, especially near corotation. It is difficult to test these theories with current available data, and complicating the situation even more, several of them may coexist. As an example, the invariant manifold mechanism has been tested using N-body simulations of barred galaxies (Athanassoula 2012; Roca-Fàbrega et al. 2013). A detailed characterization of the motion of the particles through the arms or crossing them is required (Antoja et al. 2016). It is clear that an accurate knowledge of the kinematics is mandatory to distinguish between theories (see Sect. 1.5).

Recently, Monguió et al. (2015) published the first detection of the field star overdensity in the Perseus arm towards the Galactic anticentre. Using a young population of B- and A-type stars, the authors placed the arm at 1.6 ± 0.2 kpc from the Sun, estimating its stellar density amplitude to about 10%. Moreover, these authors show how its location matches a variation in the dust distribution congruent with a dust layer in front of the arm. This favours the assumption that the Perseus arm is placed inside the corotation radius of the Milky Way spiral pattern. The obtained heliocentric distance of the Perseus arm is slightly smaller than the 2.0 kpc recently proposed by Reid et al. (2014). These authors used VLBI trigonometric parallaxes and proper motions of masers, thus star-forming regions, to accurately locate many arm segments in the Galactic disk. The outer spiral arm seems to be located at 13.0 ± 0.3 kpc from the Galactic centre, with an amplitude 0.62 ± 0.18 kpc and a pitch angle of $13.8 \pm 3.3°$. Another critical issue is the age dependence of the radial scale length. Several determinations can be found in the literature. Recently, Monguió et al. (2015) derived it using the Galactic disk young population. They obtained values of 2.9 ± 0.1 kpc for B4-A1 type stars and 3.5 ± 0.5 kpc for B4-A0 stars. Unfortunately, the uncertainty associated with these data still prevents a direct application of these results in discriminating between inside-out or out-inside formation scenarios.

1.3.2 The Galactic Warp and Flare

It is widely accepted that warps of disk galaxies are a common phenomenon (as common as spiral structure), yet warps are still not fully understood (see García-Ruiz et al. (2002) for a historical review). From the time when the first 21 cm observations of our Galaxy became available, the large-scale warp in the HI gas disk has been apparent (Burke 1957; Westerhout 1957). More than 50 years later, Levine et al. (2006) re-examined the outer HI distribution, proposing a more complex structure with the gaseous warp well described by two Fourier modes. The warp seems to start already within the Solar circle. Reylé et al. (2009), using 2MASS infrared data, found the stellar component to be well modelled by an S-shaped warp with a significantly smaller slope that the one seen in the HI warp. Since then, several authors have tried to set up the morphology of the Galactic warp, reaching

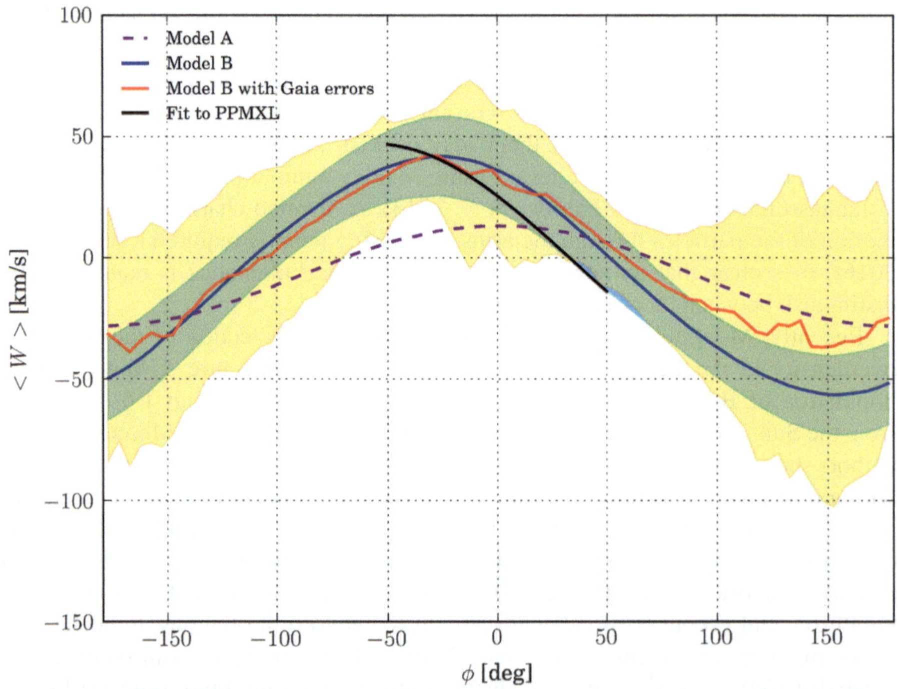

Fig. 1.4 The effect of the Galactic warp on the distribution of the mean heliocentric vertical velocity component (W) as a function of galactocentric azimuth for Red Clump stars in the galactocentric ring $13 < R < 14$ kpc. Two warp models (A and B) described in López-Corredoira et al. (2014) are plotted here (*dashed purple* and *solid blue lines*) and compared to the best fit to PPMXL proper motion data (in *black*). See López-Corredoira et al. (2014) for details

no clear conclusion. More importantly, the current uncertainties do not allow us to disentangle which mechanisms can explain it. Critical issues are the kinematics of the warped population and basic properties such as its stellar age dependence. Data available currently (PPMXL proper motions) allowed López-Corredoira et al. (2014) (see Fig. 1.4) to perform a vertical motion analysis of the warp towards the Galactic anticentre. They point out that whereas the main S-shaped structure of the warp is a long-lived feature, the perturbation that produces an irregularity in the southern part is most likely a transient phenomenon. Again, this is a complex kinematic feature from which a definitive scenario is difficult to constrain. We refer to the recent work of Abedi et al. (2014) to know the capabilities of *Gaia* in detecting and characterizing the kinematic properties of the warp. Accurate proper motions and radial velocities (see Sect. 1.5) will certainly add a new dimension to this study.

Lozinskaya and Kardashev (1963) were the first to describe the observational evidence of a flare in the HI gas in the Milky Way. Since then, the question of whether the stellar component takes part in this flaring has been a matter of debate. Momany et al. (2006) presented a first comparison of the thickness of the stellar

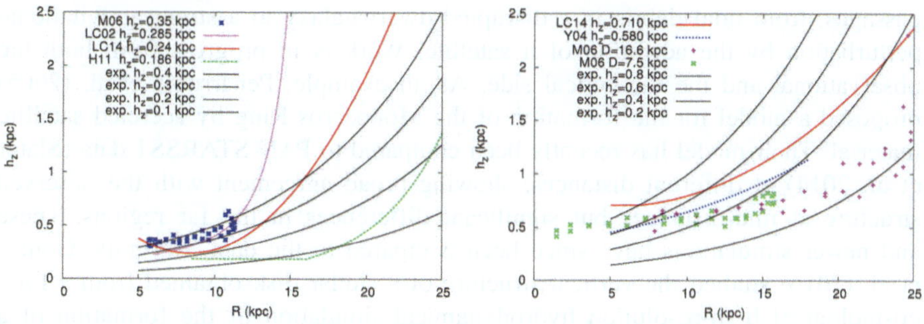

Fig. 1.5 Comparison by Kalberla et al. (2014) of the current observational data (*points*) and exponential scale height fits (*continuous lines*) to quantify the flare of the Galactic thin (*left*) and thick (*right*) components. See details in their paper

disk, neutral hydrogen gas layer and molecular clouds. The stellar flare was traced using Red Clump and red giant stars. These authors found that the variation of the disk thickness (flaring)—treating a mixture of thin and thick stellar populations—starts at $R = 15$ kpc and increases gradually until reaching a mean scale height of $\sim$1.5 kpc at $R = 23$ kpc. More recently, Kalberla et al. (2014) confirmed these results, showing strong evidence for a common flaring of gas and stars in the Milky Way. Several sources such as HI gas, Cepheids, 2MASS, SDSS and pulsar data show an increase of the scale height, growing with galactocentric radius (see Fig. 1.5). Even more, it has been proposed that flaring at large galactocentric distances could be stronger for the thin than for the thick disk. Although this is still a matter of debate, with *Gaia* we will have the opportunity not only to analyse the structure but also the kinematic properties of these outer populations.

Several models have been proposed to account for this flare. Whereas Kalberla et al. (2007) explored several mass models reflecting different dark matter distributions, others, such as Minchev et al. (2012), by preassembling *N*-body disks, showed that purely secular evolution could lead to flared disks. Others, like Roškar et al. (2010), proposed misaligned gas infall to provoke the flaring of the Galactic disk. Recent observational analysis of SDSS data points towards a smooth stellar distribution (López-Corredoira and Molgó 2014), thus supporting a continuous structure for the flare and not a combination of a Galactic disk plus some component of extragalactic origin.

1.3.3 Gravitational Interaction with Satellites

Several approaches have been followed in the last years to characterize the effects on the Galactic disks induced by dynamical perturbations by satellites (e.g. Purcell et al. 2011; Gómez et al. 2013, among others). All of them reinforce the picture that the Galactic disk can exhibit complex structure in response to close satellite passages, from tidal debris in a disrupted dwarf galaxy to a strong gravitational perturbation by the accretion of a satellite. Work is in progress from both the observational and the theoretical side. As an example, Peñarrubia et al. (2005) proposed a model for the formation of the Monoceros Ring by accreted satellite material. Their model has recently been compared to PAN-STARSS1 data (Slater et al. 2014) at different distances, showing broad agreement with the observed structure at mid-distances but significant differences in the far regions. These and newer simulations have since been compared to the data. Recently, Gómez et al. (2016) studied the vertical structure of a stellar disk obtained from a fully cosmological high-resolution hydrodynamical simulation of the formation of a Milky Way-like galaxy. The disk's mean vertical height can have amplitudes as large as 3 kpc in its outer regions as a result of a satellite-host halo-disk interaction. The simulations reproduce, qualitatively, many of the observable properties of the Monoceros Ring. Nonetheless, as pointed out by Slater et al. (2014), the crucial question for future simulations is whether such Monoceros Ring-like features can be created without causing such an unrealistically large distortion of the disk; maybe less massive satellites or particular infall trajectories could be more favourable.

1.3.4 Dynamics of the Vertical Blending and Breathing Modes

We currently have observational evidence of oscillations of the Milky Way's stellar disk in the direction perpendicular to the Galactic midplane. Some of these are possibly related to the Monoceros and Triangulum-Andromeda overdensities in the outer disk (Sect. 1.2.2), while others are more local, within $\sim$2 kpc of the Sun. Widrow et al. (2012), using SDSS/SEGUE, RAVE and LAMOST data, found a Galactic north-south asymmetry in the number density and bulk velocity of Solar neighbourhood stars which showed a gradual trend across the Galactic midplane and thus the appearance of a wavelike perturbation. This perturbation has the characteristics of a breathing mode, with compression and rarefaction motions and with the displacements and peculiar velocities having opposite signs above and below the plane. Widrow et al. (2014) demonstrate that both breathing and bending modes can be generated by a passing satellite or dark matter subhalo, with the nature of the perturbation being controlled by the satellite's vertical velocity relative to the disk (a slow-moving satellite would induce a bending mode, whereas higher vertical velocities would induce breathing modes). More recently, Monari et al. (2016) discussed that breathing modes can be induced by several effects such as bar or spiral perturbation, spiral instabilities or a possible bombarding by satellites. Even the existence of a dark matter substructure could play a role. These authors have derived explicit expressions for the full perturbed density function of a thin disk stellar population in the presence of non-axisymmetric structures such as spiral arms and bar. Undoubtedly, upcoming data will yield a more accurate and complete map of bulk motions in the stellar disk, up to about 3–4 kpc from the Sun (Sect. 1.5).

1.4 Towards a Chemodynamical Model of the Galactic Disk

Prantzos (2008, 2011) are two excellent and pedagogical papers to review the basic principles and hot topics for the development of galactic chemical evolutionary models. The state of the art is well summarized there. These models shall assume, among other factors, the evolution of several chemical elements up to the iron peak and the different compositions for the infalling material. Accurate observations are required to test these models, the products of which are usually expressed in terms of radial abundance gradients of several elements (C, N, O, Ne, Mg, Al, Si, S, Ar, Fe) and their time evolution. Thanks to the extent of present and future large surveys (see Sect. 1.5), any formation model must be able to account not only for local but also for radial and vertical large-scale chemical element distributions.

Radial migration has been firmly accepted to be an inseparable part of disk evolution in numerical simulations (Sellwood and Binney 2002). In radial migration models (e.g. Schönrich and Binney 2009), metal-poor stars born in the outer disk move inwards to the Solar neighbourhood, while metal-rich stars born in the inner disk migrate outwards. These authors suggest two mechanisms for this motion[1]: *blurring*, due to the scattering and subsequent increase of eccentricities over time, and *churning*, mostly triggered by resonant scattering at corotation. These effects would produce a large heterogeneity in the chemical abundance in the Solar neighbourhood and its environment.

1.4.1 Age-Metallicity-Kinematics Relations

It is well established that the velocity dispersion of the thin disk population increases with age, a well-known phenomenon often referred to as disk heating (Binney and Tremaine 1987). Aumer and Binney (2009) used a power law to fit the thin disk age-velocity relation (AVR) in the Solar neighbourhood, using the most accurate data at that time: the Hipparcos astrometry and photometric and spectroscopic data from the Geneva-Copenhagen survey (Holmberg et al. 2007). They favour continuous heating, but as proposed also by Seabroke and Gilmore (2007), a saturation at ages $\geq$ 4.5 Gyr could not be excluded. Concerning the age-metallicity distribution (AMD), it is important to mention the work done by Haywood (2006). Biases were evaluated in detail by this author, and a new AMD with a mean increase limited to about a factor of two in Z over the disk age was proposed. Again, it was emphasized that dynamical effects and complexity in the AMD clearly dominate. All the above relations have been established using data in the Solar neighbourhood. Nonetheless, and as pointed out by Casagrande et al. (2011), the Solar neighbourhood is not only

[1]Following the Schönrich and Binney (2009) terminology, churning implies a change of guiding-centre radii, while blurring means a steady increase of the oscillation amplitude around the guiding centre.

assembled from local stars but also from stars born in the inner and outer Galactic disk that migrated to their present positions (Schönrich and Binney 2009). New surveys (see Sect. 1.5) will open a new window and different approaches to these two key and fundamental relations.

In this direction, a new stellar chemo-kinematic relation has recently been derived by Minchev et al. (2014) using RAVE and deeper surveys such as SEGUE. Stars with [Mg/Fe] $\geq$ 0.4 dex show a peculiar kinematic behaviour. These authors used this index as a proxy of the stellar age to identify the oldest stars in the sample. These stars, born during the first years of the Galaxy's life, have velocity dispersions too large to be accounted for by internal disk heating. Minchev et al. (2014) showed that a chemo-dynamical model incorporating massive mergers in the early Universe and a subsequent radial migration of cool stars could explain the observed trends. More imprints such as the ones reported there should be expected in the outer regions of the Galactic disk, so this work is a good example of the chemo-kinematic relations that the combination of future *Gaia* spectroscopic surveys will provide. They will surely bring new constraints to the formation scenarios of galactic disks.

1.4.2 The Galactic Thick Disk

The characterization of the thick disk is an important milestone when trying to understand the assembly of disk galaxies. Despite the increasing amounts of observational data in our local environment, to date we have lacked the means to discriminate among different thick disk formation scenarios for the Milky Way. A first key and basic question which arises is: Do the thin and thick disks have a different origin? Two approaches are being used to answer this question, observational evidence and galaxy modelling. Observational evidence at present is, in some sense, highly inconsistent. Whereas Bovy et al. (2012), using SDSS data, clearly favoured a vertical structure composed of a smooth continuum of disk thicknesses with no discontinuity between the thin and thick disk, a bimodal distribution in the ([Fe/H],[α/Fe]) relation with two sequences of high and low [α/Fe] seems well established (Adibekyan et al. 2012; Nidever et al. 2014) with high [α/Fe] values more prominent in the inner disk and lower values dominating the outer parts. In this line, Recio-Blanco et al. (2014) point towards a clear kinematic distinction between thin and thick disk (two distinct populations). Concerning the structure, the thick disk occupies (in agreement with its hotter kinematics) a larger vertical volume around the midplane (e.g. Jurić et al. 2008), perhaps at a shorter scale length than the thin disk (Bensby et al. 2011; Robin et al. 2014) and a clear uncertainty in the radial metallicity gradient, as will become evident in Sect. 1.4.3.2. Nonetheless, as discussed by Bland-Hawthorn and Gerhard (2016), one of the strongest pieces of evidence of the existence of the thick disk will be a robust statistical confirmation of its unique chemistry (see the important work by Bensby et al. 2014 in this line).

Several models have been proposed to explain the existence of the thick disk: (1) a heating process of the Galactic disk due to satellite mergers (e.g. Abadi et al. 2003); (2) the formation of a puffed-up structure by mere radial migration (Sellwood and Binney 2002); (3) a so-called "upside-down" disk formation, where old stars were formed in a relatively thick component, while younger populations form in successively thinner disks (e.g. Bird et al. 2013); or (4) in situ formation by early accretion of gas. The scenario for the Milky Way thick disk is presently unclear. Several observational constraints have to be fixed. As an example, some of these models predict a vertical gradient in [α/Fe]; others do not. The in situ formation or the direct accretion of small satellites would be ruled out if α-gradients are observed (Recio-Blanco et al. 2014). Progress in this field is further discussed in Sect. 1.4.3.2. Other future constraints could be, as pointed out by B. Lemasle (private communication), the ratio of classical versus Type II Cepheids. The galactocentric radial change of these ratios, available from *Gaia*, would help us to analyse the extent of the thin/thick disk structures.

1.4.3 The Radial Abundance Gradients

Examples of two reference papers on chemical evolutionary models that show different but complementary approaches are those of Chiappini et al. (1997) and Pilkington et al. (2012a). Chiappini et al. (1997) developed a new model for the Galaxy assuming two main infall episodes for the formation of the halo-thick disk and thin disk. The model also predicts the evolution of the gas mass, the star formation rate, the supernova rates and the abundances of 16 chemical elements as functions of time and galactocentric distance. In summary, a long list of detailed model results can be constrained when compared with observational data. The approach of Pilkington et al. (2012a) is different and complementary. These authors used cosmological hydrodynamical simulations of dwarf disk galaxies to analyse the distribution of metals. Both approaches require the comparison of models and data and hence the selection of good stellar tracers, as discussed below.

The differences between the models concern effects such as the efficiency of the enrichment processes in the inner and outer regions and the nature of the material (primordial or pre-enriched falling from the halo onto the disk). These questions can help answer fundamental questions such as whether our Galactic disk has a flattening or a steepening radial metallicity gradient with time. Open clusters and Cepheids are proposed as good tools to derive the time evolution of the metallicity gradients as the need to answer this question is pressing. As an example, in the future WEAVE spectroscopic survey (Sect. 1.5), the requirement on the accuracy of the open cluster metallicity is 0.1 dex over the full age and metallicity range.

Tracers of Chemical Abundance Gradients As mentioned by Boissier and Prantzos (1999), all the existing observational data only inform us on the present properties of the Galactic disk, but not on its past history, except for the tentative

Table 1.1 Some of the ongoing and future ground-based surveys for chemodynamical studies of field stars in the outer regions of the Milky Way

Survey	Dates	Magnitude	Resolution	Trace populations
Gaia-ESO	2011–2016	$V \sim 16$–20	∼20,000	Red Clump giants
Gaia-ESO	2011–2016	$V < 16$	∼45,000	Thick disk population
LAMOST	2012–	$r \sim 19$–20	1800	North hemisphere
APOGEE	2011–	$H < 12.2$	22,500	Red giants and subgiants
EMIR	2017–	$J, K \sim 18$	4000	Red giants
WEAVE	2018–2022	$V = 16$–20	4000	Disk OBA and Red Clump giants
WEAVE	2018–2022	$V < 16$–17	20,000	Anti-centre OBA and RC disk
4MOST	2022–	$V \leq 20$	>2000	From Sun-like to RGBs

indications obtained from the study of abundances in planetary nebulae or in open clusters. Alibes et al. (2001) reviewed the main observables for the disk of the Milky Way, which can be compared with the products of a chemical evolutionary model. As emphasized by these authors, homogeneous samples are required to derive radial gradients. The young population (age <1 Gyr) provides information on the present-day abundances in the Galactic disk, the most important tracers being the HII regions and the B-type stars. HII regions have the drawback of large uncertainties due to the derivation of electron temperature. The B-type stars are bright, thus observed up to large distances, with a photospheric composition which supposedly reflects the composition of the material from which they formed, except for high rotators where there could have been some mixing of atmospheric and core material. Tracers older than 1 Gyr provide insight into the past evolution of the abundance gradients and thus into the evolution in abundance profiles. Among these are the planetary nebulae, with large uncertainties in distances, ages and masses; open clusters, which can be as old as 8 Gyr; and the FGK stars, dwarfs and giants, a population selected as tracer in most of the ongoing large spectroscopic surveys (see Table 1.1).

1.4.3.1 Cepheids as Tracers of Galactic Abundance Gradients

Since the early work by Harris (1981), the determinations of the Galactic [Fe/H] gradient from Galactic Cepheids have been remarkably consistent, all of them reporting values of ~ -0.06 dex/kpc (Lemasle et al. 2013). From open clusters, a linear gradient of approximately −0.06 dex/kpc has been proposed (e.g. Bragaglia et al. 2008) but with a flattening between 10 and 14 kpc, that is, in the outer regions of the Galactic disk. This possible flattening of the abundance gradients is a long-standing question. Whereas it is not observed from Galactic Cepheids (Lemasle et al. 2013), a clear flat (bimodal) gradient in the outer disk is derived from open clusters. Lemasle et al. (2013) measured the Galactic abundance gradient of 11 chemical elements using high-resolution spectra of more than 60 Galactic Cepheids and focussed their analysis on the outer Galactic disk. They confirm the existence of a gradient for all the heavy elements and noted that current data are not supporting a flattening of the gradient in the outer disk. Later on, the same team (Genovali et al. 2014) compiled homogeneous data for a sample of 450 Cepheids deriving a linear metallicity gradient over a broad range of galactocentric distances ($R \sim 5$–19 kpc). These authors pointed out clear evidence for the increase in the intrinsic scatter when moving towards the outer disk. It cannot be ruled out that this scatter is due to contamination by Type II Cepheids misclassified as classical Cepheids (Sect. 1.2.2).

1.4.3.2 Recent Outcomes of the *Gaia*-ESO Survey

From current surveys (SEGUE, RAVE, APOGEE, GALAH, *Gaia*-ESO Survey, among others), the large-scale characteristics of the thin and thick disks have started to emerge. It is difficult to summarize here the huge amount of new results published in the last 5–10 years. In this section, as it is impossible to make an exhaustive review, we will take one of them as an example, in this case the effort being done by the European *Gaia* research community (more than 300 scientists) to build the *Gaia*-ESO Survey (hereafter GES). The survey, planned for 5 years (2011–2016) and still ongoing, uses the FLAMES/GIRAFFE and UVES spectrographs on the Very Large Telescope (ESO), and important preliminary results have already been published. Two approaches have been followed to derive the vertical and radial trends of the chemical gradient, from open clusters and from FGK main-sequence field stars.

Open clusters are, apart from Cepheids (Sect. 1.4.3.1), the best tracers to derive metallicity over large ranges in galactocentric distance. However, accurate and homogeneous observations of open clusters are required to properly trace the chemical abundance distribution in the Galactic disk. Recently, Jacobson et al. (2016) showed that the GES open clusters at galactocentric distances $[5.5, 8]$ kpc exhibit an [Fe/H] radial metallicity gradient of -0.10 ± 0.02 dex/kpc, consistent with values obtained by Netopil et al. (2016) with a larger open cluster sample (172 open clusters) at distances $R = [6, 14]$ kpc compiled using different criteria. Both datasets are showing us no evidence for a steepening of the inner disk metallicity gradient inside the Solar circle. Furthermore, Netopil et al. (2016) compared open clusters and Cepheids with the Galactic chemical model by Minchev et al. (2013) (see Fig. 1.6). Open clusters show a much flatter gradient than Cepheids or these model predictions. Possible explanations proposed for this flattening are an underestimation of radial migration in the models or other not yet included mechanisms acting at large galactocentric radius. It is evident that more open cluster data are needed in the outer Galactic disk. Open clusters are introducing an important constraint on the chemical evolutionary models and their time evolution scenarios. Note that other independent data, in this case from APOGEE DR10 and with high-resolution metallicity measurements, reproduce interesting features such as the lack of gradient in the logarithmic α-elements-to-metal ratio, $[\alpha/M]$, across the range $7.9 < R < 14.5$ kpc (Frinchaboy et al. 2013).

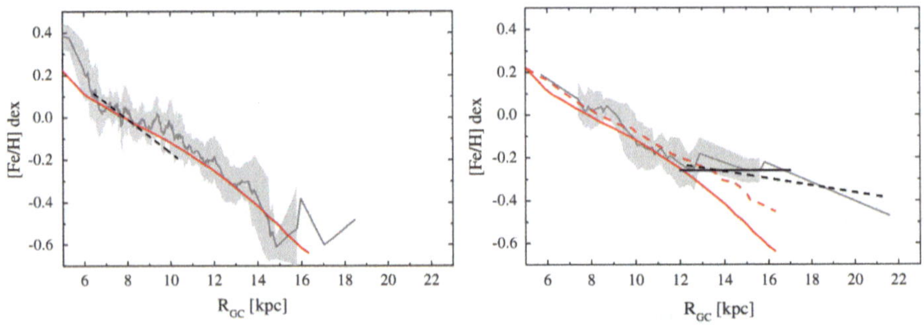

Fig. 1.6 Radial metallicity gradient derived from Cepheids and open cluster data compiled, respectively, by Genovali et al. (2014) and Netopil et al. (2016). The *left panel* compares the youngest populations of these tracers with models. The *grey line* and the *light grey area* show the running average and the spread in iron of the Cepheid data. The *black dashed line* shows the gradient derived from a subset of young open clusters with ages less than 0.5 Gyr. For comparison, the *red line* indicates the gradient for the gas from the Galactic chemical model by Minchev et al. (2014). The *right panel* presents the metallicity gradient derived from an intermediate population. The *grey area* shows the running average obtained from open clusters with ages between 0.5 and 3.0 Gyr. The *horizontal black solid line* is the mean value of those clusters located in the outer disk, with galactocentric distances larger than 12 kpc. The *black dashed line* is the gradient derived for all clusters in the outer disk. For comparison, the *red dashed line* shows the Galactic chemical model by Minchev et al. (2013) that includes the dynamical effects of the Galactic disk after 2 Gyr. This Figure is taken from Netopil et al. (2016); see this paper for details

Concerning field stars, the GES FGK dwarf and giant survey allows, for the first time, to extend high-resolution spectroscopic surveys for field stars further away from the Solar vicinity. Mikolaitis et al. (2014) published a first radial metallicity abundance gradient of eight elements (Mg, Al, Si, Ca, Ti, Fe, Cr, Ni and Y) from this survey. The bimodality observed in the [Mg/M] distribution was attributed to thick and thin disk populations. Both radial (4–12 kpc) and vertical (0–3.5 kpc) gradients in metallicity were derived for the thin and the thick disk separately. The interpretation of this huge amount of data is complex, with the thin disk showing a positive radial [α/M] gradient in contrast to the flat gradient derived for the thick disk population. Concerning vertical gradients, the thick disk hosts a shallower negative vertical metallicity gradient than the thin disk in the Solar cylinder. These results and upcoming new analyses of GES, APOGEE and SEGUE data, among others, are being examined in the context of new and complex models: from detailed analytic local disk models (e.g. Just and Rybizki 2016) to *N*-body simulations of Milky Way-like galaxies (e.g. Kawata et al. 2017; Roca-Fàbrega et al. 2016). In all cases, kinematic data will play a fundamental role here (see, e.g. Kordopatis et al. 2016).

1.4.4 The "Outside-In" Versus "Inside-Out" Disk Formation Scenarios

The comparison of observations and models involves global properties of the Milky Way such as rotation curves, scale lengths, masses (gas, stars and dark matter), gas flows and stellar abundances. Moreover, dynamical aspects also play a critical role. They can be analysed and taken into account through analytical or semi-analytical models or from cosmological N-body plus hydrodynamical models, the latter ones needed to assess the effect of the environment and satellite accretion on shaping the outer disks. Several efforts have been conducted up to now in this direction. As an example of the heterogeneity in the obtained results, we can mention the works of Cescutti et al. (2007), Pilkington et al. (2012b), Haywood et al. (2013) and Minchev et al. (2014). Whereas the first three proposed an inside-out disk formation scenario, Haywood et al. (2013) put forward an outside-in scenario to explain Solar neighbourhood data. These models are all only partially compatible with GES data (Sect. 1.4.3.2), and others have only been tested against external galaxy data (e.g. Pilkington et al. 2012a). The addition of upcoming new observational constraints to these models is complex. Although a general prediction of inside-out formation would be a fast decrease in [α/Fe] with increasing radius for stars in a narrow metallicity range, different slopes for the thin and thick disk population (GES results; see Mikolaitis et al. 2014) increase the complexity. This research field will be a key priority for the next decade. It will be possible to combine *Gaia* and spectroscopic data from the Milky Way with extragalactic work such as that done in the context of the CALIFA or MaNGA surveys.

1.5 Large Surveys in the Next Decade

The beginning of the twenty-first century is, without doubt, a new golden age for Galactic astronomy. Data coming from both space missions and ground-based surveys would be a dream in 1962, when Eggen, Lynden-Bell and Sandage derived the first model of galaxy formation and evolution (Eggen et al. 1962). Astrometric (*Gaia*, the Large Synoptic Survey Telescope, LSST) and spectroscopic ground-based surveys are fully complementary, providing both local and global measures of the six-dimensional phase-space distribution function, needed to encode valuable dynamical information. The status and products of the *Gaia* mission (2014–2022) are detailed in Sect. 1.5.1. In the coming years, more than 10^9 stars up to *Gaia* apparent magnitude $G \sim 20$ will have an astrometric accuracy never envisaged up to now. These data will be complemented with the data collected by the LSST. This telescope, expected to be fully operational in 2022, will measure proper motions with an accuracy of about 1 mas/year to a limit 4 mag fainter than the *Gaia* survey. Geometric parallaxes will also be provided by LSST, with an accuracy similar to that of the faintest stars in *Gaia* ($\sim$0.3 mas) at $V = 20$ and up to $\sim$3 mas at $V = 24$.

The acquisition of good signal-to-noise spectra at intermediate and high resolution is mandatory to complement these accurate astrometric data with good radial velocities and chemical abundances for a significant number of stars. In Table 1.1 we present a non-exhaustive list of the most promising spectroscopic surveys ongoing or planned for the near future. A detailed description of the capabilities of each of these surveys can be found in their respective web pages. Here, we will focus on one of them, the WEAVE[2] spectroscopic survey. This survey, to be undertaken in the Northern hemisphere, will open a new window to the Galactic anticentre and thus to the outskirts of the Milky Way.

WEAVE is a multiobject spectroscopic instrument using about 1000 fibres that will be operated at the William Hershel Telescope (WHT, Canary Islands). Its science case and survey plan are almost fixed. During the first 5 years of operation (2018–2022), several Galactic plane radial velocity surveys will be executed using as kinematic tracer disk Red Clump and young OBA-type stars. WEAVE's low-resolution mode ($R = 5000$) will provide radial velocities for millions of stars up to the *Gaia* limiting magnitude. We estimate to obtain radial velocities for about one million Red Clump stars with an accuracy better than $\sim$5 km/s. Expected accuracies for stars brighter than $V = 18$ are even better, with errors as small as 1–3 km/s. Complementarily, several chemical tagging programmes for Galactic archaeology will be undertaken using the high-resolution mode of WEAVE ($R = 20{,}000$). Both kinds of data are critical ingredients to establish the chemical gradients of the Galactic thin and thick disk (Sect. 1.4.3) and the dynamics of the non-axisymmetric components of the outer regions of the MW (Sect. 1.3). A promising survey of well-selected open clusters (ages >100 Myr) will significantly contribute to the derivation of the radial metallicity gradient from these excellent tracers in the outer regions of the Milky Way.

1.5.1 The Gaia *Mission*

The high-quality astrometric and photometric data already collected by *Gaia* during its first almost 2 years of successful scientific operation allow us to anticipate that, in a few years, *Gaia* will revolutionize our understanding of our own Galaxy, the Milky Way, and its surroundings. In this section we will briefly describe the *Gaia* products to be published in the first and end-of-mission data releases (end of 2022). More in detail, we will quantify the expected accuracy in position, transversal and radial velocity and stellar astrophysical parameters of some of the key stellar tracers of the outer regions of the Milky Way.

[2]See http://www.ing.iac.es/confluence/display/WEAV/The+WEAVE+Project.

***Gaia* Products** The *Gaia* data release scenario, based on the current status of data processing, is described in the *Gaia* web page.[3] The first *Gaia* data release (*Gaia* DR1) has been made public on September 14, 2016. This release, ready for the first *Gaia* science exploitation tasks, has provided to the open community the first 3D map of the Solar neighbourhood, with parallaxes and proper motions with unprecedented accuracy for more than two million Tycho sources. Estimated astrometric errors are discussed in detail by Michalik et al. (2015), and validation tasks indicate that these expectations are being accomplished. The accuracy in astrometric data for Hipparcos sources ($\sim 10^5$ stars) will improve by a significant factor. Parallax accuracy will be improved by a factor of 2–10 and proper motions by a factor of more than 25–30. For Tycho sources ($\sim 2 \times 10^6$ stars) we will have an astrometric accuracy as good as that published for the Hipparcos sources in 1997. It is not until the second *Gaia* data release (*Gaia* DR2) that the 22 months of mission operation will provide the five astrometric parameters (position, parallax and proper motions) for all sources up to $G \sim 20$ (only single-star behaviour). As described below, with this accuracy we will have good distances for some of our stellar tracers (see Sect. 1.2.1) up to 3–4 kpc from the position of the Sun. Also in this release, expected for late 2017, integrated photometry from the BP and RP instruments, basic stellar astrophysical parameters and mean radial velocities for bright sources will be published. The publication of BP/RP spectra (Jordi et al. 2010) for well-behaved objects, orbital solutions for binaries and object classification and astrophysical parameters will need to wait until 2018–2019. The final release is expected for 2022. The *Gaia* Archive core system web interface[4] has been open from the date of the first release. Its continuous updates, both for tools and data, will be a reference for upcoming studies. In the following, we present some examples of the *Gaia* data capabilities. They have been derived using the Fortran code first used in Romero-Gómez et al. (2015).[5]

Parallax Horizon Accuracy (Tracers) In Fig. 1.7 we show the accuracy expected for the *Gaia* trigonometric parallax data to be published in the second release (late 2017). We can see how Red Clump stars placed as far as 3–4 kpc towards the Galactic anticentre will have distance accuracies better than 20%. As discussed before (see Sect. 1.2.1), Red Clump giant and supergiant stars are good trace populations for both Galactic structure and kinematic studies.

Transversal Velocity Accuracy Antoja et al. (2016) analysed the *Gaia* capabilities to characterize the Milky Way spiral arms. In Fig. 1.8 we reproduce the expected accuracies in the *Gaia* end-off-mission tangential velocities for several disk populations. As discussed by Antoja et al. (2016), errors in the median transverse velocity shall be <1 km/s to distinguish spiral features, and, furthermore, errors

[3] *Gaia* Science Performance web page: http://www.cosmos.esa.int/web/gaia/science-performance.

[4] *Gaia* Archive web page: http://gaia.esac.esa.int/archive/.

[5] The code was released at the 2nd *Gaia* challenge workshop and is publicly available at https://github.com/mromerog/Gaia-errors.

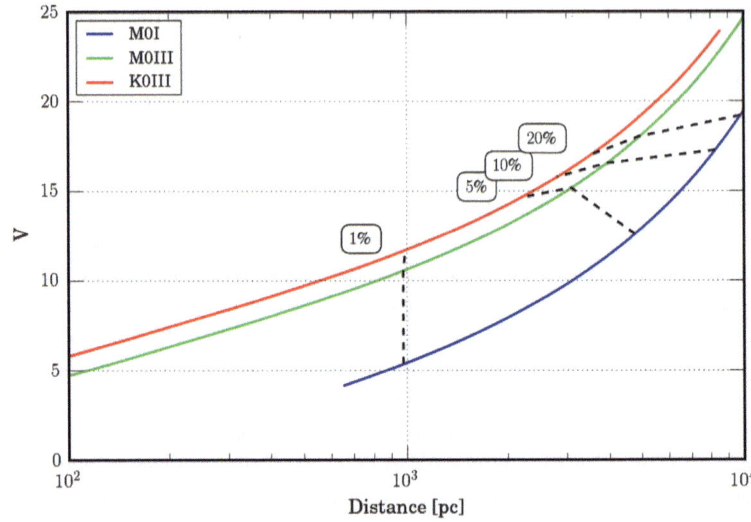

Fig. 1.7 Mean relative parallax accuracy horizons for red giants and supergiants. The plot of visual apparent magnitude versus heliocentric distance has been made assuming an extinction of 1 mag/kpc, a good approximation for the interstellar absorption towards the Galactic anticentre. *Dashed lines* represent the constant mean relative parallax accuracy expected for *Gaia* DR2 expected for the end of 2017 (code courtesy of A. Brown)

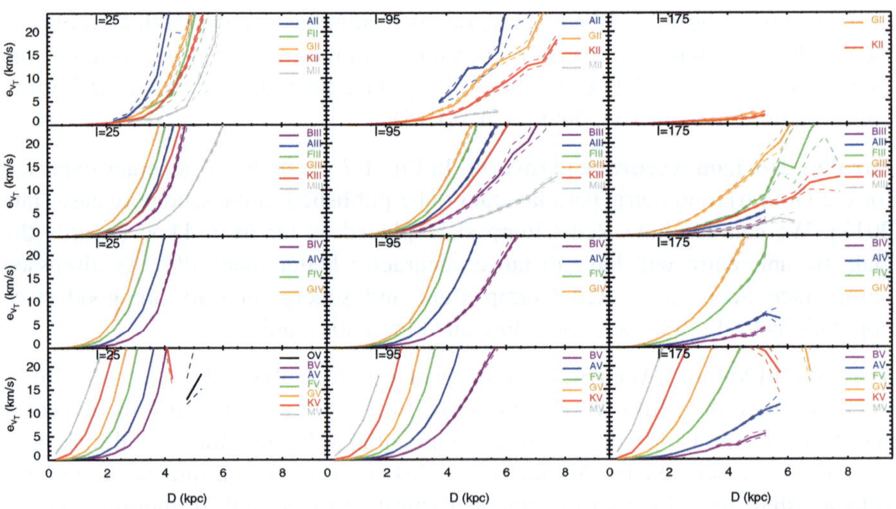

Fig. 1.8 Median error in transverse velocity (V_T) as a function of heliocentric distance expected from end-off-mission *Gaia* data. The stellar populations shown are those of the *Gaia* Universe Model Simulator (Robin et al. 2012). Three different directions in the Galactic plane have been considered. The *dashed lines* show the 75% confidence limit of the median. The locations of the main resonances CR, ILR 2:1, ILR 4:1, OLR 4:1 and OLR 2:1 of the tight-winding approximation spiral arms are shown with *black horizontal lines* (*solid*, *dashed*, *dotted*, *dashed-dotted*, *long dashed*, respectively). The rotation of the Galaxy is towards the *left* (courtesy of T. Antoja)

larger than 20% could introduce biases in the spiral pattern characterization. A careful statistical treatment of *Gaia* data will be indispensable, and as much as possible, it is recommended to work directly in the space of the observables (see Romero-Gómez et al. 2006; Abedi et al. 2014).

Radial Velocities and Chemistry The *Gaia* spectrophotometers BP and RP (Jordi et al. 2010) provide low-resolution spectra, with resolutions $R = 40$–25 and $R = 130$–70 in the blue and red bands, respectively. Being conservative, one can expect to derive end-off-mission astrophysical parameters with an accuracy of about 75–250 K in effective temperature, 0.2–0.5 dex in log(g), 0.1–0.3 dex in [Fe/H] and 0.06–0.15 mag in A_V for a star with *Gaia* magnitude $G \sim 15$ (Bailer-Jones et al. 2013). The *Gaia* intermediate-resolution spectrograph ($R \sim 11{,}000$) is expected to yield radial velocities for Red Clump stars with $V < 15$, excellent tracers on the Milky Way outskirts (see Sect. 1.2.1), with an accuracy of ~5 km/s. We estimate *Gaia* will provide us with about 10^6 of such objects. The expected radial velocity error will increase to ~10 km/s for the 2×10^6 Red Clump stars to be observed by *Gaia* with $15 < V < 16$ (see Romero-Gómez et al. 2015 for details).

New Insights into Spiral Arm Nature Up to now, most studies have been done considering only radial velocity data, proper motions at large distances being too uncertain to be used. Recently, Antoja et al. (2016) have proposed a new independent method to disentangle the nature of the spiral arm using only astrometric data from *Gaia*. This method is based on the comparison of stellar kinematics of symmetric Galactic longitude regions in the Galactic plane. In Fig. 1.9

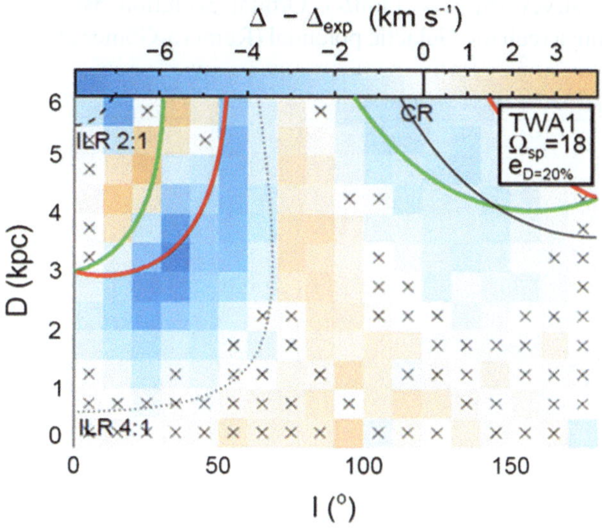

Fig. 1.9 Residuals in the tangential velocities expected from a simulated realistic density wave spiral arm perturbation. See text and Antoja et al. (2016) for a detailed explanation (courtesy of T. Antoja)

we show the differences in tangential velocities expected when comparing two regions symmetric in Galactic longitude. In this case a density wave spiral arm model has been considered, and the expected contribution of the axisymmetric galactocentric motion has been subtracted. It is important to note that the analysis is done using only *Gaia* data. This strategy allows to more clearly identify possible biases introduced when propagating errors or when combining variables. As can be seen, typical kinematic trends induced only by the spiral arms are of the order of ~2–10 km/s in tangential velocity. As expected, these differences are very small in the Galactic anticentre telling us that it will be hard to directly derive the kinematic perturbations at large distances using only *Gaia* data. Fortunately, *Gaia* astrometric data will be combined with spectroscopic data from ground-based surveys (see Sect. 1.5). Nonetheless, the precision on median transverse velocity obtained exclusively with *Gaia* data is ~1 km/s up to ~4–6 kpc for some giant stars and ~0.5 km/s up to ~2–4 kpc for subgiants and dwarfs (Antoja et al. 2016).

Orbital Analysis Back in Time In Fig. 1.10 we show the orbital trajectory back in time of two simulated test particles. On the left we simulated a Red Clump star placed near the Perseus arm and towards the Galactic anticentre as seen from the position of the Sun (galactocentric coordinates $(X, Y) = (-10.5, 0)$ kpc). Its apparent magnitude would be $V \sim 15$. The second particle, in the right figure, simulates the same target but placed inside the Sun's galactocentric radius, with coordinates $(X, Y) = (-6.5, -1.0)$ kpc, that is, near the Sagittarius arm, with an apparent magnitude $V \sim 16$. One hundred realizations of each particle have been done assuming a Gaussian distribution of the astrometric proper motion errors as expected from *Gaia* DR2 and a radial velocity error as expected from the WEAVE spectroscopic survey ($\sigma_{V_r} \sim 1$ km/s). Orbital evolution back in time has been performed using a realistic Galactic potential (Romero-Gómez et al. 2015). In black,

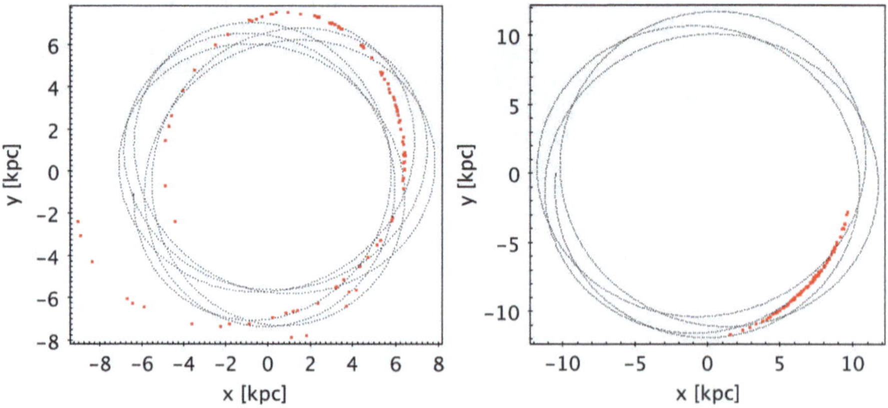

Fig. 1.10 Quantitative evaluation of the *Gaia*-DR2 plus WEAVE capabilities to derive the birth positions of two hypothetical Red Clump stars, one placed near the Perseus arm towards the anticentre (*left*) and the other near the Sagittarius arm (*right*). See text for a detailed explanation

we show the trajectories of these particles and as points (in red) the final positions of these 100 realizations after evolving $\Delta t = 1$ Gyr on the Galactic potential. The dispersion of these points gives an indication of the accuracy we will have when looking, for example, at the birth position of a particle in the Perseus or in the Sagittarius arm. We have confirmed that the WEAVE contribution to improve the radial velocity component is crucial for this detailed analysis of the birth positions of field stars or open clusters.

1.6 Conclusions

The high-quality astrometric and photometric data already collected by *Gaia* during its almost 2 years of successful scientific operation to date allow us to anticipate that, in a few years, *Gaia* and the upcoming ground-based spectroscopic surveys will revolutionize our understanding of the Milky Way. We have shown the capabilities we will have to characterize the kinematic and chemical properties of the resolved stellar populations in the outer regions of the Galactic disk. A detailed and robust statistical treatment of all these data will provide us with a local and global measure of the six-dimensional phase-space distribution function, essential to encode valuable dynamical information, whereas the spectroscopic surveys will provide us with the chemical characterization of the stellar populations. These are the ingredients needed to constrain new and powerful chemodynamical models for the Milky Way. Novel approaches in the theoretical and modelling work are required to treat this huge and unprecedentedly accurate amount of data. It is mandatory to use both bottom-up and top-down dynamical approaches. Furthermore, the *Gaia* archive, designed to maximize the scientific return from the *Gaia* mission, will be a dynamic platform to implement new and powerful software which would support scientific research in this field. These efforts and the huge improvements in the extragalactic domain (e.g. by the CALIFA or MaNGA surveys) will definitively provide new insights on the formation and evolution of galactic disks.

Acknowledgements This work was supported by the MICINN (Spanish Ministry of Science and Innovation)-FEDER through grant AYA2012-39551-C02-01 and ESP2013-48318-C2-1-R and by the European Community's Seventh Framework Programme (FP7/2007-2013) under grant agreement GENIUS, FP7-606740. I thank my colleagues at the University of Barcelona and specially Mercè Romero-Gómez, Teresa Antoja, Maria Monguió, Santi Roca-Fàbrega and Roger Mor for our common research activities in this exciting *Gaia* Era.

References

Abadi, M.G., Navarro, J.F., Steinmetz, M., Eke, V.R.: Simulations of galaxy formation in a Λ cold dark matter universe. II. The fine structure of simulated galactic disks. Astrophys. J. **597**, 21–34 (2003). doi:10.1086/378316, astro-ph/0212282

Abedi, H., Mateu, C., Aguilar, L.A., Figueras, F., Romero-Gómez, M.: Characterizing the Galactic warp with Gaia – I. The tilted ring model with a twist. Mon. Not. R. Astron. Soc. **442**, 3627–3642 (2014). doi:10.1093/mnras/stu1035, 1405.6237

Adibekyan, V.Z., Sousa, S.G., Santos, N.C., Delgado Mena, E., González Hernández, J.I., Israelian, G., Mayor, M., Khachatryan, G.: Chemical abundances of 1111 FGK stars from the HARPS GTO planet search program. Galactic stellar populations and planets. Astron. Astrophys. **545**, A32 (2012). doi:10.1051/0004-6361/201219401, 1207.2388

Alibes, A., Labay, J., Canal, R.: Chemical evolution and abundance gradients in the Milky Way. e-prints. ArXiv astro-ph/0107016

Antoja, T., Roca-Fàbrega, S., de Bruijne, J., Prusti, T.: Kinematics of symmetric Galactic longitudes to probe the spiral arms of the Milky Way with Gaia. Astron. Astrophys. **589**, A13 (2016). doi:10.1051/0004-6361/201628200, 1602.07687

Athanassoula, E.: Manifold-driven spirals in N-body barred galaxy simulations. Mon. Not. R. Astron. Soc. **426**, L46–L50 (2012). doi:10.1111/j.1745-3933.2012.01320.x, 1207.4590

Aumer, M., Binney, J.J.: Kinematics and history of the solar neighbourhood revisited. Mon. Not. R. Astron. Soc. **397**, 1286–1301 (2009); doi:10.1111/j.1365-2966.2009.15053.x, 0905.2512

Bailer-Jones, C.A.L., Andrae, R., Arcay, B., Astraatmadja, T., Bellas-Velidis, I., Berihuete, A., Bijaoui, A., Carrión, C., Dafonte, C., Damerdji, Y., Dapergolas, A., de Laverny, P., Delchambre, L., Drazinos, P., Drimmel, R., Frémat, Y., Fustes, D., García-Torres, M., Guédé, C., Heiter, U., Janotto, A.M., Karampelas, A., Kim, D.W., Knude, J., Kolka, I., Kontizas, E., Kontizas, M., Korn, A.J., Lanzafame, A.C., Lebreton, Y., Lindstrøm, H., Liu, C., Livanou, E., Lobel, A., Manteiga, M., Martayan, C., Ordenovic, C., Pichon, B., Recio-Blanco, A., Rocca-Volmerange, B., Sarro, L.M., Smith, K., Sordo, R., Soubiran, C., Surdej, J., Thévenin, F., Tsalmantza, P., Vallenari, A., Zorec, J.: The Gaia astrophysical parameters inference system (Apsis). Pre-launch description. Astron. Astrophys. **559**, A74 (2013). doi:10.1051/0004-6361/201322344, 1309.2157

Bensby, T., Alves-Brito, A., Oey, M.S., Yong, D., Meléndez, J.: A first constraint on the thick disk scale length: differential radial abundances in K giants at galactocentric radii 4, 8, and 12 kpc. Astrophys. J. Lett. **735**, L46 (2011). doi:10.1088/2041-8205/735/2/L46, 1106.1914

Bensby, T., Feltzing, S., Oey, M.S.: Exploring the Milky Way stellar disk. A detailed elemental abundance study of 714 F and G dwarf stars in the solar neighbourhood. Astron. Astrophys. **562**, A71 (2014). doi:10.1051/0004-6361/201322631, 1309.2631

Binney, J., Tremaine, S.: Galactic Dynamics. Princeton University Press, Princeton (1987)

Bird, J.C., Kazantzidis, S., Weinberg, D.H., Guedes, J., Callegari, S., Mayer, L., Madau, P.: Inside out and upside down: tracing the assembly of a simulated disk galaxy using mono-age stellar populations. Astrophys. J. **773**, 43 (2013). doi:10.1088/0004-637X/773/1/43, 1301.0620

Bland-Hawthorn, J., Gerhard, O.: The galaxy in context: structural, kinematic, and integrated properties. Annu. Rev. Astron. Astrophys. **54**, 529–596 (2016). doi:10.1146/annurev-astro-081915-023441, 1602.07702

Boissier, S., Prantzos, N.: Chemo-spectrophotometric evolution of spiral galaxies – I. The model and the Milky Way. Mon. Not. R. Astron. Soc. **307**, 857–876 (1999). doi:10.1046/j.1365-8711.1999.02699.x, astro-ph/9902148

Bono, G., Marconi, M., Cassisi, S., Caputo, F., Gieren, W., Pietrzynski, G.: Classical Cepheid pulsation models. X. The Period-Age relation. Astrophys. J. **621**, 966–977 (2005). doi:10.1086/427744, astro-ph/0411756

Bovy, J., Rix, H.W., Hogg, D.W., Beers, T.C., Lee, Y.S., Zhang, L.: The vertical motions of mono-abundance sub-populations in the Milky Way disk. Astrophys. J. **755**, 115 (2012). doi:10.1088/0004-637X/755/2/115, 1202.2819

Bragaglia, A., Sestito, P., Villanova, S., Carretta, E., Randich, S., Tosi, M.: Old open clusters as key tracers of Galactic chemical evolution. II. Iron and elemental abundances in NGC 2324, NGC 2477 NGC 2660, NGC 3960, and Berkeley 32. Astron. Astrophys. **480**, 79–90 (2008). doi:10.1051/0004-6361:20077904

Burke, B.F.: Systematic distortion of the outer regions of the galaxy. Astron. J. **62**, 90 (1957). doi:10.1086/107463

Carraro, G., Seleznev, A.F., Baume, G., Turner, D.G.: The complex stellar populations in the background of open clusters in the third Galactic quadrant. Mon. Not. R. Astron. Soc. **455**, 4031–4045. doi:10.1093/mnras/stv2663, 1511.03182

Casagrande, L., Schönrich, R., Asplund, M., Cassisi, S., Ramírez, I., Meléndez, J., Bensby, T., Feltzing, S.: New constraints on the chemical evolution of the solar neighbourhood and Galactic disc(s). Improved astrophysical parameters for the Geneva-Copenhagen Survey. Astron. Astrophys. **530**, A138 (2011). doi:10.1051/0004-6361/201016276, 1103.4651

Cescutti, G., Matteucci, F., François, P., Chiappini, C.: Abundance gradients in the Milky Way for α elements, iron peak elements, barium, lanthanum, and europium. Astron. Astrophys. **462**, 943–951 (2007). doi:10.1051/0004-6361:20065403, astro-ph/0609813

Chiappini, C., Matteucci, F., Gratton, R.: The chemical evolution of the galaxy: the two-infall model. Astrophys. J. **477**, 765–780 (1997). astro-ph/9609199

Czekaj, M.A., Robin, A.C., Figueras, F., Luri, X., Haywood, M.: The Besançon Galaxy model renewed. I. Constraints on the local star formation history from Tycho data. Astron. Astrophys. **564**, A102 (2014). doi:10.1051/0004-6361/201322139, 1402.3257

Drimmel, R., Cabrera-Lavers, A., López-Corredoira, M.: A three-dimensional Galactic extinction model. Astron. Astrophys. **409**, 205–215 (2003). doi:10.1051/0004-6361:20031070, astro-ph/0307273

Eggen, O.J., Lynden-Bell, D., Sandage, A.R.: Evidence from the motions of old stars that the Galaxy collapsed. Astrophys. J. **136**, 748 (1962). doi:10.1086/147433

Feast, M.W., Menzies, J.W., Matsunaga, N., Whitelock, P.A.: Cepheid variables in the flared outer disk of our galaxy. Nature **509**, 342–344 (2014). doi:10.1038/nature13246, 1406.7660

Frinchaboy, P.M., Thompson, B., Jackson, K.M., O'Connell, J., Meyer, B., Zasowski, G., Majewski, S.R., Chojnowksi, S.D., Johnson, J.A., Allende Prieto, C., Beers, T.C., Bizyaev, D., Brewington, H., Cunha, K., Ebelke, G., García Pérez, A.E., Hearty, F.R., Holtzman, J., Kinemuchi, K., Malanushenko, E., Malanushenko, V., Marchante, M., Mészáros, S., Muna, D., Nidever, D.L., Oravetz, D., Pan, K., Schiavon, R.P., Schneider, D.P., Shetrone, M., Simmons, A., Snedden, S., Smith, V.V., Wilson, J.C.: The open cluster chemical analysis and mapping survey: local Galactic metallicity gradient with APOGEE using SDSS DR10. Astrophys. J. Lett. **777**, L1 (2013). doi:10.1088/2041-8205/777/1/L1, 1308.4195

García-Ruiz, I., Kuijken, K., Dubinski, J.: The warp of the Galaxy and the large magellanic cloud. Mon. Not. R. Astron. Soc. **337**, 459–469 (2002). doi:10.1046/j.1365-8711.2002.05923.x

Genovali, K., Lemasle, B., Bono, G., Romaniello, M., Fabrizio, M., Ferraro, I., Iannicola, G., Laney, C.D., Nonino, M., Bergemann, M., Buonanno, R., François, P., Inno, L., Kudritzki, R.P., Matsunaga, N., Pedicelli, S., Primas, F., Thévenin, F.: On the fine structure of the Cepheid metallicity gradient in the Galactic thin disk. Astron. Astrophys. **566**, A37 (2014). doi:10.1051/0004-6361/201323198, 1403.6128

Gómez, F.A., Helmi, A., Cooper, A.P., Frenk, C.S., Navarro, J.F., White, S.D.M.: Streams in the Aquarius stellar haloes. Mon. Not. R. Astron. Soc. **436**, 3602–3613 (2013). doi:10.1093/mnras/stt1838, 1307.0008

Gómez, F.A., White, S.D.M., Marinacci, F., Slater, C.T., Grand, R.J.J., Springel, V., Pakmor, R.: A fully cosmological model of a Monoceros-like ring. Mon. Not. R. Astron. Soc. **456**, 2779–2793 (2016). doi:10.1093/mnras/stv2786, 1509.08459

Harris, H.C.: Photometric abundances of classical Cepheids and the gradient in the galactic disk. Astron. J. **86**, 707–718 (1981). doi:10.1086/112936

Haywood, M.: Revisiting two local constraints of the Galactic chemical evolution. Mon. Not. R. Astron. Soc. **371**, 1760–1776 (2006). doi:10.1111/j.1365-2966.2006.10802.x, astro-ph/0609523

Haywood, M., Di Matteo, P., Lehnert, M.D., Katz, D., Gómez, A.: The age structure of stellar populations in the solar vicinity. Clues of a two-phase formation history of the Milky Way disk. Astron. Astrophys. **560**, A109 (2013). doi:10.1051/0004-6361/201321397, 1305.4663

Holmberg, J., Nordström, B., Andersen, J.: The Geneva-Copenhagen survey of the Solar neighbourhood II. New uvby calibrations and rediscussion of stellar ages, the G dwarf problem, age-metallicity diagram, and heating mechanisms of the disk. Astron. Astrophys. **475**, 519–537 (2007). doi:10.1051/0004-6361:20077221, 0707.1891

Jacobson, H.R., Friel, E.D., Jílková, L., Magrini, L., Bragaglia, A., Vallenari, A., Tosi, M., Randich, S., Donati, P., Cantat-Gaudin, T., Sordo, R., Smiljanic, R., Overbeek, J.C., Carraro, G., Tautvaišiene, G., San Roman, I., Villanova, S., Geisler, D., Muñoz, C., Jiménez-Esteban, F., Tang, B., Gilmore, G., Alfaro, E.J., Bensby, T., Flaccomio, E., Koposov, S.E., Korn, A.J., Pancino, E., Recio-Blanco, A., Casey, A.R., Costado, M.T., Franciosini, E., Heiter, U., Hill, V., Hourihane, A., Lardo, C., de Laverny, P., Lewis, J., Monaco, L., Morbidelli, L., Sacco, G.G., Sousa, S.G., Worley, C.C., Zaggia, S.: The Gaia-ESO survey: probes of the inner disk abundance gradient. Astron. Astrophys. **591**, A37 (2016). doi:10.1051/0004-6361/201527654, 1605.04899

Jordi, C., Gebran, M., Carrasco, J.M., de Bruijne, J., Voss, H., Fabricius, C., Knude, J., Vallenari, A., Kohley, R., Mora, A.: Gaia broad band photometry. Astron. Astrophys. **523**, A48 (2010). doi:10.1051/0004-6361/201015441, 1008.0815

Jurić, M., Ivezić, Ž., Brooks, A., Lupton, R.H., Schlegel, D., Finkbeiner, D., Padmanabhan, N., Bond, N., Sesar, B., Rockosi, C.M., Knapp, G.R., Gunn, J.E., Sumi, T., Schneider, D.P., Barentine, J.C., Brewington, H.J., Brinkmann, J., Fukugita, M., Harvanek, M., Kleinman, S.J., Krzesinski, J., Long, D., Neilsen, E.H. Jr., Nitta, A., Snedden, S.A., York, D.G.: The Milky Way tomography with SDSS. I. Stellar number density distribution. Astrophys. J. **673**, 864–914 (2008). doi:10.1086/523619, astro-ph/0510520

Just, A., Rybizki, J.: Dynamical and chemical evolution of the thin disc. Astron. Nachr. **337**, 880 (2016). doi:10.1002/asna.201612390, 1512.05091

Kalberla, P.M.W., Dedes, L., Kerp, J., Haud, U.: Dark matter in the Milky Way. II. The HI gas distribution as a tracer of the gravitational potential. Astron. Astrophys. **469**, 511–527 (2007). doi:10.1051/0004-6361:20066362, 0704.3925

Kalberla, P.M.W., Kerp, J., Dedes, L., Haud, U.: Does the stellar distribution flare? A comparison of stellar scale heights with LAB H I data. Astrophys. J. **794**, 90 (2014). doi:10.1088/0004-637X/794/1/90, 1408.5334

Kawata, D., Grand, R.J.J., Gibson, B.K., Casagrande, L., Hunt, J.A.S., Brook, C.B.: Impacts of a flaring star-forming disc and stellar radial mixing on the vertical metallicity gradient. Mon. Not. R. Astron. Soc. **464**, 702–712 (2017). doi:10.1093/mnras/stw2363, 1604.07412

Kordopatis, G., Wyse, R.F.G., Binney, J.: Chemodynamics of the Milky Way and disc formation history: insight from the RAVE and Gaia-ESO surveys. Astron. Nachr. **337**, 904 (2016). doi:10.1002/asna.201612395, 1601.02673

Lemasle, B., François, P., Genovali, K., Kovtyukh, V.V., Bono, G., Inno, L., Laney, C.D., Kaper, L., Bergemann, M., Fabrizio, M., Matsunaga, N., Pedicelli, S., Primas, F., Romaniello, M.: Galactic abundance gradients from Cepheids. α and heavy elements in the outer disk. Astron. Astrophys. **558**, A31 (2013). doi:10.1051/0004-6361/201322115, 1308.3249

Levine, E.S., Blitz, L., Heiles, C.: The vertical structure of the outer Milky Way H I disk. Astrophys. J. **643**, 881–896 (2006). doi:10.1086/503091, astro-ph/0601697

Lin, C.C., Shu, F.H.: On the spiral structure of disk galaxies. Astrophys. J. **140**, 646 (1964). doi:10.1086/147955

López-Corredoira, M., Molgó, J.: Flare in the Galactic stellar outer disc detected in SDSS-SEGUE data. Astron. Astrophys. **567**, A106 (2014). doi:10.1051/0004-6361/201423706, 1405.7649

López-Corredoira, M., Abedi, H., Garzón, F., Figueras, F.: Vertical velocities from proper motions of red clump giants. Astron. Astrophys. **572**, A101 (2014). doi:10.1051/0004-6361/201424573, 1409.6222

Lozinskaya, T.A., Kardashev, N.S.: The thickness of the gas disk of the galaxy from 21-cm observations. Sov. Astron. **7**, 161 (1963)

Masset, F., Tagger, M.: Non-linear coupling of spiral waves in disk galaxies: a numerical study. Astron. Astrophys. **322**, 442–454 (1997). astro-ph/9902125

Michalik, D., Lindegren, L., Hobbs, D.: The Tycho-Gaia astrometric solution . How to get 2.5 million parallaxes with less than one year of Gaia data. Astron. Astrophys. **574**, A115 (2015). doi:10.1051/0004-6361/201425310, 1412.8770

Mikolaitis, Š., Hill, V., Recio-Blanco, A., de Laverny, P., Allende Prieto, C., Kordopatis, G., Tautvaišiene, G., Romano, D., Gilmore, G., Randich, S., Feltzing, S., Micela, G., Vallenari, A., Alfaro, E.J., Bensby, T., Bragaglia, A., Flaccomio, E., Lanzafame, A.C., Pancino, E., Smiljanic, R., Bergemann, M., Carraro, G., Costado, M.T., Damiani, F., Hourihane, A., Jofré, P., Lardo, C., Magrini, L., Maiorca, E., Morbidelli, L., Sbordone, L., Sousa, S.G., Worley, C.C., Zaggia, S.: The Gaia-ESO Survey: the chemical structure of the Galactic discs from the first internal data release. Astron. Astrophys. **572**, A33 (2014). doi:10.1051/0004-6361/201424093, 1408.6687

Minchev, I., Famaey, B., Quillen, A.C., Dehnen, W., Martig, M., Siebert, A.: Radial migration does little for Galactic disc thickening. Astron. Astrophys. **548**, A127 (2012). doi:10.1051/0004-6361/201219714, 1205.6475

Minchev, I., Chiappini, C., Martig, M.: Chemodynamical evolution of the Milky Way disk. I. The solar vicinity. Astron. Astrophys. **558**, A9 (2013). doi:10.1051/0004-6361/201220189, 1208.1506

Minchev, I., Chiappini, C., Martig, M., Steinmetz, M., de Jong, R.S., Boeche, C., Scannapieco, C., Zwitter, T., Wyse, R.F.G., Binney, J.J., Bland-Hawthorn, J., Bienaymé, O., Famaey, B., Freeman, K.C., Gibson, B.K., Grebel, E.K., Gilmore, G., Helmi, A., Kordopatis, G., Lee, Y.S., Munari, U., Navarro, J.F., Parker, Q.A., Quillen, A.C., Reid, W.A., Siebert, A., Siviero, A., Seabroke, G., Watson, F., Williams, M.: A new Stellar chemo-kinematic relation reveals the merger history of the Milky Way disk. Astrophys. J. Lett. **781**, L20 (2014). doi:10.1088/2041-8205/781/1/L20, 1310.5145

Minniti, D., Saito, R.K., Alonso-García, J., Lucas, P.W., Hempel, M.: The edge of the Milky Way Stellar disk revealed using clump giant stars as distance indicators. Astrophys. J. Lett. **733**, L43 (2011). doi:10.1088/2041-8205/733/2/L43, 1105.3151

Momany, Y., Zaggia, S., Gilmore, G., Piotto, G., Carraro, G., Bedin, L.R., de Angeli, F.: Outer structure of the Galactic warp and flare: explaining the Canis Major over-density. Astron. Astrophys. **451**, 515–538 (2006). doi:10.1051/0004-6361:20054081, astro-ph/0603385

Monari, G., Famaey, B., Siebert, A.: Modelling the Galactic disc: perturbed distribution functions in the presence of spiral arms. Mon. Not. R. Astron. Soc. **457**, 2569–2582 (2016). doi:10.1093/mnras/stw171, 1601.04714

Monguió, M., Grosbøl, P., Figueras, F.: First detection of the field star overdensity in the Perseus arm. Astron. Astrophys. **577**, A142 (2015). doi:10.1051/0004-6361/201424896, 1503.04600

Netopil, M., Paunzen, E., Heiter, U., Soubiran, C.: On the metallicity of open clusters. III. Homogenised sample. Astron. Astrophys. **585**, A150 (2016). doi:10.1051/0004-6361/201526370, 1511.08884

Newberg, H.J., Yanny, B., Rockosi, C., Grebel, E.K., Rix, H.W., Brinkmann, J., Csabai, I., Hennessy, G., Hindsley, R.B., Ibata, R., Ivezić, Z., Lamb, D., Nash, E.T., Odenkirchen, M., Rave, H.A., Schneider, D.P., Smith, J.A., Stolte, A,, York, D.G.: The ghost of sagittarius and lumps in the halo of the Milky Way. Astrophys. J. **569**, 245–274 (2002). doi:10.1086/338983, astro-ph/0111095

Nidever, D.L., Bovy, J., Bird, J.C., Andrews, B.H., Hayden, M., Holtzman, J., Majewski, S.R., Smith, V., Robin, A.C., García Pérez, A.E., Cunha, K., Allende Prieto, C., Zasowski, G., Schiavon, R.P., Johnson, J.A., Weinberg, D.H., Feuillet, D., Schneider, D.P., Shetrone, M., Sobeck, J., García-Hernández, D.A., Zamora, O., Rix, H.W., Beers, T.C., Wilson, J.C., O'Connell, R.W., Minchev, I., Chiappini, C., Anders, F., Bizyaev, D., Brewington, H., Ebelke, G., Frinchaboy, P.M., Ge, J., Kinemuchi, K., Malanushenko, E., Malanushenko, V., Marchante, M., Mészáros, S., Oravetz, D., Pan, K., Simmons, A., Skrutskie, M.F.: Tracing chemical evolution over the extent of the Milky Way's Disk with APOGEE Red Clump Stars. Astrophys. J. **796**, 38 (2014). doi:10.1088/0004-637X/796/1/38, 1409.3566

Patsis, P.A.: The stellar dynamics of spiral arms in barred spiral galaxies. Mon. Not. R. Astron. Soc. **369**, L56–L60 (2006). doi:10.1111/j.1745-3933.2006.00174.x

Peñarrubia, J., Martínez-Delgado, D., Rix, H.W., Gómez-Flechoso, M.A., Munn, J., Newberg, H., Bell, E.F., Yanny, B., Zucker, D., Grebel, E.K.: A comprehensive model for the Monoceros Tidal Stream. Astrophys. J. **626**, 128–144 (2005). doi:10.1086/429903, astro-ph/0410448

Pilkington, K., Few, C.G., Gibson, B.K., Calura, F., Michel-Dansac, L., Thacker, R.J., Mollá, M., Matteucci, F., Rahimi, A., Kawata, D., Kobayashi, C., Brook, C.B., Stinson, G.S., Couchman, H.M.P., Bailin, J., Wadsley, J.: Metallicity gradients in disks. Do galaxies form inside-out? Astron. Astrophys. **540**, A56 (2012a). doi:10.1051/0004-6361/201117466, 1201.6359

Pilkington, K., Gibson, B.K., Brook, C.B., Calura, F., Stinson, G.S., Thacker, R.J., Michel-Dansac, L., Bailin, J., Couchman, H.M.P., Wadsley, J., Quinn, T.R., Maccio, A.: The distribution of metals in cosmological hydrodynamical simulations of dwarf disc galaxies. Mon. Not. R. Astron. Soc. **425**, 969–978 (2012b). doi:10.1111/j.1365-2966.2012.21353.x, 1205.4796

Prantzos, N.: An Introduction to Galactic Chemical Evolution. In: Charbonnel, C., Zahn, J.P. (eds.) Stellar Nucleosynthesis: 50 Years After B^2FH. EAS Publications Series, vol. 32, pp. 311–356 (2008). doi:10.1051/eas:0832009, 0709.0833

Prantzos, N.: Topics on Galactic Chemical Evolution (2011). ArXiv e-prints 1101.2108

Purcell, C.W., Bullock, J.S., Tollerud, E.J., Rocha, M., Chakrabarti, S.: The Sagittarius impact as an architect of spirality and outer rings in the Milky Way. Nature **477**, 301–303 (2011). doi:10.1038/nature10417, 1109.2918

Recio-Blanco, A., de Laverny, P., Kordopatis, G., Helmi, A., Hill, V., Gilmore, G., Wyse, R., Adibekyan, V., Randich, S., Asplund, M., Feltzing, S., Jeffries, R., Micela, G., Vallenari, A., Alfaro, E., Allende Prieto, C., Bensby, T., Bragaglia, A., Flaccomio, E., Koposov, S.E., Korn, A., Lanzafame, A., Pancino, E., Smiljanic, R., Jackson, R., Lewis, J., Magrini, L., Morbidelli, L., Prisinzano, L., Sacco, G., Worley, C.C., Hourihane, A., Bergemann, M., Costado, M.T., Heiter, U., Joffre, P., Lardo, C., Lind, K., Maiorca, E.: The Gaia-ESO Survey: the Galactic thick to thin disc transition. Astron. Astrophys. **567**, A5 (2014). doi:10.1051/0004-6361/201322944, 1403.7568

Reid, M.J., Menten, K.M., Brunthaler, A., Zheng, X.W., Dame, T.M., Xu, Y., Wu, Y., Zhang, B., Sanna, A., Sato, M., Hachisuka, K., Choi, Y.K., Immer, K., Moscadelli, L., Rygl, K.L.J., Bartkiewicz, A.: Trigonometric parallaxes of high mass star forming regions: the structure and kinematics of the Milky Way. Astrophys. J. **783**, 130 (2014). doi:10.1088/0004-637X/783/2/130, 1401.5377

Reylé, C., Marshall, D.J., Robin, A.C., Schultheis, M.: The Milky Way's external disc constrained by 2MASS star counts. Astron. Astrophys. **495**, 819–826 (2009). doi:10.1051/0004-6361/200811341, 0812.3739

Robin, A.C., Creze, M., Mohan, V.: The edge of the Galactic disk. Astrophys. J. Lett. **400**, L25–L27 (1992). doi:10.1086/186640, astro-ph/9210001

Robin, A.C., Luri, X., Reylé, C., Isasi, Y., Grux, E., Blanco-Cuaresma, S., Arenou, F., Babusiaux, C., Belcheva, M., Drimmel, R., Jordi, C., Krone-Martins, A., Masana, E., Mauduit, J.C., Mignard, F., Mowlavi, N., Rocca-Volmerange, B., Sartoretti, P., Slezak, E., Sozzetti, A.: Gaia Universe model snapshot. A statistical analysis of the expected contents of the Gaia catalogue. Astron. Astrophys. **543**, A100 (2012). doi:10.1051/0004-6361/201118646, 1202.0132

Robin, A.C., Reylé, C., Fliri, J., Czekaj, M., Robert, C.P., Martins, A.M.M.: Constraining the thick disc formation scenario of the Milky Way. Astron. Astrophys. **569**, A13 (2014). doi:10.1051/0004-6361/201423415, 1406.5384

Roca-Fàbrega, S., Valenzuela, O., Figueras, F., Romero-Gómez, M., Velázquez, H., Antoja, T., Pichardo, B.: On galaxy spiral arms' nature as revealed by rotation frequencies. Mon. Not. R. Astron. Soc. **432**, 2878–2885 (2013). doi:10.1093/mnras/stt643, 1302.6981

Roca-Fàbrega, S., Valenzuela, O., Colín, P., Figueras, F., Krongold, Y., Velázquez, H., Avila-Reese, V., Ibarra-Medel, H.: GARROTXA cosmological simulations of Milky Way-sized Galaxies: general properties, hot-gas distribution, and missing baryons. Astrophys. J. **824**, 94 (2016). doi:10.3847/0004-637X/824/2/94, 1504.06261

Rocha-Pinto, H.J., Majewski, S.R., Skrutskie, M.F., Crane, J.D.: Tracing the Galactic anticenter stellar stream with 2MASS M giants. Astrophys. J. Lett. **594**, L115–L118 (2003). doi:10.1086/378668

Romero-Gómez, M., Masdemont, J.J., Athanassoula, E., García-Gómez, C.: The origin of rR_1 ring structures in barred galaxies. Astron. Astrophys. **453**, 39–45 (2006). doi:10.1051/0004-6361:20054653, astro-ph/0603124

Romero-Gómez, M., Figueras, F., Antoja, T., Abedi, H., Aguilar, L.: The analysis of realistic stellar Gaia mock catalogues – I. Red clump stars as tracers of the central bar. Mon. Not. R. Astron. Soc. **447**, 218–233 (2015). doi:10.1093/mnras/stu2457, 1411.6389

Roškar, R., Debattista, V.P., Brooks, A.M., Quinn, T.R., Brook, C.B., Governato, F., Dalcanton, J.J., Wadsley, J.: Misaligned angular momentum in hydrodynamic cosmological simulations: warps, outer discs and thick discs. Mon. Not. R. Astron. Soc. **408**, 783–796 (2010). doi:10.1111/j.1365-2966.2010.17178.x, 1006.1659

Schönrich, R., Binney, J.: Chemical evolution with radial mixing. Mon. Not. R. Astron. Soc. **396**, 203–222 (2009). doi:10.1111/j.1365-2966.2009.14750.x, 0809.3006

Seabroke, G.M., Gilmore, G.: Revisiting the relations: Galactic thin disc age-velocity dispersion relation. Mon. Not. R. Astron. Soc. **380**, 1348–1368 (2007). doi:10.1111/j.1365-2966.2007.12210.x, 0707.1027

Sellwood, J.A., Binney, J.J.: Radial mixing in galactic discs. Mon. Not. R. Astron. Soc. **336**, 785–796 (2002). doi:10.1046/j.1365-8711.2002.05806.x, astro-ph/0203510

Sellwood, J.A., Carlberg, R.G.: Spiral instabilities provoked by accretion and star formation. Astrophys. J. **282**, 61–74 (1984). doi:10.1086/162176

Slater, C.T., Bell, E.F., Schlafly, E.F., Morganson, E., Martin, N.F., Rix, H.W., Peñarrubia, J., Bernard, E.J., Ferguson, A.M.N., Martinez-Delgado, D., Wyse, R.F.G., Burgett, W.S., Chambers, K.C., Draper, P.W., Hodapp, K.W., Kaiser, N., Magnier, E.A., Metcalfe, N., Price, P.A., Tonry, J.L., Wainscoat, R.J., Waters, C.: The complex structure of stars in the outer Galactic Disk as revealed by Pan-STARRS1. Astrophys. J. **791**, 9 (2014). doi:10.1088/0004-637X/791/1/9, 1406.6368

Toomre, A.: What amplifies the spirals. In: Fall, S.M., Lynden-Bell, D. (eds.) Structure and Evolution of Normal Galaxies, pp. 111–136. Cambridge University Press, Cambridge (1981)

van der Kruit, P.C., Freeman, K.C.: Galaxy disks. Annu. Rev. Astron. Astrophys. **49**, 301–371 (2011). doi:10.1146/annurev-astro-083109-153241, 1101.1771

van der Kruit, P.C., Searle, L.: Surface photometry of edge-on spiral galaxies. I – a model for the three-dimensional distribution of light in galactic disks. Astron. Astrophys. **95**, 105–115 (1981)

Voglis, N., Tsoutsis, P., Efthymiopoulos, C.: Invariant manifolds, phase correlations of chaotic orbits and the spiral structure of galaxies. Mon. Not. R. Astron. Soc. **373**, 280–294 (2006). doi:10.1111/j.1365-2966.2006.11021.x, astro-ph/0607174

Watson, L.C., Koda, J.: Molecular gas in the outskirts. In: Knapen, J.H., Lee, J.C., Gil de Paz, A. (eds.) Outskirts of Galaxies, vol. 434. Springer, Cham (2017). doi:10.1007/978-3-319-56570-5

Westerhout, G.: The distribution of atomic hydrogen in the outer parts of the Galactic System. Bull. Astron. Inst. Neth. **13**, 201 (1957)

Widrow, L.M., Gardner, S., Yanny, B., Dodelson, S., Chen, H.Y.: Galactoseismology: discovery of vertical waves in the Galactic Disk. Astrophys. J. Lett. **750**, L41 (2012). doi:10.1088/2041-8205/750/2/L41, 1203.6861

Widrow, L.M., Barber, J., Chequers, M.H., Cheng, E.: Bending and breathing modes of the Galactic disc. Mon. Not. R. Astron. Soc. **440**, 1971–1981 (2014). doi:10.1093/mnras/stu396, 1404.4069

Xu, Y., Newberg, H.J., Carlin, J.L., Liu, C., Deng, L., Li, J., Schönrich, R., Yanny, B.: Rings and radial waves in the disk of the Milky Way. Astrophys. J. **801**, 105 (2015). doi:10.1088/0004-637X/801/2/105, 1503.00257

Chapter 2
Resolved Stellar Populations as Tracers of Outskirts

Denija Crnojević

Abstract Galaxy haloes contain fundamental clues about the galaxy formation and evolution process: hierarchical cosmological models predict haloes to be ubiquitous and to be (at least in part) the product of past merger and/or accretion events. The advent of wide-field surveys in the last two decades has revolutionized our view of our own Galaxy and its closest "sister", Andromeda, revealing copious tidal streams from past and ongoing accretion episodes, as well as doubling the number of their known faint satellites. The focus shall now be shifted to galaxy haloes beyond the Local Group: resolving individual stars over significant areas of galaxy haloes will enable estimates of their ages, metallicities and gradients. The valuable information collected for galaxies with a range of masses, morphologies and within diverse environments will ultimately test and quantitatively inform theoretical models of galaxy formation and shed light onto the many challenges faced by simulations on galactic scales.

2.1 The Importance of Haloes

Our understanding of galaxy formation and evolution has dramatically evolved in the past 50 years. The first and simplest idea for the formation scenario of our own Milky Way (MW) Galaxy was put forward by Eggen et al. (1962), who proposed the bulk of a stellar halo to be formed in a rapid collapse of gas in the protogalaxy. This scenario, often referred to as "monolithic" collapse, is a dissipative process and takes place on dynamical timescales of the order of $\sim 10^8$ years. This process gives birth to a metal-poor stellar component in the halo outer regions, while the inner regions end up being more metal rich due to the reprocessing of the gas as it collapses deeper into the protogalaxy potential well. This idea was later challenged by an alternative explanation, based on the observation that globular clusters (GCs) at different galactocentric distances have a wide range of metallicities. In this scenario, the halo

D. Crnojević (✉)
Department of Physics and Astronomy, Texas Tech University, Box 41051, Lubbock, TX 79409, USA
e-mail: denija.crnojevic@ttu.edu

J.H. Knapen et al. (eds.), *Outskirts of Galaxies*, Astrophysics and Space Science Library 434, DOI 10.1007/978-3-319-56570-5_2

is formed on longer timescales ($\sim 10^9$ years), and instead of being a self-contained system, it comes together as the product of several protogalactic fragments (Searle and Zinn 1978). These fragments can be pre-enriched before they are accreted. While both scenarios are capable of explaining many observed quantities of the Galactic halo, they cannot individually give a comprehensive picture (Norris and Ryan 1991; Chiba and Beers 2000), which has led to the development of hybrid "two-phase" models. In the latter, the inner Galaxy regions are formed in a first phase as a result of a monolithic-like process, while the outer halo regions are built up over the Galaxy's lifetime through dissipationless accretion events (Freeman and Bland-Hawthorn 2002).

In the past couple of decades, the most widely accepted paradigm of the hierarchical Lambda-Cold Dark Matter (ΛCDM) structure formation model has prevailed, favouring the predominance of merger and accretion events in the build-up of galactic haloes (White and Frenk 1991; Bullock and Johnston 2005; Springel et al. 2006; Johnston et al. 2008). These models predict the ubiquitous presence of haloes, which are characterized by old and metal-poor populations and often show signs of recent interactions, in contrast with the smooth haloes predicted by dissipative models (Bullock and Johnston 2005; Abadi et al. 2006; Font et al. 2006). The interaction events provide a mine of information on the assembly of haloes: dynamical timescales become relatively long (up to several Gyr) in the outer regions of a galaxy, and thus accretion/merger events that occurred a long time ago are often still visible as coherent structures like disrupting galaxies or streams, which readily testify the past assembly history of their host. The assembly itself depends on a variety of factors, such as number, mass, stellar content and structural properties of the accreted satellites, as well as orbital properties, timing and energy of the accretion event. Even when the progenitor is completely dissolved in the host's halo (which is particularly true in the inner halo regions where dynamical timescales are relatively short), its stripped stellar content still retains a characteristic coherence in velocity space as well as in metallicity content, thus giving important clues about the progenitor's properties. Observing the stellar "fossils" that populate galaxy haloes thus offers a unique opportunity to reconstruct the modes, timing and statistics of the halo formation process.

Besides being taletellers of their host system's merger history, the shape and size of haloes also hold vital clues to the process of galaxy formation. In particular, they can teach us about the primordial power spectrum of density fluctuations at the smallest scales, about the reionization process that shall lead to faint and concentrated haloes for an early suppression of star formation in low-mass dark matter (DM) subhaloes or about the triaxiality of DM haloes, which are predicted to be more flattened for dissipationless formation scenarios (Abadi et al. 2006). Despite only accounting for a mere $\sim 1\%$ of a galaxy's total mass (e.g. Morrison 1993), extended haloes are clearly extremely valuable to test and refine theoretical predictions on the halo assembly process. Due to their extreme faintness, however, haloes have not been as fully exploited as they should have been as key tests of galaxy formation models: they are not easily detected above the sky level, i.e.

surface brightness values of $\mu_V \sim 25\,\mathrm{mag\,arcsec}^{-2}$, posing a serious observing challenge to their investigation. Cosmological simulations predict the majority of past and ongoing accretion events to have surface brightness values well below this value (e.g. Bullock and Johnston 2005). According to some models, reaching a surface brightness of $\mu_V \sim 29\,\mathrm{mag\,arcsec}^{-2}$ should allow the detection of at least one stream per observed galaxy (Johnston et al. 2008; Cooper et al. 2010). How is it then possible to extract the information locked in the faint outskirts of galaxies?

2.1.1 Resolved Stellar Populations

The best method to study faint haloes and their substructure in nearby galaxies is to resolve individual stars. Even when sparse and faint, resolved stars can be individually counted, and a stellar number density can easily be converted into a surface brightness value. When the Galactic extinction presents a high degree of spatial inhomogeneity (possibly mimicking faint irregular substructures), and the sky level is higher than the integrated light signal coming from extremely faint sources, resolved populations provide a very powerful means to trace them. This method is not free from complications: there will always be contamination coming both from foreground Galactic stars and from background unresolved galaxies. This can be accounted for statistically, by observing "field" regions away from the main target and quantifying the contaminants, while a direct confirmation of a star's membership requires spectroscopy. At the same time, resolving individual stars poses constraints on the inherent nature and on the distance of the putative targets: for systems where the stellar density is so high that stars fall on top of each other on the sky, the "crowding" prevents the resolution of individual objects. This can of course occur also in the case of a relatively sparse galaxy which has a large line-of-sight distance, so that the stars are packed in a small region of the sky. Distance is also the principal enemy of depth: the larger the distance, the brighter the detection limit, i.e. the absolute magnitude/surface brightness that we can reach for a fixed apparent magnitude. Nonetheless, resolved stellar populations are able to deliver powerful information for galaxies located within $\sim$10 Mpc, i.e. within the so-called Local Volume.

The discovery of the Sagittarius dwarf galaxy by Ibata et al. (1994) from the identification of a co-moving group of stars opened the door to the era of halo studies and their substructure: a galaxy resembling the properties of classical dwarf spheroidals was clearly in the process of being disrupted by its giant host, our own MW. This evidence was the first to support theoretical predictions for the hierarchical assembly models and the existence of observable accretion events. Soon thereafter, stellar density maps allowed the discovery of a prominent low surface brightness stream around the MW's closest giant spiral Andromeda (M31), the so-called giant stellar stream (Ibata et al. 2001). This feature, invisible to the naked eye, is a clear example of the elusive nature of haloes and their substructure: the surface brightness of the giant stellar stream is $\mu_V \sim 30\,\text{mag arcsec}^{-2}$, which is prohibitive for integrated light images.

As challenging as it is, the mere detection of haloes and their substructures is not enough to provide quantitative constraints on models of galaxy evolution. From the stars' photometry and thus position in the colour-magnitude diagram (CMD), i.e. the observational counterpart of the Hertzsprung-Russell diagram, it is possible to characterize the properties of the considered stellar system. First and foremost, in contrast to integrated light, accurate distance measurements can be obtained from CMD features that act as standard candles, e.g. the luminosity of the tip of the red giant branch (TRGB) or of the horizontal branch (HB). Another key advantage of resolved populations is the possibility to constrain ages and metallicities more tightly than with integrated light alone. The CMD is used to quantify the star formation rate as a function of lookback time and thus derive the star formation history (SFH) of a composite stellar population (e.g. Gallart et al. 2005, and references therein). Spectroscopy of individual stars is the ultimate method to constrain their metallicity content and kinematical properties, such as radial velocity and proper motion, which allows for the full six-dimensional phase space to be investigated. The latter cannot, for the moment, be achieved beyond the LG limits and still only occasionally for M31.

Besides giving precious insights into galaxy haloes and their accretion histories, resolved stellar populations can help us in characterizing the "surviving" low-mass galaxies that have not been accreted to date and reside in the outskirts of giant hosts.

2.1.2 The Low-Mass End of the Galaxy Luminosity Function

The low-mass end of the galaxy luminosity function (LF) is of no less interest than haloes themselves. Besides the MW and M31, the Local Group (LG) contains tens of smaller galaxies which can be studied in detail due to their proximity (see Tolstoy et al. 2009 for a review). While the ΛCDM cosmological model has provided a convincing match to the large-scale structures observed in the high-redshift Universe, it falls short at the smallest, galactic scales, indicating an incomplete understanding of the physics involved in the evolution of galaxies: for example, the "missing-satellite problem" has been highlighted for the first time by Moore et al. (1999) and Klypin et al. (1999). Briefly, the number of DM subhaloes predicted in simulations exceeds the observed number of MW satellites by almost two orders of magnitude. The shape of the DM profile in the innermost regions of dwarf galaxies is also a matter of debate (the "cusp-core" problem; Walker and Peñarrubia 2011). In addition, the more massive among the MW satellites are less dense than what is expected from simulations, which is puzzling because they should be affected by fewer observational biases than their smaller, sparser siblings (the "too-big-to-fail" problem; Boylan-Kolchin et al. 2011). In addition, the fact that many of the

MW and M31 satellites are distributed along planes does not have a straightforward explanation in ΛCDM models (e.g. Pawlowski et al. 2014).

From the theoretical point of view, the inclusion of baryonic physics in DM-only simulations is key to reconcile predictions with observations of the smallest galaxies. In particular, effects such as supernova feedback, stellar winds, cosmic reionization and tidal/ram pressure stripping all concur to reduce star formation efficiency in the least massive DM haloes. Tremendous progress is being made on this front, taking into account realistic physics as well as increasing the resolution of simulations (e.g. Stinson et al. 2009; Brooks et al. 2013; Sawala et al. 2016; Wetzel et al. 2016). At the same time, new observational discoveries keep offering intriguing challenges at the smallest galactic scales, as further described in Sects. 2.2.1.3 and 2.2.2.3.

2.2 Local Group

The galaxies closest to us give us the most detailed information because of the large number of stars that can be resolved. Here, I will summarize what we have learnt in the past two decades about our own Galaxy (even though an extensive picture of the MW outskirts goes beyond the scope of this contribution and can be found in Figueras 2017), about its closest spiral neighbour M31 and about their lower-mass satellites.

2.2.1 Milky Way

The MW is traditionally divided into discrete components, i.e. the bulge, the disks (thin and thick) and the halo. The spheroidal portion of the MW is given by the central bulge, which consists mainly of metal-rich populations, and an extended diffuse component, which has a lower mean metallicity. Overall, stars and GCs in the halo have ages $\sim$ 11–13 Gyr (Carollo et al. 2007). The halo can be further deconstructed into an inner halo and an outer halo, even though the distinction could partly arise from observational biases (Schönrich et al. 2014). The inner and outer haloes also seem to have different chemical composition ([Fe/H]~ -1.6 and [Fe/H]~ -2.2, respectively; Ryan and Norris 1991; Carollo et al. 2007). According to simulations, the two halo components should also have formed on different timescales: the inner halo (<20 kpc) is partly constituted by early-formed in situ stars, partly due to a violent relaxation process and partly assembled from early, massive merging events that provide metal-rich populations (Abadi et al. 2006; Font et al. 2011; Tissera et al. 2013; Pillepich et al. 2014; Cooper et al. 2015); the outer halo is assembled more recently, with its mass beyond $\sim$30 kpc being mainly accreted in the past $\sim$8 Gyr (Bullock and Johnston 2005; Cooper et al. 2010). These predictions are, however, still not sufficient at a quantitative level and unconstrained as to the exact ratio of accreted stars versus in situ populations. At the same time, observations of the MW halo with better statistics and precision are needed to inform them.

Our position within the MW puts us at a clear disadvantage for global studies of its outskirts: the distant and sparse halo stars are observed from within the substantial disk component, which produces contamination both in terms of extinction and numerous disk stars along the line of sight, which completely "obscure" the sky at low Galactic latitudes. Nonetheless, thanks to the advent of wide-field imagers, the past two decades have revolutionized the large-scale view of our Galaxy. Several stellar tracers can be used to dig into the MW halo at different distance ranges: old main sequence turnoff (MSTO) stars are identified mostly out to ∼20 kpc and brighter RGB stars out to ∼40–50 kpc, while RR Lyrae and blue horizontal branch (BHB) stars can be detected out to 100 kpc. Spatial clustering of these stellar components indicates non-mixed substructure, which is often confirmed to be kinematically coherent.

2.2.1.1 The Emergence of Streams

After the cornerstone discovery of the disrupting Sagittarius dwarf, it became clear that substructure is not only present in the MW halo, but it also might constitute a big portion of it. To put it in Majewski's words (Majewski 1999),

> There is good reason to believe that within a decade we will have a firm handle on the contribution of satellite mergers in the formation of the halo, as we move observationally from serendipitous discoveries of circumstantial evidence to more systematic surveys for the fossils left behind by the accretion process.

In the following decade, several streamlike features have indeed emerged from a variety of multiband photometric and spectroscopic surveys, and the Sloan Digital Sky Survey (SDSS) proved to be an especially prolific mine for such discoveries around the northern Galactic cap. The Sagittarius Stream has been traced further, including in the Galactic anti-centre direction (e.g. Mateo 1998; Majewski et al. 2003), and independent substructures have been uncovered (Ivezić et al. 2000; Yanny et al. 2000; Newberg et al. 2002; Grillmair 2006; Jurić et al. 2008), most notably the Monoceros Ring, the Virgo Overdensity, the Orphan Stream and the Hercules-Aquila Cloud. Some of these have later been confirmed to be coherent with radial velocities (Duffau et al. 2006). Note that most of these substructures are discovered at galactocentric distances > 15 kpc, while the inner halo is smooth due to its shorter dynamical timescales.

During the past decade, one of the most stunning visualizations of the ongoing accretion events in the MW halo was provided by the Field of Streams (Belokurov et al. 2006), reproduced in Fig. 2.1. The stunning stellar density map is derived from SDSS data of stars around the old MSTO at the distance of Sagittarius, with a range of magnitudes to account for a range in distances. This map not only shows the Sagittarius Stream and its distance gradient but also a plethora

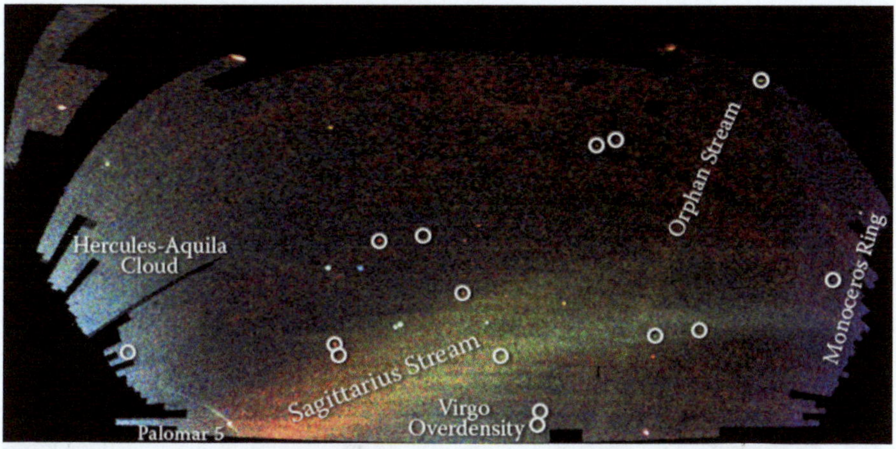

Fig. 2.1 Spatial density of SDSS stars around the Galactic cap, binned in 0.5×0.5 deg^2; the colour scale is such that *blue* indicates the nearest stars, while *red* is for the furthest ones. *Labelled* are the main halo substructures, which are in some cases streams associated with a GC or a dwarf galaxy; the *circles* show some newly discovered dwarf satellites of the MW. Plot adapted from Belokurov et al. (2006) (http://www.ast.cam.ac.uk/~vasily/sdss/field_of_streams/dr6/)

of less massive streams, as well as an abundance of previously unknown dwarf satellites (see Sect. 2.2.1.3). The Field of Streams has been now complemented with results from the latest state-of-the-art surveys, most notably the all-sky Panoramic Survey Telescope and Rapid Response System (Pan-STARRS), which covers an area significantly larger than that of SDSS. In Fig. 2.2 the first stellar density maps from Pan-STARRS are shown, obtained in a similar way as the Field of Streams (Bernard et al. 2016). The map highlights the fact that the deeper and wider we look at the Galaxy halo, the more substructures can be uncovered and used to constrain its past accretion history and the underlying DM halo properties. From this kind of maps, for example, the halo stellar mass that lies in substructures can be estimated, amounting to $\sim$2–3 $\times$ $10^8 M_\odot$ (see Belokurov 2013). Using SDSS, Bell et al. (2008) also highlight the predominant role of accretion in the formation of the MW's halo based on MSTO star counts, adding to up to $\sim$40% of the total halo stellar mass (note that, however, different tracers could indicate much smaller values, e.g. Deason et al. 2011).

Many of the known halo streams arise from tidally disrupting GCs, of which Palomar 5 is one of the most obvious examples (Odenkirchen et al. 2001). This demonstrates the possible role of GCs, besides dwarf satellites, in building up the halo stellar population and additionally implies that some of the halo GCs may be stripped remnants of nucleated accreted satellites (see Freeman and Bland-Hawthorn 2002, and references therein). In order to discern between a dwarf and a cluster origin of halo stars, we need to perform chemical "tagging", i.e. obtain spectroscopic abundances for tens of elements for these stars (e.g. Martell and Grebel 2010): stars born within the same molecular cloud will retain the same

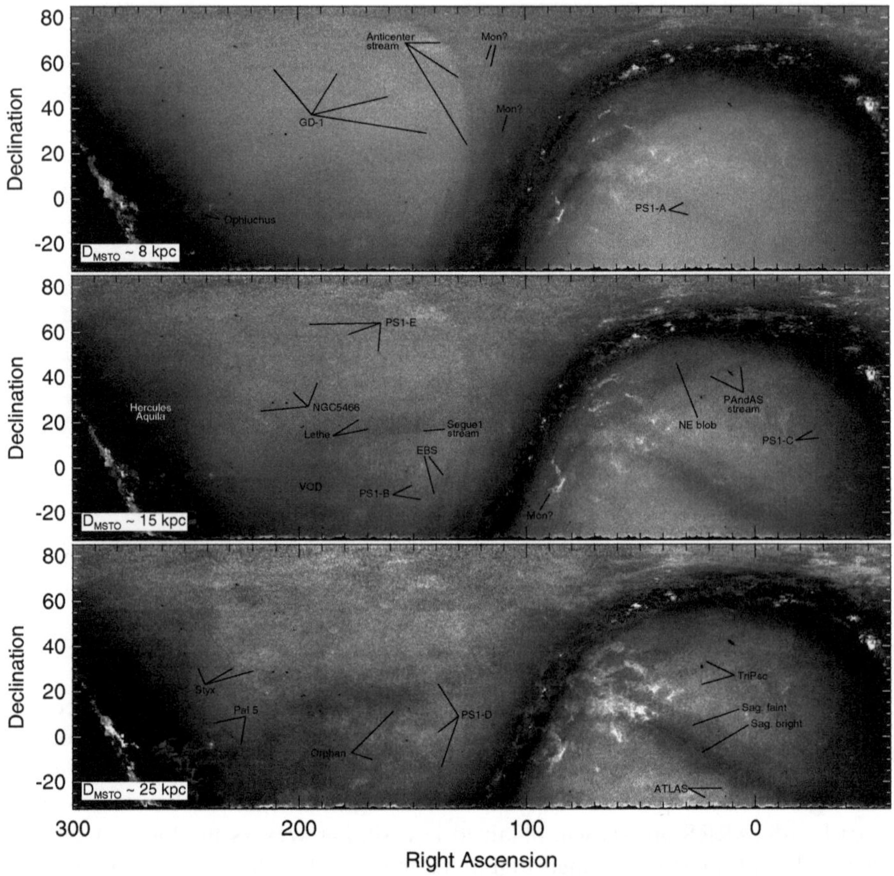

Fig. 2.2 Stellar density maps of the whole Pan-STARRS footprint, obtained by selecting MSTO stars at a range of heliocentric distances (as indicated in each panel). The map is on a logarithmic scale, with *darker areas* indicating higher stellar densities. The many substructures are highlighted in each panel. Reproduced from Bernard et al. (2016), their Fig. 1, with permission of MNRAS

chemical composition and allow us to trace the properties of their birthplace. A number of ambitious ongoing and upcoming spectroscopic surveys (SEGUE, APOGEE, Gaia-ESO, GALAH, WEAVE, 4MOST) are paving the path for this promising research line, even though theoretical models still struggle to provide robust predictions for the fraction of GC stars lost to the MW halo (e.g. Schiavon et al. 2016, and references therein).

2.2.1.2 The Smooth Halo Component

Once the substructure in the halo is detected, it is important that it is “cut out” in order to gain insights into the smooth, in situ stellar component (note that, however, the latter will inevitably suffer from residual contamination from accreted material that is now fully dissolved). The stellar profile of the Galactic halo is, in fact, not smooth at all: several studies have found a break at a radius $\sim$ 25 kpc, with a marked steepening beyond this value (Watkins et al. 2009; Sesar et al. 2013), in qualitative agreement with halo formation models. Some of the explanations put forward suggest that a density break in the halo stellar profile is the likely consequence of a massive accretion event, corresponding to the apocentre of the involved stars (Deason et al. 2013).

The kinematics of halo stars, of GCs and of satellite galaxies, as well as the spatial distribution of streams and tidal features in satellites, can be further used as mass tracers for the DM halo. The total MW mass is to date still poorly constrained, given the difficulty of evaluating it with a broad range of different tracers. The general consensus is for a virial mass value of $\sim 1.3 \pm 0.3 \times 10^{12} M_{\odot}$, even though values discrepant up to a factor of two have recently been suggested (see Bland-Hawthorn and Gerhard 2016 for a compilation of estimates). Besides providing estimates for the total MW mass, studies of SDSS kinematical data, of the Sagittarius Stream and of GC tidal streams have provided discording conclusions on the shape of the MW DM halo: nearly spherical from the modelling of streams or strongly oblate from SDSS kinematics at galactocentric distances < 20 kpc, while nearly spherical and oblate based on stream geometry or prolate from kinematical arguments for distances as large as $\sim$100 kpc (see Bland-Hawthorn and Gerhard 2016 for details). These constraints need a substantial improvement in the future to be able to inform cosmological models: the latter predict spherical/oblate shapes once baryons are included in DM-only flattened haloes (see Read 2014).

2.2.1.3 Dwarf Satellites

As mentioned above, the SDSS has revolutionized our notions of dwarf satellites of the MW. Bright enough to be easily recognized on photographic plates, a dozen "classical" MW dwarf satellites has been known for many decades before the advent of wide-field surveys (Mateo 1998; Grebel et al. 2000). Starting with the SDSS, an entirely new class of objects has started to emerge with properties intermediate between the classical dwarfs and GCs (see Willman 2010, and references therein). The so-called ultra-faint satellites have magnitudes higher than $M_V \sim -8$ and surface brightness values so low that the only way to find them is to look for spatial overdensities of resolved main sequence/BHB stars. Their discovery 10 years ago doubled the number of known MW satellites and revealed the most DM-dominated galaxies in the Universe, with mass-to-light ratios of up to several times $10^3 M_{\odot}/L_{\odot}$ (Simon and Geha 2007).

More recently, the interest in the low end of the galaxy LF has been revitalized once again with deep, wide-field surveys performed with CTIO/DECam, VST/OmegaCAM and Pan-STARRS: these have led to the discovery of more than 20 southern dwarfs in less than 2 years (Bechtol et al. 2015; Koposov et al. 2015; Kim et al. 2015; Drlica-Wagner et al. 2015; Torrealba et al. 2016, and references therein). Some of these discoveries represent extremes in the properties of MW satellites, with surface brightness values as low as $\sim$30 mag arcsec^{-2}, total luminosities of only a few hundred $L_{\odot}$ and surprisingly low stellar density regimes. One of the perhaps most intriguing properties of the newly discovered dwarfs is that many of them appear to be clustered around the Large Magellanic Cloud (LMC): this might be the smoking gun for the possible infall of a group of dwarfs onto the MW, which is predicted by simulations (D'Onghia and Lake 2008; Sales et al. 2016). Low-mass galaxies are expected to have satellites on their own and to provide a large fraction of a giant galaxy's dwarf companions (e.g. Wetzel et al. 2015). The properties of the possible LMC satellites will give us a glimpse onto the conditions of galaxy formation and evolution in an environment much different from the LG as we know it today.

These faintest galaxies, or their accreted and fully dispersed counterparts, are also excellent testbeds to look for the very most metal-poor stars and to investigate the star formation process in the early stages of the Universe (e.g. Frebel and Norris 2015). The study of the lowest mass galaxies holds the promise to challenge our knowledge of galaxy physics even further and pushes us to explore unexpected and exciting new limits.

2.2.2 M31 (Andromeda)

Our nearest giant neighbour has received growing attention in the past decade. Having a remarkable resemblance with the MW and a comparable mass (e.g. Veljanoski et al. 2014), it is a natural ground of comparison for the study of spiral haloes. In terms of a global perspective, the M31 halo is arguably known better than that of the MW: our external point of view allows us to have a panoramic picture of the galaxy and its surrounding regions. The other side of the medal is that, at a distance of $\sim$780 kpc, we can only resolve the brightest evolved stars in M31, and we are mostly limited to a two-dimensional view of its populations. Its proximity also implies a large angular size on the sky, underlining the need for wide field-of-view imagers to cover its entire area.

At the distance of M31, ground-based observations are able to resolve at best the uppermost $\sim$3–4 magnitudes below the TRGB, which is found at a magnitude $i \sim 21$. The RGB is an excellent tracer for old (>1 Gyr) populations, but suffers from a degeneracy in age and metallicity: younger, metal-rich stars overlap in magnitude and colour with older, metal-poor stars (Koch et al. 2006). Despite this, the RGB colour is often used as a photometric indicator for metallicity, once a fixed old age is assumed (VandenBerg et al. 2006; Crnojević et al. 2010). This assumption is justified as long as a prominent young and intermediate-age population seems to be absent (i.e. as judged from the lack of luminous main sequence and asymptotic giant branch, AGB, stars), and it shows very good agreement with spectroscopic metallicity values where both methods have been applied.

The very first resolved studies of M31's halo introduced the puzzling evidence that the M31 halo stellar populations along the minor axis have a higher metallicity than that of the MW at similar galactocentric distances (e.g. Mould and Kristian 1986). This was further confirmed by several studies targeting projected distances from 5 to 30 kpc and returning an average value of [Fe/H] ~ -0.8: in particular, Durrell et al. (2001) study a halo region at a galactocentric distance of $\sim$20 kpc and underline the difference between the properties of M31 and of the MW, suggesting that our own Galaxy might not represent the prototype of a typical spiral. In fact, it has later been suggested that the MW is instead fairly atypical based on its luminosity, structural parameters and the metallicity of its halo stars when compared to spirals of similar mass (Hammer et al. 2007). This result was interpreted as the consequence of an abnormally quiet accretion history for the MW, which apparently lacked a major merger in its recent past.

The wide-area studies of M31's outskirts were pioneered $\sim$15 years ago with an Isaac Newton Telescope survey mapping $\sim$ 40 deg^2 around M31, reaching significantly beyond its disk out to galactocentric distances of $\sim$55 kpc (Ibata et al. 2001; Ferguson et al. 2002). As mentioned before, the southern Giant Stream was first uncovered with this survey, and the halo and its substructures could be studied with a dramatically increased detail. A metal-poor halo component ([Fe/H]~ -1.5) was finally uncovered for regions beyond 30 kpc and out to 160 kpc (Irwin et al. 2005; Kalirai et al. 2006; Chapman et al. 2006), similar to what had been observed for the MW both in terms of metallicity and for its stellar density profile. These studies do not detect a significant gradient in metallicity across the covered radial range. Nonetheless, the properties of the inner halo remained a matter of debate: while Chapman et al. (2006) found a metal-poor halo population within 30 kpc above the disk, Kalirai et al. (2006) analysed a kinematically selected sample of stars within 20 kpc along the minor axis and derived a significantly higher value of [Fe/H]~ -0.5. At the same time, Brown et al. (2006) used deep, pencil-beam *Hubble Space Telescope* (*HST*) pointings in M31's inner halo to conclude that a significant fraction of its stellar populations have an intermediate age with an overall high metallicity. These results were later interpreted by Ibata et al. (2007) in light of their wider-field dataset: the samples from Kalirai et al. (2006) and Brown et al. (2006) are simply part of regions dominated by an extended disk component and with a high contamination from various accretion events, respectively. This underlines, once again, the importance of wide-field observations to reach a global understanding of halo properties.

The M31 INT survey was further extended out to 150 kpc (200 kpc in the direction of the low-mass spiral M33) with the Canada-France-Hawaii Telescope/Megacam and dubbed Pan-Andromeda Archaeological Survey (PAndAS; Ibata et al. 2007; McConnachie et al. 2009). This survey contiguously covered an impressive 380 deg^2 around M31, reaching 4 mag below the TRGB. The PAndAS RGB stellar density map (see Fig. 2.3) is a striking example of an active accretion history, with a copious amount of tidal substructure at both small and large galactocentric radii. PAndAS also constituted a mine for the discovery of a number of very faint satellites and GCs (see below; Richardson et al. 2011; Huxor et al. 2014; Martin et al. 2016).

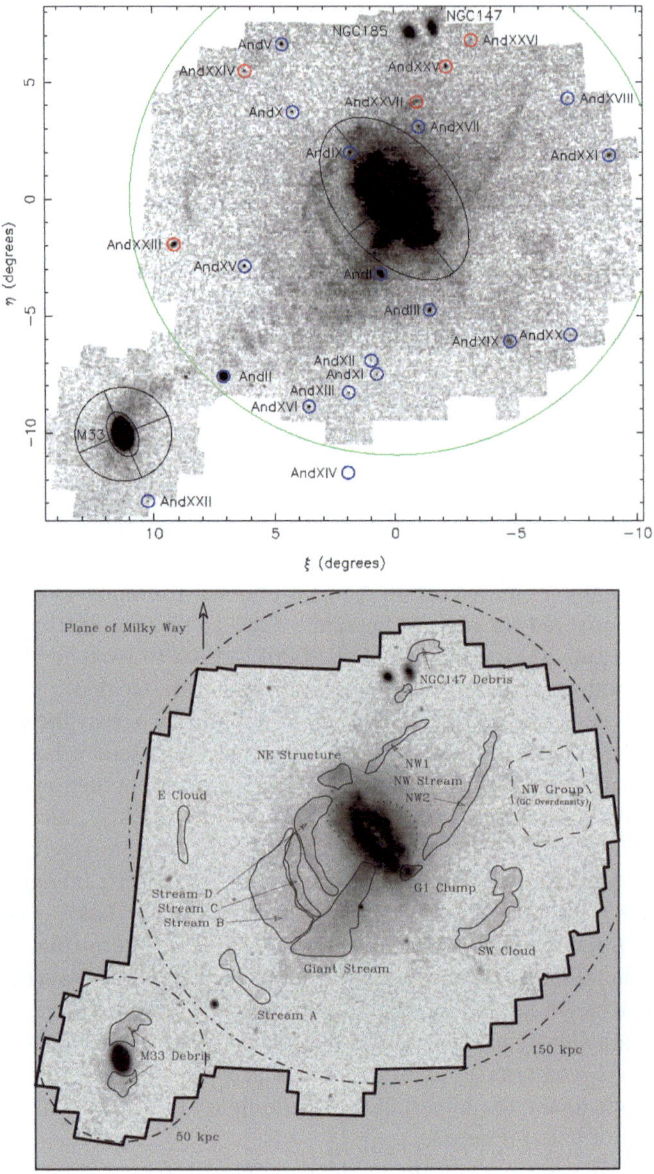

Fig. 2.3 Stellar density maps of metal-poor RGB populations at the distance of M31, as derived from the PAndAS survey. The *large circles* lie at projected radii of 150 kpc and 50 kpc from M31 and M33, respectively. *Upper panel*: The Andromeda satellites are visible as clear overdensities and are marked with *circles*. The vast majority of them were uncovered by the PAndAS survey. Reproduced by permission of the AAS from Richardson et al. (2011), their Fig. 1. *Lower panel*: The main substructures around M31 are highlighted, showcasing a broad range of morphologies and likely progenitor type. Tidal debris is also present in the vicinities of the low-mass satellites M33 and NGC 147, indicating an ongoing interaction with M31. Reproduced by permission of the AAS from Lewis et al. (2013), their Fig. 1

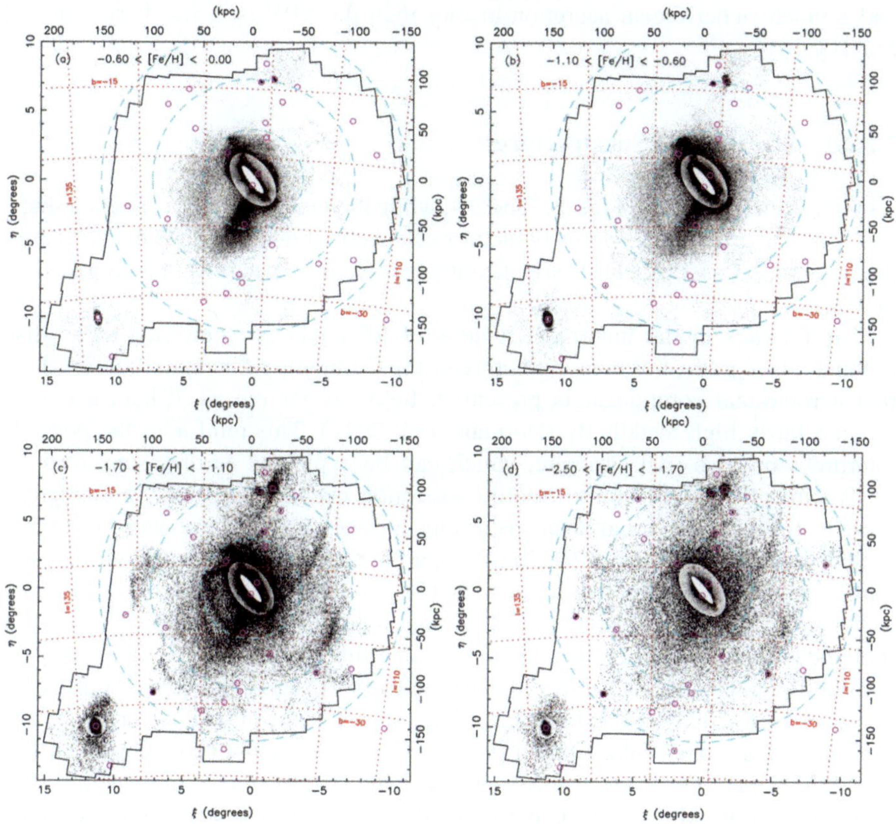

Fig. 2.4 Stellar density map of M31 (akin to Fig. 2.3), this time subdivided into photometric metallicity bins (as indicated in each subpanel). The *upper panels* show high metallicity cuts, where the Giant Stream and Stream C are the most prominent features; note that the shape of the Giant Stream changes as a function of metallicity. The *lower panels* show lower metallicity cuts: the *lower left* panel is dominated by substructure at large radii, while the most metal-poor panel (*lower right*) is smoother and believed to mostly contain in situ populations. Reproduced by permission of the AAS from Ibata et al. (2014), their Fig. 9

Figure 2.4 further shows the RGB stellar map broken into bins of photometric metallicity. The parallel Spectroscopic and Photometric Landscape of Andromeda's Stellar Halo (SPLASH) survey (Guhathakurta et al. 2006; Kalirai et al. 2006) provides a comparison dataset with both photometric and spectroscopic information, the latter obtained with Keck/DEIMOS. The SPLASH pointings are significantly smaller than the PAndAS ones but strategically cover M31 halo regions out to ~225 kpc. Deeper, pencil-beam photometric follow-up studies have further made use of the *HST* to target some of the substructures uncovered in M31's outskirts, resolving stars down to the oldest MSTO (e.g. Brown et al. 2006; Bernard et al. 2015). These observations reveal a high complexity in the stellar populations in M31, hinting at a high degree of mixing in its outskirts. Overall, M31 has evidently had a much richer recent accretion history than the MW (see also Ferguson and Mackey 2016).

2.2.2.1 Streams and Substructures

As seen from the maps in Figs. 2.3 and 2.4, while the inner halo has a flattened shape and contains prominent, relatively metal-rich substructures (e.g. the Giant Stream), the outer halo (>50 kpc) hosts significantly less extended, narrow, metal-poor tidal debris.

The features in the innermost regions of M31 can be connected to its disk populations (e.g. the north-east structure or the G1 clump): kinematic studies show that a rotational component is present in fields as far out as 70 kpc, and they retain a fairly high metallicity (Dorman et al. 2013). This reinforces the possible interpretation as a vast structure, which can be explained as disk stars torn off or dynamically heated due to satellite accretion events. Deep *HST* pointings of these features indeed reveal relatively young populations, likely produced from pre-enriched gas in a continuous fashion, comparable to the outer disk (Ferguson et al. 2005; Brown et al. 2006; Bernard et al. 2015).

The most prominent feature in M31's outer halo, the Giant Stream, was initially thought to originate from the disruption of either M32 or NGC 205, the two dwarf ellipticals located at only ~25–40 kpc from M31's centre (Ibata et al. 2001; Ferguson et al. 2002). While both these dwarfs show signs of tidal distortion, it was soon clear that none of them could produce the vast structure extending ~100 kpc into M31's halo. Great effort has been spent into mapping this substructure both photometrically and spectroscopically, in order to trace its orbit and define its nature: a gradient in its line-of-sight distance was first highlighted by McConnachie et al. (2003), who found the outer stream regions to be located behind M31, the innermost regions at about the distance of M31 and an additional stream component on the opposite (northern) side of M31 to be actually in front of M31. The stream presents a metallicity gradient, with the core regions being more metal-rich and the envelope more metal-poor (see also Fig. 2.4), as well as a very narrow velocity dispersion, with the addition of a puzzling second kinematic component (Gilbert et al. 2009); possible interpretations for the latter may be a wrap or bifurcation in the stream, as well as a component from M31's populations.

A number of increasingly sophisticated theoretical studies have tried to reproduce the appearance of the Giant Stream and picture its progenitor, which is undetected to date. The general consensus seems to be that a relatively massive ($\sim 10^9 M_\odot$) satellite, possibly with a rotating disk, impacted M31 from behind with a pericentric passage around 1–2 Gyr ago (most recently, Fardal et al. 2013; Sadoun et al. 2014). In particular, simulations can reproduce the current extension and shape of the stream and predict the progenitor to be located to the north-east of M31, just beyond its disk (Fardal et al. 2013). This study also concludes that some of the substructures linked to M31's inner regions are likely to have arisen from the same accretion event, i.e. the north-east structure and the G1 clump (Fig. 2.3): these shelf features would trace the second and third passage around M31, which is also supported by their radial velocities. CMDs of the Giant Stream populations are in agreement with these predictions: its stellar populations have mixed properties, consistent with both disk and streamlike halo features (Ferguson et al. 2005; Richardson et al. 2008). Detailed reconstruction of its SFH indicates that most star formation occurred at early ages and was possibly quenched at the time of infall in M31's potential (around 6 Gyr ago) (Bernard et al. 2015). Again, these studies deduce a likely origin of these populations as a dwarf elliptical or a spiral bulge.

Besides the Giant Stream, the only other tidal feature with a relatively high metallicity is Stream C (see Figs. 2.3 and 2.4), which appears in the metal-poor RGB maps as well. The origin of this feature is obscure, even though it is tempting to speculate that it could be part of the Giant Stream event. The lower left panel of Fig. 2.4, showing metal-poor populations, encompasses all of the narrow streams and arcs beyond 100 kpc, which extend for up to several tens of kpc in length. All these substructures are extremely faint ($\mu_V \sim 31.5$ mag arcsec^{-2}), and their origin is mostly unknown because of the difficulty in following up such faint and sparse populations. As part of the *HST* imaging of these features, Bernard et al. (2015) find that their populations are mainly formed at early ages and undergo a more rapid chemical evolution with respect to the disk populations. Despite the metal-poor nature of these features, the hypothesis of a single accretion event producing most of the tidal features observed in the outer halo is not that unlikely, given the metallicity gradient present in the Giant Stream itself.

An efficient alternative to investigate the nature of these streams is to study the halo GC population: the wide-field surveys of M31 have allowed to uncover a rich population of GCs beyond a radius of ~25 kpc (e.g. Huxor et al. 2014, and references therein), significantly more numerous than that of the MW halo. Mackey et al. (2010) first highlighted a high spatial correlation between the streams in M31's halo and the GC population, which would be extremely unlikely in a uniform distribution. Following the hypothesis that the disrupting satellites might be providing a high fraction of M31's halo GCs, Veljanoski et al. (2014) obtained spectroscopic follow-up: they were able to confirm that streams and GCs often have correlated velocities and remarkably cold kinematics. This exciting result gives hope for studies of more distant galaxies, where halo populations cannot be resolved and GCs could be readily used to trace possible substructure.

2.2.2.2 Smooth Halo

One of the first spatially extended datasets to investigate the halo of M31 in detail is described in Tanaka et al. (2010): their Subaru/SuprimeCam photometry along the minor axis in both directions is deeper, even though less extended, than PAndAS. The stellar density profile derived in this study extends out to 100 kpc and shows a consistent power law for both directions. The authors also suggest that, given the inhomogeneities in the stellar populations, the M31 halo is likely not fully mixed.

In the most metal-poor (lower right) panel of Fig. 2.4, the substructures in the outer halo fade away, displaying a smoother component that can be identified with the in situ M31 halo. Once the substructures are decoupled based on the lack of obvious spatial correlation and with an additional photometric metallicity cut, Ibata et al. (2014) derive a stellar density profile out to 150 kpc. Again, the profile follows a power law, which turns out to be steeper when increasingly more metal-rich populations are considered. Ibata et al. (2014) also conclude that only 5% of M31's total halo luminosity lies in its smooth halo, and the halo mass is as high as $\sim 10^{10} M_{\odot}$, significantly larger than what is estimated for the MW.

The SPLASH survey extends further out than PAndAS and benefits from kinematical information that is crucial to decontaminate the studied stellar samples from foreground stars and decreases the scatter in the radial profiles. Based on this dataset, Gilbert et al. (2012) find that the halo profile does not reveal any break out to 175 kpc. This is somewhat surprising given the prediction from simulations that accreted M31-sized stellar haloes should exhibit a break beyond a radius of $\sim$100 kpc (Bullock and Johnston 2005; Cooper et al. 2010). Beyond a radius of 90 kpc, significant field-to-field variations are identified in their data, which suggests that the outer halo regions are mainly comprised of stars from accreted satellites, in agreement with previous studies. At the outermost radii probed by SPLASH ($\sim$230 kpc), there is a tentative detection of M31 stars, but this is hard to confirm given the high contamination fraction. Finally, the Gilbert et al. (2012) stellar halo profile suggests a prolate DM distribution, which is also consistent with being spherical, in agreement with Ibata et al. (2014).

Both Ibata et al. (2014) and Gilbert et al. (2014) investigate the existence of a metallicity gradient in the smooth halo of M31: they found a steady decrease in metallicity of about 1 dex from the very inner regions out to 100 kpc. This might indicate the past accretion of (at least) one relatively massive satellite. At the same time, a large field-to-field metallicity variation could mean that the outer halo has been mainly built up by the accretion of several smaller progenitors.

2.2.2.3 Andromeda Satellites

Similarly to the boom of satellite discoveries around the MW, the vast majority of dwarfs in M31's extended halo have been uncovered by the SDSS, PAndAS and Pan-STARRS surveys in the past decade (see Martin et al. 2016, and references therein). The M31 satellites follow the same relations between luminosity, radius and metallicity defined by MW satellites, with the exception of systems that are likely undergoing tidal disruption (Collins et al. 2014). Once more, the characterization of the lowest-mass galaxies raises new, unexpected questions: from the analysis of accurate distances and kinematics, Ibata et al. (2013) conclude that half of the M31 satellites lie in a vast ($\sim$200 kpc) and thin ($\sim$12 kpc) corotating plane and share the same dynamical orbital properties. The extreme thinness of the plane is very hard to reconcile with ΛCDM predictions, where such structures should not survive for a Hubble time. While several theoretical interpretations have been offered (e.g. Fernando et al. 2016), none is conclusive, and this reinforces the allure of mystery surrounding low-mass satellites.

2.2.3 Low-Mass Galaxies In and Around the Local Group

Besides the detailed studies of the two LG spirals, increasing attention is being paid to lower-mass galaxies and their outskirts. Given the self-similar nature of DM, low-mass galaxies should naively be expected to possess haloes and satellites of their own; however, our difficulty in constraining star formation efficiency and physical processes affecting galaxy evolution at these scales blurs these expectations. In the last couple of years, the increasing resolution of cosmological simulations has allowed to make quantitative predictions about the halo and substructures in sub-MW-mass galaxies and about the number of satellites around them (Wheeler et al. 2015; Dooley et al. 2016). Observations are thus much needed to test these predictions.

Since the late 1990s, numerous studies of star-forming dwarfs within or just beyond the LG have claimed the detection of an RGB component extending beyond the blue, young stars (see Stinson et al. 2009, and references therein), hinting at a generic mode of galaxy formation independent on galaxy size. Such envelopes, however, were not characterized in detail and in fact could not be identified uniquely as the product of hierarchical merging without, e.g. accurate age and metallicity estimates.

The presence of extended haloes in the most luminous satellites of the MW and M31, i.e. the irregular LMC and the low-mass spiral M33, respectively, has not been confirmed to date despite the availability of exquisite datasets. Gallart et al. (2004) demonstrate how, out to a galactocentric distance of 7 kpc, the stellar density profile of the LMC disk does not show a clear break, in contrast to previous tentative claims. Clearly, the question is complicated by the fact that the LMC is undergoing tidal disruption, and stripped stellar material could easily be misinterpreted as a halo component. Nonetheless, McMonigal et al. (2014) suggest to have found a sparse LMC halo population from a wide-field dataset around the nearby dwarf galaxy Carina, at galactocentric distances as large as 20 deg. The question might be settled in the near future with the help of wide-field surveys such as the Survey of MAgellanic Stellar History (Martin et al. 2015). With regard to possible low-mass satellites, there is now tantalizing indication that the LMC might have fallen onto the MW with its own satellite system, as mentioned in Sect. 2.2.1.3. As part of the PAndAS survey, deep imaging of M33 has revealed prominent substructure in its outer disk reminiscent of a tidal disturbance and a faint, diffuse substructure possibly identified as a halo component (Cockcroft et al. 2013). This result was, however, carefully reconsidered by McMonigal et al. (2016), who claim that a definitive sign of a halo structure cannot be confirmed, and if present it must have a surface brightness below $\mu_V \sim 35\,\mathrm{mag\,arcsec^{-2}}$.

Besides the investigation of haloes and satellites, deep and wide-field views of low-mass galaxies are crucial to, for example, assess the presence of tidal disturbances, which in turn are key to estimate mass values and constrain DM profiles (e.g. Sand et al. 2012). As demonstrated by Crnojević et al. (2014), a striking similarity in the global properties (luminosity, average metallicity, size) of two low-mass galaxies, such as the M31 satellites NCG 185 and NGC 147, can be quite misleading: once deep imaging was obtained around these galaxies (within PAndAS), NCG 147 revealed extended, symmetric tidal tails, returning a much larger extent and luminosity for this dwarf than what was previously thought. This dataset further showed a flat metallicity gradient for NGC 147, in contrast with the marked gradient found in NGC 185. All these pieces of evidence point at an ongoing interaction of NGC 147 with M31. Large-scale studies of LG dwarfs also provide useful insights into their evolutionary history: by studying CMDs reaching below the MSTO, Hidalgo et al. (2013) trace significant age gradients that advocate an outside-in mode of star formation for dwarf galaxies.

Clearly, systematic deep searches are needed to detect and characterize the outskirts of low-mass satellites. With this goal in mind, wide-field surveys of nearby (<3 Mpc) dwarfs have started to be pursued. The first of these efforts targets NGC 3109, a sub-LMC-mass dwarf located just beyond the boundaries of the LG: several candidate satellites of NGC 3109 are identified from a CTIO/DECam survey targeting regions out to its virial radius (Sand et al. 2015). One of them, confirmed to be at the distance of NGC 3109, is relatively bright ($M_V \sim -10$) and is already in excess of the predicted number by Dooley et al. (2016) for this system. Other ongoing surveys are similarly looking for halo substructures and satellites in several relatively isolated dwarfs, e.g. the Solitary Local Dwarfs survey (Higgs et al. 2016) and the Magellanic Analog Dwarf Companions and Stellar Halos survey (Carlin et al. 2016), by using wide-field imagers on large telescopes such as CFHT/MegaCam, Magellan/Megacam, CTIO/DECam and Subaru/HyperSuprimeCam. These datasets will constitute a mine of information to constrain the role of baryonic processes at the smallest galactic scales.

2.3 Beyond the Local Group

The ground-breaking photometric and kinematic surveys carried out in the past two decades have significantly advanced our knowledge of haloes and their substructures within LG galaxies. Nonetheless, the MW and M31 may not be representative of generic MW-sized haloes, given the stochasticity of the hierarchical assembly process: several marked differences in the stellar populations of their haloes underline the need for observations of a statistically significant sample of galaxy haloes with different morphologies, with surveys targeting large portions of their haloes.

Cosmological simulations of MW-mass analogues show a wide variation in the properties of their haloes. As already mentioned, the relative contribution of in situ star formation and disrupted satellites remains unclear: depending on the models (e.g. full hydrodynamical simulations, *N*-body models with particle tagging), they can vary from a negligible number of accretion events for a MW-sized halo to making up for most of a stellar halo content (e.g. Lu et al. 2014; Tissera and Scannapieco 2014). Even within the same set of simulations, the number, mass ratio and morphology of accretion and merger events span a wide range of possible values (Bullock and Johnston 2005; Johnston et al. 2008; Garrison-Kimmel et al. 2014). The chemical content of extended haloes can provide useful insights into their assembly history: mergers or accretion events of similar-mass satellites will generally tend to produce mild to flat gradients; in situ populations will feature increasingly metal-poor populations as a function of increasing galactocentric radius, similarly to the accretion of one or two massive companions (e.g. Cooper et al. 2010; Font et al. 2011). More extended merger histories are also expected to return younger and relatively metal-rich populations with respect to those coming from a shorter assembly and to produce more massive stellar haloes, with the final result that the mean halo metallicities of MW-mass spirals can range by up to 1 dex (e.g. Renda et al. 2005).

Comprehensive observational constraints are key to guide future simulations of galaxy haloes: the past decade has seen a dramatic increase in the observational census of resolved galaxy haloes beyond the LG, thanks to deep imaging obtained with space facilities, as well as to the advent of wide-field imagers on large ground-based telescopes.

While the increasing target distance means that it is easier to survey larger portions of their haloes, the drawback is that the depth of the images decreases dramatically, and thus we are only able to detect the brightest surface brightness features in the haloes, i.e. the uppermost $\sim$2–3 mag below the TRGB in terms of resolved stars (see Fig. 6 in Radburn-Smith et al. (2011) for a schematic visualization of the different stellar evolutionary phases recognizable in such shallow CMDs). A number of studies have surveyed relatively nearby and more distant galaxy haloes in integrated light despite the serious challenges posed by sky subtraction at such faint magnitudes, masking of bright stars, flat fielding and scattered light effects, point spread function modelling and/or spatially variable Galactic extinction.

A few early studies have been able to uncover a halo component and tidal debris in the target galaxies (e.g. Malin et al. 1983; Morrison et al. 1994; Sackett et al. 1994), without, however, settling the questions about their existence, nature or ubiquity. Different approaches have been adopted to detect haloes and their substructures, i.e. targeting either individual galaxies (e.g. Zheng et al. 1999; Pohlen et al. 2004; Jablonka et al. 2010; Janowiecki et al. 2010; Martínez-Delgado et al. 2010; Adams et al. 2012; Atkinson et al. 2013) or stacking the images of thousands of objects (e.g. Zibetti et al. 2004; van Dokkum 2005; Tal et al. 2009). A precise quantification of the occurrence of faint substructure in the outskirts of nearby galaxies seems as uncertain as it can be, ranging from a few percent to $\sim$70% (see, e.g. Atkinson et al. 2013, and references therein). This is perhaps unsurprising given the heterogeneity of methods used, target galaxy samples and surface brightness limits in such studies. Besides the identification of such features, the characterization of unresolved halo stellar populations constitutes an even harder challenge: integrated colours and spectra can at most reach a few effective radii, thus missing the outer haloes. Even for the available datasets, the degeneracies between age, metallicity and extinction are generally challenging to break (e.g. de Jong et al. 2007); in addition, tidal features can rarely tell us about the mass ratio of a merger event or its orbit (with the exception of tails). Here, we do not intend to discuss the detection of haloes and the variety of fractions and morphologies for tidal features observed in integrated light studies; Knapen and Trujillo (2017) treat this topic in detail, while this contribution focusses on resolved populations.

Obtaining resolved photometry beyond the LG is a daunting task as well, due to the very faint luminosities involved—the brightest RGB stars for galaxies at ~4–10 Mpc have magnitudes of $I \sim 24$–28.5, and thus this approach is so far really limited to the Local Volume. Early attempts to perform photometry of individual stars in the outskirts of nearby galaxies have been made using large photographic plates and the first CCDs (e.g. Humphreys et al. 1986; Davidge and Jones 1989; Georgiev et al. 1992). The brightest populations (i.e. the youngest) could often be reconciled with being members of the parent galaxy, but the critical information on the faint, old stars was still out of reach. With the advent of wide-format CCDs in the mid-1990s, photometry finally became robust enough to open up new perspectives on the resolved stellar content of our closest neighbours.

The first studies of this kind date back to 20 years ago and mainly focus on the inner regions of the target galaxies, most commonly their disks or inner haloes, with the goal of studying their recent star formation and of deriving TRGB distances (see, e.g. Soria et al. 1996 for CenA, Sakai and Madore 1999 for M81 and M82). Elson (1997), in particular, resolved individual stars in the halo of the S0 galaxy NGC 3115 with *HST*. By analysing the uppermost 1.5 mag of the RGB at a galactocentric distance of 30 kpc, they derived a distance of ~11 Mpc and additionally discovered for the first time a bimodality in the photometric metallicity distribution function of this early-type galaxy. Tikhonov et al. (2003) studied for the first time the resolved content of the nearest (~3.5 Mpc) S0 galaxy NGC 404 with combined ground-based and *HST* imaging. Their furthermost *HST* pointings (~20 kpc in projection) contain RGB stars that are clearly older than the main disk population, with similar colour (metallicity). The authors conclude that the disk of NGC 404 extends out to this galactocentric distance, but they do not mention a halo component.

Beyond these early studies of individual galaxies, the need for systematic investigations of resolved stellar haloes was soon recognized. Next we describe the design and results of some systematic surveys targeting samples of galaxies in the Local Volume.

2.3.1 Systematic Studies

A decade ago, Mouhcine et al. (2005a,b,c) started an effort to systematically observe the haloes of eight nearby (<7 Mpc) spiral galaxies with the resolution of *HST*. In particular, they utilized WFPC2 to target fields off of the galaxies' disks (2–13 kpc in projection along the minor axis) with the goal of investigating their stellar populations and obtaining accurate distance estimates as well as photometric metallicity distribution functions, to gain insights into the halo formation process. Mouhcine et al. (2005c) find the haloes to predominantly contain old populations, with no younger components and little to no intermediate-age populations. Interestingly, Mouhcine et al. (2005b) find a correlation between luminosity and metallicity for the target galaxies, where the metallicity is derived from the mean colour of the resolved RGB. Both the spiral galaxies from their sample (NGC 253, NGC 4244, NGC 4945, NGC 4258, NGC 55, NGC 247, NGC 300 and NGC 3031 or M81) and the two ellipticals (NGC 3115 and NGC 5128 or Centaurus A, included in their comparison from previous literature data) fall on the same relation, indicating that haloes might have a common origin regardless of the galaxy morphological type. Interestingly enough, the MW halo turns out to be substantially more metal-poor than those of the other galaxies of comparable luminosity, based on kinematically selected pressure-supported halo stars within ~10 kpc above the disk (see also Sect. 2.2.2). This relation is consistent with a scenario where halo field stars form in the potential well of the parent galaxy in a gradual way from pre-enriched gas. Moreover, the relatively high metallicities of the target haloes seem to suggest that they likely originate from the disruption of intermediate-mass galaxies, rather than smaller metal-poor dwarf galaxies (Mouhcine et al. 2005c).

Interestingly, the dataset presented and studied in Mouhcine et al. (2005a,b,c) is further analysed by Mouhcine (2006) to find that each spiral of the sample presents a bimodal metallicity distribution. In particular, both a metal-poor and a metal-rich component are present in the outskirts of the target galaxies, and both components correlate with the host's luminosity. This is taken as a hint that these populations are born in subgalactic fragments that were already embedded in the dark haloes of the host galaxy; the metal-poor component additionally has a broader dispersion than that of the metal-rich population. These properties show similarities with GC subpopulations in the haloes of early-type galaxies (e.g. Peng et al. 2006). Mouhcine (2006) argues that the metal-poor component may arise from the accretion of low-mass satellites, while the metal-rich one could be linked to the formation of the bulge or the disk.

The shortcoming of this ambitious study is, however, twofold: first, the limited field of view (FoV) of *HST* hampers global conclusions on the galaxies' haloes, and the stellar populations at even larger radii may have different properties than those in the observed fields; second, perhaps most importantly, it is not obvious what structure of the galaxy is really targeted, i.e. the halo, the outer bulge or disk or a mixture of these.

Along the same lines of these studies, Radburn-Smith et al. (2011) present an even more ambitious *HST* survey of 14 nearby disk galaxies within 17 Mpc, with a range of luminosities, inclinations and morphological types. The Galaxy Halos, Outer disks, Substructure, Thick disks, and Star clusters (GHOSTS) survey aims at investigating radial light profiles, axis ratios, metallicity distribution functions (MDFs), SFHs, possible tidal streams and GC populations, all to be considered as a function of galaxy type and position within the galaxies. The 76 ACS pointings of the survey are located along both major and minor axes for most of the targets, and reach ~2–3 mag below the TRGB, down to surface brightness values of $V \sim 30\,\mathrm{mag\,arcsec^{-2}}$. This dataset thus represents a very valuable resource for testing hierarchical halo formation models. Monachesi et al. (2016) investigate six of the galaxies in this sample (NGC 253, NGC 891, M81, NGC 4565, NGC 4945 and NGC 7814) and conclude that all of them contain a halo component out to 50 kpc, and two of them out to 70 kpc along their minor axis. The colour (i.e. photometric metallicity) distribution of RGB stars in the target haloes is analysed and reveals a non-homogeneity which likely indicates the presence of non-mixed populations from accreted objects. The average metallicity out to the largest radii probed remains relatively high when compared to the values of the MW halo; metallicity gradients are also detected in half of the considered galaxies. Surprisingly, and in contrast to the results presented by Mouhcine et al. (2005b), the spiral galaxies in this sample do not show a strong correlation between the halo metallicity and the total mass of the galaxies, highlighting instead the stochasticity inherent to the halo formation process through accretion events (e.g. Cooper et al. 2010). The advantage of the GHOSTS dataset over the one from Mouhcine et al. (2005b) is that the GHOSTS fields are deeper, there are several pointings per galaxy, and they reach significantly larger galactocentric distances, thus offering a more global view of the haloes of the targets.

In an effort to increase the sample of nearby galaxies for which stellar haloes are resolved and characterized, several groups have individually targeted Local Volume objects with either ground-based or space-borne facilities: the low-mass spirals NGC 2403 (Barker et al. 2012, with Subaru/SuprimeCam), NGC 300 (Vlajić et al. 2009, with Gemini/GMOS) and NGC 55 (Tanaka et al. 2011, with Subaru/SuprimeCam,), the ellipticals NGC 3379 (Harris et al. 2007b, with *HST*) and NGC 3377 (Harris et al. 2007a, with *HST*) and the lenticular NGC 3115 (Peacock et al. 2015, with *HST*). In most of these galaxies, a resolved faint halo (or at least an extended, faint and diffuse component) has been detected and is characterized by populations more metal-poor than the central/disk regions. Most of these haloes also show signs of substructure, pointing at past accretion/merger events as predicted by a hierarchical galaxy formation model. Even galaxies as far as the central elliptical of the Virgo cluster, M87 (~16 Mpc), are starting to be targeted with *HST*, although pushing its resolution capabilities to the technical limits (Bird et al. 2010).

While spectroscopically targeting individual RGB stars to obtain radial velocity and metallicity information is still prohibitive beyond the LG (see Sect. 2.2.3), some cutting-edge studies have pushed the limits of spectroscopy for dwarf galaxies within ~1.5 Mpc (e.g. Kirby et al. 2012, and references therein). At the same time, novel spectroscopic techniques are being developed to take full advantage of the information locked into galaxy haloes. One example is the use of co-added spectra of individual stars, or stellar blends, to obtain radial velocities, metallicities and possibly gradients in galaxies within ~4 Mpc, as robustly demonstrated by Toloba et al. (2016a). The development of new analysis methods and the advent of high-resolution spectrographs will soon allow for systematic spectroscopic investigations of nearby galaxy haloes which will importantly complement the available photometric studies, similarly to the studies of LG galaxies.

Besides the systematic studies presented here, which mostly involve deep space observations, an increasing effort is being invested in producing spatial density maps of outer haloes in some of the closest galaxies with ground-based observations, akin to the panoramic view of M31 offered by PAndAS. In the following section, we describe some of these efforts.

2.3.2 Panoramic Views of Individual Galaxies

Panoramic views of nearby galaxies can be obtained with the use of remarkable ground-based wide-field imagers such as Subaru/SuprimeCam and HyperSuprime-Cam and CFTH/MegaCam in the northern hemisphere, and Magellan/Megacam, CTIO/DECam and VISTA/VIRCAM in the southern hemisphere. Clearly, such CMDs cannot reach the depth of those obtained for M31; these studies nevertheless represent cornerstones for our investigation of global halo properties and serve as precursor science cases for the next generation of telescopes that will open new perspectives for this kind of studies to be performed on a significantly larger sample of galaxies. As mentioned in Sect. 2.2.3, the haloes of low-mass galaxies are also starting to be systematically investigated, to gain a more complete picture of galaxy formation at all mass scales. Here we further describe the few examples of spatially extended imaging obtained to date for some of the closest spiral and elliptical galaxies.

2.3.2.1 NGC 891

Despite its relatively large distance (∼9 Mpc, Radburn-Smith et al. 2011), the "MW-twin" NCG 891 (van der Kruit 1984) is one of the first spirals to be individually investigated in resolved light. Its high inclination and absence of a prominent bulge make it an appealing target for halo studies.

Mouhcine et al. (2007) exploit three *HST* pointings located approximately 10 kpc above the disk of NGC 891 to investigate the properties of this galaxy's halo. The broad observed RGB indicates a wide range of metallicities in this population, with metal-rich peaks and extended metal-poor tails. The three fields also show a decreasing mean metallicity trend as a function of increasing distance along the major axis. The mean metallicity of this sample of RGB stars ([Fe/H]~ -1) falls on the halo metallicity-galaxy luminosity relation pointed out by Mouhcine et al. (2005b): this, together with the gradient mentioned before, is in contrast with the lower metallicities and absence of a gradient for non-rotating stars in the inner haloes of the MW and M31 (Chapman et al. 2006; Kalirai et al. 2006). Mouhcine et al. (2007) thus suggest that not all massive galaxies' outskirts are dominated by metal-poor, pressure-supported stellar populations (because of the inclination and absence of a bulge, the studied RGB sample is thought to be representative of the true halo population). A possible explanation is suggested with the presence of two separate populations: a metal-rich one that is present in the most massive galaxies' outskirts and one constituting the metal-poor, pressure-supported halo, coming from the accretion of moderate-mass satellites. For smaller-mass galaxies, the halo would instead be dominated by debris of small satellites with lower metallicities.

Follow-up analysis on the same *HST* dataset has been carried out by Ibata et al. (2009) and Rejkuba et al. (2009). After careful accounting for the internal reddening of the galaxy, a mild metallicity gradient is confirmed in NGC 891's spheroidal component, which is surveyed out to ∼20 kpc (assuming elliptical radii), and suggested to arise from the presence of a distinct outer halo, similarly to the MW (Ibata et al. 2009). Most importantly, and for the first time, this refined analysis reveals a substantial amount of substructure not only in the RGB spatial distribution but also as metallicity fluctuations in the halo of NGC 891. This evidence points at multiple small accretion events that have not fully blended into the smooth halo.

Motivated by these studies, Mouhcine et al. (2010) provide the first attempt to derive a PAndAS-like map of a MW-analogue beyond the LG: their wide-field map of NGC 891's halo is shown in Fig. 2.5. The panoramic survey, performed contiguously with Subaru/SuprimeCam, covers an impressive $\sim 90 \times 90\,\mathrm{kpc}^2$ in the halo of NGC 891 with the *V* and *i* filters, reaching ∼2 mag below the TRGB. Among

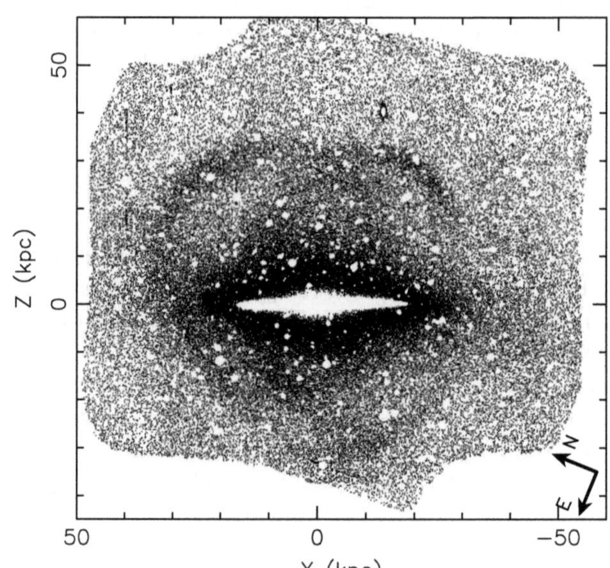

Fig. 2.5 Surface density map of RGB stars in the halo of NGC 891, obtained with Subaru/SuprimeCam. The overdensities of old RGB stars reveal a large complex of arcing streams that loops around the galaxy, tracing the remnants of an ancient accretion. The second spectacular morphological feature is the *dark cocoon-like structure* enveloping the high surface brightness disk and bulge. Figure 1 from Mouhcine et al. (2010), reproduced by permission of the AAS

the abundant substructures uncovered by the RGB map around NGC 891, a system of arcs/streams reaches out some ~50 kpc into the halo, including the first giant stream detected beyond the LG with ground-based imaging. The latter's shape does not rule out a single accretion event origin, but a possible progenitor cannot be identified as a surviving stellar overdensity. These structures appear to be old, given the absence of corresponding overdensities in the luminous AGB (i.e. intermediate-age populations) maps. Another surprising feature highlighted by the RGB map is a flattened, superthick envelope surrounding the disk and bulge of NGC 891, which does not seem to constitute a simple extension of its thick disk but is instead believed to generate from the tidal disruption of satellites given its non-smooth nature (Ibata et al. 2009).

2.3.2.2 M81

Located at a distance of 3.6 Mpc (Radburn-Smith et al. 2011) and with a dynamical mass inside 20 kpc of $\sim 10^{11}\, M_{\odot}$, M81 is one of the closest MW-analogues and has thus been among the first targets for extended halo studies beyond the LG. The earliest HI imaging of the galaxy group dominated by this spiral unambiguously shows a spectacular amount of substructure, most prominently a bridge of gas between M81 and its brightest companions NGC 3077 and M82, located at a projected distance of ~60 kpc (van der Hulst 1979; Yun et al. 1994).

Given the high level of interaction and HI substructure in a group that can be considered as a LG-analogue, it is natural to pursue the investigation of this complex environment even further. The intergalactic gas clouds embedding this environment are traced by young stellar systems identified in resolved stellar studies (Durrell et al. 2004; Davidge 2008; de Mello et al. 2008). Some of them are classified as tidal dwarf galaxies, such as Holmberg IX and the Garland (Makarova et al. 2002; Karachentsev et al. 2004; Sabbi et al. 2008; Weisz et al. 2008), characterized by a predominance of young stellar populations. This type of galaxy has no counterpart in our own LG, and it is believed to be DM-free (see, e.g. Duc et al. 2000).

The first detailed look into the resolved populations in the outskirts of M81 is through the eye of *HST*: the predominantly old halo RGB stars show a broad range of metallicities and a radial gradient (Tikhonov et al. 2005; Mouhcine et al. 2005c). The radial stellar counts (along several different directions) also reveal a break at a radius of ~25 kpc, which is interpreted as the transition point between thick disk and halo (Tikhonov et al. 2005). In a similar fashion, the ground-based wide-field imager Subaru/SuprimeCam has been used to uncover a faint and extended component beyond M81's disk with a flat surface brightness profile extending out to ~0.5 deg (or ~30 kpc) to the north of M81 (Barker et al. 2009). This low surface brightness feature (~28 mag arcsec^{-2}) traced by the brightest RGB star counts appears bluer than the disk, suggesting a metallicity lower than that of M81's main body, but its true nature remains unclear. The authors suggest this component to have intermediate properties between the MW's halo and its thick disk, but the limited surveyed area (0.3 deg^2) precludes any robust conclusions.

As part of a campaign to obtain panoramic views of nearby galaxy haloes, Mouhcine and Ibata (2009) present a $0.9 \times 0.9\,\mathrm{deg}^2$ view of M81's surroundings obtained with the CFHT/MegaCam imager. The images resolve individual RGB stars down to $\sim$2 mag below the TRGB, but this study focusses on the younger, bright populations such as massive main sequence stars and red supergiants, which reveal further young systems tracing the HI tidal distribution between M81 and its companions. These systems are younger than the estimated dynamical age of the large-scale interaction and do not have an old population counterpart, suggesting that they are not simply being detached from the main body of the primary galaxies but are instead formed within the HI clouds.

Durrell et al. (2010) recently conducted a deeper, albeit spatially limited, *HST* study of a field at a galactocentric distance of $\sim$20 kpc. This field reveals an [M/H]~ -1.15 population with an approximate old age of $\sim$9 Gyr. This field thus contains the most metal-poor stars found in M81's halo to that date, which led the authors to the conclusion that they were dealing with an authentic halo component. This study is extended by Monachesi et al. (2013) with the *HST* GHOSTS dataset (see Sect. 2.3.1): they construct a colour profile out to a radius of $\sim$50 kpc, and this dataset does not show a significant gradient. The mean photometric metallicity derived is [Fe/H]~ -1.2, similarly to Durrell et al. (2010). This result is found to be in good agreement with simulations, and the authors suggest that the halo of M81 could have been assembled through an early accretion of satellites with comparable mass (e.g. Cooper et al. 2010; Font et al. 2006).

As a further step in the investigation of M81's halo, the Barker et al. (2009) and Mouhcine and Ibata (2009) ground-based imaging of M81 is being improved by means of the Subaru/HyperSuprimeCam. The first $\sim 2 \times 2\,\mathrm{deg}^2$ ($\sim 100 \times 115\,\mathrm{kpc}^2$) resolved stellar maps from different subpopulations (upper main sequence, red supergiants, RGB and AGB stars) are presented in Okamoto et al. (2015) and constitute a preview of an even wider-field effort to map the extended halo of this group. These first maps (see Fig. 2.6) confirm a high degree of substructure, most interestingly: the youngest populations nicely trace the HI gas content, confirming previous small FoV studies; the RGB distributions are smoother and significantly more extended than the young component and show streamlike overlaps between the dominant group galaxies, e.g. M82's stars clearly being stripped by M81; a redder RGB distribution is detected for M81 and NGC 3077 compared to M82, indicating a lower metallicity in the latter; and, in addition, M82 and NGC 3077's outer regions present S-shaped morphologies, a smoking gun of the tidal interaction with M81 and typical of interacting dwarf galaxies with larger companions (e.g. Peñarrubia et al. 2009).

Not less importantly, the widest-field survey to date ($\sim 65\,\mathrm{deg}^2$) of the M81 group has been performed by Chiboucas et al. (2009) with CFHT/MegaCam, although with only one filter. The main goal of this survey was to identify new, faint dwarf galaxies and investigate the satellite LF in a highly interacting group environment as compared to the LG. This is the first survey to systematically search for faint dwarfs beyond the LG. Resolved spatial overdensities consistent with candidate dwarfs have been followed up with two-band *HST*/ACS and *HST*/WFPC2 observations.

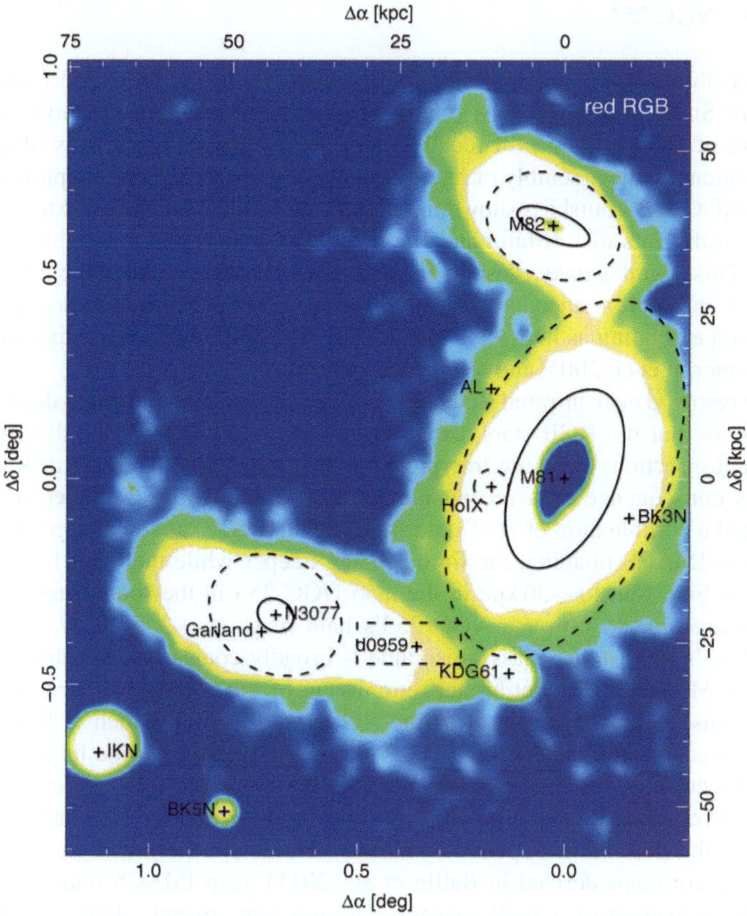

Fig. 2.6 Isodensity contour map of red RGB stars in the M81 group, as observed by Subaru/HyperSuprimeCam. Structures up to 20σ above the background level are visible; the *cross marks* represent the centres of known M81 group members, while *solid lines* are R_{25} of galaxies. The high degree of substructure underlines the ongoing tidal interactions in this group; note in particular the S-shape of the outer regions in NGC 3077 and M82. Figure 4 from Okamoto et al. (2015), reproduced by permission of the AAS

Fourteen of the 22 candidates turned out to be real satellites of M81 based on their CMDs and TRGB distances, extending the previously known galaxy LF in this group by three orders of magnitude down to $M_r \sim -9.0$ (Chiboucas et al. 2013), with an additional possibly ultra-faint member at $M_r \sim -7.0$. The measured slope of the LF in the M81 group appears to be flatter than cosmological predictions ($\alpha \sim -1.27$, in contrast to the theoretical value of $\alpha \sim -1.8$), similar to what has been found for the MW and M31 satellites.

2.3.2.3 NGC 253

Another obvious MW-mass spiral target for halo studies is NCG 253 ($\sim$3.5 Mpc, Radburn-Smith et al. 2011). Its role of brightest object within the loose Sculptor filament of galaxies makes it ideally suited to investigate the effects of external environment on the assembly of haloes. As already apparent from old photographic plates, NGC 253's outskirts show faint perturbation signs, such as an extended shelf to the south of its disk (Malin and Hadley 1997), pointing at a possible accretion event. This spiral galaxy, despite its relative isolation, is experiencing a recent starburst and a pronounced nuclear outflow: the latter is believed to host local star formation extending as high as $\sim$15 kpc above the disk in the minor axis direction (see Comerón et al. 2001, and references therein).

The resolved near-infrared study of Davidge (2010) allowed them to detect bright AGB stars, but not RGB stars, extending out to $\sim$13 kpc from the disk plane in the south direction: these are interpreted as being expelled from the disk into the halo as consequence of a recent interaction. Subsequently, Bailin et al. (2011) exploited a combination of *HST* data from the GHOSTS survey and ground-based Magellan/IMACS imaging, the former being deeper while the latter have a more extended FoV (out to $\sim$30 kpc in the halo NGC 253 in the south direction). The authors are able to estimate NGC 253's halo mass as $\sim 2 \times 10^9 M_\odot$ or 6% of the galaxy's total stellar mass: this value is broadly consistent with those derived from the MW and M31 but higher reminiscent of the halo-to-halo scatter seen in simulations. A power law is fit to the RGB radial profile which is found to be slightly steeper than that of the two LG spirals and appears to be flattened in the same direction as the disk component. This is one of the few studies to date to quantitatively measure such parameters for a halo beyond the LG, and it sets the stage for the possibilities opened by similar studies of other nearby galaxies. The RGB density maps derived in Bailin et al. (2011) from IMACS imaging confirm the early detection of a shelf structure and uncover several additional kpc-scale substructures in the halo of this spiral.

A more recent wide-field study of NGC 253 is presented by Greggio et al. (2014), who exploit the near-infrared VISTA/VIRCAM imager to study the RGB and AGB stellar content of this galaxy out to $\sim$40–50 kpc, covering also the northern portion which was not included in previous studies. This portion, in particular, reveals an RGB substructure symmetric (and likely connected) to the one in the south. A prominent arc ($\sim$20 kpc in length) to the north-west of the disk is detected and estimated to arise from a progenitor with a stellar mass of roughly $\sim 7 \times 10^6 M_\odot$. The RGB radial density profile shows a break at a radius of $\sim$25 kpc, indicative of the transition from disk to halo. The elongated halo component already discussed in Bailin et al. (2011) is confirmed here, but is considered to be an inner halo: an outer, more spherical and homogeneous component extends at least out to the galactocentric distances covered by this survey. Intriguingly, the AGB density map reveals that 25% of this intermediate-age (i.e. up to a few Gyr old) population is spread out to $\sim$30 kpc above the disk: this component cannot easily be explained with either an in situ or an accreted origin.

NGC 253 is also one of the two targets of the Panoramic Imaging Survey of Centaurus and Sculptor (PISCeS), recently initiated with the wide-field imager Magellan/Megacam. This ambitious survey aims at obtaining RGB stellar maps of this galaxy and of the elliptical Centaurus A (Cen A; see next section) out to galactocentric radii of ∼150 kpc, similarly to the PAndAS survey of M31. Early results from this survey include the discovery of two new faint satellites of NGC 253, one of which is clearly elongated and in the process of being disrupted by its host (Sand et al. 2014; Toloba et al. 2016b).

2.3.2.4 NGC 5128 (Centaurus A)

It is important to target galaxies of different morphologies and environments to thoroughly investigate the assembly of haloes. The closest (∼3.8 Mpc; Harris et al. 2010) elliptical galaxy is Centaurus A (Cen A; technically speaking, Maffei 1 is slightly closer, but it lies behind the Galactic disk and is thus heavily reddened; see Wu et al. 2014). Cen A is the dominant galaxy of a rich and dense group, which also has a second subgroup component centred on the spiral M83 (e.g. Karachentsev et al. 2007).

Despite having often been referred to as a peculiar galaxy, due to its pronounced radio activity, its central dust lanes and a perturbed morphology, the luminosity of Cen A is quite typical of field elliptical galaxies: a recent (<1 Gyr) merger event is believed to be the culprit for its peculiar features (see Israel 1998, and references therein). Besides this main merger event, Peng et al. (2002) uncover a system of faint shells and an arc within ∼25 kpc of Cen A's centre from integrated light observations; the arc is believed to have been produced by the infall of a low-mass, star-forming galaxy around ∼300 Myr ago.

This elliptical galaxy has been the subject of a systematic study conducted with *HST*/ACS and *HST*/WFPC2 throughout the past couple of decades: a number of pointings at increasingly large galactocentric radii (from a few out to ∼150 kpc) have been used to investigate the properties and gradients of Cen A's halo populations (Rejkuba et al. 2014, and references therein). The considered pointings out to 40 kpc reveal metal-rich populations ([Fe/H]> -1.0), not dissimilar to what has been observed for the haloes of spiral galaxies. The deepest CMD to date of this elliptical is presented by Rejkuba et al. (2011) for the *HST* field at 40 kpc: this study concludes that the vast majority of Cen A's halo population is old (∼12 Gyr), with a younger (∼2–4 Gyr) component accounting for ∼20% of the total population.

The first wide-field study of Cen A was performed with the ground-based VLT/VIMOS imager, reaching out to ∼85 kpc along both minor and major axes (Crnojević et al. 2013). Cen A's halo population seems to extend all the way out to this large radius. This study confirms the relatively high metallicity for halo populations found by the *HST* studies, although with a considerable presence of metal-poor stars at all radii; the authors also highlight the absence of a strong metallicity gradient from a ∼30 kpc radius out to the most distant regions probed.

This study suggests that the outer regions of Cen A's halo show an increase in ellipticity as a function of radius, which could, however, be interpreted as the presence of substructure contaminating the observed fields. A subsequent study exploits additional *HST* pointings out to a remarkably large radius of ~150 kpc: the edge of Cen A's halo is not reached even by this study (Rejkuba et al. 2014). This dataset, analysed together with the previous *HST* pointings, confirms that a very mild metallicity gradient is present, with median metallicities remaining high out to the largest distances probed. Rejkuba et al. (2014), however, also detect a significant pointing-to-pointing variation in both the RGB star counts and the median metallicity, which is likely indicative of non-mixed accreted populations.

Recently, the PISCeS survey (see previous section) has sketched a PAndAS-like picture of Cen A's halo out to ~150 kpc: the RGB stellar density map derived from a mosaic of Magellan/Megacam images is presented in Fig. 2.7. This map, very much like the ones obtained for M31 and NGC 891, uncovers a plethora of faint substructures, both in the inner regions of the target galaxy and in its outskirts. The morphological variety of these features is reminiscent of that observed in PAndAS,

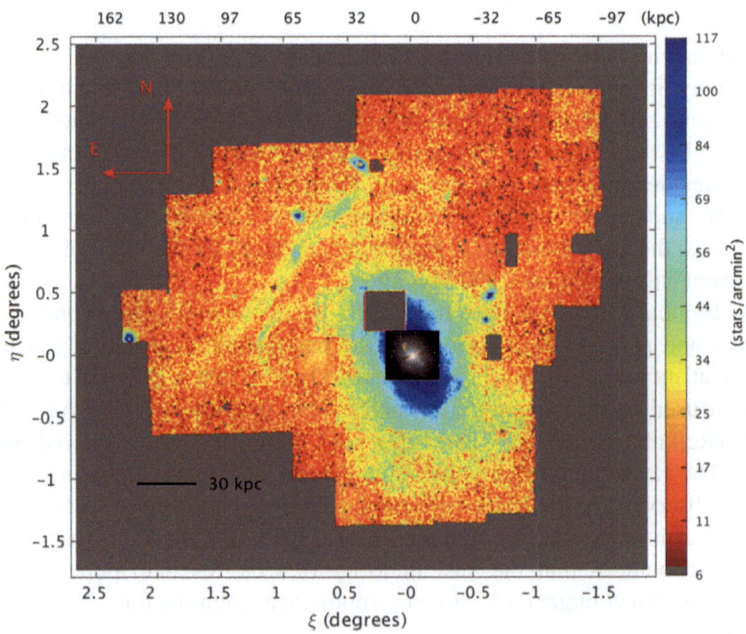

Fig. 2.7 Surface density map of RGB stars in the halo of Cen A, obtained with Magellan/Megacam as part of the PISCeS survey. The map extends out to a radius of 150 kpc in the north and east directions (physical and density scales are reported). Several tidal features are easily recognized, including a stunning disrupting dwarf with tails 2 deg long in the outer halo, an extended sparse cloud to the south of the galaxy, as well as arcs and plumes around the inner regions, tracing both ongoing and past accretion events. Figure 3 from Crnojević et al. (2016), reproduced by permission of the AAS

with shells, plumes, an extended cloud and long tidal streams. In particular, one of the newly discovered dwarf satellites of Cen A is clearly in the process of being disrupted, with ~2 deg long tails: taking into account the stellar content of these tails, this galaxy's pre-disruption luminosity could have been similar to that of Sagittarius in the LG. This survey also led to the discovery of nine (confirmed) dwarf satellites down to $M_V \sim -7$. Their properties are consistent with those of faint LG satellites, but some of them lie at the faint/diffuse end of the LG luminosity/surface brightness/radius distribution: this indicates that we might be looking at previously unexplored physical regimes for these faintest satellites, which opens new exciting perspectives for future studies.

2.4 Summary and Future Prospects

In a ΛCDM hierarchical model, all galaxies are predicted to have experienced mergers, of which many should be recognizable as debris/streams that make up for a large fraction of their haloes. Haloes and their substructures thus provide a unique glimpse into the assembly history of galaxies and can inform the models at the smallest galactic scales, where they still fall short in reproducing observations. The time is now ripe for in-depth systematic studies of the resolved stellar populations in galaxy haloes, which will dramatically increase our understanding of galaxy evolution over the next decade.

The challenges for this type of studies are of a different nature: for our own Galaxy, state-of-the-art results on its halo shape, profile and mass inevitably suffer from assumptions on underlying density models and extrapolations of the available data to radii larger than observed. The major current limitation of MW halo studies lies in observational biases due to small field-of-view samples, which preclude the identification of possible substructure contamination. Future surveys hold the promise to advance the knowledge of our Galaxy by obtaining significantly larger samples of tracers, especially in areas so far not covered. Most notably, the astrometric *Gaia* mission (which will provide unprecedented six-dimensional phase space information for two billion stars out to the inner MW halo) and the Large Synoptic Survey Telescope (LSST, designed to provide a southern sky counterpart to SDSS and reaching ~4 magnitudes fainter than its predecessor for a total sample of tens of billions of stars) are going to revolutionize our view of the MW. At the same time, the current and future generation of high-resolution spectrographs will follow up these surveys from the ground, providing comprehensive kinematic and chemical information to assess the origin of halo stars and characterize their birthplaces (see also Figueras 2017).

The pioneering studies of an increasing number of haloes beyond the LG, and across a range of masses, will soon be extended by the next generation of ground-based extremely large telescopes (E-ELT, GMT, TMT), as well as space-borne missions (*JWST*, *Euclid*, *WFIRST*). The PAndAS survey of M31 has extensively demonstrated that only the synergy of wide-field ground-based observations, deep (but spatially limited) observations from space and spectroscopy can return a truly global understanding of haloes made up of a complex mixture of in situ and accreted populations. The aforementioned facilities will open new perspectives with wide-field optical and infrared imagers in concert with high-resolution spectrographs, which will allow us to systematically survey hundreds of galaxies within tens of Mpc in the next decade or two. For example, with the E-ELT/MICADO and *JWST*/NIRCam imagers (the former having higher resolving power and the latter a wider field of view), we should resolve stars down to the HB within ~10 Mpc, thus identifying and characterizing the SFHs of streams and faint satellites; derive radial profiles, MDFs and stellar population gradients in haloes within 20 Mpc from the uppermost few magnitudes of the RGB; and trace the halo shape and possible overdensities down to $\mu_V \sim 33$ mag arcsec^{-2} from the uppermost ~0.5 mag of the RGB out to 50 Mpc (Greggio et al. 2016).

These observational constraints will be crucial to inform increasingly sophisticated theoretical models and ultimately answer intriguing open questions (as well as possibly unexpected ones that will likely be raised by these observations themselves), such as:

- Do all galaxies have haloes?
- What are the relative fractions of in situ versus accreted populations in galaxy haloes, and how does this depend on galactocentric distance, galaxy morphology and environment?
- What are the properties of the objects currently being accreted, i.e. mass, chemical content, SFH and orbital properties, and how do they relate to those of the present-day low-mass satellites?
- Do low-mass galaxies possess haloes/satellites of their own, and what is their fate and contribution upon infall onto a massive galaxy?
- How extended really are the haloes of massive galaxies?
- What is the shape and mass of the DM haloes underlying galaxies?
- What is the relation between the outer halo and the bulge/disk of a galaxy?
- What is the role of internal versus external processes in shaping a galaxy's properties, especially at the low-mass end of the galaxy LF?
- What is the relation between the present-day haloes/satellites and their unresolved, high-redshift counterparts?

The era of resolved populations in galaxy haloes has just begun, and it holds the promise to be a golden one.

Acknowledgements I would like to thank the organizers for a lively and stimulating conference. I am indebted to S. Pasetto for his advice and support throughout the preparation of this contribution. I acknowledge the hospitality of the Carnegie Observatories during the completion of this work.

References

Abadi, M.G., Navarro, J.F., Steinmetz, M.: Stars beyond galaxies: the origin of extended luminous haloes around galaxies. Mon. Not. R. Astron. Soc. **365**, 747–758 (2006). doi:10.1111/j.1365-2966.2005.09789.x, astro-ph/0506659

Adams, S.M., Zaritsky, D., Sand, D.J., Graham, M.L., Bildfell, C., Hoekstra, H., Pritchet, C.: The environmental dependence of the incidence of galactic tidal features. Astron. J. **144**, 128 (2012). doi:10.1088/0004-6256/144/5/128, 1208.4843

Atkinson, A.M., Abraham, R.G., Ferguson, A.M.N.: Faint tidal features in galaxies within the Canada-France-Hawaii Telescope legacy survey wide fields. Astrophys. J. **765**, 28 (2013). doi:10.1088/0004-637X/765/1/28, 1301.4275

Bailin, J., Bell, E.F., Chappell, S.N., Radburn-Smith, D.J., de Jong, R.S.: The resolved stellar halo of NGC 253. Astrophys. J. **736**, 24 (2011). doi:10.1088/0004-637X/736/1/24, 1105.0005

Barker, M.K., Ferguson, A.M.N., Irwin, M., Arimoto, N., Jablonka, P.: Resolving the stellar outskirts of M81: evidence for a faint, extended structural component. Astron. J. **138**, 1469–1484 (2009). doi:10.1088/0004-6256/138/5/1469, 0909.1430

Barker, M.K., Ferguson, A.M.N., Irwin, M.J., Arimoto, N., Jablonka, P.: Quantifying the faint structure of galaxies: the late-type spiral NGC 2403. Mon. Not. R. Astron. Soc. **419**, 1489–1506 (2012). doi:10.1111/j.1365-2966.2011.19814.x, 1109.2625

Bechtol, K., Drlica-Wagner, A., Balbinot, E., Pieres, A., Simon, J.D., Yanny, B., Santiago, B., Wechsler, R.H., Frieman, J., Walker, A.R., Williams, P., Rozo, E., Rykoff, E.S., Queiroz, A., Luque, E., Benoit-Lévy, A., Tucker, D., Sevilla, I., Gruendl, R.A., da Costa, L.N., Fausti Neto, A., Maia, M.A.G., Abbott, T., Allam, S., Armstrong, R., Bauer, A.H., Bernstein, G.M., Bernstein, R.A., Bertin, E., Brooks, D., Buckley-Geer, E., Burke, D.L., Carnero Rosell, A., Castander, F.J., Covarrubias, R., D'Andrea, C.B., DePoy, D.L., Desai, S., Diehl, H.T., Eifler, T.F., Estrada, J., Evrard, A.E., Fernandez, E., Finley, D.A., Flaugher, B., Gaztanaga, E., Gerdes, D., Girardi, L., Gladders, M., Gruen, D., Gutierrez, G., Hao, J., Honscheid, K., Jain, B., James, D., Kent, S., Kron, R., Kuehn, K., Kuropatkin, N., Lahav, O., Li, T.S., Lin, H., Makler, M., March, M., Marshall, J., Martini, P., Merritt, K.W., Miller, C., Miquel, R., Mohr, J., Neilsen, E., Nichol, R., Nord, B., Ogando, R., Peoples, J., Petravick, D., Plazas, A.A., Romer, A.K., Roodman, A., Sako, M., Sanchez, E., Scarpine, V., Schubnell, M., Smith, R.C., Soares-Santos, M., Sobreira, F., Suchyta, E., Swanson, M.E.C., Tarle, G., Thaler, J., Thomas, D., Wester, W., Zuntz, J., DES Collaboration: Eight new milky way companions discovered in first-year dark energy survey data. Astrophys. J. **807**, 50 (2015). doi:10.1088/0004-637X/807/1/50, 1503.02584

Bell, E.F., Zucker, D.B., Belokurov, V., Sharma, S., Johnston, K.V., Bullock, J.S., Hogg, D.W., Jahnke, K., de Jong, J.T.A., Beers, T.C., Evans, N.W., Grebel, E.K., Ivezić, Ž., Koposov, S.E., Rix, H.W., Schneider, D.P., Steinmetz, M., Zolotov, A.: The accretion origin of the milky way's stellar halo. Astrophys. J. **680**, 295–311 (2008). doi:10.1086/588032, 0706.0004

Belokurov, V.: Galactic archaeology: the dwarfs that survived and perished. NewAR **57**, 100–121 (2013). doi:10.1016/j.newar.2013.07.001, 1307.0041

Belokurov, V., Zucker, D.B., Evans, N.W., Gilmore, G., Vidrih, S., Bramich, D.M., Newberg, H.J., Wyse, R.F.G., Irwin, M.J., Fellhauer, M., Hewett, P.C., Walton, N.A., Wilkinson, M.I., Cole, N., Yanny, B., Rockosi, C.M., Beers, T.C., Bell, E.F., Brinkmann, J., Ivezić, Ž., Lupton, R.: The field of streams: sagittarius and its siblings. Astrophys. J. Lett. **642**, L137–L140 (2006). doi:10.1086/504797, astro-ph/0605025

Bernard, E.J., Ferguson, A.M.N., Richardson, J.C., Irwin, M.J., Barker, M.K., Hidalgo, S.L., Aparicio, A., Chapman, S.C., Ibata, R.A., Lewis, G.F., McConnachie, A.W., Tanvir, N.R.: The nature and origin of substructure in the outskirts of M31 - II. Detailed star formation histories. Mon. Not. R. Astron. Soc. **446**, 2789–2801 (2015). doi:10.1093/mnras/stu2309, 1406.2247

Bernard, E.J., Ferguson, A.M.N., Schlafly, E.F., Martin, N.F., Rix, H.W., Bell, E.F., Finkbeiner, D.P., Goldman, B., Martínez-Delgado, D., Sesar, B., Wyse, R.F.G., Burgett, W.S., Chambers, K.C., Draper, P.W., Hodapp, K.W., Kaiser, N., Kudritzki, R.P., Magnier, E.A., Metcalfe, N.,

Wainscoat, R.J., Waters, C.: A synoptic map of halo substructures from the Pan-STARRS1 3π Survey. Mon. Not. R. Astron. Soc. **463**, 1759–1768 (2016). doi:10.1093/mnras/stw2134, 1607.06088

Bird, S., Harris, W.E., Blakeslee, J.P., Flynn, C.: The inner halo of M 87: a first direct view of the red-giant population. Astron. Astrophys. **524**, A71 (2010). doi:10.1051/0004-6361/201014876, 1009.3202

Bland-Hawthorn, J., Gerhard, O.: The galaxy in context: structural, kinematic, and integrated properties. Annu. Rev. Astron. Astrophys. **54**, 529–596 (2016). doi:10.1146/annurev-astro-081915-023441, 1602.07702

Boylan-Kolchin, M., Bullock, J.S., Kaplinghat, M.: Too big to fail? The puzzling darkness of massive Milky Way subhaloes. Mon. Not. R. Astron. Soc. **415**, L40–L44 (2011). doi:10.1111/j.1745-3933.2011.01074.x, 1103.0007

Brooks, A.M., Kuhlen, M., Zolotov, A., Hooper, D.: A baryonic solution to the missing satellites problem. Astrophys. J. **765**, 22 (2013). doi:10.1088/0004-637X/765/1/22, 1209.5394

Brown, T.M., Smith, E., Ferguson, H.C., Rich, R.M., Guhathakurta, P., Renzini, A., Sweigart, A.V., Kimble, R.A.: The detailed star formation history in the spheroid, outer disk, and tidal stream of the andromeda galaxy. Astrophys. J. **652**, 323–353 (2006). doi:10.1086/508015, astro-ph/0607637

Bullock, J.S., Johnston, K.V.: Tracing galaxy formation with stellar halos. I. Methods. Astrophys. J. **635**, 931–949 (2005). doi:10.1086/497422, arXiv:astro-ph/0506467

Carlin, J.L., Sand, D.J., Price, P., Willman, B., Karunakaran, A., Spekkens, K., Bell, E.F., Brodie, J.P., Crnojević, D., Forbes, D.A., Hargis, J., Kirby, E., Lupton, R., Peter, A.H.G., Romanowsky, A.J., Strader, J.: First results from the MADCASH survey: a faint dwarf galaxy companion to the low-mass spiral galaxy NGC 2403 at 3.2 Mpc. Astrophys. J. Lett. **828**, L5 (2016). doi:10.3847/2041-8205/828/1/L5, 1608.02591

Carollo, D., Beers, T.C., Lee, Y.S., Chiba, M., Norris, J.E., Wilhelm, R., Sivarani, T., Marsteller, B., Munn, J.A., Bailer-Jones, C.A.L., Fiorentin, P.R., York, D.G.: Two stellar components in the halo of the Milky Way. Nature **450**, 1020–1025 (2007). doi:10.1038/nature06460, 0706.3005

Chapman, S.C., Ibata, R., Lewis, G.F., Ferguson, A.M.N., Irwin, M., McConnachie, A., Tanvir, N.: A kinematically selected, metal-poor stellar halo in the outskirts of M31. Astrophys. J. **653**, 255–266 (2006). doi:10.1086/508599, astro-ph/0602604

Chiba, M., Beers, T.C.: Kinematics of metal-poor stars in the galaxy. III. Formation of the stellar halo and thick disk as revealed from a large sample of nonkinematically selected stars. Astron. J. **119**, 2843–2865 (2000). doi:10.1086/301409, astro-ph/0003087

Chiboucas, K., Karachentsev, I.D., Tully, R.B.: Discovery of new dwarf galaxies in the M81 group. Astron. J. **137**, 3009–3037 (2009). doi:10.1088/0004-6256/137/2/3009, 0805.1250

Chiboucas, K., Jacobs, B.A., Tully, R.B., Karachentsev, I.D.: Confirmation of faint dwarf galaxies in the M81 group. Astron. J. **146**, 126 (2013). doi:10.1088/0004-6256/146/5/126, 1309.4130

Cockcroft, R., McConnachie, A.W., Harris, W.E., Ibata, R., Irwin, M.J., Ferguson, A.M.N., Fardal, M.A., Babul, A., Chapman, S.C., Lewis, G.F., Martin, N.F., Puzia, T.H.: Unearthing foundations of a cosmic cathedral: searching the stars for M33's halo. Mon. Not. R. Astron. Soc. **428**, 1248–1262 (2013). doi:10.1093/mnras/sts112, 1210.4114

Collins, M.L.M., Chapman, S.C., Rich, R.M., Ibata, R.A., Martin, N.F., Irwin, M.J., Bate, N.F., Lewis, G.F., Peñarrubia, J., Arimoto, N., Casey, C.M., Ferguson, A.M.N., Koch, A., McConnachie, A.W., Tanvir, N.: The masses of local group dwarf spheroidal galaxies: the death of the universal mass profile. Astrophys. J. **783**, 7 (2014). doi:10.1088/0004-637X/783/1/7, 1309.3053

Comerón, F., Torra, J., Méndez, R.A., Gómez, A.E.: Possible star formation in the halo of NGC 253. Astron. Astrophys. **366**, 796–810 (2001). doi:10.1051/0004-6361: 20000336

Cooper, A.P., Cole, S., Frenk, C.S., White, S.D.M., Helly, J., Benson, A.J., De Lucia, G., Helmi, A., et al.: Galactic stellar haloes in the CDM model. Mon. Not. R. Astron. Soc. **406**, 744–766 (2010). doi:10.1111/j.1365-2966.2010.16740.x, 0910.3211

Cooper, A.P., Parry, O.H., Lowing, B., Cole, S., Frenk, C.: Formation of in situ stellar haloes in Milky Way-mass galaxies. Mon. Not. R. Astron. Soc. **454**, 3185–3199 (2015). doi:10.1093/mnras/stv2057, 1501.04630

Crnojević, D., Grebel, E.K., Koch, A.: A close look at the Centaurus A group of galaxies. I. Metallicity distribution functions and population gradients in early-type dwarfs. Astron. Astrophys. **516**, A85 (2010). doi:10.1051/0004-6361/200913429, 1002.0341

Crnojević, D., Ferguson, A.M.N., Irwin, M.J., Bernard, E.J., Arimoto, N., Jablonka, P., Kobayashi, C.: The outer halo of the nearest giant elliptical: a VLT/VIMOS survey of the resolved stellar populations in Centaurus A to 85 kpc. Mon. Not. R. Astron. Soc. **432**, 832–847 (2013). doi:10.1093/mnras/stt494, 1303.4736

Crnojević, D., Ferguson, A.M.N., Irwin, M.J., McConnachie, A.W., Bernard, E.J., Fardal, M.A., Ibata, R.A., Lewis, G.F., Martin, N.F., Navarro, J.F., Noël, N.E.D., Pasetto, S.: A PAndAS view of M31 dwarf elliptical satellites: NGC 147 and NGC 185. Mon. Not. R. Astron. Soc. **445**, 3862–3877 (2014). doi:10.1093/mnras/stu2003, 1409.7065

Crnojević, D., Sand, D.J., Spekkens, K., Caldwell, N., Guhathakurta, P., McLeod, B., Seth, A., Simon, J.D., Strader, J., Toloba, E.: The extended halo of Centaurus A: uncovering satellites, streams, and substructures. Astrophys. J. **823**, 19 (2016). doi:10.3847/0004-637X/823/1/19, 1512.05366

Davidge, T.J.: An arc of young stars in the Halo of M82. Astrophys. J. Lett. **678**, L85 (2008). doi:10.1086/588551, 0803.3613

Davidge, T.J.: Shaken, not stirred: the disrupted disk of the starburst Galaxy NGC 253. Astrophys. J. **725**, 1342–1365 (2010). doi:10.1088/0004-637X/725/1/1342, 1011.3006

Davidge, T.J., Jones, J.H.: The evolved stellar content of Holmberg IX. Astron. J. **97**, 1607–1613 (1989). doi:10.1086/115102

de Jong, R.S., Seth, A.C., Radburn-Smith, D.J., Bell, E.F., Brown, T.M., Bullock, J.S., Courteau, S., Dalcanton, J.J., Ferguson, H.C., Goudfrooij, P., Holfeltz, S., Holwerda, B.W., Purcell, C., Sick, J., Zucker, D.B.: Stellar populations across the NGC 4244 truncated galactic disk. Astrophys. J. Lett. **667**, L49–L52 (2007). doi:10.1086/522035, 0708.0826

de Mello, D.F., Smith, L.J., Sabbi, E., Gallagher, J.S., Mountain, M., Harbeck, D.R.: Star formation in the H I bridge between M81 and M82. Astron. J. **135**, 548–554 (2008). doi:10.1088/0004-6256/135/2/548, 0711.2685

Deason, A.J., Belokurov, V., Evans, N.W.: The Milky Way stellar halo out to 40 kpc: squashed, broken but smooth. Mon. Not. R. Astron. Soc. **416**, 2903–2915 (2011). doi:10.1111/j.1365-2966.2011.19237.x, 1104.3220

Deason, A.J., Belokurov, V., Evans, N.W., Johnston, K.V.: Broken and unbroken: the milky way and M31 stellar Halos. Astrophys. J. **763**, 113 (2013). doi:10.1088/0004-637X/763/2/113, 1210.4929

D'Onghia, E., Lake, G.: Small dwarf galaxies within larger dwarfs: why some are luminous while most go dark. Astrophys. J. Lett. **686**, L61 (2008). doi:10.1086/592995, 0802.0001

Dooley, G.A., Peter, A.H.G., Yang, T., Willman, B., Griffen, B.F., Frebel A.: An observer's guide to the (Local Group) dwarf galaxies: predictions for their own dwarf satellite populations (2016). ArXiv e-prints 1610.00708

Dorman, C.E., Widrow, L.M., Guhathakurta, P., Seth, A.C., Foreman-Mackey, D., Bell, E.F., Dalcanton, J.J., Gilbert, K.M., Skillman, E.D., Williams, B.F.: A new approach to detailed structural decomposition from the SPLASH and PHAT surveys: kicked-up disk stars in the andromeda galaxy? Astrophys. J. **779**, 103 (2013). doi:10.1088/0004-637X/779/2/103, 1310.4179

Drlica-Wagner, A., Bechtol, K., Rykoff, E.S., Luque, E., Queiroz, A., Mao, Y.Y., Wechsler, R.H., Simon, J.D., Santiago, B., Yanny, B., Balbinot, E., Dodelson, S., Fausti Neto, A., James, D.J., Li, T.S., Maia, M.A.G., Marshall, J.L., Pieres, A., Stringer, K., Walker, A.R., Abbott, T.M.C., Abdalla, F.B., Allam, S., Benoit-Lévy, A., Bernstein, G.M., Bertin, E., Brooks, D., Buckley-Geer, E., Burke, D.L., Carnero Rosell, A., Carrasco Kind, M., Carretero, J., Crocce, M., da Costa, L.N., Desai, S., Diehl, H.T., Dietrich, J.P., Doel, P., Eifler T.F., Evrard, A.E., Finley, D.A., Flaugher, B., Fosalba, P., Frieman, J., Gaztanaga, E., Gerdes, D.W., Gruen, D., Gruendl,

R.A., Gutierrez, G., Honscheid, K., Kuehn, K., Kuropatkin, N., Lahav, O., Martini, P., Miquel, R., Nord, B., Ogando, R., Plazas, A.A., Reil, K., Roodman, A., Sako, M., Sanchez, E., Scarpine, V., Schubnell, M., Sevilla-Noarbe, I., Smith, R.C., Soares-Santos, M., Sobreira, F., Suchyta, E., Swanson, M.E.C., Tarle, G., Tucker, D., Vikram, V., Wester, W., Zhang, Y., Zuntz, J., DES Collaboration: Eight ultra-faint galaxy candidates discovered in year two of the dark energy survey. Astrophys. J. **813**, 109 (2015). doi:10.1088/0004-637X/813/2/109, 1508.03622

Duc, P., Brinks, E., Springel, V., Pichardo, B., Weilbacher, P., Mirabel, I.F.: Formation of a tidal dwarf galaxy in the interacting system Arp 245 (NGC 2992/93). Astron. J. **120**, 1238–1264 (2000). doi:10.1086/301516, arXiv:astro-ph/0006038

Duffau, S., Zinn, R., Vivas, A.K., Carraro, G., Méndez, R.A., Winnick, R., Gallart, C.: Spectroscopy of QUEST RR lyrae variables: the new virgo stellar stream. Astrophys. J. Lett. **636**, L97–L100 (2006). doi:10.1086/500130, astro-ph/0510589

Durrell, P.R., Harris, W.E., Pritchet, C.J.: Photometry and the metallicity distribution of the outer halo of M31. Astron. J. **121**, 2557–2571 (2001). doi:10.1086/320403, arXiv:astro-ph/0101436

Durrell, P.R., Decesar, M.E., Ciardullo, R., Hurley-Keller, D., Feldmeier, J.J.: A CFH12K survey of red giant stars in the M81 Group. In: Duc, P.A., Braine, J., Brinks, E. (eds.) Recycling Intergalactic and Interstellar Matter, IAU Symposium, vol. 217, p. 90 (2004). astro-ph/0311130

Durrell, P.R., Sarajedini, A., Chandar, R.: Deep HST/ACS Photometry of the M81 Halo. Astrophys. J. **718**, 1118–1127 (2010). doi:10.1088/0004-637X/718/2/1118, 1006.2036

Eggen, O.J., Lynden-Bell, D., Sandage, A.R.: Evidence from the motions of old stars that the Galaxy collapsed. Astrophys. J. **136**, 748 (1962). doi:10.1086/147433

Elson, R.A.W.: Red giants in the halo of the S0 galaxy NGC 3115: a distance and a bimodal metallicity distribution. Mon. Not. R. Astron. Soc. **286**, 771–776 (1997). doi:10.1093/mnras/286.3.771, astro-ph/9612037

Fardal, M.A., Weinberg, M.D., Babul, A., Irwin, M.J., Guhathakurta, P., Gilbert, K.M., Ferguson, A.M.N., Ibata, R.A., Lewis, G.F., Tanvir, N.R., Huxor, A.P.: Inferring the Andromeda Galaxy's mass from its giant southern stream with Bayesian simulation sampling. Mon. Not. R. Astron. Soc. **434**, 2779–2802 (2013). doi:10.1093/mnras/stt1121, 1307.3219

Ferguson, A.M.N., Mackey, A.D.: Substructure and tidal streams in the Andromeda Galaxy and its satellites. In: Newberg, H.J., Carlin, J.L. (eds.) Astrophysics and Space Science Library, vol. 420, p. 191 (2016). doi:10.1007/978-3-319-19336-6_8, 1603.01993

Ferguson, A.M.N., Irwin, M.J., Ibata, R.A., Lewis, G.F., Tanvir, N.R.: Evidence for stellar substructure in the halo and outer disk of M31. Astron. J. **124**, 1452–1463 (2002). doi:10.1086/342019, arXiv:astro-ph/0205530

Ferguson, A.M.N., Johnson, R.A., Faria, D.C., Irwin, M.J., Ibata, R.A., Johnston, K.V., Lewis, G.F., Tanvir, N.R.: The stellar populations of the M31 Halo substructure. Astrophys. J. Lett. **622**, L109–L112 (2005). doi:10.1086/429371, astro-ph/0501511

Fernando, N., Arias, V., Guglielmo, M., Lewis, G.F., Ibata, R.A., Power C.: On the stability of satellite planes I: effects of mass, velocity, halo shape and alignment (2016). ArXiv e-prints 1610.05393

Figueras, F.: Outer regions of the Milky Way. In: Knapen, J.H., Lee, J.C., Gil de Paz, A. (eds.) Outskirts of Galaxies, vol. 434. Springer, Cham (2017). doi:10.1007/978-3-319-56570-5

Font, A.S., Johnston, K.V., Bullock, J.S., Robertson, B.E.: Chemical abundance distributions of galactic halos and their satellite systems in a ΛCDM universe. Astrophys. J. **638**, 585–595 (2006). doi:10.1086/498970, astro-ph/0507114

Font, A.S., McCarthy, I.G., Crain, R.A., Theuns, T., Schaye, J., Wiersma, R.P.C., Dalla Vecchia, C.: Cosmological simulations of the formation of the stellar haloes around disc galaxies. Mon. Not. R. Astron. Soc. **416**, 2802–2820 (2011). doi:10.1111/j.1365-2966.2011.19227.x, 1102.2526

Frebel, A., Norris, J.E.: Near-field cosmology with extremely metal-poor stars. Annu. Rev. Astron. Astrophys. **53**, 631–688 (2015). doi:10.1146/annurev-astro-082214-122423, 1501.06921

Freeman, K., Bland-Hawthorn, J.: The new galaxy: signatures of its formation. Annu. Rev. Astron. Astrophys. **40**, 487–537 (2002). doi:10.1146/annurev.astro.40.060401.093840, astro-ph/0208106

Gallart, C., Stetson, P.B., Hardy, E., Pont, F., Zinn, R.: Surface brightness and stellar populations at the outer edge of the large magellanic cloud: no stellar halo yet. Astrophys. J. Lett. **614**, L109–L112 (2004). doi:10.1086/425866, astro-ph/0409023

Gallart, C., Zoccali, M., Aparicio, A.: The adequacy of stellar evolution models for the interpretation of the color-magnitude diagrams of resolved stellar populations. Annu. Rev. Astron. Astrophys. **43**, 387–434 (2005). doi:10.1146/annurev.astro.43.072103.150608

Garrison-Kimmel, S., Boylan-Kolchin, M., Bullock, J.S., Kirby, E.N.: Too big to fail in the Local Group. Mon. Not. R. Astron. Soc. **444**, 222–236 (2014). doi:10.1093/mnras/stu1477, 1404.5313

Georgiev, T.B., Bilkina, B.I., Tikhonov, N.A.: The distribution of blue and red stars around the M81 galaxy. Astron. Astrophys. Suppl. Ser. **96**, 569–581 (1992)

Gilbert, K.M., Guhathakurta, P., Kollipara, P., Beaton, R.L., Geha, M.C., Kalirai, J.S., Kirby, E.N., Majewski, S.R., Patterson, R.J.: The splash survey: a spectroscopic portrait of Andromeda's giant southern stream. Astrophys. J. **705**, 1275–1297 (2009). doi:10.1088/0004-637X/705/2/1275, 0909.4540

Gilbert, K.M., Guhathakurta, P., Beaton, R.L., Bullock, J., Geha, M.C., Kalirai, J.S., Kirby, E.N., Majewski, S.R., Ostheimer, J.C., Patterson, R.J., Tollerud, E.J., Tanaka, M., Chiba, M.: Global properties of M31's stellar Halo from the SPLASH survey. I. Surface brightness profile. Astrophys. J. **760**, 76 (2012). doi:10.1088/0004-637X/760/1/76, 1210.3362

Gilbert, K.M., Kalirai, J.S., Guhathakurta, P., Beaton, R.L., Geha, M.C., Kirby, E.N., Majewski, S.R., Patterson, R.J., Tollerud, E.J., Bullock, J.S., Tanaka, M., Chiba, M.: Global properties of M31's stellar halo from the SPLASH survey. II. Metallicity profile. Astrophys. J. **796**, 76 (2014). doi:10.1088/0004-637X/796/2/76, 1409.3843

Grebel, E.K., Seitzer, P., Dolphin, A., Geisler, D., Guhathakurta, P., Hodge, P., Karachentseva, I., Sarajedini, A.: A dwarf galaxy survey in the local volume. In: Alloin, D., Olsen, K., Galaz, G. (eds.) Stars, Gas and Dust in Galaxies: Exploring the Links, Astronomical Society of the Pacific Conference Series, vol. 221, p. 147 (2000)

Greggio, L., Rejkuba, M., Gonzalez, O.A., Arnaboldi, M., Iodice, E., Irwin, M., Neeser, M.J., Emerson, J.: A panoramic VISTA of the stellar halo of NGC 253. Astron. Astrophys. **562**, A73 (2014). doi:10.1051/0004-6361/201322759, 1401.1665

Greggio, L., Falomo, R., Uslenghi, M.: Studying stellar halos with future facilities. In: Bragaglia, A., Arnaboldi, M., Rejkuba, M., Romano, D. (eds.) The General Assembly of Galaxy Halos: Structure, Origin and Evolution, IAU Symposium, vol. 317, pp. 209–214 (2016). doi:10.1017/S1743921315007024, 1510.03181

Grillmair, C.J.: Detection of a 60deg-long dwarf galaxy debris stream. Astrophys. J. Lett. **645**, L37–L40 (2006). doi:10.1086/505863, astro-ph/0605396

Guhathakurta, P., Rich, R.M., Reitzel, D.B., Cooper, M.C., Gilbert, K.M., Majewski, S.R., Ostheimer, J.C., Geha, M.C., Johnston, K.V., Patterson, R.J.: Dynamics and stellar content of the giant southern stream in M31. I. Keck spectroscopy of red giant stars. Astron. J. **131**, 2497–2513 (2006). doi:10.1086/499562, astro-ph/0406145

Hammer, F., Puech, M., Chemin, L., Flores, H., Lehnert, M.D.: The milky way, an exceptionally quiet galaxy: implications for the formation of spiral galaxies. Astrophys. J. **662**, 322–334 (2007). doi:10.1086/516727, astro-ph/0702585

Harris, W.E., Harris, G.L.H., Layden, A.C., Stetson, P.B.: Hubble space telescope photometry for the halo stars in the leo elliptical NGC 3377. Astron. J. **134**, 43–55 (2007a). doi:10.1086/518233, 0706.1997

Harris, W.E., Harris, G.L.H., Layden, A.C., Wehner, E.M.H.: The leo elliptical NGC 3379: a metal-poor halo emerges. Astrophys. J. **666**, 903–918 (2007b). doi:10.1086/520799, 0706.1995

Harris, G.L.H., Rejkuba, M., Harris, W.E.: The distance to NGC 5128 (Centaurus A). Publ. Astron. Soc. Aust. **27**, 457–462 (2010). doi:10.1071/AS09061, 0911.3180

Hidalgo, S.L., Monelli, M., Aparicio, A., Gallart, C., Skillman, E.D., Cassisi, S., Bernard, E.J., Mayer, L., Stetson, P., Cole, A., Dolphin, A.: The ACS LCID Project. IX. Imprints of the early universe in the radial variation of the star formation history of dwarf galaxies. Astrophys. J. **778**, 103 (2013). doi:10.1088/0004-637X/778/2/103, 1309.6130

Higgs, C.R., McConnachie, A.W., Irwin, M., Bate, N.F., Lewis, G.F., Walker, M.G., Côté, P., Venn, K., Battaglia, G.: Solo dwarfs I: survey introduction and first results for the Sagittarius dwarf irregular galaxy. Mon. Not. R. Astron. Soc. **458**, 1678–1695 (2016). doi:10.1093/mnras/stw257, 1602.01881

Humphreys, R.M., Aaronson, M., Lebofsky, M., McAlary, C.W., Strom, S.E., Capps, R.W.: The luminosities of M supergiants and the distances to M101, NGC 2403, and M81. Astron. J. **91**, 808–821 (1986). doi:10.1086/114061

Huxor, A.P., Mackey, A.D., Ferguson, A.M.N., Irwin, M.J., Martin, N.F., Tanvir, N.R., Veljanoski, J., McConnachie, A., Fishlock, C.K., Ibata, R., Lewis, G.F.: The outer halo globular cluster system of M31 - I. The final PAndAS catalogue. Mon. Not. R. Astron. Soc. **442**, 2165–2187 (2014). doi:10.1093/mnras/stu771, 1404.5807

Ibata, R.A., Gilmore, G., Irwin, M.J.: A dwarf satellite galaxy in Sagittarius. Nature **370**, 194–196 (1994). doi:10.1038/370194a0

Ibata, R., Irwin, M., Lewis, G., Ferguson, A.M.N., Tanvir, N.: A giant stream of metal-rich stars in the halo of the galaxy M31. Nature **412**, 49–52 (2001). astro-ph/0107090

Ibata, R., Martin, N.F., Irwin, M., Chapman, S., Ferguson, A.M.N., Lewis, G.F., McConnachie, A.W.: The haunted halos of andromeda and triangulum: a panorama of galaxy formation in action. Astrophys. J. **671**, 1591–1623 (2007). doi:10.1086/522574, 0704.1318

Ibata, R., Mouhcine, M., Rejkuba, M.: An HST/ACS investigation of the spatial and chemical structure and sub-structure of NGC 891, a Milky Way analogue. Mon. Not. R. Astron. Soc. **395**, 126–143 (2009). doi:10.1111/j.1365-2966.2009.14536.x, 0903.4209

Ibata, R.A., Lewis, G.F., Conn, A.R., Irwin, M.J., McConnachie, A.W., Chapman, S.C., Collins, M.L., Fardal, M., Ferguson, A.M.N., Ibata, N.G., Mackey, A.D., Martin, N.F., Navarro, J., Rich, R.M., Valls-Gabaud, D., Widrow, L.M.: A vast, thin plane of corotating dwarf galaxies orbiting the Andromeda galaxy. Nature **493**, 62–65 (2013). doi:10.1038/nature11717, 1301.0446

Ibata, R.A., Lewis, G.F., McConnachie, A.W., Martin, N.F., Irwin, M.J., Ferguson, A.M.N., Babul, A., Bernard, E.J., Chapman, S.C., Collins, M., Fardal, M., Mackey, A.D., Navarro, J., Peñarrubia, J., Rich, R.M., Tanvir, N., Widrow, L.: The large-scale structure of the halo of the andromeda galaxy. I. Global stellar density, morphology and metallicity properties. Astrophys. J. **780**, 128 (2014). doi:10.1088/0004-637X/780/2/128, 1311.5888

Irwin, M.J., Ferguson, A.M.N., Ibata, R.A., Lewis, G.F., Tanvir, N.R.: A minor-axis surface brightness profile for M31. Astrophys. J. Lett. **628**, L105–L108 (2005). doi:10.1086/432718, astro-ph/0505077

Israel, F.P.: Centaurus A - NGC 5128. Astron. Astrophys. Rev. **8**, 237–278 (1998). doi:10.1007/s001590050011, arXiv:astro-ph/9811051

Ivezić, Ž., Goldston, J., Finlator, K., Knapp, G.R., Yanny, B., McKay, T.A., Amrose, S., Krisciunas, K., Willman, B., Anderson, S., Schaber, C., Erb, D., Logan, C., Stubbs, C., Chen, B., Neilsen, E., Uomoto, A., Pier, J.R., Fan, X., Gunn, J.E., Lupton, R.H., Rockosi, C.M., Schlegel, D., Strauss, M.A., Annis, J., Brinkmann, J., Csabai, I., Doi, M., Fukugita, M., Hennessy, G.S., Hindsley, R.B., Margon, B., Munn, J.A., Newberg, H.J., Schneider, D.P., Smith, J.A., Szokoly, G.P., Thakar, A.R., Vogeley, M.S., Waddell, P., Yasuda, N., York, D.G., SDSS Collaboration: Candidate RR lyrae stars found in sloan digital sky survey commissioning data. Astron. J. **120**, 963–977 (2000). doi:10.1086/301455, astro-ph/0004130

Jablonka, P., Tafelmeyer, M., Courbin, F., Ferguson, A.M.N.: Direct detection of galaxy stellar halos: NGC 3957 as a test case. Astron. Astrophys. **513**, A78 (2010). doi:10.1051/0004-6361/200913320, 1001.3067

Janowiecki, S., Mihos, J.C., Harding, P., Feldmeier, J.J., Rudick, C., Morrison H Diffuse Tidal Structures in the Halos of Virgo Ellipticals. Astrophys. J. **715**, 972–985 (2010). doi:10.1088/0004-637X/715/2/972, 1004.1473

Johnston, K.V., Bullock, J.S., Sharma, S., Font, A., Robertson, B.E., Leitner, S.N.: Tracing galaxy formation with stellar halos. II. Relating substructure in phase and abundance space to accretion histories. Astrophys. J. **689**, 936–957 (2008). doi:10.1086/592228, 0807.3911

Jurić, M., Ivezić, Ž., Brooks, A., Lupton, R.H., Schlegel, D., Finkbeiner, D., Padmanabhan, N., Bond, N., Sesar, B., Rockosi, C.M., Knapp, G.R., Gunn, J.E., Sumi, T., Schneider, D.P.,

Barentine, J.C., Brewington, H.J., Brinkmann, J., Fukugita, M., Harvanek, M., Kleinman, S.J., Krzesinski, J., Long, D., Neilsen, E.H. Jr., Nitta, A., Snedden, S.A., York, D.G.: The milky way tomography with SDSS. I. Stellar number density distribution. Astrophys. J. **673**, 864–914 (2008). doi:10.1086/523619, astro-ph/0510520

Kalirai, J.S., Gilbert, K.M., Guhathakurta, P., Majewski, S.R., Ostheimer, J.C., Rich, R.M., Cooper, M.C., Reitzel, D.B., Patterson, R.J.: The metal-poor halo of the andromeda spiral galaxy (M31)1. Astrophys. J. **648**, 389–404 (2006). doi:10.1086/505697, astro-ph/0605170

Karachentsev, I.D., Karachentseva, V.E., Huchtmeier, W.K., Makarov, D.I.: A catalog of neighboring galaxies. Astron. J. **127**, 2031–2068 (2004). doi:10.1086/382905

Karachentsev, I.D., Tully, R.B., Dolphin, A., Sharina, M., Makarova, L., Makarov, D., Sakai, S., Shaya, E.J., et al.: The hubble flow around the Centaurus A/M83 galaxy complex. Astron. J. **133**, 504–517 (2007). doi:10.1086/510125, arXiv:astro-ph/0603091

Kim, D., Jerjen, H., Mackey, D., Da Costa, G.S., Milone, A.P.: A hero's dark horse: discovery of an ultra-faint milky way satellite in pegasus. Astrophys. J. Lett. **804**, L44 (2015). doi:10.1088/2041-8205/804/2/L44, 1503.08268

Kirby, E.N., Cohen, J.G., Bellazzini, M.: The dynamics and metallicity distribution of the distant dwarf galaxy VV124. Astrophys. J. **751**, 46 (2012). doi:10.1088/0004-637X/751/1/46, 1203.4561

Klypin, A., Kravtsov, A.V., Valenzuela, O., Prada, F.: Where are the missing galactic satellites? Astrophys. J. **522**, 82–92 (1999). doi:10.1086/307643, arXiv:astro-ph/9901240

Knapen, J.H., Trujillo, I.: Ultra-deep imaging: structure of disks and haloes. In: Knapen, J.H., Lee, J.C., Gil de Paz, A. (eds.) Outskirts of Galaxies, vol. 434. Springer, Cham (2017). doi:10.1007/978-3-319-56570-5

Koch, A., Grebel, E.K., Wyse, R.F.G., Kleyna, J.T., Wilkinson, M.I., Harbeck, D.R., Gilmore, G.F., Evans, N.W.: Complexity on small scales: the metallicity distribution of the carina dwarf spheroidal galaxy. Astron. J. **131**, 895–911 (2006). doi:10.1086/499490, arXiv:astro-ph/0511087

Koposov, S.E., Belokurov, V., Torrealba, G., Evans, N.W.: Beasts of the southern wild: discovery of nine ultra faint satellites in the vicinity of the magellanic clouds. Astrophys. J. **805**, 130 (2015). doi:10.1088/0004-637X/805/2/130, 1503.02079

Lewis, G.F., Braun, R., McConnachie, A.W., Irwin, M.J., Ibata, R.A., Chapman, S.C., Ferguson, A.M.N., Martin, N.F., Fardal, M., Dubinski, J., Widrow, L., Mackey, A.D., Babul, A., Tanvir, N.R., Rich, M.: PAndAS in the mist: the stellar and gaseous mass within the halos of M31 and M33. Astrophys. J. **763**, 4 (2013). doi:10.1088/0004-637X/763/1/4, 1211.4059

Lu, Z., Mo, H.J., Lu, Y., Katz, N., Weinberg, M.D., van den Bosch, F.C., Yang, X.: An empirical model for the star formation history in dark matter haloes. Mon. Not. R. Astron. Soc. **439**, 1294–1312 (2014). doi:10.1093/mnras/stu016, 1306.0650

Mackey, A.D., Huxor, A.P., Ferguson, A.M.N., Irwin, M.J., Tanvir, N.R., McConnachie, A.W., Ibata, R.A., Chapman, S.C., Lewis, G.F.: Evidence for an accretion origin for the outer halo globular cluster system of M31. Astrophys. J. Lett. **717**, L11–L16 (2010). doi:10.1088/2041-8205/717/1/L11, 1005.3812

Majewski, S.R.: The role of accretion in the formation of the halo: observational view. In: Gibson, B.K., Axelrod, R.S., Putman, M.E. (eds.) The Third Stromlo Symposium: The Galactic Halo, Astronomical Society of the Pacific Conference Series, vol. 165, p. 76 (1999)

Majewski, S.R., Skrutskie, M.F., Weinberg, M.D., Ostheimer, J.C.: A two micron all sky survey view of the sagittarius dwarf galaxy. I. Morphology of the sagittarius core and tidal arms. Astrophys. J. **599**, 1082–1115 (2003). doi:10.1086/379504, astro-ph/0304198

Makarova, L.N., Grebel, E.K., Karachentsev, I.D., Dolphin, A.E., Karachentseva, V.E., Sharina, M.E., Geisler, D., Guhathakurta, P., Hodge, P.W., Sarajedini, A., Seitzer, P.: Tidal dwarfs in the M81 group: the second generation? Astron. Astrophys. **396**, 473–487 (2002). doi:10.1051/0004-6361 20021426

Malin, D., Hadley, B.: HI in shell galaxies and other merger remnants. Publ. Astron. Soc. Aust. **14**, 52–58 (1997). doi:10.1071/AS97052

Malin, D.F., Quinn, P.J., Graham, J.A.: Shell structure in NGC 5128. Astrophys. J. Lett. **272**, L5–L7 (1983). doi:10.1086/184106

Martell, S.L., Grebel, E.K.: Light-element abundance variations in the Milky Way halo. Astron. Astrophys. **519**, A14 (2010). doi:10.1051/0004-6361/201014135, 1005.4070

Martin, N.F., Nidever, D.L., Besla, G., Olsen, K., Walker, A.R., Vivas, A.K., Gruendl, R.A., Kaleida, C.C., Muñoz, R.R., Blum, R.D., Saha, A., Conn, B.C., Bell, E.F., Chu, Y.H., Cioni, M.R.L., de Boer, T.J.L., Gallart, C., Jin, S., Kunder, A., Majewski, S.R., Martinez-Delgado, D., Monachesi, A., Monelli, M., Monteagudo, L., Noël, N.E.D., Olszewski, E.W., Stringfellow, G.S., van der Marel, R.P., Zaritsky, D.: Hydra II: a faint and compact milky way dwarf galaxy found in the survey of the magellanic stellar history. Astrophys. J. Lett. **804**, L5 (2015). doi:10.1088/2041-8205/804/1/L5, 1503.06216

Martin, N.F., Ibata, R.A., Lewis, G.F., McConnachie, A., Babul, A., Bate, N.F., Bernard, E., Chapman, S.C., Collins, M.M.L., Conn, A.R., Crnojević, D., Fardal, M.A., Ferguson, A.M.N., Irwin, M., Mackey, A.D., McMonigal, B., Navarro, J.F., Rich, R.M.: The PAndAS view of the Andromeda satellite system - II. Detailed properties of 23 M31 dwarf spheroidal galaxies (2016). ArXiv e-prints 1610.01158

Martínez-Delgado, D., Gabany, R.J., Crawford, K., Zibetti, S., Majewski, S.R., Rix, H., Fliri, J., et al.: Stellar tidal streams in spiral galaxies of the local volume: a pilot survey with modest aperture telescopes. Astron. J. **140**, 962–967 (2010). doi:10.1088/0004-6256/140/4/962, 1003.4860

Mateo, M.L.: Dwarf galaxies of the local group. Annu. Rev. Astron. Astrophys. **36**, 435–506 (1998). doi:10.1146/annurev.astro.36.1.435, arXiv:astro-ph/9810070

McConnachie, A.W., Irwin, M.J., Ibata, R.A., Ferguson, A.M.N., Lewis, G.F., Tanvir N.: The three-dimensional structure of the giant stellar stream in Andromeda. Mon. Not. R. Astron. Soc. **343**, 1335–1340 (2003). doi:10.1046/j.1365-8711.2003.06785.x, astro-ph/0305160

McConnachie, A.W., Irwin, M.J., Ibata, R.A., Dubinski, J., Widrow, L.M., Martin, N.F., Côté, P., Dotter, A.L., et al.: The remnants of galaxy formation from a panoramic survey of the region around M31. Nature **461**, 66–69 (2009). doi:10.1038/nature08327, 0909.0398

McMonigal, B., Bate, N.F., Lewis, G.F., Irwin, M.J., Battaglia, G., Ibata, R.A., Martin, N.F., McConnachie, A.W., Guglielmo, M., Conn, A.R.: Sailing under the magellanic clouds: a DECam view of the Carina dwarf. Mon. Not. R. Astron. Soc. **444**, 3139–3149 (2014). doi:10.1093/mnras/stu1659, 1408.2907

McMonigal, B., Lewis, G.F., Brewer, B.J., Irwin, M.J., Martin, N.F., McConnachie, A.W., Ibata, R.A., Ferguson, A.M.N., Mackey, A.D., Chapman, S.C.: The elusive stellar halo of the Triangulum galaxy. Mon. Not. R. Astron. Soc. **461**, 4374–4388 (2016). doi:10.1093/mnras/stw1657, 1607.02190

Monachesi, A., Bell, E.F., Radburn-Smith, D.J., Vlajić, M., de Jong, R.S., Bailin, J., Dalcanton, J.J., Holwerda, B.W., Streich, D.: Testing galaxy formation models with the GHOSTS survey: the color profile of M81's stellar halo. Astrophys. J. **766**, 106 (2013). doi:10.1088/0004-637X/766/2/106, 1302.2626

Monachesi, A., Bell, E.F., Radburn-Smith, D.J., Bailin, J., de Jong, R.S., Holwerda, B., Streich, D., Silverstein, G.: The GHOSTS survey - II. The diversity of halo colour and metallicity profiles of massive disc galaxies. Mon. Not. R. Astron. Soc. **457**, 1419–1446 (2016). doi:10.1093/mnras/stv2987, 1507.06657

Moore, B., Ghigna, S., Governato, F., Lake, G., Quinn, T., Stadel, J., Tozzi P.: Dark matter substructure within galactic halos. Astrophys. J. Lett. **524**, L19–L22 (1999). doi:10.1086/312287, arXiv:astro-ph/9907411

Morrison, H.L.: The local density of halo giants. Astron. J. **106**, 578–590 (1993). doi:10.1086/116662

Morrison, H.L., Boroson, T.A., Harding, P.: Stellar populations in edge-on galaxies from deep CCD surface photometry, 1: NGC 5907. Astron. J. **108**, 1191–1208 (1994). doi:10.1086/117148

Mouhcine, M.: The outskirts of spiral galaxies: evidence for multiple stellar populations. Astrophys. J. **652**, 277–282 (2006). doi:10.1086/504104, astro-ph/0603191

Mouhcine, M., Ibata, R.: A panoramic view of M81: new stellar systems in the debris field. Mon. Not. R. Astron. Soc. **399**, 737–743 (2009). doi:10.1111/j.1365-2966.2009.15135.x

Mouhcine, M., Ferguson, H.C., Rich, R.M., Brown, T.M., Smith, T.E.: Halos of spiral galaxies. I. The tip of the red giant branch as a distance indicator. Astrophys. J. **633**, 810–820 (2005a). doi:10.1086/468177, astro-ph/0510253

Mouhcine, M., Ferguson, H.C., Rich, R.M., Brown, T.M., Smith, T.E.: Halos of spiral galaxies. II. Halo metallicity-luminosity relation. Astrophys. J. **633**, 821–827 (2005b). doi:10.1086/468178, astro-ph/0510254

Mouhcine, M., Rich, R.M., Ferguson, H.C., Brown, T.M., Smith T.E.: Halos of spiral galaxies. III. Metallicity distributions. Astrophys. J. **633**, 828–843 (2005c). doi:10.1086/468179, arXiv:astro-ph/0510255

Mouhcine, M., Rejkuba, M., Ibata, R.: The stellar halo of the edge-on galaxy NGC 891. Mon. Not. R. Astron. Soc. **381**, 873–880 (2007). doi:10.1111/j.1365-2966.2007.12291.x

Mouhcine, M., Ibata, R., Rejkuba, M.: A panoramic view of the milky way analog NGC 891. Astrophys. J. Lett. **714**, L12–L15 (2010). doi:10.1088/2041-8205/714/1/L12, 1002.0461

Mould, J., Kristian, J.: The stellar population in the halos of M31 and M33. Astrophys. J. **305**, 591–599 (1986). doi:10.1086/164273

Newberg, H.J., Yanny, B., Rockosi, C., Grebel, E.K., Rix, H.W., Brinkmann, J., Csabai, I., Hennessy, G., Hindsley, R.B., Ibata, R., Ivezić, Z., Lamb, D., Nash, E.T., Odenkirchen, M., Rave, H.A., Schneider, D.P., Smith, J.A., Stolte, A., York, D.G.: The ghost of sagittarius and lumps in the halo of the milky way. Astrophys. J. **569**, 245–274 (2002). doi:10.1086/338983, astro-ph/0111095

Norris, J.E., Ryan, S.G.: Population studies. XI - The extended disk, halo configuration. Astrophys. J. **380**, 403–418 (1991). doi:10.1086/170599

Odenkirchen, M., Grebel, E.K., Rockosi, C.M., Dehnen, W., Ibata, R., Rix, H.W., Stolte, A., Wolf, C., Anderson, J.E. Jr., Bahcall, N.A., Brinkmann, J., Csabai, I., Hennessy, G., Hindsley, R.B., Ivezić, Ž., Lupton, R.H., Munn, J.A., Pier, J.R., Stoughton, C., York, D.G.: Detection of massive tidal tails around the globular cluster Palomar 5 with sloan digital sky survey commissioning data. Astrophys. J. Lett. **548**, L165–L169 (2001). doi:10.1086/319095, astro-ph/0012311

Okamoto, S., Arimoto, N., Ferguson, A.M.N., Bernard, E.J., Irwin, M.J., Yamada, Y., Utsumi, Y.: A hyper suprime-cam view of the interacting galaxies of the M81 group. Astrophys. J. Lett. **809**, L1 (2015). doi:10.1088/2041-8205/809/1/L1, 1507.04889

Pawlowski, M.S., Famaey, B., Jerjen, H., Merritt, D., Kroupa, P., Dabringhausen, J., Lüghausen, F., Forbes, D.A., Hensler, G., Hammer, F., Puech, M., Fouquet, S., Flores, H., Yang, Y.: Co-orbiting satellite galaxy structures are still in conflict with the distribution of primordial dwarf galaxies. Mon. Not. R. Astron. Soc. **442**, 2362–2380 (2014). doi:10.1093/mnras/stu1005, 1406.1799

Peñarrubia, J., Navarro, J.F., McConnachie, A.W., Martin, N.F.: The signature of galactic tides in local group dwarf spheroidals. Astrophys. J. **698**, 222–232 (2009). doi:10.1088/0004-637X/698/1/222, 0811.1579

Peacock, M.B., Strader, J., Romanowsky, A.J., Brodie, J.P.: Detection of a distinct metal-poor stellar halo in the early-type galaxy NGC 3115. Astrophys. J. **800**, 13 (2015). doi:10.1088/0004-637X/800/1/13, 1412.2752

Peng, E.W., Ford, H.C., Freeman, K.C., White, R.L.: A young blue tidal stream in NGC 5128. Astron. J. **124**, 3144–3156 (2002). doi:10.1086/344308, arXiv:astro-ph/0208422

Peng, E.W., Jordán, A., Côté, P., Blakeslee, J.P., Ferrarese, L., Mei, S., West, M.J., Merritt, D., Milosavljević, M., Tonry, J.L.: The ACS virgo cluster survey. IX. The color distributions of globular cluster systems in early-type galaxies. Astrophys. J. **639**, 95–119 (2006). doi:10.1086/498210, astro-ph/0509654

Pillepich, A., Vogelsberger, M., Deason, A., Rodriguez-Gomez, V., Genel, S., Nelson, D., Torrey, P., Sales, L.V., Marinacci, F., Springel, V., Sijacki, D., Hernquist, L.: Halo mass and assembly history exposed in the faint outskirts: the stellar and dark matter haloes of Illustris galaxies. Mon. Not. R. Astron. Soc. **444**, 237–249 (2014). doi:10.1093/mnras/stu1408, 1406.1174

Pohlen, M., Martínez-Delgado, D., Majewski, S., Palma, C., Prada, F., Balcells, M.: Tidal streams around external galaxies. In: Prada, F., Martinez Delgado, D., Mahoney, T.J. (eds.) Satellites

and Tidal Streams, Astronomical Society of the Pacific Conference Series, vol. 327, p. 288 (2004). astro-ph/0308142
Radburn-Smith, D.J., de Jong, R.S., Seth, A.C., Bailin, J., Bell, E.F., Brown, T.M., Bullock, J.S., Courteau, S., Dalcanton, J.J., Ferguson, H.C., Goudfrooij, P., Holfeltz, S., Holwerda, B.W., Purcell, C., Sick, J., Streich, D., Vlajic, M., Zucker, D.B.: The GHOSTS Survey. I. Hubble space telescope advanced camera for surveys data. Astrophys. J. Suppl. Ser. **195**, 18 (2011). doi:10.1088/0067-0049/195/2/18
Read, J.I.: The local dark matter density. J. Phys. G: Nucl. Part. Phys. **41**(6), 063101 (2014). doi:10.1088/0954-3899/41/6/063101, 1404.1938
Rejkuba, M., Mouhcine, M., Ibata, R.: The stellar population content of the thick disc and halo of the Milky Way analogue NGC 891. Mon. Not. R. Astron. Soc. **396**, 1231–1246 (2009). doi:10.1111/j.1365-2966.2009.14821.x
Rejkuba, M., Harris, W.E., Greggio, L., Harris, G.L.H.: How old are the stars in the halo of NGC 5128 (Centaurus A)? Astron. Astrophys. **526**, A123 (2011). doi:10.1051/0004-6361/201015640, 1011.4309
Rejkuba, M., Harris, W.E., Greggio, L., Harris, G.L.H., Jerjen, H., Gonzalez, O.A.: Tracing the Outer Halo in a Giant Elliptical to 25 R_{eff}. Astrophys. J. Lett. **791**, L2 (2014). doi:10.1088/2041-8205/791/1/L2, 1406.4627
Renda, A., Gibson, B.K., Mouhcine, M., Ibata, R.A., Kawata, D., Flynn, C., Brook, C.B.: The stellar halo metallicity-luminosity relationship for spiral galaxies. Mon. Not. R. Astron. Soc. **363**, L16–L20 (2005). doi:10.1111/j.1745-3933.2005.00075.x, astro-ph/0507281
Richardson, J.C., Ferguson, A.M.N., Johnson, R.A., Irwin, M.J., Tanvir, N.R., Faria, D.C., Ibata, R.A., Johnston, K.V., Lewis, G.F., McConnachie, A.W., Chapman, S.C.: The nature and origin of substructure in the outskirts of M31. I. Surveying the stellar content with the hubble space telescope advanced camera for surveys. Astron. J. **135**, 1998–2012 (2008). doi:10.1088/0004-6256/135/6/1998, 0803.2614
Richardson, J.C., Irwin, M.J., McConnachie, A.W., Martin, N.F., Dotter, A.L., Ferguson, A.M.N., Ibata, R.A., Chapman, S.C., Lewis, G.F., Tanvir, N.R., Rich, R.M.: PAndAS' progeny: extending the M31 dwarf galaxy cabal. Astrophys. J. **732**, 76 (2011). doi:10.1088/0004-637X/732/2/76, 1102.2902
Ryan, S.G., Norris, J.E.: Subdwarf Studies. III. The halo metallicity distribution. Astron. J. **101**, 1865–1878 (1991). doi:10.1086/115812
Sabbi, E., Gallagher, J.S., Smith, L.J., de Mello, D.F., Mountain, M.: Holmberg IX: the nearest young galaxy. Astrophys. J. Lett. **676**, L113 (2008). doi:10.1086/587548, 0802.4446
Sackett, P.D., Morrisoni, H.L., Harding, P., Boroson, T.A.: A faint luminous halo that may trace the dark matter around spiral galaxy NGC5907. Nature **370**, 441–443 (1994). doi:10.1038/370441a0, astro-ph/9407068
Sadoun, R., Mohayaee, R., Colin, J.: A single-merger scenario for the formation of the giant stream and the warp of M31. Mon. Not. R. Astron. Soc. **442**, 160–175 (2014). doi:10.1093/mnras/stu850, 1307.5044
Sakai, S., Madore, B.F.: Detection of the red giant branch stars in M82 using the hubble space telescope. Astrophys. J. **526**, 599–606 (1999). doi:10.1086/308032, astro-ph/9906484
Sales, L.V., Navarro, J.F., Kallivayalil, N., Frenk, C.S.: Identifying true satellites of the Magellanic Clouds (2016). ArXiv e-prints 1605.03574
Sand, D.J., Strader, J., Willman, B., Zaritsky, D., McLeod, B., Caldwell, N., Seth, A., Olszewski, E.: Tidal signatures in the faintest milky way satellites: the detailed properties of Leo V, Pisces II, and Canes Venatici II. Astrophys. J. **756**, 79 (2012). doi:10.1088/0004-637X/756/1/79, 1111.6608
Sand, D.J., Crnojević, D., Strader, J., Toloba, E., Simon, J.D., Caldwell, N., Guhathakurta, P., McLeod, B., Seth, A.C.: Discovery of a new faint dwarf galaxy associated with NGC 253. Astrophys. J. Lett. **793**, L7 (2014). doi:10.1088/2041-8205/793/1/L7, 1406.6687
Sand, D.J., Spekkens, K., Crnojević, D., Hargis, J.R., Willman, B., Strader, J., Grillmair, C.J.: Antlia B: a faint dwarf galaxy member of the NGC 3109 association. Astrophys. J. Lett. **812**, L13 (2015). doi:10.1088/2041-8205/812/1/L13, 1508.01800

Sawala, T., Frenk, C.S., Fattahi, A., Navarro, J.F., Bower, R.G., Crain, R.A., Dalla Vecchia, C., Furlong, M., Helly, J.C., Jenkins, A., Oman, K.A., Schaller, M., Schaye, J., Theuns, T., Trayford, J., White, S.D.M.: The APOSTLE simulations: solutions to the Local Group's cosmic puzzles. Mon. Not. R. Astron. Soc. **457**, 1931–1943 (2016). doi:10.1093/mnras/stw145, 1511.01098

Schiavon, R.P., Zamora, O., Carrera, R., Lucatello, S., Robin, A.C., Ness, M., Martell, S.L., Smith, V.V., García-Hernández, D.A., Manchado, A., Schönrich, R., Bastian, N., Chiappini, C., Shetrone, M., Mackereth, J.T., Williams, R.A., Mészáros, S., Allende Prieto, C., Anders, F., Bizyaev, D., Beers, T.C., Chojnowski, S.D., Cunha, K., Epstein, C., Frinchaboy, P.M., García Pérez, A.E., Hearty, F.R., Holtzman, J.A., Johnson, J.A., Kinemuchi, K., Majewski, S.R., Muna, D., Nidever, D.L., Nguyen, D.C., O'Connell, R.W., Oravetz, D., Pan, K., Pinsonneault, M., Schneider, D.P., Schultheis, M., Simmons, A., Skrutskie, M.F., Sobeck, J., Wilson, J.C., Zasowski, G.: Chemical tagging with APOGEE: discovery of a large population of N-rich stars in the inner Galaxy. Mon. Not. R. Astron. Soc. (2016). doi:10.1093/mnras/stw2162, 1606.05651

Schönrich, R., Asplund, M., Casagrande, L.: Does SEGUE/SDSS indicate a dual Galactic halo? Astrophys. J. **786**, 7 (2014). doi:10.1088/0004-637X/786/1/7, 1403.0937

Searle, L., Zinn, R.: Compositions of halo clusters and the formation of the galactic halo. Astrophys. J. **225**, 357–379 (1978). doi:10.1086/156499

Sesar, B., Ivezić, Ž., Stuart, J.S., Morgan, D.M., Becker, A.C., Sharma, S., Palaversa, L., Jurić, M., Wozniak, P., Oluseyi, H.: Exploring the variable sky with LINEAR. II. Halo structure and substructure traced by RR Lyrae Stars to 30 kpc. Astron. J. **146**, 21 (2013). doi:10.1088/0004-6256/146/2/21, 1305.2160

Simon, J.D., Geha, M.: The kinematics of the ultra-faint milky way satellites: solving the missing satellite problem. Astrophys. J. **670**, 313–331 (2007). doi:10.1086/521816, 0706.0516

Soria, R., Mould, J.R., Watson, A.M., Gallagher, J.S. III, Ballester, G.E., Burrows, C.J., Casertano, S., Clarke, J.T., et al.: Detection of the tip of the red giant branch in NGC 5128. Astrophys. J. **465**, 79 (1996). doi:10.1086/177403

Springel, V., Frenk, C.S., White, S.D.M.: The large-scale structure of the Universe. Nature **440**, 1137–1144 (2006). doi:10.1038/nature04805, astro-ph/0604561

Stinson, G.S., Dalcanton, J.J., Quinn, T., Gogarten, S.M., Kaufmann, T., Wadsley, J.: Feedback and the formation of dwarf galaxy stellar haloes. Mon. Not. R. Astron. Soc. **395**, 1455–1466 (2009). doi:10.1111/j.1365-2966.2009.14555.x, 0902.0775

Tal, T., van Dokkum, P.G., Nelan, J., Bezanson, R.: The frequency of tidal features associated with nearby luminous elliptical galaxies from a statistically complete sample. Astron. J. **138**, 1417–1427 (2009). doi:10.1088/0004-6256/138/5/1417, 0908.1382

Tanaka, M., Chiba, M., Komiyama, Y., Guhathakurta, P., Kalirai, J.S., Iye, M.: Structure and population of the andromeda stellar halo from a subaru/suprime-cam survey. Astrophys. J. **708**, 1168–1203 (2010). doi:10.1088/0004-637X/708/2/1168, 0908.0245

Tanaka, M., Chiba, M., Komiyama, Y., Guhathakurta, P., Kalirai, J.S.: Structure and population of the NGC 55 Stellar halo from a subaru/suprime-cam survey. Astrophys. J. **738**, 150 (2011). doi:10.1088/0004-637X/738/2/150, 1107.0911

Tikhonov, N.A., Galazutdinova, O.A., Aparicio, A.: Stellar content of NGC 404 - The nearest S0 Galaxy. Astron. Astrophys. **401**, 863–872 (2003). doi:10.1051/0004-**6361**, 20021819

Tikhonov, N.A., Galazutdinova, O.A., Drozdovsky, I.O.: Thick disks and halos of spiral galaxies M 81, NGC 55 and NGC 300. Astron. Astrophys. **431**, 127–142 (2005). doi:10.1051/0004-**6361**, 20047042, astro-ph/0407389

Tissera, P.B., Scannapieco, C.: Low-metallicity stellar halo populations as tracers of dark matter haloes. Mon. Not. R. Astron. Soc. **445**, L21–L25 (2014). doi:10.1093/mnrasl/slu114, 1407.5800

Tissera, P.B., Scannapieco, C., Beers, T.C., Carollo, D.: Stellar haloes of simulated Milky-Way-like galaxies: chemical and kinematic properties. Mon. Not. R. Astron. Soc. **432**, 3391–3400 (2013). doi:10.1093/mnras/stt691, 1301.1301

Toloba, E., Sand, D., Guhathakurta, P., Chiboucas, K., Crnojević, D., Simon, J.D.: Spectroscopic confirmation of the dwarf spheroidal galaxy d0994+71 as a member of the M81 group of galaxies. Astrophys. J. Lett. **830**, L21 (2016a). doi:10.3847/2041-8205/830/1/L21, 1610.03856

Toloba, E., Sand, D.J., Spekkens, K., Crnojević, D., Simon, J.D., Guhathakurta, P., Strader, J., Caldwell, N., McLeod, B., Seth, A.C.: A tidally disrupting dwarf galaxy in the halo of NGC 253. Astrophys. J. Lett. **816**, L5 (2016b). doi:10.3847/2041-8205/816/1/L5, 1512.03816

Tolstoy, E., Hill, V., Tosi, M.: Star-formation histories, abundances, and kinematics of dwarf galaxies in the local group. Annu. Rev. Astron. Astrophys. **47**, 371–425 (2009). doi:10.1146/annurev-astro-082708-101650, 0904.4505

Torrealba, G., Koposov, S.E., Belokurov, V., Irwin, M.: The feeble giant. Discovery of a large and diffuse Milky Way dwarf galaxy in the constellation of Crater. Mon. Not. R. Astron. Soc. **459**, 2370–2378 (2016). doi:10.1093/mnras/stw733, 1601.07178

van der Hulst, J.M.: The structure and kinematics of the neutral hydrogen bridge between M81 and NGC 3077. Astron. Astrophys. **75**, 97–111 (1979)

van der Kruit, P.C.: A comparison of our Galaxy to NGC 891 and NGC 4565 and the structure of their spheroids. Astron. Astrophys. **140**, 470–475 (1984)

van Dokkum, P.G.: The recent and continuing assembly of field elliptical galaxies by red mergers. Astron. J. **130**, 2647–2665 (2005). doi:10.1086/497593, astro-ph/0506661

VandenBerg, D.A., Bergbusch, P.A., Dowler, P.D.: The Victoria-Regina stellar models: evolutionary tracks and isochrones for a wide range in mass and metallicity that allow for empirically constrained amounts of convective core overshooting. Astrophys. J. Suppl. Ser. **162**, 375–387 (2006). doi:10.1086/498451, arXiv:astro-ph/0510784

Veljanoski, J., Mackey, A.D., Ferguson, A.M.N., Huxor, A.P., Côté, P., Irwin, M.J., Tanvir, N.R., Peñarrubia, J., Bernard, E.J., Fardal, M., Martin, N.F., McConnachie, A., Lewis, G.F., Chapman, S.C., Ibata, R.A., Babul A.: The outer halo globular cluster system of M31 - II. Kinematics. Mon. Not. R. Astron. Soc. **442**, 2929–2950 (2014). doi:10.1093/mnras/stu1055, 1406.0186

Vlajić, M., Bland-Hawthorn, J., Freeman, K.C.: The abundance gradient in the extremely faint outer disk of NGC 300. Astrophys. J. **697**, 361–372 (2009). doi:10.1088/0004-637X/697/1/361, 0903.1855

Walker, M.G., Peñarrubia, J.: A method for measuring (slopes of) the mass profiles of dwarf spheroidal galaxies. Astrophys. J. **742**, 20 (2011). doi:10.1088/0004-637X/742/1/20, 1108.2404

Watkins, L.L., Evans, N.W., Belokurov, V., Smith, M.C., Hewett, P.C., Bramich, D.M., Gilmore, G.F., Irwin, M.J., Vidrih, S., Wyrzykowski, Ł., Zucker, D.B.: Substructure revealed by RRLyraes in SDSS Stripe 82. Mon. Not. R. Astron. Soc. **398**, 1757–1770 (2009). doi:10.1111/j.1365-2966.2009.15242.x, 0906.0498

Weisz, D.R., Skillman, E.D., Cannon, J.M., Dolphin, A.E., Kennicutt, R.C. Jr., Lee, J., Walter, F.: The recent star formation histories of M81 group dwarf irregular galaxies. Astrophys. J. **689**, 160–183 (2008). doi:10.1086/592323, 0809.5059

Wetzel, A.R., Deason, A.J., Garrison-Kimmel, S.: Satellite dwarf galaxies in a hierarchical universe: infall histories, group preprocessing, and reionization. Astrophys. J. **807**, 49 (2015). doi:10.1088/0004-637X/807/1/49, 1501.01972

Wetzel, A.R., Hopkins, P.F., Kim Jh, Faucher-Giguère, C.A., Kereš, D., Quataert, E.: Reconciling dwarf galaxies with ΛCDM cosmology: simulating a realistic population of satellites around a milky way-mass galaxy. Astrophys. J. Lett. **827**, L23 (2016). doi:10.3847/2041-8205/827/2/L23, 1602.05957

Wheeler, C., Oñorbe, J., Bullock, J.S., Boylan-Kolchin, M., Elbert, O.D., Garrison-Kimmel, S., Hopkins, P.F., Kereš, D.: Sweating the small stuff: simulating dwarf galaxies, ultra-faint dwarf galaxies, and their own tiny satellites. Mon. Not. R. Astron. Soc. **453**, 1305–1316 (2015). doi:10.1093/mnras/stv1691, 1504.02466

White, S.D.M., Frenk, C.S.: Galaxy formation through hierarchical clustering. Astrophys. J. **379**, 52–79 (1991). doi:10.1086/170483

Willman, B.: In pursuit of the least luminous galaxies. Adv. Astron. **2010**, 285454 (2010). doi:10.1155/2010/285454, 0907.4758

Wu, P.F., Tully, R.B., Rizzi, L., Dolphin, A.E., Jacobs, B.A., Karachentsev, I.D.: Infrared tip of the red giant branch and distances to the Maffei/IC 342 group. Astron. J. **148**, 7 (2014). doi:10.1088/0004-6256/148/1/7, 1404.2987

Yanny, B., Newberg, H.J., Kent, S., Laurent-Muehleisen, S.A., Pier, J.R., Richards, G.T., Stoughton, C., Anderson, J.E. Jr., Annis, J., Brinkmann, J., Chen, B., Csabai, I., Doi, M., Fukugita, M., Hennessy, G.S., Ivezić, Ž., Knapp, G.R., Lupton, R., Munn, J.A., Nash, T., Rockosi, C.M., Schneider, D.P., Smith, J.A., York, D.G.: Identification of A-colored stars and structure in the halo of the milky way from sloan digital sky survey commissioning data. Astrophys. J. **540**, 825–841 (2000). doi:10.1086/309386, astro-ph/0004128

Yun, M.S., Ho, P.T.P., Lo, K.Y.: A high-resolution image of atomic hydrogen in the M81 group of galaxies. Nature **372**, 530–532 (1994). doi:10.1038/372530a0

Zheng, Z., Shang, Z., Su, H., Burstein, D., Chen, J., Deng, Z., Byun, Y.I., Chen, R., Chen, W.P., Deng, L., Fan, X., Fang, L.Z., Hester, J.J., Jiang, Z., Li, Y., Lin, W., Sun, W.H., Tsay, W.S., Windhorst, R.A., Wu, H., Xia, X., Xu, W., Xue, S., Yan, H., Zheng, Z., Zhou, X., Zhu, J., Zou, Z., Lu, P.: Deep intermediate-band surface photometry of NGC 5907. Astron. J. **117**, 2757–2780 (1999). doi:10.1086/300866

Zibetti, S., White, S.D.M., Brinkmann, J.: Haloes around edge-on disc galaxies in the sloan digital sky survey. Mon. Not. R. Astron. Soc. **347**, 556–568 (2004). doi:10.1111/j.1365-2966.2004.07235.x, astro-ph/0309623

Chapter 3
The Impact of Stellar Migration on Disk Outskirts

Victor P. Debattista, Rok Roškar, and Sarah R. Loebman

Abstract Stellar migration, whether due to trapping by transient spirals (churning), or to scattering by non-axisymmetric perturbations, has been proposed to explain the presence of stars in outer disks. After a review of the basic theory, we present compelling, but not yet conclusive, evidence that churning has been important in the outer disks of galaxies with Type II (down-bending) profiles, while scattering has produced the outer disks of Type III (up-bending) galaxies. In contrast, field galaxies with Type I (pure exponential) profiles appear to not have experienced substantial migration. We conclude by suggesting work that would improve our understanding of the origin of outer disks.

3.1 Introduction

By the mid-1990s the view of disks that prevailed was that they are truncated at some radius with little or no stars at larger radii. Using photographic images of three edge-on galaxies, van der Kruit (1979) had shown that disks appear sharply truncated. Based on four edge-on galaxies, van der Kruit and Searle (1982) measured truncation radii at $4.2 \pm 0.6\, R_{\rm d}$, where $R_{\rm d}$ is the scale length of the disk. In face-on galaxies, van der Kruit (1988) found closely spaced light contours in the outer disks, which he interpreted as the advent of the truncation. van der Kruit (1987) suggested these truncations are the result of detailed angular momentum conservation in the collapse of a uniformly rotating gas sphere. While HI disks are

V.P. Debattista (✉)
Jeremiah Horrocks Institute, University of Central Lancashire, Preston PR1 2HE, UK
e-mail: vpdebattista@uclan.ac.uk

R. Roškar
Research Informatics, Scientific IT Services, ETH Zürich, Weinbergstrasse 11, CH-8092 Zürich, Switzerland
e-mail: rokroskar@gmail.com

S.R. Loebman
Department of Astronomy, University of Michigan, 1085 S. University Ave., Ann Arbor, MI 48109, USA
e-mail: sloebman@umich.edu

J.H. Knapen et al. (eds.), *Outskirts of Galaxies*, Astrophysics and Space Science Library 434, DOI 10.1007/978-3-319-56570-5_3

typically larger than the break radius, van der Kruit (2007) showed that HI warps often begin at about the truncation radius. Kregel et al. (2002) found truncations in 20 of 34 galaxies with a slightly smaller truncation radius, $3.6 \pm 0.6\,R_{\rm d}$. Pohlen et al. (2000), on the basis of a sample of 31 edge-on galaxies, revised the mean truncation radius further to $2.9 \pm 0.7\,R_{\rm d}$ (truncations at 11–35 kpc).

Then the picture of disk profiles started changing. Already in NGC 4565 (Naeslund and Joersaeter 1997) and IC 5249 (Byun 1998), a transition from one exponential profile to a steeper one, rather than a sharp truncation, had been found. In his sample of four edge-on galaxies, de Grijs et al (2001) found that truncations are not perfectly sharp. Finally the deep optical imaging of 72 edge-on (Pohlen 2002) and three face-on (Pohlen et al. 2002) galaxies firmly established that disk galaxies have broken exponential profiles, rather than sharp truncations. The transition between the two exponentials has come to be known as a "break", although the term "truncation" persists. This discovery led to increased interest in disk outskirts because the presence of stars at large radii needed an explanation. The first workshop dedicated to disk outskirts, "Outer Edges of Disk Galaxies: A Truncated Perspective", was held at the Lorentz Center in 2005. That same year Erwin et al. (2005) identified a new class of disk profiles, the Type III "anti-truncated" profiles. The first pure N-body simulations aimed at producing disk profiles with breaks were presented by Debattista et al. (2006), who proposed angular momentum redistribution during bar formation as the cause, while Bournaud et al. (2007) and Foyle et al. (2008) included also the effect of gas. Roškar et al. (2008b) introduced the idea that disk outskirts are populated by stars that have migrated outwards without heating.

A decade later much progress has been made but much more remains to be understood. While generally comprising only a small fraction of the overall stellar mass, disk outskirts are an ideal laboratory for understanding the processes of galaxy formation. These are regions where the dark matter dominates and that are therefore easily perturbed, are generally inhospitable to star formation and are the interface at which gas, and some satellites, are accreted. In recent years it has become clear that the stars in these regions are generally old. Are these stars the product of ongoing inefficient star formation or the fossil of a long-ago, efficient, in situ star formation? Was mass migrated to the outer disk (by bars, clumps or spirals)? Were the stars accreted from satellites? Understanding which of these processes dominates, and where, provides crucial information on how galaxies formed. This review explores the extent to which stellar migration is responsible for populating the outer disks. In the remainder of this section, we clarify what we mean by profile breaks. In Sect. 3.2 we review the demographics of different profile types. We introduce stellar migration and angular momentum redistribution in Sect. 3.3. We then consider each of the profile types in turn in Sects. 3.4 (Type IIs), 3.5 (Type Is) and 3.6 (Type IIIs). Section 3.7 concludes with suggestions for future directions that can greatly advance our understanding of the origin of stars in outer disks.

3.1.1 Our Definition of Breaks

In the truncated picture of disks, the meaning and location of a truncation is unambiguous. Instead disks do not end abruptly but extend significantly beyond a break. While disks are often well fit by piecewise exponential profiles, the presence of bars, lenses, spirals, rings and asymmetries add bumps and wiggles to profiles. For instance, Laine et al. (2014) noted that their galaxies included breaks associated with lenses, with rings and with star formation breaks. Therefore sometimes it is difficult to assign the location of breaks, particularly if the data being fit are shallow. Such difficulties account for the significant differences in the locations of breaks assigned by different authors. For example, in NGC 4244 the resolved stellar population study of de Jong et al. (2007) found a very clear break at 420″ while Martín-Navarro et al. (2012) found one at ∼290″ and another at ∼550″. Although sharp truncations are not supported by observations, star formation is often observed to be more strongly truncated than the overall mass distribution (Bakos et al. 2008). How stars end up past this point is a very important question that gives us insight into important processes that operate in most disk galaxies. With this in mind, this review focusses not on the bumps and wiggles produced by bars, rings or spirals, but on breaks related to star formation. Thus, our primary focus is on Type II profiles, which exhibit such breaks, whereas our interest in pure exponential (Type I) profiles comes from the absence of these breaks and in anti-truncated (Type III) profiles from their ability to populate the outer disk via some other mechanism.

3.2 Demographics of Profile Type

Stellar disk profiles are piecewise exponential.[1] This does not imply that the profile of star formation itself must be exponential since even a radially constant star formation rate produces an exponential disk (e.g. Roškar et al. 2008b). Why nature prefers exponential disks is uncertain, although many studies have attacked this problem from a variety of perspectives. Whatever the reason, the exponential profile is evidently an attractor. Freeman (1970) identified two types of disk profiles. The first (Type I profile) is exponential to the last measured point, while in the second (Type II profile) an inner exponential is followed by a steeper one in the outer disk. Erwin et al. (2005) extended this classification to include Type III profiles, where the outer profile is shallower than the inner one.

Pohlen and Trujillo (2006) found that ∼60% of late-type spirals have a Type II profile with the break at $2.5 \pm 0.6\,R_\mathrm{d}$ (at $\mu_r \sim 23.5 \pm 0.8$ mag arcsec^{-2}), while 30% have a Type III profile, with the break at larger radii, $4.9 \pm 0.6\,R_\mathrm{d}$ (at $\mu_r \sim 24.7 \pm 0.8$ mag arcsec^{-2}). The mix of profile types varies with the Hubble type, with Type IIs

[1] Bulges at the centres of galaxies contribute additional light above an exponential.

Table 3.1 The fraction of Type I, II and III profiles among unbarred (SA), weakly barred (SAB) and barred (SB) galaxies

	I	II	III
S0-Sb galaxies			
SA	0.26 ± 0.07	0.14 ± 0.06	0.69 ± 0.08
SAB	0.18 ± 0.07	0.45 ± 0.09	0.48 ± 0.09
SB	0.36 ± 0.07	0.49 ± 0.07	0.20 ± 0.06
Sbc-Sd galaxies			
SA	0.17 ± 0.08	0.52 ± 0.10	0.30 ± 0.10
SAB	0.12 ± 0.07	0.67 ± 0.10	0.21 ± 0.08
SB	0.09 ± 0.06	0.83 ± 0.08	0.09 ± 0.06

Data for S0-Sb galaxies are from Erwin et al. (2008) and Gutiérrez et al. (2011). The data for Sbc-Sd galaxies are from Pohlen and Trujillo (2006)

more frequent in late-type galaxies, while Type IIIs are more common in early-type galaxies. Erwin et al. (2008) found a breakdown of 27%:42%:27% for Types I:II:III for 66 early-type galaxies, with ~6% having a composite Type II+III profile: a Type II profile at small radii and a Type III further out. Gutiérrez et al. (2011), studying 47 face-on early-type unbarred galaxies, found that the Type II profiles are more common in late-type galaxies, while the Type I and III profiles are more common in early types. By combining all the published studies and accounting for differences in sample definitions, they found global (S0-Sd) frequencies of 21%:50%:38% for Types I:II:III, with 8% of galaxies having composite II+III profiles. Table 3.1 presents the fraction of each profile type (with Type II+III profiles counted twice) from the combination of these three studies.

3.2.1 The Role of Environment

Erwin et al. (2012) and Roediger et al. (2012) found a strong dichotomy between field and Virgo Cluster lenticulars. Both studies found that the fraction of Type I profiles rises at the expense of Type II profiles. Gutiérrez et al. (2011) found Type I profiles in one third of lenticular galaxies, with Type II profiles uncommon among them. Pranger et al. (2016) used the Sloan Digital Sky Survey to construct field and cluster samples of ~100 galaxies each, in a narrow mass range (1–$4 \times 10^{10} M_\odot$). They found that Type I galaxies are three times more frequent in clusters than in the field. Maltby et al. (2012a) came to a very different conclusion, finding roughly half of both their field and cluster samples having a Type I profile and only 10% having a Type II profile. However, because they concentrated on outer disks through a surface brightness cut at $24 < \mu < 26.5$ mag arcsec^{-2}, they missed most Type II breaks, which are typically brighter. Nevertheless, their fraction of Type III profiles is consistent with that of the field sample of Pohlen and Trujillo (2006).

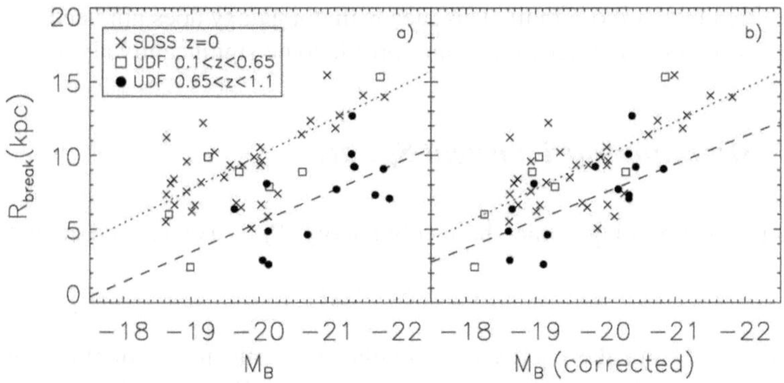

Fig. 3.1 Break radius as a function of rest-frame *B*-band absolute magnitude for galaxies at different redshift, as indicated in the inset at *left*. (**a**) The observed relation. (**b**) The relation after correcting the absolute magnitude for the mean surface brightness evolution since $z = 1$. The *dashed lines* show the best-fitting relation for the distant sample while the *dotted line* shows the local relation. Reproduced with permission from Trujillo and Pohlen (2005)

3.2.2 *Redshift Evolution*

Pérez (2004) found six breaks in a sample of 16 galaxies at $0.6 \leq z \leq 1.0$; the average break radius was $1.8\,R_d$, smaller than in local samples although the detection was biassed to smaller radii. In a study of 36 galaxies to $z \sim 1$, shown in Fig. 3.1, Trujillo and Pohlen (2005) found that the location of the breaks, corrected for the mean surface brightness evolution, has increased by 1-3 kpc (25%), while the $\sim 10\times$ larger sample of Azzollini et al. (2008b) found that, at fixed mass, the break radius has increased by a factor 1.3 ± 0.1 since $z = 1$.

3.3 Stellar Migration

The breaks in disk profiles are generally at $2 - 5\,R_d$ (Pohlen and Trujillo 2006). At these large radii, the surface density is low, and the total mass in the outer disk is small. Bakos et al. (2008) found that the mass of stars in the outer disk constitutes $14.7{\pm}1.2\%$ in Type II galaxies and $9.2{\pm}1.4\%$ in Type III galaxies. Therefore even a quite small transfer of mass from small to large radii can have a significant effect on the overall profile shape. For a pure exponential profile, the total mass outside $4\,R_d$ is less than 10% of the total. If just one quarter of the mass between R_d and $2\,R_d$ (i.e. $\sim$8% of the total disk mass) is redistributed to radii outside $4\,R_d$, this nearly doubles the mass at those radii. This means that a galaxy does not need substantial mass redistribution in order for its outer profile to be significantly altered.[2]

3.3.1 Migration via Transient Spirals

The energy in a rotating frame, the Jacobi energy, E_J, is a conserved quantity:

$$E_J = E - \Omega_p L_z, \tag{3.1}$$

where E and L_z are the energy and angular momenta in the inertial frame, and Ω_p is the angular frequency of the rotating frame (Binney and Tremaine 1987). E_J is a conserved quantity even when the system is not axisymmetric provided the system is viewed in the frame in which the potential is stationary. For a bar or a spiral perturbation, the stationary frame corresponds to the one where Ω_p is the pattern speed of the perturbation. Conservation of E_J requires that changes in angular momentum, ΔL_z, and in energy, ΔE, are related as $\Delta E = \Omega_p \Delta L_z$. For most stars ΔL_z averages to zero over an orbit; the exception is for stars at resonances (Lynden-Bell and Kalnajs 1972). The change in angular momentum is accompanied by a change in the radial action, J_R; in the epicyclic approximation, this change is given by

$$\Delta J_R = \frac{\Omega_p - \Omega}{\kappa} \Delta L_z, \tag{3.2}$$

where Ω and κ are the angular and radial frequencies of a star. Sellwood and Binney (2002) realized that changes in energy and angular momentum at the corotation resonance (CR), where $\Omega_p = \Omega$, are not accompanied by any changes in radial action and result in no radial heating.

Stars trapped by a spiral at its CR change their energy and angular momentum at fixed E_J without changing J_R. With spirals being transient (Sellwood and Carlberg 1984; Sellwood 2011), trapping by a spiral is short-lived, typically lasting for one rotation period. When the spiral decays, trapped stars are released with a different angular momentum than they started with. This is the basis of migration due to corotation-trapping by transient spirals, which Sellwood and Binney (2002) refer to as "churning". Figure 3.2 presents an example of churning in a disk with a single

[2]This also poses a challenge for simulations studying the evolution of disk outskirts, which need to minimize noise; for instance initial conditions that are even slightly out of equilibrium can significantly alter the profile at large radii. The most widely used technique for producing initial conditions for simulations assumes Gaussian velocity distributions, which is only an approximate equilibrium. Kazantzidis et al. (2004) presented a spectacular example of how this approximation leads to dramatically wrong conclusions in the case of satellite galaxy disruption. The freely available GALACTICS initial conditions code (Kuijken and Dubinski 1995; Widrow and Dubinski 2005; Widrow et al. 2008) instead uses the correct distribution function method for setting up galaxies. Likewise, the GALAXY package (Sellwood 2014) includes software for setting up disk galaxies in proper equilibrium.

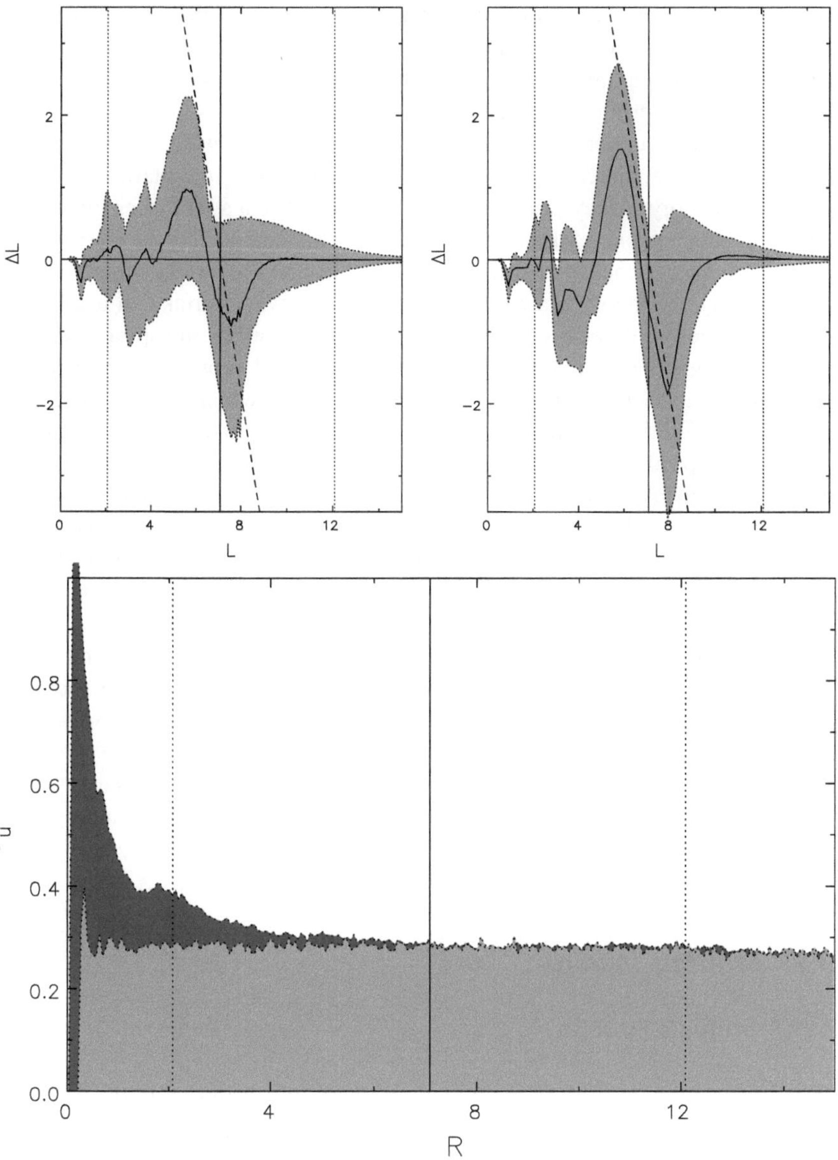

Fig. 3.2 *Top*: Exchanges of angular momentum driven by a single spiral. The *left panel* shows angular momentum exchanges for all orbits while the *right panel* shows exchanges for nearly circular orbits. The *solid vertical line* indicates the angular momentum of a circular orbit corotating with the spiral, while the *dotted vertical lines* show the ILR and OLR. Strong angular momentum exchanges occur at the CR. The *shaded region* shows the 20–80% interval at each angular momentum, with the *solid line* showing the mean change in angular momentum. *Bottom*: The radial velocity dispersion before (*light grey*) and after (*dark grey*) the lifetime of the spiral. The CR (*solid line*), and the ILR and OLR (*dotted lines*) are indicated. Although the spiral has driven substantial migration at CR, as shown in the *left panels*, the radial heating there is negligible. Reproduced with permission from Sellwood and Binney (2002)

strong spiral. The vertical solid lines in the top panels show the location of the CR. In spite of the substantial amount of angular momentum exchanges taking place around the CR, the radial velocity dispersion, shown in the bottom panel, at the CR is barely altered after the spiral has died down.

A simple way to understand this behaviour is to consider what happens as a spiral density wave forms, increasing the local density in some region. Stars that have $\Omega > \Omega_{\rm p}$ will gain energy as they catch up with it, but then lose it again when they overtake it. The net change in energy for such stars is minimal. Conversely the stars with $\Omega < \Omega_{\rm p}$ first lose energy, but then gain it again as the spiral overtakes them, also with little net change in energy. Stars that have $\Omega \simeq \Omega_{\rm p}$ never overtake, or are overtaken by, the spiral before it decays. These particles are trapped by the spiral, exchanging energy and angular momentum with it. Once the spiral subsides, their net change in angular momentum depends on the relative phase at which they escape. The top panel of Fig. 3.3 shows the location of stars that are about to be trapped and migrated by a growing spiral. Stars that will gain angular momentum are located behind the spiral's peak density, while those that will lose angular momentum are ahead of the spiral. The angular phase of the peak density demarcates a separation between the gainers and the losers. Stars are not trapped at just the corotation radius but up to $\sim$1 kpc from it. Daniel and Wyse (2015) showed that the trapping region (example shown in Fig. 3.4) is broadened and not limited to a single radius.

The efficiency of trapping declines rapidly with radial random motion (Sellwood and Binney 2002; Daniel and Wyse 2015; Solway et al. 2012), but the vertical random motion reduces the trapping efficiency much more slowly (Solway et al. 2012). The left panel of Fig. 3.5 plots the angular momentum change of stars as a function of orbital eccentricity, showing that angular momentum changes become smaller as the orbital eccentricity increases. The right panel shows that migration is not strongly affected by the vertical amplitude of orbits.

3.3.2 *Multiple Patterns*

3.3.2.1 Multiple Spirals

Typically disks support multiple spirals with different pattern speeds. The unconstrained simulation of Sellwood and Binney (2002) supported many separate spiral frequencies, which resulted in migration at locations across the disk. Because the difference between initial and final radii of stars was much larger than the epicyclic radius, particularly at large radii, they concluded that migration by churning was driving this migration. The simulations of Roškar et al. (2012) also contained multiple spirals. The bottom panel of Fig. 3.3 shows a typical example: at least three separate strong spirals with different pattern speeds are clearly recognizable at this particular time. The left panel of Fig. 3.6 shows the angular momentum exchanges

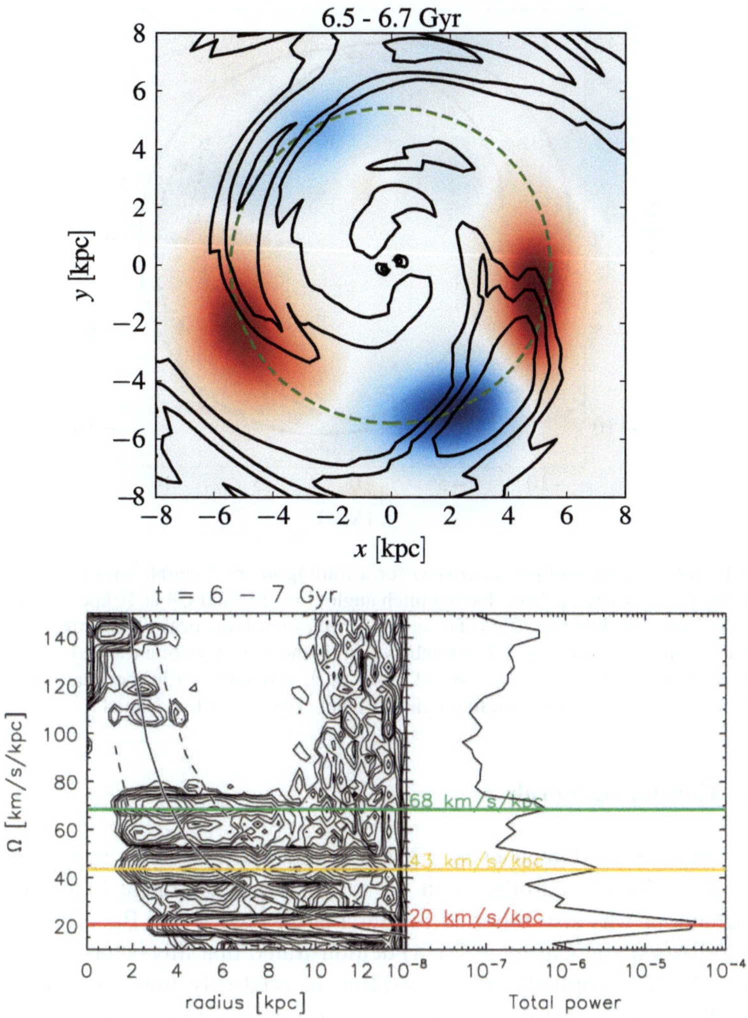

Fig. 3.3 *Top*: The location of stars relative to a growing spiral shortly before it reaches peak amplitude, trapping stars and migrating them. The *red* (*blue*) *shading* indicates the density of stars that lose (gain) angular momentum, while the contours show the overall density distribution. *Bottom*: The spectrum of spiral structure during the time interval 6–7 Gyr for the same simulation. Three prominent spirals can be seen (with pattern speeds indicated by the *horizontal coloured lines*) as well as a number of weaker ones. Reproduced with permission from Roškar et al. (2012)

occurring during this time. The presence of multiple spirals increases the number of locations at which stars are migrating (compare with the upper left panel of Fig. 3.2). For a small, random sample of stars that had large migrations, Roškar et al. (2012) showed explicitly that their orbits are trapped by the spiral at the CR.

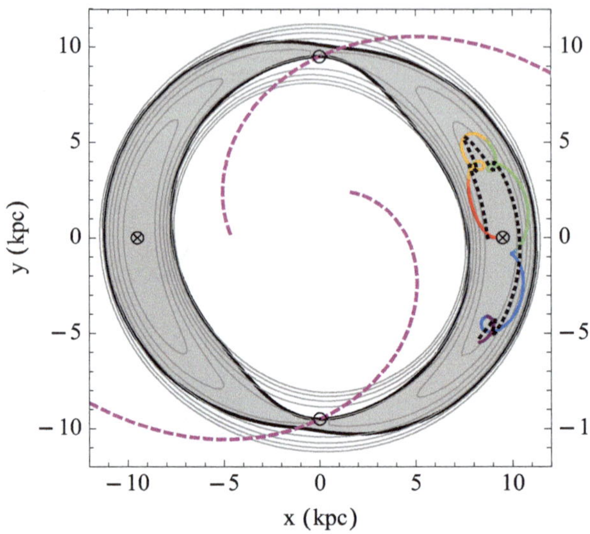

Fig. 3.4 Effective potential (*grey contours*) for a trailing $m = 2$ spiral, with peak perturbation indicated by *dashed magenta lines*, having pitch angle $\theta = 25°$ and CR at 10 kpc. The potential is described in Daniel and Wyse (2015). The *crossed circles* mark the peak of the effective potential. The capture region is shaded *grey*. The rainbow path is the trajectory of a trapped star with initial phase-space coordinates $(x, y, v_x, v_y) = (9.1\,\text{kpc}, 0, 0, -10\,\text{km s}^{-1})$ in the rotating frame. The *black* (*dotted*) *curve* is the star's guiding radius. Figure courtesy of K.J. Daniel

3.3.2.2 Corotating Spirals

In their simulations, Grand et al. (2012) found multiple short spirals such that most of them locally corotate with the stars at a large range of radii. Similar spiral behaviour was also reported by Baba et al. (2013) and Roca-Fàbrega et al. (2013). Sellwood and Carlberg (2014) demonstrated that this behaviour is a result of several high multiplicity spirals residing in relatively low-mass disks. Grand et al. (2012) showed that these spirals trapped a large fraction of the stars at CR, leading to migration across a large extent of the disk. The extreme migrators in their simulations retain nearly circular orbits, demonstrating that the mechanism is still that of churning.

3.3.2.3 Bars and Chaotic Scattering

The presence of multiple patterns also introduces new physics. Daniel and Wyse (2015) showed that a star trapped at CR can escape if its guiding radius approaches an inner or an outer Lindblad resonance (ILR or OLR) of another pattern, without the spiral amplitude changing. This changes the rate of migration; unlike trapping at the CR, however, this is a scattering process, raising the random motion of stars.

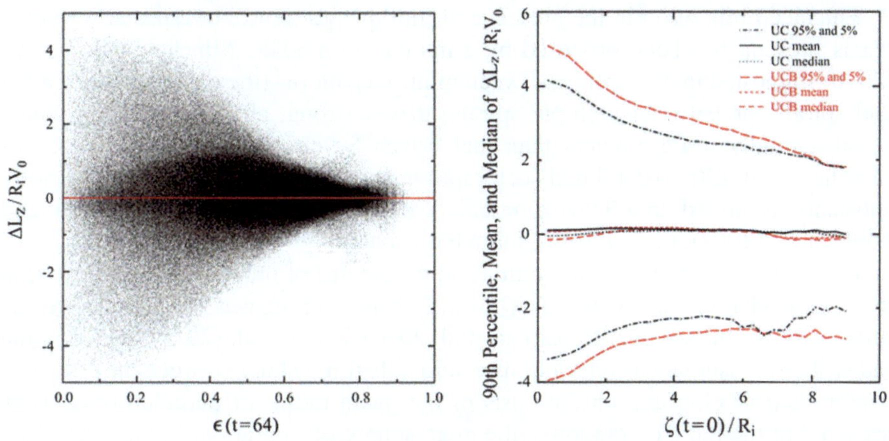

Fig. 3.5 *Left*: The effect of orbital eccentricity on migration efficiency. On the *horizontal axis* is plotted the initial eccentricity of individual stellar orbits, while on the *vertical axis* is shown the normalized change in angular momentum of the stars. Migration is most efficient for stars on nearly circular orbits. *Right*: The effect of a bar on migration. On the *horizontal axis* ζ is a proxy for the vertical amplitude of the initial orbits. The mean, median and 5–95% of the distributions of angular momentum changes are shown by different lines as indicated at *top right*. Simulation UC (*black lines*) did not form a bar, while simulation UCB, with the same physical model (with $Q \simeq 1.5$) but different random initialization, formed a bar. The difference is computed over ~6 Gyr, during which time a bar formed in UCB within the first Gyr. The effect of the bar on the net migration is small compared to that driven by spirals. Both simulations also show that the effect of height on migration is relatively weak. Reproduced with permission from Solway et al. (2012)

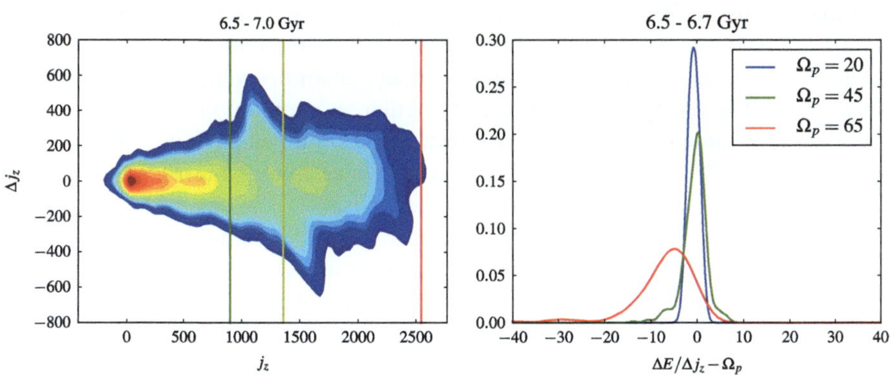

Fig. 3.6 *Left*: Distribution of changes in angular momentum as a function of starting angular momentum at around the same time as in Fig. 3.3. The *vertical dashed lines* mark the locations of the CR of some of the prominent spirals. *Right*: For stars exchanging angular momentum around each of the main spiral CRs, the distribution of $\Delta E/\Delta j_z$ is strongly peaked around Ω_p. Typical uncertainties on measuring the pattern speeds are $\sim 5\,\mathrm{km\,s^{-1}\,kpc^{-1}}$. Thus the Jacobi energy is a conserved quantity, rather than evolving chaotically, despite the presence of multiple patterns. This implies that large angular momentum exchanges are dominated by the spiral corotation trapping (churning). Reproduced with permission from Roškar et al. (2012)

Enhanced migration in the presence of multiple patterns, but especially when a bar is present, has been proposed by a number of studies. Minchev and Famaey (2010) suggested that coupling between multiple patterns (the combination of a bar and spirals, or between multiple spirals) drives strong, chaotic migration which is substantially more efficient than that driven by churning. The simulations of Minchev et al. (2011) exhibited such rapid migration that stellar populations were substantially mixed on a 3 Gyr timescale. Contrary to Debattista et al. (2006) and Foyle et al. (2008), they suggested that bar-driven migration is persistent. Brunetti et al. (2011) also found fast migration (on timescales of order a rotation period) in the case of a bar forming in a cool ($Q \sim 1$) disk, but mixing was much reduced if the initial disk is hot (see also Debattista et al. 2006). Halle et al. (2015) studied radial redistribution using a bar-unstable simulation, distinguishing between the effects of heating and of churning on the basis of the mean radius of particle orbits. They argued that the bar's corotation is the main source of migration. In the outer disk, they found heating dominates, rather than migration, by a factor of ~2–8.

Other studies have questioned whether such strong migration occurs as a result of multiple patterns. A quantitative assessment of the impact of bars on migration was obtained by Solway et al. (2012). The right panel of Fig. 3.5 compares the degree of angular momentum changes in two models, which represent the same physical initial conditions, but, because of the stochasticity inherent in bar formation (Sellwood and Debattista 2009), in one a strong bar formed, while in the other no bar formed. The bar formed during the first Gyr; after ~6 Gyr of evolution, the change in angular momentum is larger when a bar is present, but not by the order of magnitude predicted by Minchev et al. (2011).

Nor is there evidence for strong scattering resulting from the presence of multiple spirals. Roškar et al. (2012) tested explicitly for chaotic migration in their simulations with multiple spiral arms (see Fig. 3.3). They found smoothly evolving orbital frequencies, a characteristic of regular, not chaotic, evolution. Moreover they showed that the changes in energy and angular momentum conserve the Jacobi energy in the frame of the spirals, as shown in the right panel of Fig. 3.6. E_J would not be conserved if the chaotic interaction of two or more patterns was scattering the stars, since no stationary closed frame then exists.

The observational evidence also does not favour extremely efficient mixing driven by bars. For instance Sánchez-Blázquez et al. (2014) measured stellar age and metallicity gradients in a sample of barred and unbarred galaxies. They found no significant difference in their gradients, arguing that migration by bars is not strong enough to erase gradients.

3.3.2.4 Churning Versus Scattering

The word "migration" has come to mean the motion of stars from one radius to another either because of trapping at corotation by spirals (churning) or by a variety of heating mechanisms. As Sellwood and Binney (2002) noted, the measured velocity dispersion of old stars allows very little room for random motions in the Solar neighbourhood to have moved stars by more than $\Delta R \sim 1.3$ kpc.

The key difference between churning and scattering is in the random motion created, with churning not increasing random motions appreciably while scattering leads to disks becoming hotter.

3.3.3 Evidence for Migration in the Milky Way

Because we can study it in much greater detail than other galaxies, thus far the evidence for migration has been strongest in the Milky Way:

- The age-metallicity relation of the Milky Way is flat and broad (Haywood et al. 2013; Bergemann et al. 2014; Rebassa-Mansergas et al. 2016), contrary to the expected evolution from traditional chemical evolution models. Sellwood and Binney (2002) argued that these are characteristics of migration, as has been shown by models (e.g. Roškar et al. 2008a; Schönrich and Binney 2009).
- The radial [Fe/H] gradient decreases with stellar age (Yu et al. 2012). Since the metallicity gradient is expected to be steeper at earlier times, this flattening must result from migration smearing the metallicity profile of older stars (Roškar et al. 2008a; Schönrich and Binney 2009).
- Bovy et al. (2016) showed that low-[α/Fe], mono-abundance populations flare radially, which is the expected behaviour for populations that migrate while conserving the vertical action (Solway et al. 2012; Vera-Ciro and D'Onghia 2016).
- The skewness of the metallicity distribution function varies across the disk, changing from skewed to low [Fe/H] in the inner disk to skewed to high [Fe/H] at larger radii (Hayden et al. 2015). This pattern arises because at larger radii, stellar populations include an increasing fraction of metal-rich stars that have migrated outwards from small radii (Hayden et al. 2015; Loebman et al. 2016).

This partial list demonstrates that a broad range of observations favour migration having occurred in the Milky Way. Whether this is because of churning or scattering is still not decided. The realization that metal-rich stars, which are probably outward migrators, become proportionally more common at radii past the Solar cylinder hints that the outer disk is a repository of migrated stars.

3.4 Type II Profiles

The Type II profile is the most common among star-forming galaxies. Two obvious potential causes for a steeper exponential profile in the outer disk are obscuration by dust and a slowly declining star formation rate (SFR). In an Hα spectroscopic study of 15 edge-on galaxies, Christlein et al. (2010) excluded both of these possibilities.

Their sample was free of surface brightness enhancements, and the data clearly showed a cut-off in the SFR much steeper than that of the stellar continuum. Therefore another mechanism is required to explain the mass in the outer disks.

Pohlen and Trujillo (2006) and Erwin et al. (2008) classified Type II profiles into various classes based on the break radius and the origin which this suggests. They distinguished between inner and outer breaks, the latter separated into breaks arising from asymmetric disks, classical breaks at large radius and, in barred galaxies, ones at roughly twice the bar radius, which they associated with the bar's OLR. Pohlen and Trujillo (2006) found classical breaks in 40% of Sc and later-type galaxies.

3.4.1 Theoretical Models

Thus Type II breaks are probably due to a number of mechanisms; here we concentrate on classical breaks which are associated with star formation breaks. In the standard paradigm of galaxy formation, galaxies grow by accreting gas and forming stars (Fall and Efstathiou 1980; Mo et al. 1998). Because gas accreted at later times has higher angular momentum, it lands at larger radii (Guo et al. 2011). Galaxies should therefore form stars at increasingly large radii, which is termed "inside-out formation". Inside-out formation predicts that young stars should be present in the outskirts of galaxies (Larson 1976; Matteucci and Francois 1989; Chiappini et al. 1997; Naab and Ostriker 2006), as is observed inside the break (de Jong 1996; Bell and de Jong 2000; MacArthur et al. 2004; Muñoz-Mateos et al. 2007).

3.4.1.1 Star Formation Breaks

Star formation in disk galaxies typically drops abruptly at a few R_d, even though the atomic gas extends beyond this point (e.g. Kennicutt 1989; Martin and Kennicutt 2001; Bigiel et al. 2010). Efforts to associate stellar breaks with star formation breaks started early (e.g. van der Kruit 1987; Ferguson et al. 1998). Schaye (2004) proposed that disk breaks occur as a result of star formation thresholds caused by the phase transition of gas from warm ($T \sim 10^4$ K) to cold ($T \sim 10^2$ K). Once this thermal instability sets in, it is quickly followed by a gravitational instability due to the associated sharp drop in Toomre-Q, allowing star formation to proceed. Schaye (2004) computed where the stellar breaks should occur and found a good match to observed stellar break radii, as shown in Fig. 3.7. Thus breaks in the stellar distribution show a strong relation to breaks in the present day star formation.

The model of Schaye (2004) assumed that the disk is axisymmetric. Recognizing that any perturbations that arise in the outer disk will enhance the local gas density, he predicted that the thermal instability will occur, triggering star formation in the outer disk. Thus, it is not surprising to find star formation at large radii, beyond $2\,R_{25}$ (e.g. Gil de Paz et al. 2005; Thilker et al. 2005; Zaritsky and Christlein 2007;

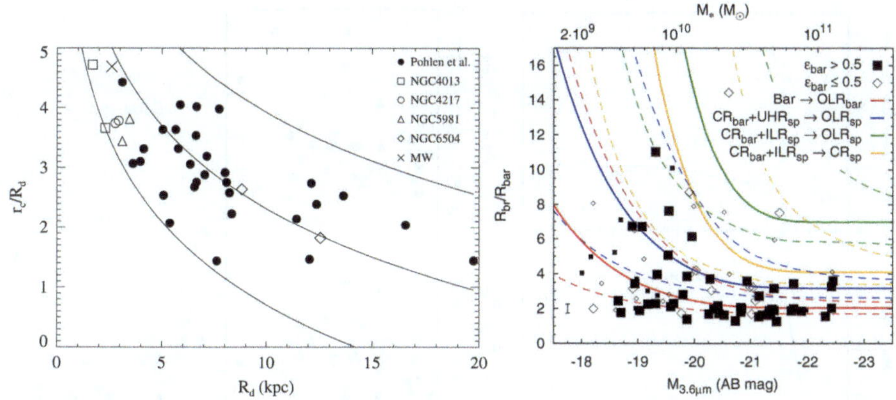

Fig. 3.7 *Left*: Break radius, r_c, as a function of R_d, in the model of Schaye (2004). The *solid lines* correspond to disks of different masses ($M_{\rm disk} = 7.5\times10^9\,M_\odot$, $3.8\times10^{10}\,M_\odot$ and $1.9\times10^{11}\,M_\odot$, from *top to bottom*) assuming a surface density of the phase transition at $\log(N_{\rm H,crit}\,{\rm cm}^2) = 20.75$. Observational datapoints from the literature are indicated. However, both sides are well fit by models at fixed galaxy mass. Reproduced with permission from Schaye (2004). *Right*: Evidence for the effect of bars on the locations of disk breaks. *Filled squares* and *open diamonds* correspond to strong and weak bars. The *red solid curve* shows the locus of the bar's OLR while the *blue solid line* corresponds to the OLR of a spiral which has its inner 4:1 resonance at the bar's CR. The *yellow and green lines* are the OLR and CR of spirals with ILR at the bar's CR. Reproduced with permission from Muñoz-Mateos et al. (2013)

Christlein and Zaritsky 2008), with extended UV (XUV) emission in about 20% of galaxies (Lemonias et al. 2011), although it is unlikely that all XUV disks can be explained this way. Schaye (2004) also neglected the effects of bars, which are efficient at modifying the distribution of gas. Bars accumulate gas at the OLR (e.g. Schwarz 1981; Byrd et al. 1994; Rautiainen and Salo 2000). A bar may also influence the gas further out via resonances of spirals coupled to the bar. Both these effects are present in the barred sample of Muñoz-Mateos et al. (2013), who showed that breaks tend to follow two loci (shown in the right panel of Fig. 3.7): one given by the OLR of the bar and the other by the OLR of a spiral that has its 4:1 resonance at the CR of the bar. In both cases the effect of bars is to alter the gas distribution and thereby the location of star formation breaks. Finally, star formation need not truncate sharply: Elmegreen and Hunter (2006) proposed a multicomponent star formation model with star formation in the outer disk predominantly driven by turbulent processes (see the review by Elmegreen and Hunter 2017).

3.4.1.2 Angular Momentum Redistribution Models

Since most stars do not form in the outer disks, stars may reach there from the inner disk. The first mechanism that was proposed to explain Type II profiles sought to transform Type I profiles via angular momentum redistribution due to coupled bars

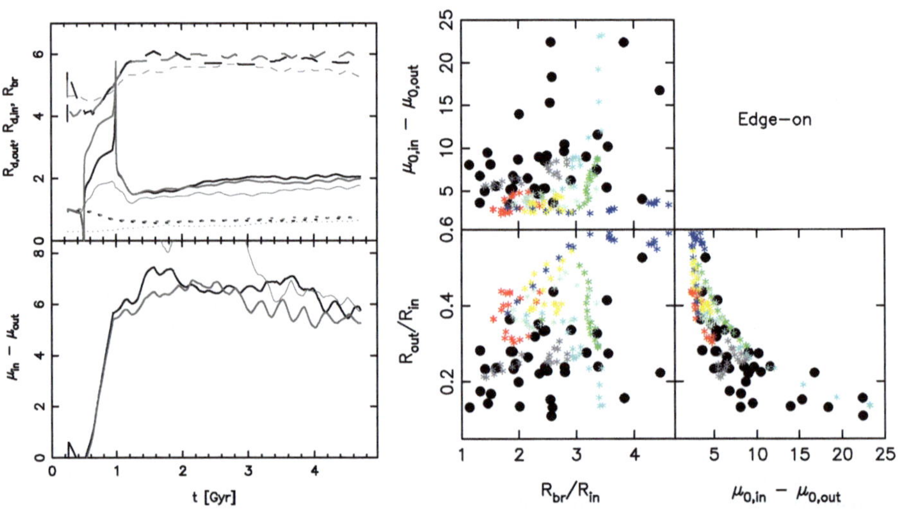

Fig. 3.8 *Left*: The evolution of break parameters in the pure N-body simulations of Debattista et al. (2006): R_{br} (the radius of the break, *dashed lines*), $R_{d,in}$ and $R_{d,out}$ (the scale lengths of the inner and outer disks, *solid and dotted lines*, respectively) and the difference in central surface brightness between the inner and outer disks, $\mu_{in} - \mu_{out}$. In these simulations the bar forms by 1 Gyr. The *black, grey and thin lines* correspond to different *initial* radial extents of the same model. *Right*: A comparison between parameters of profile breaks in observed (*solid circles*) and simulated (*coloured stars*) edge-on galaxies. The observational data are taken from Pohlen (2002). Each simulation corresponds to a different colour, and different viewing angles are shown by different points. Reproduced with permission from Debattista et al. (2006)

and spirals (Debattista et al. 2006). Their pure N-body simulations produced spirals with inner 4:1 resonance coinciding with the CR of the bar; angular momentum was transported to the OLR of the spiral, where a break in the profile developed. This angular momentum redistribution occurs *during bar formation* and decreases thereafter. The left panel of Fig. 3.8 shows that the break barely evolves after the first 1.5 Gyr, during which time the bar forms. The structural parameters of the resulting breaks are in good agreement with observations, as shown in the right panel of Fig. 3.8. Debattista et al. (2006) found that the inner disk needs to be cool ($Q \lesssim 1.6$) in order for bar formation to require enough angular momentum redistribution to significantly alter the density profile, whereas in hotter disks strong bars formed without a significant break developing. The idea was developed further by Foyle et al. (2008), who conducted a large simulation study of bulge+disk systems, including gas and star formation. They found that when breaks form, the outer profile remains unchanged while the inner disk changes substantially. They reported that Type II profiles developed when $m_d/\lambda \geq 1$, where m_d is the disk mass fraction and λ is the halo angular momentum. The inclusion of gas led to a break in star formation, but this did not always result in a break in the total profile. In agreement with Debattista et al. (2006), they found that the location of the break does not evolve much once it formed. In a similar spirit, Minchev et al. (2011)

found that, within 1–3 Gyr, as bars form, the density profile of their models extended outwards to more than 10 $R_{\rm d}$, with a flattened metallicity distribution. Minchev et al. (2012) emphasize that outer disks forming via this bar-spiral resonance overlap are radially hot.

The formation of Type II profiles by angular momentum redistribution during bar formation presents a number of difficulties; most obviously strong bars are not present in all galaxies. Minchev et al. (2011) argued that bars are a recurrent phenomenon and that the absence of a bar now does not mean that one was never present. However, bars are not easily destroyed (e.g. Shen and Sellwood 2004). Additionally bar destruction leaves a kinematically hot disk; a re-formed bar would subsequently lead to reduced angular momentum transport (Debattista et al. 2006; Brunetti et al. 2011). The formation of Type II profiles in such simulations is likely significantly affected by the very unstable initial conditions usually employed, which may be unlikely in nature.

Another way of scattering material into the outer disk was proposed by Bournaud et al. (2007), who used simulations of gas-rich disks which form clumps similar to those observed in chain galaxies. These clumps interact gravitationally with each other, flinging material outwards while sinking to form a central bulge. The material flung out forms a kinematically hot outer exponential of a Type II profile. The existence of bulgeless galaxies with Type II profiles (e.g. NGC 3589, NGC 6155, UGC 12709; Pohlen and Trujillo 2006) is a challenge for this model to explain Type II profiles observed at low redshift since clump formation inevitably also leads to bulge formation.

3.4.1.3 Churning

Alternatively, Roškar et al. (2008b) proposed that the outer disk in Type II profiles results from stars that have migrated past the star formation break via churning. Stellar density breaks formed in this way are coincident with a drop in the star formation rate. In their simulations, stars are formed from gas continuously cooling from the corona rather than being introduced into the simulation as part of the initial conditions. Because of this, the disk heats as it is forming, remaining at $Q \simeq 2$ through most of its extent and never becomes violently unstable, while still being cool enough to continuously support multiple spirals. Roškar et al. (2008b) found that $\sim$85% of the stars that end up beyond the break formed interior to it, yet are on nearly circular orbits, with epicyclic radii half of the average radial excursion, which is because spiral trapping is most efficient for nearly circular orbits (Fig. 3.9).

In the churning model of Type II profiles, while the break radius grows with time, churning forces the break to occur at the same radius for stars of all ages, and all heights above the plane, as seen in Fig. 3.10. The model also predicts that the mean age of stars beyond the star formation break increases outwards, as shown in Fig. 3.9. This age upturn occurs because churning, being stochastic, requires increasingly long times to populate regions further from where stars form.

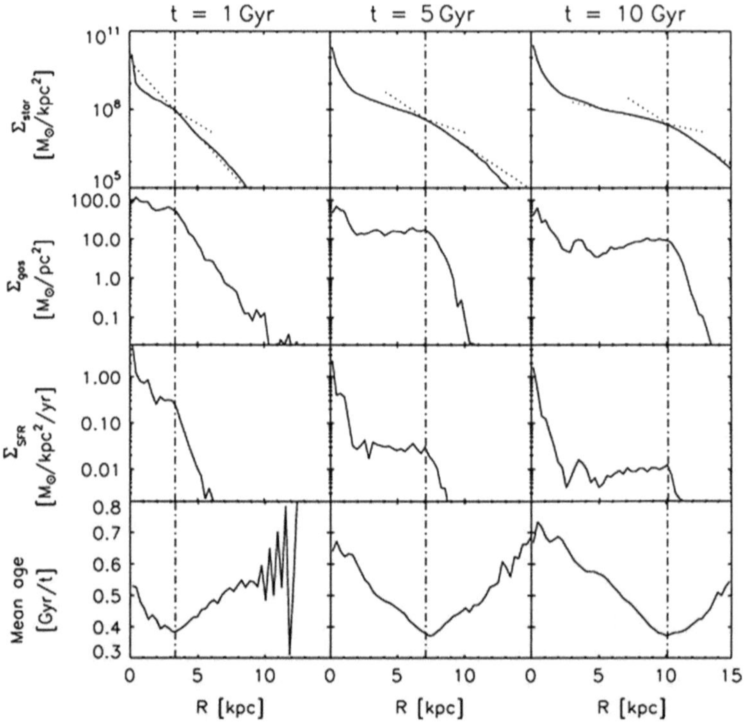

Fig. 3.9 Evolution of the system with stars forming from gas cooling off a hot corona. The *top row* shows the stellar density profile, the *second row* shows the surface density of cool gas, the *third row* shows the star formation rate surface density and the *bottom row* shows the average age of stars, normalized to the age of the galaxy at that time. From *left* to *right*, the columns show the system after 1, 5 and 10 Gyr. The *vertical dot-dashed lines* show the stellar break radius while the *dotted lines* in the top row show the inner and outer exponential disk fits

3.4.1.4 Insight from Cosmological Simulations

Sánchez-Blázquez et al. (2009) studied breaks using a cosmological simulation. The break in the simulation was caused by a drop in the star formation due to a warp, which led to a drop in the volume density of the gas. The stars beyond the break were predominantly (57%) formed interior to the break, with the remainder coming equally from in situ formation and from satellite accretion. They interpreted the age upturn in their simulation as resulting from a low but constant star formation at large radii, rather than to migration. Ruiz-Lara et al. (2016a) also found age upturns in their cosmological simulations, which were caused by satellites accreted onto the outer disks. An observational signature of such accretion is that past the break the age upturn is followed within $\sim 1\,R_{\rm d}$ by an age plateau.

However, cosmological simulations are still plagued by excessively hot disks. Sánchez-Blázquez et al. (2009) dealt with this by identifying the disk component via a kinematic cut on stellar orbits to retain stars on nearly circular orbits. This resulted

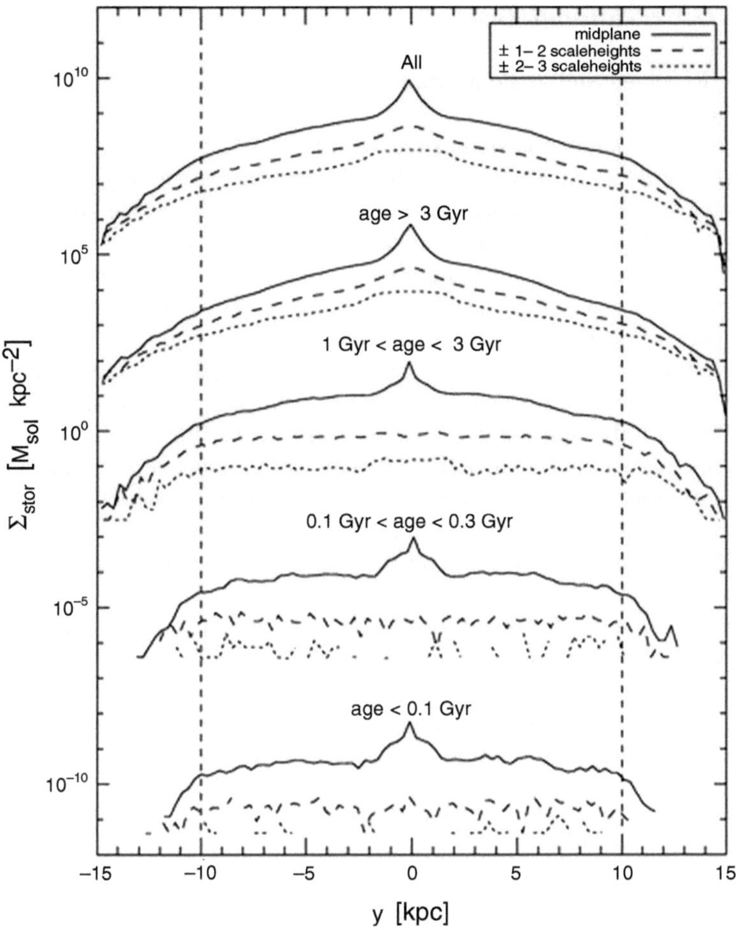

Fig. 3.10 The density distribution of stars for the galaxy on the left at 10 Gyr, observed edge-on. Different age bins are shown, as indicated, with the top profiles showing the profiles of all the stars. In each set of profiles, three heights above the mid-plane are shown, as indicated in the inset. An arbitrary offset has been applied to each set of profiles, for clarity. Reproduced with permission from Roškar et al. (2008b)

in a meagre 37% of stars being sufficiently cool to be considered as disk stars. Likewise Ruiz-Lara et al. (2016a) applied a kinematic cut to the stellar particles, but were left with a hot system with even young populations being as hot as the old stars in the Solar neighbourhood. Roškar et al. (2010) stressed that however the disk is defined, one is left with a high Toomre-Q system which substantially inhibits spiral structure (e.g. Binney and Tremaine 2008), suppressing churning; therefore, these simulations cannot yet give a clear indication of the relative contribution of accreted, scattered and migrated by churning stars in outer disks.

3.4.2 Observational Tests

3.4.2.1 Resolved Stellar Population Studies

Resolved stellar population studies are ideal for assessing the populations of outer disks because they can be observed directly and little modelling is necessary. An early study of stellar populations across the break came from the GHOSTS survey (Radburn-Smith et al. 2011) for the edge-on galaxy NGC 4244 (de Jong et al. 2007). They separated stars into four age bins at different heights above the midplane. They found that stars of all ages have a break at the same radius, regardless of whether they are in the midplane or above it, as seen in Fig. 3.11. Unless the break formed very recently (for no apparent reason), or formed once and has not evolved since (whereas observations show that break radii have evolved since $z \sim 1$, e.g. Pérez 2004; Trujillo and Pohlen 2005; Azzollini et al. 2008b), the most plausible explanation is that the break radius evolves with time, with something forcing all stars to adopt a common break radius. The churning model predicts exactly this behaviour: as seen in Fig. 3.9, the break radius in the simulations of Roškar et al. (2008b) evolves with time, yet the location of the break is identical for all populations.

A second example of a Type II profile with a common break radius for all ages is the low-inclination flocculent spiral galaxy NGC 7793 (Radburn-Smith et al. 2012). Here, the break is just inside the break of both the HI gas and of the star formation. Beyond the break, R_d increases with the age of the stellar population. Figure 3.11 shows the scale length of different age bins for NGC 7793 and for a simulation of a galaxy of comparable mass and size which experienced significant churning.

M33 also has a Type II profile with a break at ~8 kpc (Ferguson et al. 2007), which is present also in the surface *mass* density (Barker et al. 2011); stars past the break exhibit a positive age gradient (Barker et al. 2007; Williams et al. 2009; Barker et al. 2011). Unlike NGC 4244 and NGC 7793, M33 has had a strong interaction in the past 1–3 Gyr (Braun and Thilker 2004; Putman et al. 2009; Davidge and Puzia 2011). Barker et al. (2011) found that about half of all stars just inside the break have ages between 2.5 and 4.5 Gyr, with less than 14% of stars older. While the star formation within 8 kpc does not appear to have changed much in the past few Myr, the main sequence luminosity function of the outer disk indicates that the recent star formation rate in the outer disk has declined (Davidge and Puzia 2011), suggesting enhanced star formation in the outer disk or contamination by young stars. In the past 10 Gyr R_d has increased by a factor of ~2. Williams et al. (2013) contrasted M33 with its near equals in mass, NGC 300 and NGC 2403, both relatively isolated bulgeless late-type galaxies with Type I profiles. They concluded that the environment has played a role in the formation of the break in M33.

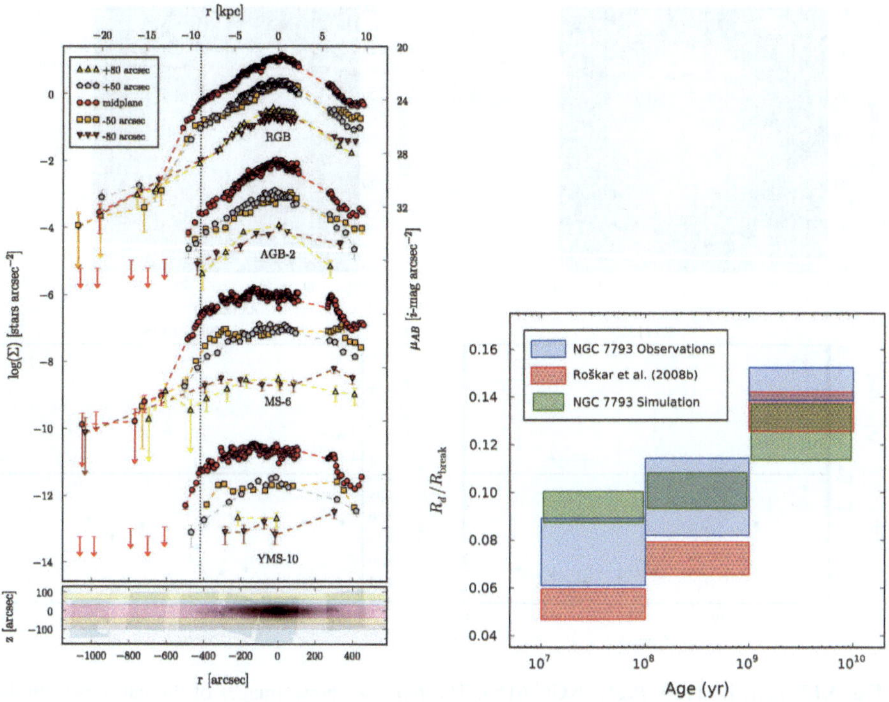

Fig. 3.11 *Left*: Radial density profiles for resolved stellar populations in the edge-on NGC 4244 (shown at *bottom*). The different populations are young main sequence (YMS) stars (<100 Myr), main sequence (MS) stars (100–300 Myr), asymptotic giant branch (AGB) stars, (older than 0.3 Gyr and peaking at 1–3 Gyr with a tail to 10 Gyr) and metal poor red giant branch (RGB) stars (older than 5 Gyr). Five strips are used, as indicated at *top left*, at three different offsets from the midplane. The AGB, MS and YMS profiles have been offset vertically by −2, −6 and −10, respectively, for clarity. This can be compared with Fig. 3.10. Reproduced with permission from de Jong et al. (2007). *Right*: The outer disk exponential scale lengths, normalized by the break radius, as a function of age for NGC 7793 (*shaded blue box*) and two simulations, the Milky Way mass one of Roškar et al. (2008a) (*dotted red box*) and another of the same mass and size as NGC 7793 (*hatched green box*). The height of the boxes represents the uncertainty on the scale length. Reproduced with permission from Radburn-Smith et al. (2012)

3.4.2.2 Spectroscopic Studies

Yoachim et al. (2012) modelled integral field data for six Type II profile galaxies, obtaining their star formation history. In all cases the average age profile has a positive gradient beyond the break. The modelling fitted an exponentially declining SFR, $\psi(t) \propto e^{-t/\tau}$, where the timescale, τ, was allowed to take both positive (declining SFR) and negative (increasing SFR) values. In three of the six galaxies, τ was positive, as expected if the stars in this region migrated there. The other three galaxies had negative τ, indicating an increasing SFR, which is not expected if the outer disk formed exclusively from migrated stars. Figure 3.12 presents an example of each behaviour. NGC 6155, shown on the right, is a galaxy where the stellar

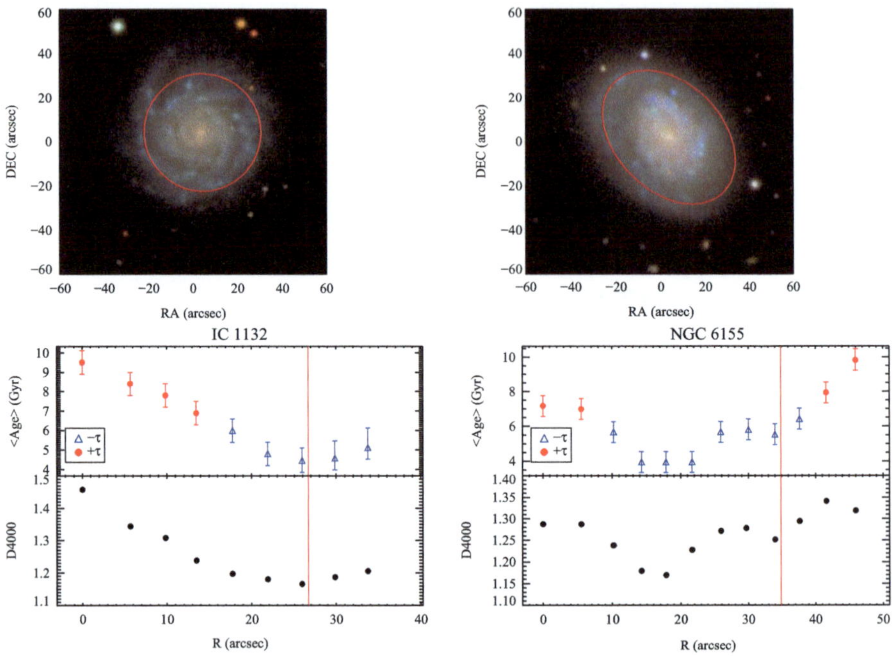

Fig. 3.12 *Left*: IC 1132. *Right*: NGC 6155. The *top row* shows images of the galaxies with the location of the break indicated by the *red ellipses*. The *bottom row* shows the mean age profiles (*top panel*) and the D4000 index (*bottom panel*). In the mean age profiles, *blue points* correspond to $\tau < 0$ (exponentially increasing SFR), while the *red points* correspond to $\tau > 0$ (exponentially decreasing SFR). Reproduced with permission from Yoachim et al. (2012)

populations past the break have a radially increasing average age and, except for the innermost bin, temporally decreasing star formation rate. While the average stellar age increases past the break, the minimum age is not at the break itself, but halfway between the centre and the break; inspection of the top panel of Fig. 3.12 shows that, at the location of the age minimum, the galaxy has a strongly star-forming ring encircling a bar.

Churning predicts that the mean age increases past the break. However, the break need not be the location of the minimum in average age, which happens only if inside-out growth is exact.

Another example of an age minimum before the break radius is M95, a barred galaxy with a ring at ~8 kpc previously classified as having an OLR Type II profile (Erwin et al. 2008). However, the star formation break is at ~13 kpc, beyond which the colour gradient reverses, extending to ~25 kpc (Watkins et al. 2014).

For three galaxies, Yoachim et al. (2012) found that the star formation timescale remains negative (increasing SFR) to the last measured point. The left panels of Fig. 3.12 show an example, IC 1132. What appears to be happening in each of these galaxies is that new stars are forming in situ due to spirals extending past the break radius, as envisaged by Schaye (2004). Since the SFR does not terminate at the break, either in the observations or in the simulations (Roškar et al. 2008b found that 15% of stars formed in situ), it should not be surprising to find hints of recent star formation past the break. Moreover, such galaxies are easier to measure spectroscopically and so may be overrepresented in a small sample.

Ruiz-Lara et al. (2016b) studied the stellar populations of 44 face-on spiral galaxies from the CALIFA survey. They found that ∼40% of these show age upturns beyond the break in the luminosity-weighted spectra. They found very flat mass-weighted mean age profiles, with less than 1 Gyr variation in NGC 551 and NGC 4711. Moreover, they found that luminosity-weighted mean age upturns are present in both Type II and Type I profiles. Roediger et al. (2012) had also seen age upturns in Type I profiles for galaxies in the Virgo Cluster. This suggests that age upturns in Type I galaxies may be a result of environment (see Sect. 3.5 below). Their presence in Type I profiles led Ruiz-Lara et al. (2016b) to propose that age upturns are the result of the early formation of the entire disk followed by a gradual inside-out quenching of star formation.

3.4.2.3 Photometric Methods

Because of the age-metallicity degeneracy (e.g. Worthey 1994), colour profiles are less constraining in probing the stellar populations of outer disks. Azzollini et al. (2008a) stacked colour (approximating rest-frame $u-g$) profiles of different profile types to $z \sim 1.1$. They scaled the radius by the break radius in Type II and III galaxies and to $2\,R_\mathrm{d}$ in Type I galaxies. In the Type I galaxies, the resulting colour profiles are flat or slightly rising, with no features. Likewise the Type III galaxies showed a range of gradients ranging from negative to positive with increasing redshift. Type II galaxies instead become increasingly red past the break at all redshifts. For a sample of nearby late-type galaxies, Bakos et al. (2008) carried out a similar stacking analysis, shown in Fig. 3.13; they used the colour profiles to obtain mass-to-light ratios and thereby recovered the mass profiles. They found that galaxies with Type II profiles have much weaker breaks in the overall *mass* distribution than in the light. Zheng et al. (2015) stacked a larger sample of 698 galaxies spanning a wide range of mass and morphology. In the bluer bands, in the majority of galaxies they found Type II profiles, becoming Type I in red bands, in contradiction to Martín-Navarro et al. (2012), who found numerous examples of Type II profiles in red bands, albeit weaker than in the blue. However, the latter authors stacked galaxies at fixed r_{90} rather than to the break radius, which probably smears any weak breaks.

3.4.3 Synthesis and Outlook

Evidence that the outer disks of Type II profiles are due to migration comes from the increasing age of stars past the break; in contrast an outer disk dominated by

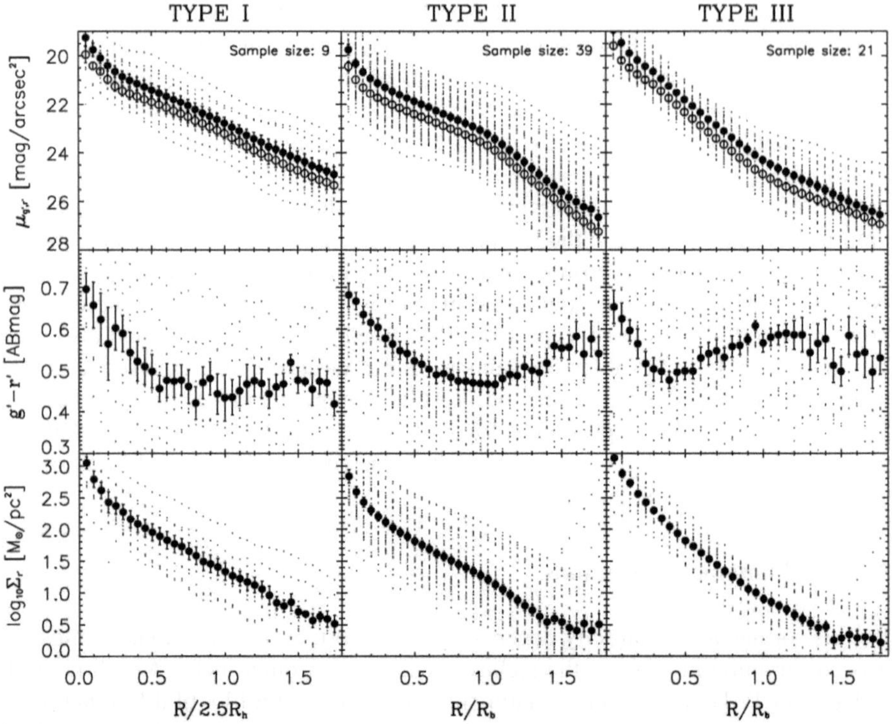

Fig. 3.13 Stacked profiles of 9 Type I (*left*), 39 Type II (*middle*) and 21 Type III (*right*) galaxies. The *top row* shows the r' (*filled circles*) and g' profiles (*open circles*). The *middle row* shows the $g' - r'$ colour profiles. The *bottom row* shows the computed mass density profiles. Reproduced with permission from Bakos et al. (2008)

accretion would have a constant age profile. The coincidence of star formation breaks and stellar continuum breaks, and the common break for stars of all ages, favours the churning model. However, the evidence is not yet conclusive. Measurements of kinematics in the outer disks would provide definitive proof of whether migration is dominated by scattering or by churning. Further observational study and modelling of the role of environment are also required. The extent to which the mass profile deviates from pure exponential needs further study and may help constrain the relative importance of churning and scattering.

3.5 Type I Profiles

Type I profiles, extending to $\sim$6–8 R_d, are present in roughly 10–15% of field late-type spirals (Gutiérrez et al. 2011). Examples of galaxies with Type I profiles are sufficiently nearby that their stellar populations can be resolved and studied directly.

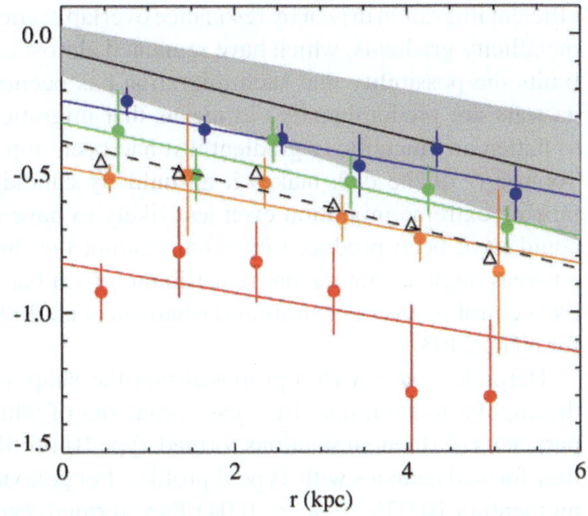

Fig. 3.14 The metallicity profiles of NGC 300. *Solid black line* with *grey shaded region*: metallicity profile of 24 young A-type supergiants from Kudritzki et al. (2008). *Blue, green, orange and red circles* are the metallicity profiles for stars with ages < 100 Myr, 1–5 Gyr, 5–10 Gyr and 10–14 Gyr, respectively. The *black triangles and dashed line* indicate the mean metallicity for the entire population. Reproduced with permission from Gogarten et al. (2010)

NGC 300 is a relatively isolated, unbarred galaxy with a Type I profile extending unbroken to 10 disk scale lengths (∼14 kpc; Bland-Hawthorn et al. 2005). Using *HST* to resolve the stellar populations of the inner ∼5 kpc, Gogarten et al. (2010) showed that the majority of its stars are old (see also Vlajić et al. 2009), with 80% of them older than 6 Gyr. R_d increases from 1.1 kpc for old stars to 1.3 kpc for young stars, indicating a modest inside-out formation, which is also evident in the broadband colours (Muñoz-Mateos et al. 2007). Bland-Hawthorn et al. (2005) estimated that everywhere beyond ∼6 kpc, the Toomre-$Q = 5 \pm 2$. Thus spiral structure in NGC 300 is unlikely to be strong, which, together with its low surface mass density, makes churning inefficient, as confirmed by simulations (Gogarten et al. 2010). In agreement with this prediction, both young stars (Kudritzki et al. 2008) and old stars (Vlajić et al. 2009; Gogarten et al. 2010) exhibit the same negative metallicity gradient outside 2 kpc, which has remained quite constant over the past ∼10 Gyr (Gogarten et al. 2010), as seen in Fig. 3.14.

NGC 300 is not unique: NGC 2403 is another nearby, isolated bulgeless galaxy with a Type I profile. Whereas NGC 300 has a strong HI warp, NGC 2403 has no warp (Williams et al. 2013). All stellar populations, including the young stars, follow the same density profile, with no break. The star formation histories are parallel at all radii to 11 R_d, with the surface density of star formation following the same exponential profile as the overall density profile (Williams et al. 2013).

3.5.1 Origin of Type I Profiles in Isolated Galaxies

These observational results help constrain the origin of Type I profiles. Minchev et al. (2011) proposed that the Type I profile of NGC 300 was produced by very efficient migration driven by resonance overlap scattering. The presence of the radial metallicity gradients, which have remained almost constant over ~10 Gyr, severely limits the possibility that such migration has occurred in NGC 300. The fact that its stars are predominantly old means that migration would have had ample time to flatten any metallicity gradient if it had been important. The low-mass and large Toomre-Q of the disk makes it an unlikely candidate for strong spiral structure, making extreme migration even less likely to have occurred. The large Toomre-Q could have been produced by a bar heating the disk, but NGC 300 is bulgeless, whereas angular momentum redistribution by a bar would have produced a bulge-like central mass concentration (Debattista et al. 2006). A similar case can be made for NGC 2403.

Herpich et al. (2015a) proposed that the shapes of density profiles depend on the angular momentum of the gas corona out of which the disk forms. At low spin parameter, λ, their simulations formed Type III profiles (see below), while at high λ they formed galaxies with Type II profiles. For galaxies with an intermediate angular momentum ($0.035 \leq \lambda \leq 0.04$) they formed Type I profiles. This model very naturally accounts for the relative rarity of the Type I profile. Furthermore, because it does not require stars to migrate, this model is consistent with the metallicity gradients in NGC 300 and in NGC 2403. The absence of a break in the profiles of stars younger than 200 Myr in NGC 300 also supports this model. Lastly, this scenario is also consistent with the fact that the *inner* profiles of Type II and Type I galaxies are consistent with each other (Gutiérrez et al. 2011).

3.5.2 Type I Profiles in Cluster Lenticulars

While Type I profiles are rare in the field, they are common among cluster lenticulars (see Sect. 3.2.1). Since a narrow range of λs cannot account for a dominant population, a different explanation for Type I profiles in cluster lenticulars is required. Clarke et al. (2017) evolved the model of Roškar et al. (2008b) (with $\lambda = 0.065$) in a cluster environment. Their model galaxy was quenched by ram pressure stripping and transformed into a lenticular with very little spiral structure at late times. At the time of peak ram pressure stripping, the disk had a break radius at ~6 kpc; therefore, the galaxy never gets a chance to form many stars beyond this radius. As shown in Fig. 3.9, the same model evolved in isolation resulted in a Type II profile with a break at 10 kpc. Nonetheless stars in the cluster galaxy moved to large radii through migration via strong, tidally induced, spirals, but not through heating of the disk. As a result, a Type I density profile developed, as shown by the top panel of Fig. 3.15. This combination of quenching by ram pressure stripping and churning by tidally induced spirals accounts for the high incidence of Type I profiles among cluster lenticulars. Because of the strong churning into the outer disk, Clarke et al. (2017) predicted that the age profile of cluster lenticulars with Type I profiles will be flat or slightly rising past the point where the galaxy had a break radius before star formation is quenched, as shown in the bottom panel of Fig. 3.15. Indeed

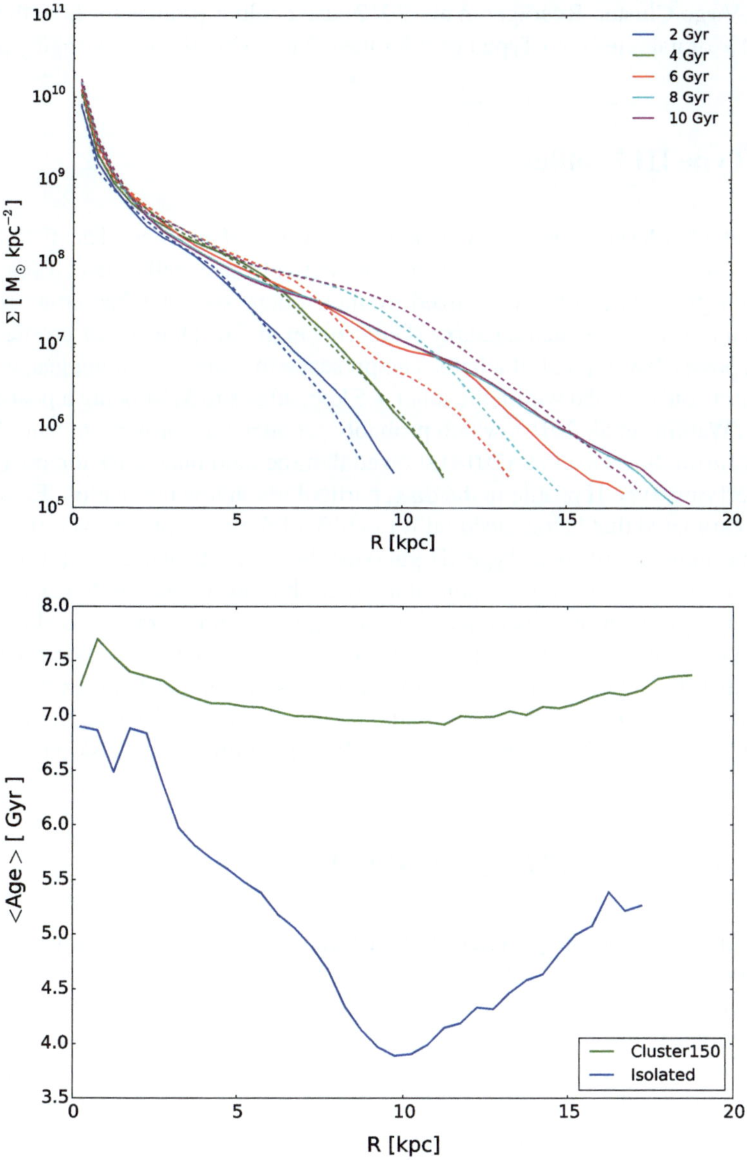

Fig. 3.15 *Top*: Evolution of surface density profile of the model of Roškar et al. (2008b), when evolved in isolation (*dashed lines*) and when it falls into a cluster, reaching pericentre at 5 Gyr (*solid lines*). Whereas the isolated simulation develops a Type II profile, the same system evolving in the cluster is transformed into a Type I profile via the joint action of ram pressure stripping and tidally induced spirals. *Bottom*: The final mean age profiles of the isolated and cluster models. The isolated model has an increasing average age past the break, whereas the cluster model has a nearly flat age distribution. Reproduced with permission from Clarke et al. (2017)

for the Virgo Cluster, Roediger et al. (2012) used colour profiles to show that more than 70% of galaxies with Type I profiles have flat or slightly positive age gradients.

3.6 Type III Profiles

Erwin et al. (2005) identified two kinds of Type III profiles. The first kind is produced by the excess light above an exponential from a stellar halo. They argued that such galaxies can be recognized by their outer isophotes becoming rounder. M64 is an example of such a galaxy (Gutiérrez et al. 2011); its outer profile is well fit by a power law, typical of a halo. The presence of counter-rotating gas, with star formation concentrated within the inner 4.5 kpc, attests to M64 being a post-merger galaxy (Watkins et al. 2016), which probably produced the prominent halo. In such cases, Martín-Navarro et al. (2014) showed that the halo may mask the presence of an underlying Type II profile in the disk, particularly at low inclination. Erwin et al. (2005) estimated that haloes account for $\sim$30% of their sample of Type III profiles.

In the majority of their Type III galaxies, however, Erwin et al. (2005) found sharp transitions from one exponential to a shallower one, with isophotes not becoming significantly rounder and, in some cases, hosting spiral arms. Maltby et al. (2012b) showed that in $\sim$85% of cases, the extrapolation of the bulge profile to large radii is insufficient to account for the excess light at the break. Thus, except in cases where the light profile is flattened by the point spread function (see the review by Knapen and Trujillo 2017), the majority of Type III profiles are a disk phenomenon.

3.6.1 Formation of Type III Disk Profiles

Breaks in Type III profile galaxies are not associated with a colour change (Azzollini et al. 2008a; Bakos et al. 2008; Zheng et al. 2015). Changes in stellar populations and star formation are therefore not the origin of these profiles.

Anecdotally, Erwin et al. (2005) noted that a number of galaxies with Type III profiles exhibit asymmetries in the outer disks, while Pohlen and Trujillo (2006) found evidence of recent interactions among Type III galaxies. It is unsurprising therefore that many models of Type III profile formation rely on mergers and interactions. Laurikainen and Salo (2001) used simulations to show that interactions result in long-lasting shallow outer profiles. Peñarrubia et al. (2006) proposed that extended outer disks are composed of tidally disrupted dwarf satellites accreted directly onto the plane of the disk. They showed that the degree of rotational support of the accreted material depends on the satellite orbit, with dispersion increasing with orbital eccentricity. Instead of the outer disk being accreted, Younger et al. (2007) proposed that it forms from the main disk itself after a minor merger forces gas to the centre, causing the galaxy's inner region to contract and its outer region to expand. In the absence of gas, the pure N-body simulations of Kazantzidis et al. (2009) produced Type III profiles when substructures excited angular momentum redistribution within the disk. Borlaff et al. (2014) demonstrated that, in simulations, gas-rich major mergers, which destroy the main disk then build a new one at large radius, develop into S0 galaxies with Type III profiles.

Alternatively both Minchev et al. (2012) and Herpich et al. (2015b) proposed that Type III profiles formed via the action of bars. Minchev et al. (2012) invoke resonance overlap in the presence of a bar together with gas accretion to populate the outer disk. In the model of Herpich et al. (2015b), instead, stars in the inner disk gained energy directly from the bar, getting boosted to increasingly eccentric orbits, and reaching large radii.

Observations do not yet provide very strong constraints on how Type III profiles form. The lack of an observed difference between Type III profiles in barred and unbarred galaxies (Borlaff et al. 2014; Eliche-Moral et al. 2015) and the fact that Type III profiles are *less* common among barred galaxies in the field (see Table 3.1) suggests that bars are unlikely to account for the majority of Type III profiles. In edge-on galaxies, anti-truncations appear in thick disks, supporting the view that a hot population is responsible for the outer disk (Comerón et al. 2012), contrary to the model of Borlaff et al. (2014). The most promising models therefore are the direct accretion model (Peñarrubia et al. 2006) and the heating by substructure model (Kazantzidis et al. 2009), both of which produce hot outer disks. The model of Younger et al. (2007) may also lead to hot outer disks because of the rapid change of the inner potential. Which of these models works best can be decided by measuring the kinematics of both inner and outer disks, as well as the chemistry of the outer disk. Additional constraints may be possible from comparing the properties of Type III and Type I/II. For instance, Gutiérrez et al. (2011) note that the inner exponentials of Type III galaxies have shorter scale lengths and higher central surface brightness than galaxies with Type I profiles, which argues against the direct accretion model but might favour the model of Younger et al. (2007).

3.7 Future Prospects

Much work remains to be done to better distinguish between competing scenarios of outer disk formation and to clarify what role, if any, migration has played. With the imminent launch of the *JWST*, resolved stellar population studies will be possible in a volume roughly three times larger than is accessible with *HST*, improving the statistical robustness of age and metallicity dissections of outer disks. In addition, we identify three areas that would greatly advance our understanding of disk outskirts.

- *Kinematics in the outer disks in Type II galaxies* The two classes of models for producing the outer disks of Type II profiles make very different kinematic predictions. Models based on scattering of stars from the inner disk all produce radially hot outer disks. In contrast, churning predicts a relatively cooler outer disk. The observed line-of-sight velocity dispersion, $\sigma_{\rm los}$, is given by

$$\sigma_{\rm los}^2 = \frac{1}{2}\sin^2 i \left[(\sigma_{\rm R}^2 + \sigma_\phi^2 + 2\sigma_z^2 \cot^2 i) - (\sigma_{\rm R}^2 - \sigma_\phi^2)\cos 2\phi\right], \qquad (3.3)$$

where $\sigma_{\rm R}$, σ_ϕ and σ_z are the velocity dispersions in the radial, tangential and vertical directions, respectively, i is the galaxy's inclination and ϕ is the angle from the major axis of the disk in its intrinsic plane. This can be written as $\sigma_{\rm los}^2 = a(R) + b(R)\cos 2\phi$. The first term, $a(R)$, is positive definite and independent of ϕ. The second term, $b(R)$ is proportional to $-\beta_\phi = -(1-\sigma_\phi^2/\sigma_{\rm R}^2)$; it is therefore negative in the inner disk where generally $\beta_\phi > 0$. When the outer disk is formed of material that migrated via churning, then it is kinematically relatively cool, and $b(R)$ remains negative. If instead the outer disk is comprised of stars scattered outwards, then $\beta_\phi < 0$ and $b(R)$ is positive. In the former case, $\sigma_{\rm los}$ peaks on the major axis, while it peaks on the minor axis for scattering. This provides a very promising means for distinguishing which class of models is responsible for the outer disks in Type II galaxies.

- *Kinematics of Type III galaxies* Most models of Type III galaxies predict that outer disks and, for some models, the inner disk are hot and that the outer disk is comprised of stars that have been moved out or accreted directly. Confirming this prediction would give us an important insight into the properties of galaxies that have experienced these heating or accretion processes.
- *The outer disk of the Milky Way* As with many other aspects of galaxy evolution, the Milky Way provides an important laboratory for our understanding of disk outskirts. In the era of *Gaia*, we should be able to study the ages, metallicities, abundances and 3D kinematics of stars past the break. This will allow us to test the relations between the outer disk and the inner disk, the warp and the bar. Accreted stars may also be identified. These data will result in a much more detailed picture of the origin of the Milky Way's outer disk.

While the field of disk outskirts has made huge strides in the past decade, the next decade promises to be even more exciting in terms of understanding their assembly. Confirmation of an important role for radial migration would permit an important means for assessing its *overall* action on disk galaxies.

Acknowledgements VPD is supported by the STFC Consolidated Grant #ST/M000877/1. We thank Kathryne Daniel for providing us with the unpublished Fig. 3.4. Discussions with Peter Erwin have been especially enlightening. VPD thanks the Max-Planck-Institut für Astronomie for hosting him during which time this review was started.

References

Azzollini, R., Trujillo, I., Beckman, J.E.: Color profiles of disk galaxies since z ˜ 1: probing outer disk formation scenarios. Astrophys. J. **679**, L69–L72 (2008a). doi:10.1086/589283, arXiv:0804.2336

Azzollini, R., Trujillo, I., Beckman, J.E.: Cosmic evolution of stellar disk truncations: from z ˜ 1 to the local universe. Astrophys. J. **684**, 1026–1047 (2008b). doi:10.1086/590142, arXiv:0805.2259

Baba, J., Saitoh, T.R., Wada, K.: Dynamics of non-steady spiral arms in disk galaxies. Astrophys. J. **763**, 46 (2013). doi:10.1088/0004-637X/763/1/46, arXiv:1211.5401

Bakos, J., Trujillo, I., Pohlen, M.: Color profiles of spiral galaxies: clues on outer-disk formation scenarios. Astrophys. J. **683**, L103 (2008). doi:10.1086/591671, arXiv:0807.2776

Barker, M.K., Sarajedini, A., Geisler, D., Harding, P., Schommer, R.: The stellar populations in the outer regions of M33. III. Star formation history. Astrophys. J. **133**, 1138–1160 (2007). doi:10.1086/511186, astro-ph/0611892

Barker, M.K., Ferguson, A.M.N., Cole, A.A., Ibata, R., Irwin, M., Lewis, G.F., Smecker-Hane, T.A., Tanvir, N.R.: The star formation history in the far outer disc of M33. Mon. Not. R. Astron. Soc. **410**, 504–516 (2011). doi:10.1111/j.1365-2966.2010.17458.x, arXiv:1008.0760

Bell, E.F., de Jong, R.S.: The stellar populations of spiral galaxies. Mon. Not. R. Astron. Soc. **312**, 497–520 (2000). doi:10.1046/j.1365-8711.2000.03138.x, astro-ph/9909402

Bergemann, M., Ruchti, G.R., Serenelli, A., Feltzing, S., Alves-Brito, A., Asplund, M., Bensby, T., Gruyters, P., Heiter, U., Hourihane, A., Korn, A., Lind, K., Marino, A., Jofre, P., Nordlander, T., Ryde, N., Worley, C.C., Gilmore, G., Randich, S., Ferguson, A.M.N., Jeffries, R.D., Micela, G., Negueruela, I., Prusti, T., Rix, H.W., Vallenari, A., Alfaro, E.J., Allende Prieto, C., Bragaglia, A., Koposov, S.E., Lanzafame, A.C., Pancino, E., Recio-Blanco, A., Smiljanic, R., Walton, N., Costado, M.T., Franciosini, E., Hill, V., Lardo, C., de Laverny, P., Magrini, L., Maiorca, E., Masseron, T., Morbidelli, L., Sacco, G., Kordopatis, G., Tautvaišiene, G.: The Gaia-ESO survey: radial metallicity gradients and age-metallicity relation of stars in the Milky Way disk. Astron. Astrophys. **565**, A89 (2014). doi:10.1051/0004-6361/201423456, arXiv:1401.4437

Bigiel, F., Leroy, A., Walter, F., Blitz, L., Brinks, E., de Blok, W.J.G., Madore, B.: Extremely inefficient star formation in the outer disks of nearby galaxies. Astrophys. J. **140**, 1194–1213 (2010). doi:10.1088/0004-6256/140/5/1194, arXiv:1007.3498

Binney, J., Tremaine, S.: Galactic Dynamics. Princeton University Press, Princeton (1987)

Binney, J., Tremaine, S.: In: Binney, J., Tremaine, S. (eds.) Galactic Dynamics, 2nd edn. Princeton University Press, Princeton (2008). ISBN 978-0-691-13026-2 (HB)

Bland-Hawthorn, J., Vlajić, M., Freeman, K.C., Draine, B.T.: NGC 300: an extremely faint, outer stellar disk observed to 10 scale lengths. Astrophys. J. **629**, 239–249 (2005). doi:10.1086/430512, arXiv:astro-ph/0503488

Borlaff, A., Eliche-Moral, M.C., Rodríguez-Pérez, C., Querejeta, M., Tapia, T., Pérez-González, P.G., Zamorano, J., Gallego, J., Beckman, J.: Formation of S0 galaxies through mergers. Antitruncated stellar discs resulting from major mergers . Astron. Astrophys. **570**, A103 (2014). doi:10.1051/0004-6361/201424299, arXiv:1407.5097

Bournaud, F., Elmegreen, B.G., Elmegreen, D.M.: Rapid formation of exponential disks and bulges at high redshift from the dynamical evolution of clump-cluster and chain galaxies. Astrophys. J. **670**, 237–248 (2007). doi:10.1086/522077, arXiv:0708.0306

Bovy, J., Rix, H.W., Schlafly, E.F., Nidever, D.L., Holtzman, J.A., Shetrone, M., Beers, T.C.: The stellar population structure of the galactic disk. Astrophys. J. **823**, 30 (2016). doi:10.3847/0004-637X/823/1/30, arXiv:1509.05796

Braun, R., Thilker, D.A.: The WSRT wide-field H I survey. II. Local group features. Astron. Astrophys. **417**, 421–435 (2004). doi:10.1051/0004-6361:20034423, astro-ph/0312323

Brunetti, M., Chiappini, C., Pfenniger, D.: Stellar diffusion in barred spiral galaxies. Astron. Astrophys. **534**, A75 (2011). doi:10.1051/0004-6361/201117566, arXiv:1108.5631

Byrd, G., Rautiainen, P., Salo, H., Buta, R., Crocher, D.A.: Pattern speed domains in ringed disk galaxies from observational and simulational databases. Astrophys. J. **108**, 476–490 (1994). doi:10.1086/117085

Byun, Y.I.: Surface photometry of edge-on galaxies: IC 5249 and ESO 404-G18. Chin. J. Phys. **36**, 677–692 (1998)

Chiappini, C., Matteucci, F., Gratton, R.: The chemical evolution of the galaxy: the two-infall model. Astrophys. J. **477**, 765–780 (1997). astro-ph/9609199

Christlein, D., Zaritsky, D.: The kinematic properties of the extended disks of spiral galaxies: a sample of edge-on galaxies. Astrophys. J. **680**, 1053–1071 (2008). doi:10.1086/587468, arXiv:0803.2225

Christlein, D., Zaritsky, D., Bland-Hawthorn, J.: A spectroscopic study of the Hα surface brightness profiles in the outer discs of galaxies. Mon. Not. R. Astron. Soc. **405**, 2549–2560 (2010). doi:10.1111/j.1365-2966.2010.16631.x, arXiv:1003.4872

Clarke, A.J., Debattista, V.P., Roškar, R., Quinn, T.: The origin of type I profiles in cluster lenticulars: an interplay between ram pressure stripping and tidally induced spiral migration. Mon. Not. R. Astron. Soc. **465**, L79–L83 (2017). doi:10.1093/mnrasl/slw214, arXiv:1610.04030

Comerón, S., Elmegreen, B.G., Salo, H., Laurikainen, E., Athanassoula, E., Bosma, A., Knapen, J.H., Gadotti, D.A., Sheth, K., Hinz, J.L., Regan, M.W., Gil de Paz, A., Muñoz-Mateos, J.C., Menéndez-Delmestre, K., Seibert, M., Kim, T., Mizusawa, T., Laine, J., Ho, L.C., Holwerda, B.: Breaks in thin and thick disks of edge-on galaxies imaged in the spitzer survey stellar structure in galaxies (S^4G). Astrophys. J. **759**, 98 (2012). doi:10.1088/0004-637X/759/2/98, arXiv:1209.1513

Daniel, K.J., Wyse, R.F.G.: Constraints on radial migration in spiral galaxies - I. Analytic criterion for capture at corotation. Mon. Not. R. Astron. Soc. **447**, 3576–3592 (2015). doi:10.1093/mnras/stu2683, arXiv:1412.6110

Davidge, T.J., Puzia, T.H.: The stellar archeology of the M33 disk: recent star-forming history and constraints on the timing of an interaction with M31. Astrophys. J. **738**, 144 (2011). doi:10.1088/0004-637X/738/2/144, arXiv:1107.0077

de Grijs, R., Kregel, M., Wesson, K.H.: Radially truncated galactic discs. Mon. Not. R. Astron. Soc. **324**, 1074–1086 (2001). doi:10.1046/j.1365-8711.2001.04380.x, astro-ph/0002523

de Jong, R.S.: Near-infrared and optical broadband surface photometry of 86 face-on disk dominated galaxies. IV. Using color profiles to study stellar and dust content of galaxies. Astron. Astrophys. **313**, 377–395 (1996). astro-ph/9604010

de Jong, R.S., Seth, A.C., Radburn-Smith, D.J., Bell, E.F., Brown, T.M., Bullock, J.S., Courteau, S., Dalcanton, J.J., Ferguson, H.C., Goudfrooij, P., Holfeltz, S., Holwerda, B.W., Purcell, C., Sick, J., Zucker, D.B.: Stellar populations across the NGC 4244 truncated galactic disk. Astrophys. J. **667**, L49–L52 (2007). doi:10.1086/522035, arXiv:0708.0826

Debattista, V.P., Mayer, L., Carollo, C.M., Moore, B., Wadsley, J., Quinn, T.: The secular evolution of disk structural parameters. Astrophys. J. **645**, 209–227 (2006). doi:10.1086/504147, astro-ph/0509310

Eliche-Moral, M.C., Borlaff, A., Beckman, J.E., Gutiérrez, L.: Photometric scaling relations of anti-truncated stellar discs in S0-Scd galaxies. Astron. Astrophys. **580**, A33 (2015). doi:10.1051/0004-6361/201424692, arXiv:1505.04797

Elmegreen, B.G., Hunter, D.A.: Radial profiles of star formation in the far outer regions of galaxy disks. Astrophys. J. **636**, 712–720 (2006). doi:10.1086/498082, astro-ph/0509190

Elmegreen, B.G., Hunter, D.A.: Outskirts of nearby disk galaxies: star formation and stellar populations. In: Knapen, J.H., Lee, J.C., Gil de Paz, A. (eds.) Outskirts of Galaxies, vol. 434. Springer, Cham (2017). doi:10.1007/978-3-319-56570-5

Erwin, P., Beckman, J.E., Pohlen, M.: Antitruncation of disks in early-type barred galaxies. Astrophys. J. **626**, L81–L84 (2005). doi:10.1086/431739, arXiv:astro-ph/0505216

Erwin, P., Pohlen, M., Beckman, J.E.: The outer disks of early-type galaxies. I. Surface-brightness profiles of barred galaxies. Astrophys. J. **135**, 20–54 (2008). doi:10.1088/0004-6256/135/1/20, arXiv:0709.3505

Erwin, P., Gutiérrez, L., Beckman, J.E.: A strong dichotomy in S0 disk profiles between the virgo cluster and the field. Astrophys. J. **744**, L11 (2012). doi:10.1088/2041-8205/744/1/L11, arXiv:1111.5027

Fall, S.M., Efstathiou, G.: Formation and rotation of disc galaxies with haloes. Mon. Not. R. Astron. Soc. **193**, 189–206 (1980). doi:10.1093/mnras/193.2.189

Ferguson, A.M.N., Wyse, R.F.G., Gallagher, J.S., Hunter, D.A.: Discovery of recent star formation in the extreme outer regions of disk galaxies. Astrophys. J. **506**, L19–L22 (1998). doi:10.1086/311626, astro-ph/9808151

Ferguson, A., Irwin2, M., Chapman, S., Ibata, R., Lewis, G., Tanvir, N.: Resolving the stellar outskirts of M31 and M33. Astrophys. Space Sci. Proc. **3**, 239 (2007). doi:10.1007/978-1-4020-5573-7-39, astro-ph/0601121

Foyle, K., Courteau, S., Thacker, R.J.: An N-body/SPH study of isolated galaxy mass density profiles. Mon. Not. R. Astron. Soc. **386**, 1821–1844 (2008). doi:10.1111/j.1365-2966.2008.13201.x, arXiv:0803.2716

Freeman, K.C.: On the disks of spiral and so galaxies. Astrophys. J. **160**, 811–830 (1970)

Gil de Paz, A., Madore, B.F., Boissier, S., Swaters, R., Popescu, C.C., Tuffs, R.J., Sheth, K., Kennicutt, R.C. Jr., Bianchi, L., Thilker, D., Martin, D.C.: Discovery of an extended ultraviolet disk in the nearby galaxy NGC 4625. Astrophys. J. **627**, L29–L32 (2005). doi:10.1086/432054, arXiv:astro-ph/0506357

Gogarten, S.M., Dalcanton, J.J., Williams, B.F., Roškar, R., Holtzman, J., Seth, A.C., Dolphin, A., Weisz, D., Cole, A., Debattista, V.P., Gilbert, K.M., Olsen, K., Skillman, E., de Jong, R.S., Karachentsev, I.D., Quinn, T.R.: The advanced camera for surveys nearby galaxy survey treasury. V. Radial star formation history of NGC 300. Astrophys. J. **712**, 858–874 (2010). doi:10.1088/0004-637X/712/2/858, arXiv:1002.1743

Grand, R.J.J., Kawata, D., Cropper, M.: The dynamics of stars around spiral arms. Mon. Not. R. Astron. Soc. **421**, 1529–1538 (2012). doi:10.1111/j.1365-2966.2012.20411.x, arXiv:1112.0019

Guo, Q., White, S., Boylan-Kolchin, M., De Lucia, G., Kauffmann, G., Lemson, G., Li, C., Springel, V., Weinmann, S.: From dwarf spheroidals to cD galaxies: simulating the galaxy population in a ΛCDM cosmology. Mon. Not. R. Astron. Soc. **413**, 101–131 (2011). doi:10.1111/j.1365-2966.2010.18114.x, arXiv:1006.0106

Gutiérrez, L., Erwin, P., Aladro, R., Beckman, J.E.: The outer disks of early-type galaxies. II. Surface-brightness profiles of unbarred galaxies and trends with hubble type. Astrophys. J. **142**, 145 (2011). doi:10.1088/0004-6256/142/5/145, arXiv:1108.3662

Halle, A., Di Matteo, P., Haywood, M., Combes, F.: Quantifying stellar radial migration in an N-body simulation: blurring, churning, and the outer regions of galaxy discs. Astron. Astrophys. **578**, A58 (2015). doi:10.1051/0004-6361/201525612, arXiv:1501.00664

Hayden, M.R., Bovy, J., Holtzman, J.A., Nidever, D.L., Bird, J.C., Weinberg, D.H., Andrews, B.H., Allende Prieto, C., Anders, F., Beers, T.C., Bizyaev, D., Chiappini, C., Cunha, K., Frinchaboy, P., García-Herńandez, D.A., García Pérez, A.E., Girardi, L., Harding, P., Hearty, F.R., Johnson, J.A., Majewski, S.R., Mészáros, S., Minchev, I., O'Connell, R., Pan, K., Robin, A.C., Schiavon, R.P., Schneider, D.P., Schultheis, M., Shetrone, M., Skrutskie, M., Steinmetz, M., Smith, V., Zamora, O., Zasowski, G.: Chemical cartography with APOGEE: metallicity distribution functions and the chemical structure of the milky way disk. Astrophys. J. **808**, 132 (2015). doi:10.1088/0004-637X/808/2/132, arXiv:1503.02110

Haywood, M., Di Matteo, P., Lehnert, M.D., Katz, D., Gómez, A.: The age structure of stellar populations in the solar vicinity. Clues of a two-phase formation history of the milky way disk. Astron. Astrophys. **560**, A109 (2013). doi:10.1051/0004-6361/201321397, arXiv:1305.4663

Herpich, J., Stinson, G.S., Dutton, A.A., Rix, H.W., Martig, M., Roškar, R., Macciò, A.V., Quinn, T.R., Wadsley, J.: How to bend galaxy disc profiles: the role of halo spin. Mon. Not. R. Astron. Soc. **448**, L99–L103 (2015a). doi:10.1093/mnrasl/slv006, arXiv:1501.01960

Herpich, J., Stinson, G.S., Rix, H.W., Martig, M., Dutton, A.A.: How to bend galaxy disc profiles II: stars surfing the bar in anti-truncated discs (2015b). ArXiv e-prints arXiv:1511.04442

Kazantzidis, S., Magorrian, J., Moore, B.: Generating equilibrium dark matter halos: inadequacies of the local Maxwellian approximation. Astrophys. J. **601**, 37–46 (2004). doi:10.1086/380192

Kazantzidis, S., Zentner, A.R., Kravtsov, A.V., Bullock, J.S., Debattista, V.P.: Cold dark matter substructure and galactic disks. II. Dynamical effects of hierarchical satellite accretion. Astrophys. J. **700**, 1896–1920 (2009). doi:10.1088/0004-637X/700/2/1896, 0902.1983

Kennicutt, R.C. Jr.: The star formation law in galactic disks. Astrophys. J. **344**, 685–703 (1989). doi:10.1086/167834

Knapen, J.H., Trujillo, I.: Ultra-deep imaging: structure of disks and haloes. In: Knapen, J.H., Lee, J.C., Gil de Paz, A. (eds.) Outskirts of Galaxies, vol. 434. Springer, Cham (2017). doi:10.1007/978-3-319-56570-5

Kregel, M., van der Kruit, P.C., de Grijs, R.: Flattening and truncation of stellar discs in edge-on spiral galaxies. Mon. Not. R. Astron. Soc. **334**, 646–668 (2002). doi:10.1046/j.1365-8711.2002.05556.x, astro-ph/0204154

Kudritzki, R.P., Urbaneja, M.A., Bresolin, F., Przybilla, N., Gieren, W., Pietrzyński, G.: Quantitative Spectroscopy of 24 A Supergiants in the Sculptor Galaxy NGC 300: flux-weighted gravity-luminosity relationship, metallicity, and metallicity gradient. Astrophys. J. **681**, 269–289 (2008). doi:10.1086/588647, arXiv:0803.3654

Kuijken, K., Dubinski, J.: Nearly self-consistent disc/bulge/halo models for galaxies. Mon. Not. R. Astron. Soc. **277**, 1341–1353 (1995)

Laine, J., Laurikainen, E., Salo, H., Comerón, S., Buta, R.J., Zaritsky, D., Athanassoula, E., Bosma, A., Muñoz-Mateos, J.C., Gadotti, D.A., Hinz, J.L., Erroz-Ferrer, S., Gil de Paz, A., Kim, T., Menéndez-Delmestre, K., Mizusawa, T., Regan, M.W., Seibert, M., Sheth, K.: Morphology and environment of galaxies with disc breaks in the S^4G and NIRSOS. Mon. Not. R. Astron. Soc. **441**, 1992–2012 (2014). doi:10.1093/mnras/stu628, arXiv:1404.0559

Larson, R.B.: Models for the formation of disc galaxies. Mon. Not. R. Astron. Soc. **176**, 31–52 (1976). doi:10.1093/mnras/176.1.31

Laurikainen, E., Salo, H.: BVRI imaging of M51-type interacting galaxy pairs - III. Analysis of the photometric parameters. Mon. Not. R. Astron. Soc. **324**, 685–698 (2001). doi:10.1046/j.1365-8711.2001.04347.x

Lemonias, J.J., Schiminovich, D., Thilker, D., Wyder, T.K., Martin, D.C., Seibert, M., Treyer, M.A., Bianchi, L., Heckman, T.M., Madore, B.F., Rich, R.M.: The space density of extended ultraviolet (XUV) disks in the local universe and implications for gas accretion onto galaxies. Astrophys. J. **733**, 74 (2011). doi:10.1088/0004-637X/733/2/74, arXiv:1104.4501

Loebman, S.R., Debattista, V.P., Nidever, D.L., Hayden, M.R., Holtzman, J.A., Clarke, A.J., Roškar, R., Valluri, M.: Imprints of radial migration on the milky way's metallicity distribution functions. Astrophys. J. **818**, L6 (2016). doi:10.3847/2041-8205/818/1/L6, arXiv:1511.06369

Lynden-Bell, D., Kalnajs, A.J.: On the generating mechanism of spiral structure. Mon. Not. R. Astron. Soc. **157**, 1 (1972). doi:10.1093/mnras/157.1.1

MacArthur, L.A., Courteau, S., Bell, E., Holtzman, J.A.: Structure of disk-dominated Galaxies. II. Color gradients and stellar population models. Astrophys. J. Suppl. Ser. **152**, 175–199 (2004). doi:10.1086/383525, astro-ph/0401437

Maltby, D.T., Gray, M.E., Aragón-Salamanca, A., Wolf, C., Bell, E.F., Jogee, S., Häußler, B., Barazza, F.D., Böhm, A., Jahnke, K.: The environmental dependence of the structure of outer galactic discs in STAGES spiral galaxies. Mon. Not. R. Astron. Soc. **419**, 669–686 (2012a). doi:10.1111/j.1365-2966.2011.19727.x, arXiv:1108.6206

Maltby, D.T., Hoyos, C., Gray, M.E., Aragón-Salamanca, A., Wolf, C.: Antitruncated stellar light profiles in the outer regions of STAGES spiral galaxies: bulge or disc related? Mon. Not. R. Astron. Soc. **420**, 2475–2479 (2012b). doi:10.1111/j.1365-2966.2011.20211.x, arXiv:1111.3801

Martin, C.L., Kennicutt, R.C. Jr.: Star formation thresholds in galactic disks. Astrophys. J. **555**, 301–321 (2001). doi:10.1086/321452, astro-ph/0103181

Martín-Navarro, I., Bakos, J., Trujillo, I., Knapen, J.H., Athanassoula, E., Bosma, A., Comerón, S., Elmegreen, B.G., Erroz-Ferrer, S., Gadotti, D.A., Gil de Paz, A., Hinz, J.L., Ho, L.C., Holwerda, B.W., Kim, T., Laine, J., Laurikainen, E., Menéndez-Delmestre, K., Mizusawa, T., Muñoz-Mateos, J.C., Regan, M.W., Salo, H., Seibert, M., Sheth, K.: A unified picture of breaks and truncations in spiral galaxies from SDSS and S^4G imaging. Mon. Not. R. Astron. Soc. **427**, 1102–1134 (2012). doi:10.1111/j.1365-2966.2012.21929.x, arXiv:1208.2893

Martín-Navarro, I., Trujillo, I., Knapen, J.H., Bakos, J., Fliri, J.: Stellar haloes outshine disc truncations in low-inclined spirals. Mon. Not. R. Astron. Soc. **441**, 2809–2814 (2014). doi:10.1093/mnras/stu767, arXiv:1401.3749

Matteucci, F., Francois, P.: Galactic chemical evolution - abundance gradients of individual elements. Mon. Not. R. Astron. Soc. **239**, 885–904 (1989)

Minchev, I., Famaey, B.: A new mechanism for radial migration in galactic disks: spiral-bar resonance overlap. Astrophys. J. **722**, 112–121 (2010). doi:10.1088/0004-637X/722/1/112, arXiv:0911.1794

Minchev, I., Famaey, B., Combes, F., Di Matteo, P., Mouhcine, M., Wozniak, H.: Radial migration in galactic disks caused by resonance overlap of multiple patterns: self-consistent simulations. Astron. Astrophys. **527**, A147 (2011). doi:10.1051/0004-6361/201015139, arXiv:1006.0484

Minchev, I., Famaey, B., Quillen, A.C., Di Matteo, P., Combes, F., Vlajić, M., Erwin, P., Bland-Hawthorn, J.: Evolution of galactic discs: multiple patterns, radial migration, and disc outskirts. Astron. Astrophys. **548**, A126 (2012). doi:10.1051/0004-6361/201219198, arXiv:1203.2621

Mo, H.J., Mao, S., White, S.D.M.: The formation of galactic discs. Mon. Not. R. Astron. Soc. **295**, 319–336 (1998). doi:10.1046/j.1365-8711.1998.01227.x, astro-ph/9707093

Muñoz-Mateos, J.C., Gil de Paz, A., Boissier, S., Zamorano, J., Jarrett, T., Gallego, J., Madore, B.F.: Specific star formation rate profiles in nearby spiral galaxies: quantifying the inside-out formation of disks. Astrophys. J. **658**, 1006–1026 (2007). doi:10.1086/511812, arXiv:astro-ph/0612017

Muñoz-Mateos, J.C., Sheth, K., Gil de Paz, A., Meidt, S., Athanassoula, E., Bosma, A., Comerón, S., Elmegreen, D.M., Elmegreen, B.G., Erroz-Ferrer, S., Gadotti, D.A., Hinz, J.L., Ho, L.C., Holwerda, B., Jarrett, T.H., Kim, T., Knapen, J.H., Laine, J., Laurikainen, E., Madore, B.F., Menendez-Delmestre, K., Mizusawa, T., Regan, M., Salo, H., Schinnerer, E., Seibert, M., Skibba, R., Zaritsky, D.: The impact of bars on disk breaks as probed by S^4G imaging. Astrophys. J. **771**, 59 (2013). doi:10.1088/0004-637X/771/1/59, arXiv:1304.6083

Naab, T., Ostriker, J.P.: A simple model for the evolution of disc galaxies: the Milky Way. Mon. Not. R. Astron. Soc. **366**, 899–917 (2006). doi:10.1111/j.1365-2966.2005.09807.x, astro-ph/0505594

Naeslund, M., Joersaeter, S.: Surface photometry of the edge-on spiral NGC 4565. I. V-band data and the extended optical warp. Astron. Astrophys. **325**, 915–922 (1997)

Peñarrubia, J., McConnachie, A., Babul, A.: On the formation of extended galactic disks by tidally disrupted dwarf galaxies. Astrophys. J. **650**, L33–L36 (2006). doi:10.1086/508656, astro-ph/0606101

Pérez, I.: Truncation of stellar disks in galaxies at $z \sim 1$. Astron. Astrophys. **427**, L17–L20 (2004). doi:10.1051/0004-6361:200400090, astro-ph/0410250

Pohlen, M.: The radial structure of galactic stellar disks. Ph.D thesis, Ruhr-Universität (2002)

Pohlen, M., Trujillo, I.: The structure of galactic disks. Studying late-type spiral galaxies using SDSS. Astron. Astrophys. **454**, 759–772 (2006). doi:10.1051/0004-6361:20064883, arXiv:astro-ph/0603682

Pohlen, M., Dettmar, R.J., Lütticke, R.: Cut-off radii of galactic disks. A new statistical study on the truncation of galactic disks. Astron. Astrophys. **357**, L1–L4 (2000). arXiv:astro-ph/0003384

Pohlen, M., Dettmar, R.J., Lütticke, R., Aronica, G.: Outer edges of face-on spiral galaxies. Deep optical imaging of NGC 5923, UGC 9837 and NGC 5434. Astron. Astrophys. **392**, 807–816 (2002). doi:10.1051/0004-6361:20020994

Pranger, F., Trujillo, I., Kelvin, L.S., Cebrián, M.: The effect of environment on the structure of disc galaxies. Mon. Not. R. Astron. Soc. **467**, 2127–2144 (2017). doi:10.1093/mnras/stx199. arXiv:1605.08845

Putman, M.E., Peek, J.E.G., Muratov, A., Gnedin, O.Y., Hsu, W., Douglas, K.A., Heiles, C., Stanimirovic, S., Korpela, E.J., Gibson, S.J.: The disruption and fueling of M33. Astrophys. J. **703**, 1486–1501 (2009). doi:10.1088/0004-637X/703/2/1486, arXiv:0812.3093

Radburn-Smith, D.J., de Jong, R.S., Seth, A.C., Bailin, J., Bell, E.F., Brown, T.M., Bullock, J.S., Courteau, S., Dalcanton, J.J., Ferguson, H.C., Goudfrooij, P., Holfeltz, S., Holwerda, B.W., Purcell, C., Sick, J., Streich, D., Vlajic, M., Zucker, D.B.: The GHOSTS survey. I. Hubble space telescope advanced camera for surveys data. Astrophys. J. Suppl. Ser. **195**, 18 (2011). doi:10.1088/0067-0049/195/2/18

Radburn-Smith, D.J., Roškar, R., Debattista, V.P., Dalcanton, J.J., Streich, D., de Jong, R.S., Vlajić, M., Holwerda, B.W., Purcell, C.W., Dolphin, A.E., Zucker, D.B.: Outer-disk populations in NGC 7793: evidence for stellar radial migration. Astrophys. J. **753**, 138 (2012). doi:10.1088/0004-637X/753/2/138, arXiv:1206.1057

Rautiainen, P., Salo, H.: N-body simulations of resonance rings in galactic disks. Astron. Astrophys. **362**, 465–586 (2000)

Rebassa-Mansergas, A., Anguiano, B., García-Berro, E., Freeman, K.C., Cojocaru, R., Manser, C.J., Pala, A.F., Gänsicke, B.T., Liu, X.W.: The age-metallicity relation in the solar neighbourhood from a pilot sample of white dwarf-main sequence binaries. Mon. Not. R. Astron. Soc. **463**, 1137–1143 (2016). doi:10.1093/mnras/stw2021, arXiv:1608.03064

Roca-Fàbrega, S., Valenzuela, O., Figueras, F., Romero-Gómez, M., Velázquez, H., Antoja, T., Pichardo, B.: On galaxy spiral arms' nature as revealed by rotation frequencies. Mon. Not. R. Astron. Soc. **432**, 2878–2885 (2013). doi:10.1093/mnras/stt643, arXiv:1302.6981

Roediger, J.C., Courteau, S., Sánchez-Blázquez, P., McDonald, M.: Stellar populations and radial migrations in virgo disk galaxies. Astrophys. J. **758**, 41 (2012). doi:10.1088/0004-637X/758/1/41, arXiv:1201.6361

Roškar, R., Debattista, V.P., Quinn, T.R., Stinson, G.S., Wadsley, J.: Riding the spiral waves: implications of stellar migration for the properties of galactic disks. Astrophys. J. **684**, L79–L82 (2008a). doi:10.1086/592231, arXiv:0808.0206

Roškar, R., Debattista, V.P., Stinson, G.S., Quinn, T.R., Kaufmann, T., Wadsley, J.: Beyond inside-out growth: formation and evolution of disk outskirts. Astrophys. J. **675**, L65–L68 (2008b). doi:10.1086/586734, arXiv:0710.5523

Roškar, R., Debattista, V.P., Brooks, A.M., Quinn, T.R., Brook, C.B., Governato, F., Dalcanton, J.J., Wadsley, J.: Misaligned angular momentum in hydrodynamic cosmological simulations: warps, outer discs and thick discs. Mon. Not. R. Astron. Soc. **408**, 783–796 (2010). doi:10.1111/j.1365-2966.2010.17178.x, arXiv:1006.1659

Roškar, R., Debattista, V.P., Quinn, T.R., Wadsley, J.: Radial migration in disc galaxies - I. Transient spiral structure and dynamics. Mon. Not. R. Astron. Soc. **426**, 2089–2106 (2012). doi:10.1111/j.1365-2966.2012.21860.x, arXiv:1110.4413

Ruiz-Lara, T., Few, C.G., Gibson, B.K., Pérez, I., Florido, E., Minchev, I., Sánchez-Blázquez, P.: The imprint of satellite accretion on the chemical and dynamical properties of disc galaxies. Astron. Astrophys. **586**, A112 (2016a). doi:10.1051/0004-6361/201526470, arXiv:1512.00625

Ruiz-Lara, T., Pérez, I., Florido, E., Sánchez-Blázquez, P., Méndez-Abreu, J., Lyubenova, M., Falcón-Barroso, J., Sánchez-Menguiano, L., Sánchez, S.F., Galbany, L., García-Benito, R., González Delgado, R.M., Husemann, B., Kehrig, C., López-Sánchez, Á.R., Marino, R.A., Mast, D., Papaderos, P., van de Ven, G., Walcher, C.J., Zibetti, S., CALIFA Team: No direct coupling between bending of galaxy disc stellar age and light profiles. Mon. Not. R. Astron. Soc. **456**, L35–L39 (2016b). doi:10.1093/mnrasl/slv174, arXiv:1511.03499

Sánchez-Blázquez, P., Courty, S., Gibson, B.K., Brook, C.B.: The origin of the light distribution in spiral galaxies. Mon. Not. R. Astron. Soc. **398**, 591–606 (2009). doi:10.1111/j.1365-2966.2009.15133.x, arXiv:0905.4579

Sánchez-Blázquez, P., Rosales-Ortega, F.F., Méndez-Abreu, J., Pérez, I., Sánchez, S.F., Zibetti, S., Aguerri, J.A.L., Bland-Hawthorn, J., Catalán-Torrecilla, C., Cid Fernandes, R., de Amorim, A., de Lorenzo-Caceres, A., Falcón-Barroso, J., Galazzi, A., García Benito, R., Gil de Paz, A., González Delgado, R., Husemann, B., Iglesias-Páramo, J., Jungwiert, B., Marino, R.A., Márquez, I., Mast, D., Mendoza, M.A., Mollá, M., Papaderos, P., Ruiz-Lara, T., van de

Ven, G., Walcher, C.J., Wisotzki, L.: Stellar population gradients in galaxy discs from the CALIFA survey. The influence of bars. Astron. Astrophys. **570**, A6 (2014). doi:10.1051/0004-6361/201423635, arXiv:1407.0002

Schaye, J.: Star formation thresholds and galaxy edges: why and where. Astrophys. J. **609**, 667–682 (2004). doi:10.1086/421232, arXiv:astro-ph/0205125

Schönrich, R., Binney, J.: Chemical evolution with radial mixing. Mon. Not. R. Astron. Soc. **396**, 203–222 (2009). doi:10.1111/j.1365-2966.2009.14750.x, arXiv:0809.3006

Schwarz, M.P.: The response of gas in a galactic disk to bar forcing. Astrophys. J. **247**, 77–88 (1981). doi:10.1086/159011

Sellwood, J.A.: The lifetimes of spiral patterns in disc galaxies. Mon. Not. R. Astron. Soc. **410**, 1637–1646 (2011). doi:10.1111/j.1365-2966.2010.17545.x, arXiv:1008.2737

Sellwood, J.A.: GALAXY package for N-body simulation (2014). ArXiv e-prints arXiv:1406.6606

Sellwood, J.A., Binney, J.J.: Radial mixing in galactic discs. Mon. Not. R. Astron. Soc. **336**, 785–796 (2002). doi:10.1046/j.1365-8711.2002.05806.x

Sellwood, J.A., Carlberg, R.G.: Spiral instabilities provoked by accretion and star formation. Astrophys. J. **282**, 61–74 (1984). doi:10.1086/162176

Sellwood, J.A., Carlberg, R.G.: Transient spirals as superposed instabilities. Astrophys. J. **785**, 137 (2014). doi:10.1088/0004-637X/785/2/137, arXiv:1403.1135

Sellwood, J.A., Debattista, V.P.: Stochasticity in N-body simulations of disc galaxies. Mon. Not. R. Astron. Soc. **398**, 1279–1297 (2009). doi:10.1111/j.1365-2966.2009.15219.x, arXiv:0903.3554

Shen, J., Sellwood, J.A.: The destruction of bars by central mass concentrations. Astrophys. J. **604**, 614–631 (2004)

Solway, M., Sellwood, J.A., Schönrich, R.: Radial migration in galactic thick discs. Mon. Not. R. Astron. Soc. **422**, 1363–1383 (2012). doi:10.1111/j.1365-2966.2012.20712.x, arXiv:1202.1418

Thilker, D.A., Bianchi, L., Boissier, S., Gil de Paz, A., Madore, B.F., Martin, D.C., Meurer, G.R., Neff, S.G., Rich, R.M., Schiminovich, D., Seibert, M., Wyder, T.K., Barlow, T.A., Byun, Y.I., Donas, J., Forster, K., Friedman, P.G., Heckman, T.M., Jelinsky, P.N., Lee, Y.W., Malina, R.F., Milliard, B., Morrissey, P., Siegmund, O.H.W., Small, T., Szalay, A.S., Welsh, B.Y.: Recent star formation in the extreme outer disk of M83. Astrophys. J. **619**, L79–L82 (2005). doi:10.1086/425251, arXiv:astro-ph/0411306

Trujillo, I., Pohlen, M.: Stellar disk truncations at high z: probing inside-out galaxy formation. Astrophys. J. **630**, L17–L20 (2005). doi:10.1086/491472, astro-ph/0507533

van der Kruit, P.C.: Optical surface photometry of eight spiral galaxies studied in Westerbork. Astron. Astrophys. Suppl. Ser. **38**, 15–38 (1979)

van der Kruit, P.C.: The radial distribution of surface brightness in galactic disks. Astron. Astrophys. **173**, 59–80 (1987)

van der Kruit, P.C.: The three-dimensional distribution of light and mass in disks of spiral galaxies. Astron. Astrophys. **192**, 117–127 (1988)

van der Kruit, P.C.: Truncations of stellar disks and warps of HI-layers in edge-on spiral galaxies. Astron. Astrophys. **466**, 883–893 (2007). doi:10.1051/0004-6361:20066941, arXiv:astro-ph/0702486

van der Kruit, P.C., Searle, L.: Surface photometry of edge-on spiral galaxies. IV - The distribution of light, colour, and mass in the disk and spheroid of NGC 7814. Astron. Astrophys. **110**, 79–94 (1982)

Vera-Ciro, C., D'Onghia, E.: On the conservation of the vertical action in galactic disks. Astrophys. J. **824**, 39 (2016). doi:10.3847/0004-637X/824/1/39

Vlajić, M., Bland-Hawthorn, J., Freeman, K.C.: The abundance gradient in the extremely faint outer disk of NGC 300. Astrophys. J. **697**, 361–372 (2009). doi:10.1088/0004-637X/697/1/361, arXiv:0903.1855

Watkins, A.E., Mihos, J.C., Harding, P., Feldmeier, J.J.: Searching for diffuse light in the M96 galaxy group. Astrophys. J. **791**, 38 (2014). doi:10.1088/0004-637X/791/1/38, arXiv:1406.6982

Watkins, A.E., Mihos, J.C., Harding, P.: The red and featureless outer disks of nearby spiral galaxies. Astrophys. J. **826**, 59 (2016). doi:10.3847/0004-637X/826/1/59, arXiv:1605.05183

Widrow, L.M., Dubinski, J.: Equilibrium disk-bulge-halo models for the milky way and andromeda galaxies. Astrophys. J. **631**, 838–855 (2005). doi:10.1086/432710

Widrow, L.M., Pym, B., Dubinski, J.: Dynamical blueprints for galaxies. Astrophys. J. **679**, 1239–1259 (2008). doi:10.1086/587636, arXiv:0801.3414

Williams, B.F., Dalcanton, J.J., Dolphin, A.E., Holtzman, J., Sarajedini, A.: The detection of inside-out disk growth in M33. Astrophys. J. **695**, L15–L19 (2009). doi:10.1088/0004-637X/695/1/L15, arXiv:0902.3460

Williams, B.F., Dalcanton, J.J., Stilp, A., Dolphin, A., Skillman, E.D., Radburn-Smith, D.: The ACS nearby galaxy survey treasury. XI. The remarkably undisturbed NGC 2403 disk. Astrophys. J. **765**, 120 (2013). doi:10.1088/0004-637X/765/2/120, arXiv:1301.4521

Worthey, G.: Comprehensive stellar population models and the disentanglement of age and metallicity effects. Astrophys. J. Suppl. Ser. **95**, 107–149 (1994). doi:10.1086/192096

Yoachim, P., Roškar, R., Debattista, V.P.: Spatially resolved spectroscopic star formation histories of nearby disks: hints of stellar migration. Astrophys. J. **752**, 97 (2012). doi:10.1088/0004-637X/752/2/97, 1204.0026

Younger, J.D., Cox, T.J., Seth, A.C., Hernquist, L.: Antitruncated stellar disks via minor mergers. Astrophys. J. **670**, 269–278 (2007). doi:10.1086/521976, arXiv:0707.4481

Yu, J., Sellwood, J.A., Pryor, C., Chen, L., Hou, J.: A test for radial mixing using local star samples. Astrophys. J. **754**, 124 (2012). doi:10.1088/0004-637X/754/2/124, arXiv:1205.7001

Zaritsky, D., Christlein, D.: On the extended knotted disks of galaxies. Astrophys. J. **134**, 135–141 (2007). doi:10.1086/518238, astro-ph/0703487

Zheng, Z., Thilker, D.A., Heckman, T.M., Meurer, G.R., Burgett, W.S., Chambers, K.C., Huber, M.E., Kaiser, N., Magnier, E.A., Metcalfe, N., Price, P.A., Tonry, J.L., Wainscoat, R.J., Waters, C.: The structure and stellar content of the outer disks of galaxies: a new view from the Pan-STARRS1 medium deep survey. Astrophys. J. **800**, 120 (2015). doi:10.1088/0004-637X/800/2/120, arXiv:1412.3209

Chapter 4
Outskirts of Nearby Disk Galaxies: Star Formation and Stellar Populations

Bruce G. Elmegreen and Deidre A. Hunter

Abstract The properties and star formation processes in the far-outer disks of nearby spiral and dwarf irregular galaxies are reviewed. The origin and structure of the generally exponential profiles in stellar disks is considered to result from cosmological infall combined with a non-linear star formation law and a history of stellar migration and scattering from spirals, bars and random collisions with interstellar clouds. In both spirals and dwarfs, the far-outer disks tend to be older, redder and thicker than the inner disks, with the overall radial profiles suggesting inside-out star formation plus stellar scattering in spirals and outside-in star formation with a possible contribution from scattering in dwarfs. Dwarf irregulars and the far-outer parts of spirals both tend to be gas dominated, and the gas radial profile is often non-exponential although still decreasing with radius. The ratio of Hα to far-UV flux tends to decrease with lower surface brightness in these regions, suggesting either a change in the initial stellar mass function or the sampling of that function or a possible loss of Hα photons.

4.1 Introduction

The outer parts of galaxies represent a new frontier in observational astronomy at the limits of faint surface brightness. We know little about these regions except that galaxies viewed deeply enough can usually be traced out to ten stellar scale lengths or more, without any evident edge. We do not know in detail how the stars and gas got there and whether stars actually formed there or just scattered from the inner parts. Neither do we know as much as we'd like about the properties, elemental abundances, scale heights and kinematics of outer disk stars except for a limited view in the Milky Way (e.g. Bovy et al. 2016) and the Andromeda galaxy

B.G. Elmegreen (✉)
IBM Research Division, T.J. Watson Research Center, 1101 Kitchawan Road, Yorktown Heights, NY 10598, USA
e-mail: bge@us.ibm.com

D.A. Hunter
Lowell Observatory, 1400 West Mars Hill Road, Flagstaff, AZ 86001, USA
e-mail: dah@lowell.edu

J.H. Knapen et al. (eds.), *Outskirts of Galaxies*, Astrophysics and Space Science Library 434, DOI 10.1007/978-3-319-56570-5_4

(Dalcanton et al. 2012). Yet the outer parts of disks are expected to be where galaxy growth is occurring today and where the leftover and recycled cosmological gas accretes or gets stored for later conversion into stars in the inner disk (Lemonias et al. 2011; Moffett et al. 2012). The outer parts should also show the history of a galaxy's interactions with other galaxies, as the orbital time is relatively long. A high fraction of outer disks are lopsided too, correlating with the stellar mass fraction in the outer parts (i.e. the ratio of the stellar mass to the total from the rotation curve; Zaritsky et al. 2013), perhaps because of uneven accretion, interactions or halo sloshing (Ghosh et al. 2016). This chapter reviews disk structure, star formation and stellar populations in the outer parts of nearby galaxies. General properties of these outer disks are in Sects. 4.2–4.7, and a focus on dwarf irregular galaxies (dIrrs) is in Sect. 4.8. The observational difficulties in observing the faint outer parts of disks are discussed in other chapters in this volume.

4.2 Outer Disk Structure from Collapse Models of Galaxy Formation

A fundamental property of galaxy disks is their exponential or piecewise exponential radial light profile (de Vaucouleurs 1959). Freeman (1970) noted that this profile gives a distribution of cumulative angular momentum versus radius that matches that of a flattened uniformly rotating sphere (Mestel 1963), but this match is only good for about four disk scale lengths. The problem is that an exponential disk has very little mass and a lot of angular momentum in the far-outer parts, unlike a power-law halo which has both mass and angular momentum increasing with radius in proportion (Efstathiou 2000). Nevertheless, observations show some disks with 8–10 scale lengths (Weiner et al. 2001; Bland-Hawthorn et al. 2005; Grossi et al. 2011; Hunter et al. 2011; Radburn-Smith et al. 2012; Vlajić et al. 2011; Barker et al. 2012; Hunter et al. 2013; Mihos et al. 2013; van Dokkum et al. 2014). These large extents compared to the predicted four scale lengths from pure collapse models need to be explained (Ferguson and Clarke 2001).

Thus we have a problem: if the halo collapses to about four scale lengths in a disk, then how can we get the observed eight or more scale lengths in the stars that eventually form? The answer may lie with the conversion of incoming gas into stars. In a purely gaseous medium, interstellar collapse proceeds at a rate per unit area that is proportional to the square of the mass column density, $\Sigma_{\rm gas}$ (Elmegreen 2015). One factor of $\Sigma_{\rm gas}$ accounts for the amount of fuel available for star formation, and the other factor accounts for the rate of conversion of this fuel into stars. This squared Kennicutt-Schmidt law converts four scale lengths of primordial gas into eight scale lengths of stars after they form (Sect. 4.7). Stellar scattering from clouds and other irregularities could extend or smooth out this exponential further (Elmegreen and Struck 2013, 2016).

There is an additional observation in Wang et al. (2014) that in local gas-rich galaxies, the outer gas radial profiles are all about the same when scaled to the radius where $\Sigma_{\rm HI} = 1\, M_\odot\, {\rm pc}^{-2}$. Bigiel and Blitz (2012) found a similar universality to the gas profile when normalized to R_{25}, the radius at 25 magnitudes per square arcsec in the V band. Wang et al. (2014) found that the ratio of the radius at $1\, M_\odot\, {\rm pc}^{-2}$ to the gaseous scale length in the outer disk is about four, the same as the maximum number of scale lengths in a pure halo collapse. This similarity may not be a coincidence (Sect. 4.7).

Cosmological simulations now have a high enough resolution to form individual galaxies with reasonable properties (Vogelsberger et al. 2014; Schaye et al. 2015). Zoom-in models in a cosmological environment show stellar exponential radial profiles in these galaxies (Robertson et al. 2004) even though specific angular momentum is not preserved during the collapse and feedback moves substantial amounts of gas around, especially for low-mass galaxies (El-Badry et al. 2016). For example, Aumer and White (2013) ran models with rotating halo gas aligned in various ways with respect to the dark matter symmetry axis. They found broken exponential disks with a break radius related to the maximum angular momentum of the gas in the halo, increasing with time as the outer disk cooled and formed stars. Star formation is from the inside-out. Angular momentum was redistributed through halo torques, but still the disks were approximately exponential. Aumer et al. (2013) further studied 16 simulated galaxies with various masses. All of them produced near-exponential disks.

In a systematic study of angular momentum, Herpich et al. (2015) found a transition from exponentials with up-bending outer profiles (Type III; Sect. 4.3) at low specific angular momentum (λ) to Type I (single exponential) and Type II (down-bending outer parts) at higher λ. An intermediate value of $\lambda = 0.035$, similar to what has been expected theoretically (Mo et al. 1998), corresponded to the pure exponential Type I. The reason for this change of structure with λ was that collapse at low spin parameter produces a high disk density in a small initial radius, and this leads to significant stellar scattering and a large redistribution of mass to the outer disk, making the up-bending Type III. Conversely, large λ produces a large and low-density initial disk, which does not scatter much and nearly preserves the initial down-bending profile of Type II.

4.3 Outer Disk Structure: Three Exponential Types

Galaxy radial profiles are often classified as exponential Type I, II or III according to whether the outer parts continue with the same scale length as the inner parts, continue with a shorter scale length (i.e. bend down a little) or continue with a larger scale length (bend up a little), respectively (Fig. 4.1). The Sersic profile with $n = 1$ corresponds to Type I; the other types do not have a constant Sersic index. For a review, see van der Kruit (2001), and for early surveys, see Pohlen and Trujillo (2006) and Erwin et al. (2008).

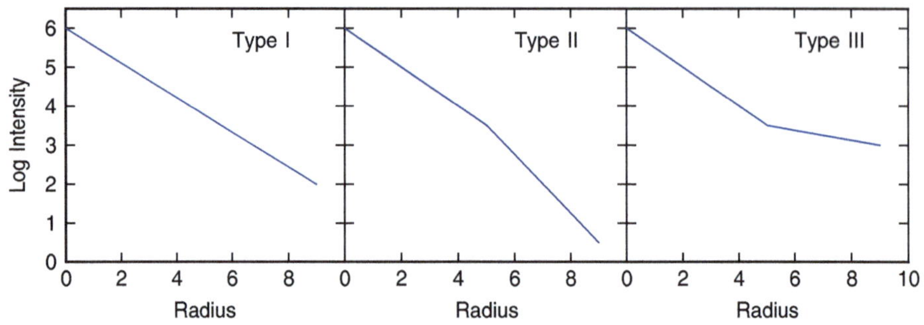

Fig. 4.1 Three types of exponential or piecewise exponential profiles

Gutiérrez et al. (2011) determined the proportion of the three exponential profile types for barred and non-barred galaxies of various Hubble types, including 183 large local face-on galaxies from three separate studies. For S0 and earlier, the three profile types are nearly evenly divided. For Sab to Sbc, Types II and III are about equal and Type I becomes relatively rare (10%). For Scd to Sdm, Type II dominates with ∼80% of the total. Herrmann et al. (2013) continued this study to dIrrs and blue compact dwarfs (BCDs; see also Sect. 4.8.4). dIrr galaxies are dominated (80%) by Type IIs, while BCDs have steep inner parts from a starburst and are usually Type III.

A general caution should be mentioned about possible contamination at faint light levels from scattered light. Sandin (2015) showed R and I band radial profiles for NGC 4102 that were fit to a single exponential model with a broad point spread function from the instrument. The usual Type III profile for this galaxy turned into a Type I when the outer excess was corrected for the instrumental profile.

4.4 Outer Disk Stellar Populations: Colour and Age Gradients

Radial profiles become more complex when changes in stellar colours and ages are considered. Bakos et al. (2008) noted that Type II light profiles tend to correspond to U-shaped $B - V$ colour profiles, which means that the inner part of the disk gets bluer with radius at first and then the outer part of the disk gets red again. This colour change presumably corresponds to a change in the mass-to-light ratio, with large ratios in the outer parts. Then the down-bending Type II in a light profile tends to straighten out and become Type I in a mass profile. That is, the outer red trend gives an increasing mass-to-light ratio, causing an increasing conversion factor from surface brightness to mass surface density. The red outer parts could be from old stars that scattered there from the inner regions (Roškar et al. 2008), as distinct from the common model of inside-out growth for spiral galaxies. A larger survey recently confirmed this result. Zheng et al. (2015) included 700 galaxies using deep images from the Pan-STARRS survey. The average g-band (peak at 5150 Å) light profile was the down-bending Type II for low-mass galaxies ($<10^{10}\,M_\odot$) and slightly less bent for high-mass galaxies ($<10^{10.5}\,M_\odot$), as usual, and the average $g - i$ colour profiles (i band peaks at 7490 Å) were U-shaped to various degrees, so the average mass profile became Type I for all galaxy masses. Muñoz-Mateos et al. (2015) made average radial profiles separated into eight mass bins for ~2400 galaxies using 3.6 μm emission from the *Spitzer* Survey of Stellar Structure in Galaxies. Such long wavelength emission is a nearly direct probe of galaxy mass, although there is some PAH emission from dust in it too. All masses showed Type II profiles on average, with a straighter trend like Type I from a central bulge in the more massive galaxies. This suggests that the mass profile for most galaxies is not exactly Type I, but still tapers off more steeply in the outer parts, beyond 1 kpc for low-mass galaxies ($<10^{9}\,M_\odot$) and beyond 10 kpc for high-mass galaxies ($>10^{10.5}\,M_\odot$).

The most telling observations are of stellar age gradients because colour gradients can be from a mixture of age gradients and metallicity gradients. Roediger et al. (2012) determined radial age profiles from photometry and stellar population models of 64 Virgo cluster disk galaxies. They found U-shaped age profiles in 15% of Type Is and also in 36% of both Types II and III. In one-third of all exponential types, the age increased steadily with radius. Yoachim et al. (2012) found about the same mixture of age profiles, measuring ages from spectra in 12 galaxies. Dale et al. (2016) determined star formation histories for 15 nearby galaxies with masses in the range $10^{8}\,M_\odot$–$10^{11}\,M_\odot$ using ultraviolet and infrared data; they also found U-shaped age profiles. These results imply that outer disks generally have old stars, although most also still have some star formation.

A recent integral field unit survey of 44 nearby spiral galaxies (CALIFA) by Ruiz-Lara et al. (2016) also found U-shaped age profiles in Types I and II when the stars were weighted by brightness, as would be the case from integrated spectra or photometry. This is in agreement with the previous surveys mentioned above. The galaxies were observed beyond their break radii or for at least three scale lengths. In contrast, Ruiz-Lara et al. (2016) found constant age profiles when the stars were weighted by mass. They suggested that the entire disk formed early with star formation stopping in the inner parts first and then quenching from inside-out. This is unlike cosmological simulations that have the outer disk form more slowly than the inner disk and also unlike models where the outer disk stars migrate there from the inner disk.

Watkins et al. (2016) viewed three nearby spirals with very deep images, covering a range of about ten magnitudes in surface brightness for B band. They found smooth and red stellar distributions with no spiral arms in the far-outer disks. For the typical colour of $B - V \sim 0.8$ mag in the outer parts, and from a lack of FUV light, they concluded that the star formation rate (SFR) had to be less than 3–$5 \times 10^{-5}\,M_\odot\,\mathrm{pc}^{-2}\,\mathrm{Myr}^{-1}$. This seemed to be too low for continuous star formation and disk building, suggesting some radial migration. However, the lack of spiral arms makes the usually invoked churning mechanism (Sellwood and Binney 2002; Roškar et al. 2008; Berrier and Sellwood 2015) inoperable. Churning is a process of stellar migration back and forth around corotation. Perhaps stellar scattering off local gas irregularities makes the outer exponential structure (Elmegreen and Struck 2013).

4.5 Mono-Age Structure of Stellar Populations

Age profiles in galaxy disks can be viewed in another way too. A series of galaxy simulations have looked at the distributions of stars of various ages in the final model. For example, Bird et al. (2013) did a simulation of the Milky Way and found that older stars in the present-day disk have shorter radial scale lengths and thicker perpendicular scale heights than younger stars. Other mono-age population studies of simulated disks are in Sánchez-Blázquez et al. (2009), Stinson et al. (2013), Martig et al. (2014), Minchev et al. (2015) and Athanassoula et al. (2016), giving the same result.

Bovy et al. (2016) found structure related to this in the Milky Way using 14,700 red clump stars. Higher metallicity populations are more centrally concentrated than lower metallicity populations (not considering the α-enhanced "thick disk" component). Each narrow metallicity range tends to have a maximum surface density of stars at a particular radius where the disk has that average metallicity. Plus, each mono-metallicity population has a perpendicular scale height that increases with radius, producing a flare.

The correspondence between metallicity and peak surface density for a population of stars suggests that star formation, feedback, halo recycling and other processes establish an equilibrium metallicity in a region that depends primarily on local conditions, such as the local mass surface density (Bovy et al. 2016). Stellar migration then broadens this distribution to produce the observed total profiles. This local equilibrium concept is consistent with the results of Rosales-Ortega et al. (2012), who found for 2000 HII regions in nearby galaxies that metallicity depends mostly on stellar mass surface density, as determined from photometry. Bresolin and Kennicutt (2015) present a similar result: that the metallicity gradients in galaxies are all the same when expressed in units of the disk scale length.

4.6 Outer Disk Structure: Environmental Effects and the Role of Bulges and Bars

Environment may also affect outer disk structure. Younger et al. (2007) showed that prograde minor mergers can drive mass inward and outward, creating a Type III profile. Borlaff et al. (2014) also suggested that Type III S0 galaxies can result from a merger. This is consistent with observations in Erwin et al. (2012) that S0 galaxies in Virgo have proportionally more Types I and III, suggesting that interactions or mergers have been important. Erwin et al. also found that bars have little effect on the proportion of exponential types. Athanassoula et al. (2016) simulated a gas-rich major merger and showed that it formed an exponential disk in the final system.

On the other hand, Maltby et al. (2012) measured the V-band radial profiles of 330 galaxies observed with *Hubble Space Telescope* over a half-degree field surrounding a galaxy supercluster at redshift 0.165. They found no dependence on environment, cluster versus field, for the ratio of the outer to the inner disk scale length or the outer scale length itself. Head et al. (2015) got a similar result looking at S0 galaxies in the Coma cluster; using a profile decomposition algorithm to remove the bulge, they found that bars are important for disk structure, correlating with Types II and III (contrast this with the Erwin et al. (2012) result above), but that location in the cluster is not important. According to Head et al. (2015), the relative proportion of Types I, II and III is the same in the core, at intermediate radii and in the outskirts of Coma (in fact, most of the S0 galaxies were Type I).

Some of the appearance of Type III could be from a bright halo or extended bulge and not from stars in the disk (Erwin et al. 2005). Maltby et al. (2015) suggested that half of the S0 Type III structures in various environments come from extended bulge light, although this fraction is only 15% in later Hubble-type spirals. This implies that disk fading can make an S0 from a spiral, preserving the scale length. Simulations by Cooper et al. (2013) also found that the outer stellar structure can be in a halo and not a disk, as a result of mergers.

Bars and spirals seem to be important in determining the break radius for down-bending (Type II) exponentials. Muñoz-Mateos et al. (2013) suggested that the break radius for Type IIs is either at the outer Lindblad resonance (OLR) of a bar or the OLR of a spiral that is outside of a bar. The spiral and bar are assumed to have their pattern speeds in a resonance with the inner 4:1 resonance of the spiral at the corotation radius of the bar (see Pohlen and Trujillo 2006 and Erwin et al. 2008). Laine et al. (2014) also found that bars and spirals are important: 94% of Type II breaks are associated with some type of feature; 48% are in early-type galaxies with an outer ring or pseudoring; 8% are with a lens, assumed to be the OLR of a bar, and if there is no outer ring, then the breaks are at 2 times the radius of an inner ring (this being the ratio of radii for outer and inner ring resonances); 14% are in late-type galaxies associated with an end to strong star formation; and 24% are at the radius where the spiral arms end. For Type III breaks studied by Laine et al. (2014), 30% are associated with inner or outer lenses or outer rings.

4.7 Outer Disk Structure: Star Formation Models

The outer disks of spiral galaxies and most dIrrs are dominated by gas in an atomic form and not stars. Because stars form in molecular gas, it is difficult to observe directly how stars form in these regions. Moreover, outer disks and dIrrs tend to be stable by the Toomre Q condition (Elmegreen and Hunter 2015). Nevertheless, star formation usually looks normal there, forming clusters and associations at low

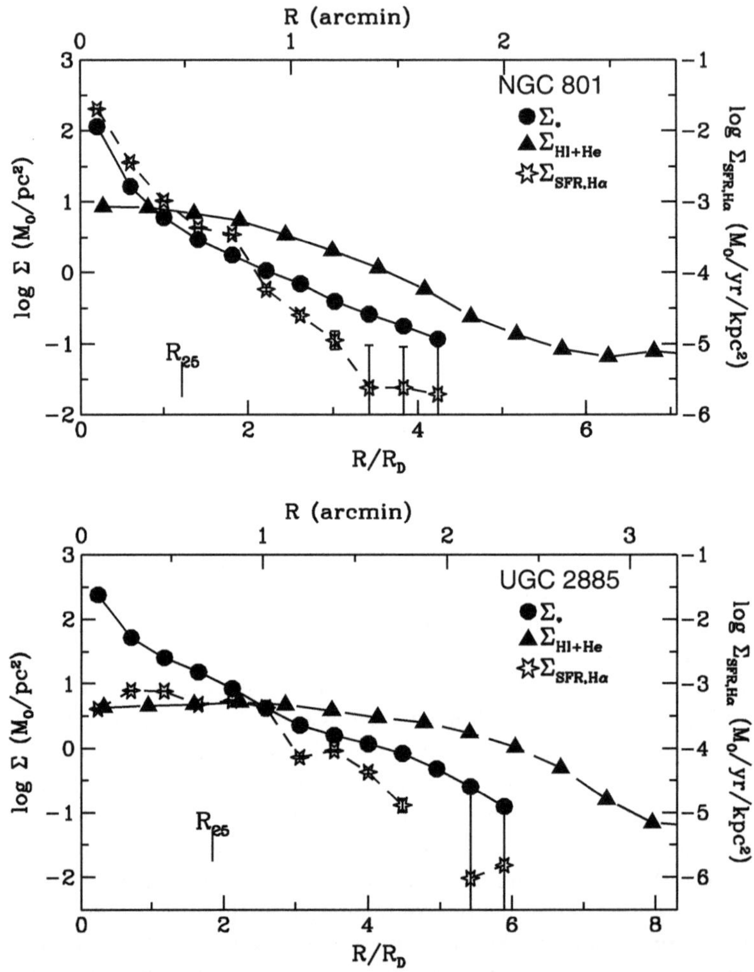

Fig. 4.2 Stellar mass surface density Σ_*, H I+He surface density $\Sigma_{\rm HI+He}$ and SFR density $\Sigma_{\rm SFR,H\alpha}$ plotted as a function of radius for two very luminous ($M_V = -22$ to -23) Sc-Type spiral galaxies, NGC 801 and UGC 2885. The radius is normalized to the optical V-band disk scale length $R_{\rm D}$. The gas and stellar mass surface densities have been corrected to face-on. The logarithmic interval is the same for all three quantities, but the SFR zero point is different. Adapted from Hunter et al. (2013)

density (Melena et al. 2009; Hunter et al. 2016), although it may stop short of the full extent of the gas disk (e.g. see Fig. 4.2).

Krumholz (2013) formulated a model for star formation in these conditions that considers the existence of a two-phase atomic medium (i.e. a warm neutral medium in pressure equilibrium with a cool neutral medium) and the molecular fraction in such a medium. He then assumed that star formation occurs in the molecular medium at a rate given by an efficiency per unit free-fall time, $\epsilon_{\rm ff} \sim 0.01$, times the molecular mass divided by the free-fall time in the molecular gas:

$$\Sigma_{\rm SFR} = \epsilon_{\rm ff}\Sigma_{\rm mol}/t_{\rm ff,mol}. \tag{4.1}$$

The free-fall time depends on the molecular cloud density, which for outer disks in their model depends on the molecular cloud mass and a fiducial value of the molecular cloud surface density, $\Sigma_{\rm GMC} = 85\,M_{\odot}\,{\rm pc}^{-2}$. The molecular cloud mass was taken to be the turbulent Jeans mass in the interstellar medium (ISM), $M_{\rm GMC} = \sigma^4/(G\Sigma_{\rm gas})$ for turbulent speed σ and average ISM gas surface density $\Sigma_{\rm gas}$. For inner disks, the free-fall time was taken to be the value for an average disk density where the Toomre Q parameter equals unity. To span the inner and outer regions, the minimum of these two free-fall times was used.

One uncertainty in the Krumholz (2013) model is the assumption that a two-phase medium is present, because it need not be present everywhere in the outer disk. But this assumption seems reasonable for star formation because cool gas greatly facilitates cloud formation (Elmegreen and Parravano 1994; Schaye 2004; Forbes et al. 2016). A second assumption is that the SFR is given only by the molecular gas mass and density and that this is related to the total density by the molecular fraction, which depends on the ratio of the radiation field to the cool cloud density. In this model, molecule formation is calculated separately as a precursor to star formation, and then whatever is calculated for molecules is used to determine the SFR.

Another model considers that the average SFR is determined by the average ISM dynamics and that molecule formation is incidental, i.e. molecule formation happens along the way but it is not a limiting factor. Star formation in predominantly atomic gas has been predicted by Glover and Clark (2012) and Krumholz (2012) and suggested by observations in Michałowski et al. (2015) and Elmegreen et al. (2016). In this model, the gas mass available for star formation is the total gas mass in all forms, even atomic gas, and the free-fall rate of this gas is given by the average midplane density, regardless of molecular content (Elmegreen 2015). Cool clouds are still required so the ISM cannot be purely warm phase. Also, because molecular hydrogen is slow to form at low density (Mac Low and Glover 2012), there could be a substantial fraction of H_2 in stagnant, diffuse clouds without significant CO emission and with little connection to star formation (Elmegreen and Hunter 2015). Such a diffuse H_2 medium was found in simulations by Hu et al. (2016) and Safranek-Shrader et al. (2016) but has not been observed yet. These diffuse H_2 clouds, along with more atom-rich clouds, would presumably come together during localized ISM collapse as a precursor to star formation. In this model the SFR per unit area is given by

$$\Sigma_{\rm SFR} = \epsilon_{\rm ff}\Sigma_{\rm gas}/t_{\rm ff,gas} \tag{4.2}$$

for midplane free-fall time t_{ff} at the density $\rho = \Sigma_{gas}/(2H)$ and scale height $H = \sigma^2/(\pi G \Sigma_{gas})$. The result is (Elmegreen 2015)

$$\Sigma_{SFR} = \epsilon_{ff}(4/3^{1/2})(G/\sigma)\Sigma_{gas}^2 = 1.7 \times 10^{-5}(\Sigma_{gas}/[1\,M_\odot/\mathrm{pc}^2])^2(\sigma/6\,\mathrm{km\,s}^{-1})^{-1} \tag{4.3}$$

where σ is the gas velocity dispersion and $\epsilon_{ff} \sim 1\%$ is the efficiency of star formation per unit free-fall time.

Both of the above models compare well with observations (Krumholz 2013; Elmegreen 2015).

Hu et al. (2016) simulated dwarf galaxies with a chemical model to form H_2, CO and other molecules, cloud self-shielding from radiation, and a SFR given by Eq. (4.2) at a threshold density of $100\,\mathrm{cm}^{-3}$ and a temperature less than 100 K. They found that Σ_{SFR} decreases faster than Σ_{gas} but not because of a flare (the extra Σ_{gas} factor in Eq. (4.3)). Rather, Σ_{SFR} follows the cold gas with a rate that scales directly with the cold gas fraction, i.e. a linear law, and this cold gas fraction decreases with radius. The linear law in their model is because of the assumed constant threshold density. Most star-forming gas is close to this fixed density, so the characteristic dynamical time is the fixed value at this density.

This point about a fixed density is similar to the explanation for the linear star formation law in Elmegreen (2015), where it was pointed out that if CO, HCN and other star formation tracers emit mostly at their fixed excitation density, as determined by the Einstein *A* coefficient, then the effective free-fall time is the fixed value at this density. The SFR then scales only with the amount of gas at or above this observationally selected density. The fixed density in this case is not because of an assumption about a star formation threshold, as there is no threshold in the Elmegreen (2015) model. There is just a continuous collapse of ISM gas at a rate given by the midplane density and a feedback return of the dense gas to a low-density form.

The existence of a fixed threshold density for star formation is something to be tested observationally. Elmegreen (2015) suggest there is no threshold density because clouds are strongly self-gravitating when $\pi G \Sigma_{cloud}^2 > P_{ISM}$ for cloud surface density Σ_{cloud} and ambient pressure P. Because the interstellar pressure varies with the square of the total surface density of gas and stars inside the gas layer, there is a large range in pressure over several exponential scale lengths in a galaxy disk—a range that may exceed a factor of 100 for dIrrs and 1000 for spirals. Thus if there is a threshold for star formation, the above equation suggests that it might be

$$\Sigma_{cloud} > (P/\pi G)^{1/2}, \tag{4.4}$$

in which case it should vary with radius.

We return to a point made in Sect. 4.1 about the gas surface density profile in the far-outer regions of gas-rich galaxies. Wang et al. (2014) noted that the gas exponential scale length beyond the radius R_1, where $\Sigma_{\rm gas} = 1\,M_\odot\,{\rm pc}^{-2}$, is always about 0.25 times this radius. We can see this here also from Eq. (4.3), which states that

$$\Sigma_{\rm SFR} = 2 \times 10^{-5}\,M_\odot\,{\rm pc}^{-2}\,{\rm Myr}^{-1} \text{ at } \Sigma_{\rm gas} = 1\,M_\odot\,{\rm pc}^{-2}. \quad (4.5)$$

After a Hubble time of 10^4 Myr, $\Sigma_{\rm stars}$ is approximately $0.2\,M_\odot\,{\rm pc}^{-2}$. According to the average disk mass profiles in Zheng et al. (2015), this outer stellar surface density is lower than that at the disk centre by $10^{-3.5}$ on average, which represents eight scale lengths in stars. But eight scale lengths in stars is four scale lengths in gas for Eq. (4.3). Thus the radius at $1\,M_\odot\,{\rm pc}^{-1}$ is about four times the scale length in the gas, as observed further out by Wang et al. (2014).

4.8 The Disks of Dwarf Irregular Galaxies

Dwarf irregular galaxies are like the outer parts of spiral galaxies in terms of gas surface density, SFR and gas consumption time. Tiny dIrrs have extended exponential disks as well. For example, Saha et al. (2010) traced the Large Magellanic Cloud (LMC) to $12\,R_{\rm D}$, an effective surface brightness of 34 mag arcsec^{-2} in I, and Sanna et al. (2010) found stars in IC 10 to $\sim 10\,R_{\rm D}$. Bellazzini et al. (2014) detected stars associated with Sextans A and Sextans B to $6\,R_{\rm D}$, and Hunter et al. (2011) measured surface brightness profiles in four nearby dIrrs and one BCD to 29.5 mag arcsec^{-2} in V, corresponding to 3–8 $R_{\rm D}$. These extended stellar disks represent extreme galactic environments for star formation and are potentially sensitive probes of galaxy evolutionary processes, and yet they are relatively unexplored. In this section we examine what is known about outer disks of dIrr galaxies.

4.8.1 Radial Trends

4.8.1.1 The Gas Disk

The H I gas often dominates the stellar component of dIrr galaxies, both in extent and mass. How much further the gas extends compared to the stars was demonstrated by Krumm and Burstein (1984) for DDO 154 where the H I was traced to $8\,R_{25}$ at a column density of 2×10^{19} atoms cm^{-2} ($0.22\,M_\odot\,{\rm pc}^{-2}$). In the LITTLE THINGS sample of 41 nearby (<10.3 Mpc), relatively isolated dIrrs (Hunter et al. 2012), most systems have gas extending to 2–4 R_{25} or 3–7 $R_{\rm D}$ at that same (face-on) column density. Some spiral galaxies also have extended H I; Portas (2010) found that the Sbc galaxy NGC 765, for example, has gas extending to $4\,R_{25}$. Large holes (up to 2.3 kpc diameter) are also sometimes found in the gas beyond $2\,R_{\rm D}$ (Dopita et al. 1985, R.N. Pokhrel, in preparation).

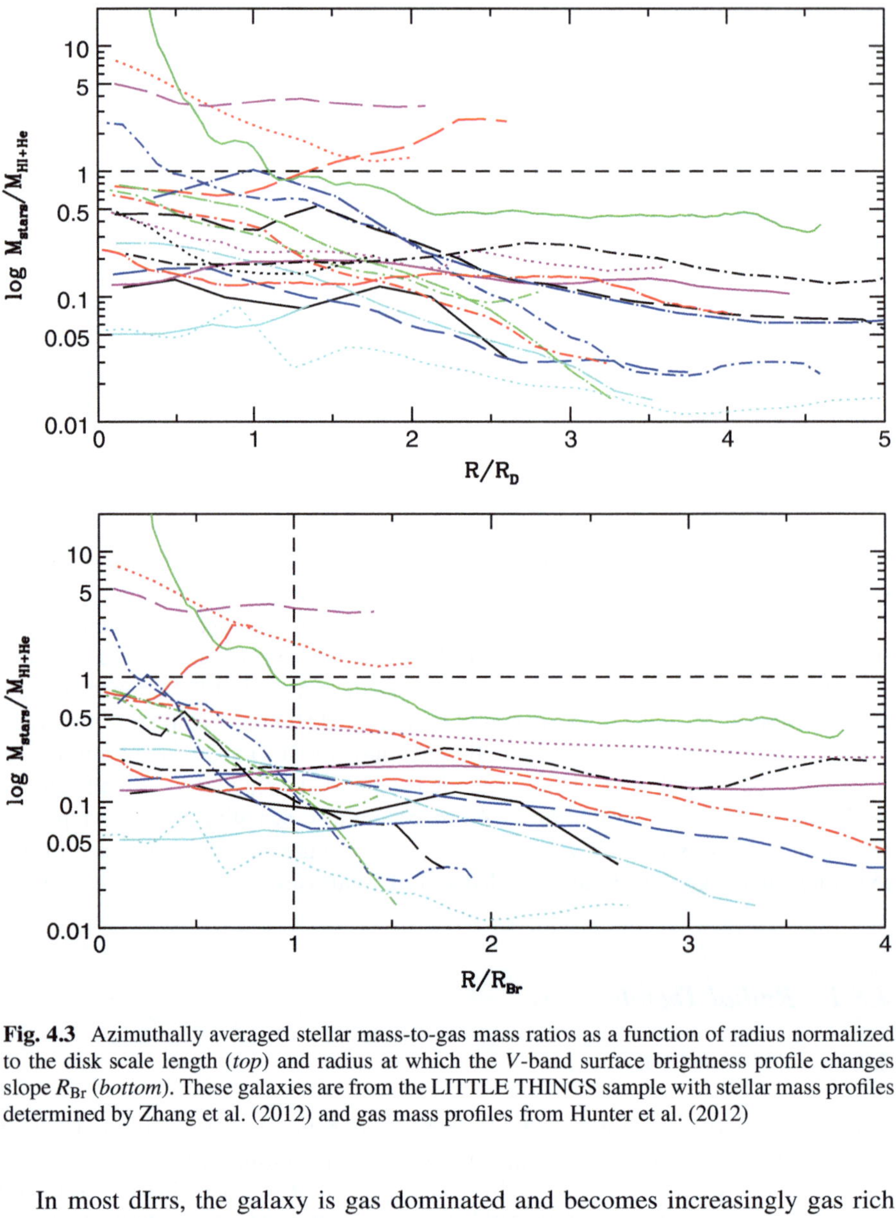

Fig. 4.3 Azimuthally averaged stellar mass-to-gas mass ratios as a function of radius normalized to the disk scale length (*top*) and radius at which the V-band surface brightness profile changes slope R_{Br} (*bottom*). These galaxies are from the LITTLE THINGS sample with stellar mass profiles determined by Zhang et al. (2012) and gas mass profiles from Hunter et al. (2012)

In most dIrrs, the galaxy is gas dominated and becomes increasingly gas rich with radius (Fig. 4.3). This implies a decreasing large-scale star formation efficiency (Leroy et al. 2008; Bigiel et al. 2010). The lack of sharp transitions in the star-to-gas ratio, including at breaks in the optical exponential surface brightness profiles, suggests that the factors dominating the drop in star formation with radius are changing relatively steadily.

The gas surface density drops off with radius usually in a non-exponential fashion. In Sects. 4.2–4.7, we have approximated the radial gas profiles of spiral galaxies as exponentials. However, especially in dIrrs, the gas profiles are rarely pure exponentials, and since dIrrs are gas dominated, the shape of the gas profile is crucial. Note, however, that in dIrrs, the radial *stellar* profiles are usually exponential in shape. In a subsample of the THINGS spirals (Walter et al. 2008), Portas (2010) found that the gas is approximately constant at 5–10 $\times$ 10^{20} atoms cm^{-2} and then drops off rapidly. A Sersic function fits the profiles with indices of $n = 0.14$–0.22. For comparison, an exponential disk has an n of 1.0. In five THINGS dIrrs the gas density dropped more shallowly with radius, and the distribution of n peaked around 0.3. For the LITTLE THINGS sample of dwarfs, the shape of the H I radial profiles varied from galaxy to galaxy, and n varied from 0.2 to 1.65 with most having values 0.2–0.8. The lack of correlations between the H I profile index n and characteristics of the stellar disk suggest that the role of the gas distribution in determining the stellar disk properties is complex.

4.8.1.2 The Stellar Disk

Zhang et al. (2012) performed spectral energy distribution fitting to azimuthally averaged surface photometry of the LITTLE THINGS galaxies. The fitting included up to 11 passbands from the FUV to the NIR. From these fits they constructed SFRs as a function of radius over three broad timescales: 100 Myr, 1 Gyr and galaxy lifetime. Zhang et al. found that the bulk star formation activity has been shrinking with radius over the lifetime of dwarf galaxies, and they adopted the term "outside-in" disk growth. Although Zhang et al. found that "outside-in" disk growth applied primarily to dIrrs with baryonic masses $< 10^8\,M_\odot$, Gallart et al. (2008) and Meschin et al. (2014) found the same phenomenon in the LMC, a more massive irregular galaxy. Similarly, Pan et al. (2015) suggested from colour profiles that the same process is occurring in a large sample of Sloan Digital Sky Survey galaxies with stellar masses up to $10^{10}\,M_\odot$. This outside-in disk growth is in contrast to the inside-out disk growth identified in spirals (White and Frenk 1991; Mo et al. 1998; Muñoz-Mateos et al. 2007; Williams et al. 2009, but see Ruiz-Lara et al. 2016).

Hunter et al. (2011) carried out an ultra-deep imaging program on four nearby dIrrs and one BCD. They measured surface photometry in this sample to 29.5 mag arcsec^{-2} in V and also obtained deep B images of three of the galaxies and deep FUV and NUV images with the *Galaxy Evolution Explorer* (*GALEX*, Martin et al. 2005). Figure 4.4 shows the V-band image and photometry of DDO 133, illustrating what they found. What does a surface brightness of 29.5 mag arcsec^{-2} mean? In DDO 133, that is a factor of $\sim$160 down in brightness from the centre. A 1 kpc-wide annulus at 29.5 mag arcsec^{-2} corresponds to a SFR of 0.0004 $M_\odot$ yr^{-1}, assuming a mass-to-light ratio from the $B-V$ colour and a constant SFR for 12 Gyr. This is roughly seven Orion nebulae every 10 Myr.

In their five dwarfs, Hunter et al. (2011) found that the stellar surface brightnesses in V and FUV continue exponentially as far as could be measured. Furthermore, the stellar disk profiles are exponential and extraordinarily regular in spite of the fact that dIrr galaxies are clumpy in gas and SFR and star formation is sporadic. Saha

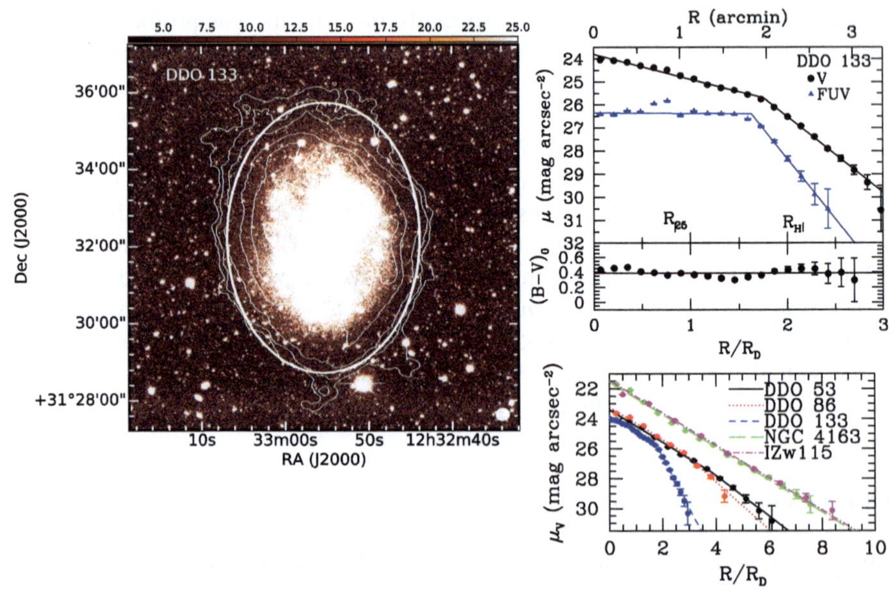

Fig. 4.4 *Left*: V-band image of DDO 133 from Hunter et al. (2011). The *white ellipse* marks the extent of the galaxy measured to 29.5 mag arcsec^{-2}. The white contours trace column densities of 5, 30, 100, 300, 500, 1000 and 3000 $\times$ 10^{18} atoms cm^{-2} in H I (Hunter et al. 2012). *Top right*: surface photometry in V, FUV and $B-V$ as a function of radius normalized to the disk scale length R_D. *Bottom right*: V-band surface photometry of all five galaxies from Hunter et al. (2011). IZw 115 is a BCD; the rest are dIrrs

et al. (2010) found the same thing for the LMC and Bellazzini et al. (2014) for Sextans B. However, Bellazzini et al. also found that, by contrast, Sextans A has a very complex surface brightness profile and suggested that that is the consequence of past outside perturbations, assuming that a regular profile is "normal" for an isolated galaxy.

4.8.2 Star Formation in Dwarfs

The deep FUV surface photometry of Hunter et al. (2011) also shows that there is a continuity of star formation with radius. The Toomre (1964) model, in which star formation is driven by two-dimensional gravitational instabilities in the gas, predicts a precipitous end to star formation where the gas surface density drops below a critical level. Nevertheless, these data show that young stars extend into the realm where the gas is a few percent of the critical gas density and should be stable against spontaneous gravitational collapse (Kennicutt 1989). Models suggest that dIrrs need to be treated as three-dimensional systems, in which case the Q parameter is not a good measure of total stability. Also, the dynamical time at the midplane density is more important than the growth time of a two-dimensional instability, which is more closely related to spiral arms than star formation (Elmegreen 2015; Elmegreen and Hunter 2015).

The presence of FUV emission in outer disks poses a stringent test of star formation models by extending measures of star formation activity to the regime of low gas densities. How low can the gas density get and still have star formation? H II regions have been found in the far-outer disks of spirals (Ferguson et al. 1998), and *GALEX* has found FUV-bright regions out to 2–3 times the optical radius of the spiral (Gil de Paz et al. 2005; Thilker et al. 2007). Bush et al. (2008) proposed that these FUV regions could be due to spiral density waves from the inner disk propagating into the outer disk and raising local gas regions above a threshold for star formation. In fact, Barnes et al. (2012) found evidence for greater instability in outer disk spirals compared to inner disk spirals in eight nearby spiral galaxies. Dwarf irregular galaxies, however, do not have spiral density waves and neither do the far-outer parts of the galaxies observed by Watkins et al. (2016), so the problem still remains of how stars form in or get scattered to extreme outer disks.

Recently, *GALEX* images have been used to identify FUV-bright knots in the outer disks of dIrrs in order to determine how far-out young star clusters are formed in situ and the nature of the star clusters found there. Hunter et al. (2016) identified the furthest-out FUV knot of emission in the LITTLE THINGS galaxies and found knots at radii of 1–8 R_D (see Fig. 4.5). Most of these outermost regions are found intermittently where the H I surface density is $\sim 2\,M_\odot\,\mathrm{pc}^{-2}$, although both the H I and dispersed old stars go out much further (also true of some spiral galaxies, e.g. Grossi et al. 2011). In a sample of 11 of the LITTLE THINGS dwarfs within 3.6 Mpc, Melena et al. (2009) identified all of the FUV knots and modelled their

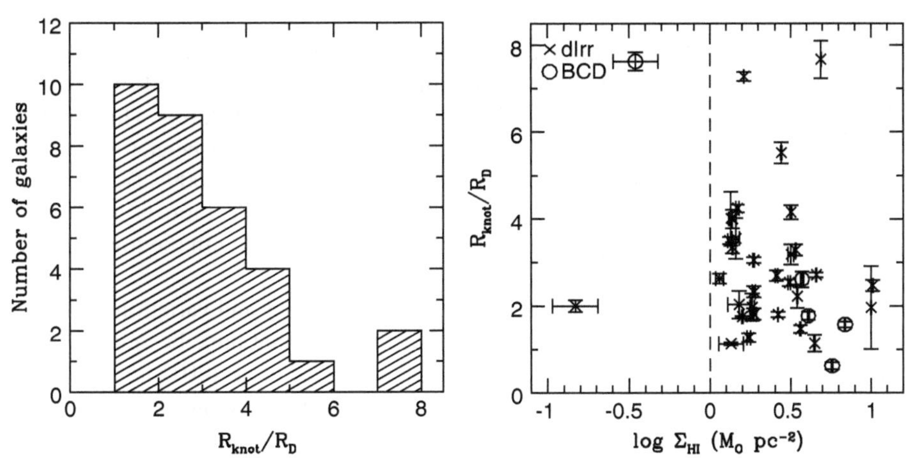

Fig. 4.5 *Left*: histogram of the distance from the centre to the furthest knot of FUV emission in the LITTLE THINGS dIrrs relative to the disk scale length R_D (Hunter et al. 2016). *Right*: distance from the galaxy centre to the FUV knot vs. average H I surface density at the radius of the knot. The Σ_{HI} have been corrected for a scaling error as described in Hunter et al. (2017)

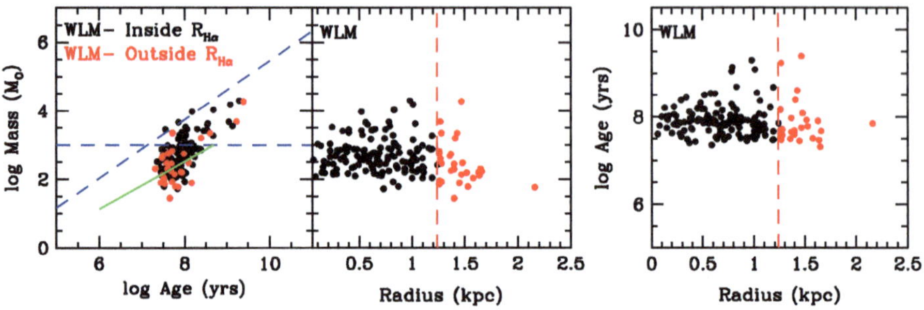

Fig. 4.6 Plots of mass vs. age, mass vs. galactocentric radius and age vs. radius for WLM, one of the galaxies from Fig. 3 of Melena et al. (2009). The radius corresponding to the extent of Hα is marked with a *vertical dashed red line*, and the regions outside the Hα extent are in *red*. The *horizontal dashed line* in the *left panel* is the mass limit for completeness to an age of 500 Myr. The *slanted dashed line* is a fit by eye to the upper envelope of the cluster distribution. The *slanted solid line* shows the slope for a fading relationship in which the minimum observable mass scales as log mass $\propto$0.69 log age

UV, optical and NIR colours to determine masses and ages. They found no radial gradients in region masses and ages (see Fig. 4.6 for an example), even beyond the realm of Hα emission, although there is an exponential decrease in the luminosity density and number density of the regions with radius. In other words, young objects in outer disks cover the same range of masses and ages as inner disk star clusters.

4.8.3 The Hα/FUV Ratio

Hα and FUV emissions are often used to trace star formation in galaxies, including dwarfs. However, commonly the Hα emission drops off faster than, and ends before, the FUV emission as one traces star formation into the outer disk (e.g. Hunter et al. 2010 and Fig. 4.2). In addition global ratios of Hα/FUV have been found to be a function of galactic surface brightness (e.g. Meurer et al. 2009; Treyer et al. 2011). Lee et al. (2009), for example, find that the Hα/FUV ratio is lower than expected by a factor of 2–10 in the nearby 11HUGS galaxies with the lowest SFRs ($<0.003\,M_\odot\,\mathrm{yr}^{-1}$).

The decline of the ratio Hα/FUV with radius in galaxies and variations between galaxies have been the subject of much debate. Causes that have been considered include variations in Meurer et al. (2009) and Bruzzese et al. (2015) or sampling issues with (Fumagalli et al. 2011; da Silva et al. 2014) the stellar initial mass function. Other explanations include variations in star formation history, loss of ionizing photons from the galaxy and undetectability of diffuse Hα emission in outer disks (e.g. Hunter et al. 2010, 2011; Lee et al. 2011; Eldridge 2012; Relaño et al. 2012; Weisz et al. 2012).

Since escape of ionizing photons from galaxies, and preferentially from small galaxies, is believed to have been responsible for the epoch of reionization in the early Universe, measuring the amount of leakage has been an important motivation for observations of Lyman continuum emission around galaxies in the nearby and more distant Universe. These observations give us the opportunity to see if leakage of ionizing photons from galaxies or outer disks could explain low Hα/FUV ratios. Lyman continuum escape fractions have been measured of order 6–13% in compact star-forming galaxies at $z \sim 0.3$ and Lyα escape fractions of order 20–40% (Izotov et al. 2016b). Rutkowski et al. (2016) have placed a limit of $\leq$2.1% on the Lyman continuum escape fraction of a sample of most star-forming dwarf galaxies at $z \sim 1$, and Izotov et al. (2016a) measured an escape fraction of order 8% in a relatively low-mass star-forming galaxy at $z \sim 0.3$. In a sample of four nearby galaxies, Leitet et al. (2013) detected Lyman continuum in one, yielding an escape fraction of 2.4%, but Zastrow et al. (2013) mapped [SIII]/[SII] in six nearby dwarf starburst galaxies and found that the fraction of emission that escapes may depend on the orientation of the galaxy to the observer, the morphology of the ISM and the age and concentration of the starburst producing the emission. Nevertheless, we see that escape fractions are not high enough to explain the lowest ratios of Hα/FUV. On the other hand, Hunter et al. (2013), in a study of two luminous spirals, suggest that the drop in Hα emission with radius is due to low gas densities in outer disks and the resulting loss of Lyman continuum photons from the vicinity of star-forming regions, making them undetectable in Hα, and not from a loss of photons out of the galaxy altogether.

Could we instead be underestimating the amount of Hα emission that is actually there? To test the idea that significant amounts of Hα emission have been missed in outer disks, Lee et al. (2016) performed very deep imaging in Hα of three nearby dwarf galaxies, reaching flux limits of order ten times lower than that of the 11HUGS/LVL survey (Kennicutt et al. 2008). Their new images (Fig. 4.7) do show emission extending up to 2.5 times further than the previous survey data, but this additional emission only contributes ~5% more Hα flux. Therefore, the additional emission found in these deep images does not account for the radial trend in Hα/FUV.

The emission measure of individual HII regions in outer disks can be very low, however, because of the extremely low average density. Following Hunter et al. (2011), one can consider the possible values for emission measure if the far-outer disks in Fig. 4.2 are completely ionized. The limits of the stellar disks in these galaxies correspond to radii of 60 kpc in NGC 801 ($R/R_D = 4.2$ in the figure) and 71 kpc in UGC 2885 ($R/R_D = 5.9$). The total surface densities at these radii can be used to determine the gas disk thicknesses assuming a velocity dispersion of 10 km s^{-1}. These thicknesses are $T = 26.2$ kpc and 11.4 kpc, respectively, if we consider thickness to be two isothermal scale heights. When combined with the HI surface densities, the corresponding average densities are only $n = 0.00052$ and 0.0031 cm^{-3}. If the entire disk thicknesses were ionized at these densities, then the emission measures would be $n^2T = 0.0069\,\mathrm{cm^{-6}\,pc}$ and 0.11 cm^{-6} pc. We convert emission measure to surface brightness as $\Sigma_{\mathrm{H}\alpha}(\mathrm{erg\,s^{-1}\,pc^{-2}}) = 7.7 \times 10^{30}\,\mathrm{EM}\,(\mathrm{cm^{-6}\,pc})$. Converting this to intensity units, we get $I_{\mathrm{H}\alpha}(\mathrm{erg\,s^{-1}\,cm^{-2}\,arcsec^{-2}}) = 1.5 \times 10^{-18}\,\mathrm{EM}\,(\mathrm{cm^{-6}\,pc})$. The limit of detection in the very deep survey by Lee et al. (2016) was $8 \times 10^{-18}\,\mathrm{erg\,s^{-1}\,cm^{-2}\,arcsec^{-2}}$, which is still too high to see the fully ionized far-outer disks in Fig. 4.2 by a factor of ~50 or more.

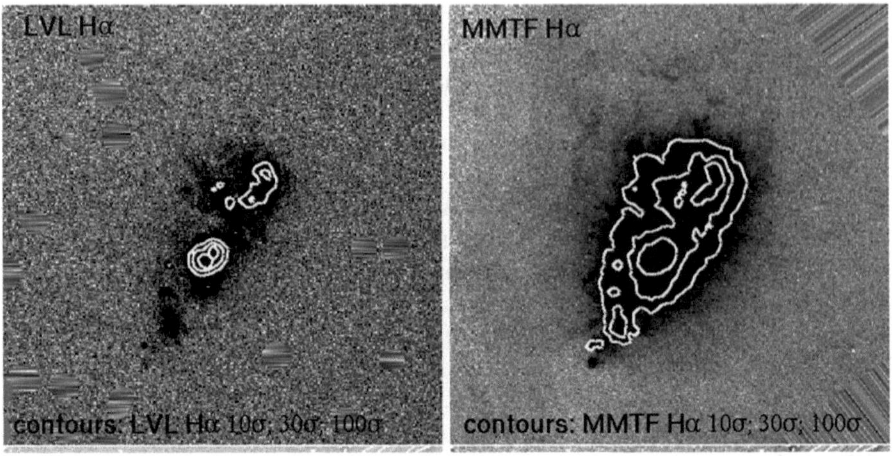

Fig. 4.7 Hα images of UGC 5456 displayed on a linear scale to emphasize emission from diffuse ionized gas. Contours are at 10σ, 30σ and 100σ above the background. On the *left* is the standard continuum subtracted narrow band image from the 11HUGS/LVL survey (Kennicutt et al. 2008) and on the *right* is the deeper image from Lee et al. (2016). Reproduced from Lee et al. (2016) with permission from the AAS

4.8.4 Breaks in Radial Profiles in dIrr Galaxies

Figure 4.8 (right) illustrates another common feature of outer dIrr disks: abrupt breaks in azimuthally averaged surface brightness profiles (Herrmann et al. 2013). Most often the profile drops in brightness into the outer disk more steeply than in the inner disk (Type II profiles; Sect. 4.3), but occasionally it drops less steeply (Type III). Surface brightness profiles without breaks (Type I) are relatively rare. Radial profile breaks are common in spirals as well and were first discovered there (van der Kruit and Shostak 1982; Shostak and van der Kruit 1984; de Grijs et al. 2001; Kregel et al. 2002; Pohlen et al. 2002; MacArthur et al. 2003; Kregel and van der Kruit 2004; Erwin et al. 2005; Pohlen and Trujillo 2006; Erwin et al. 2008; Gutiérrez et al. 2011). They are also found in high redshift disks (Pérez 2004). Bakos et al. (2008) and Ruiz-Lara et al. (2016) found that the Type II downturn at mid-radius decreases significantly in spirals when stellar mass profiles are considered instead of surface brightness. However, this is not the case for most dIrrs, as found by Herrmann et al. (2016). Thus, R_{Br} appears to represent a change in stellar population in spirals but a change in stellar surface mass density, at least in part, in dwarfs.

Herrmann et al. (2013) examined the surface brightness profiles of 141 dwarfs in up to 11 passbands, and typical Type II and Type III profiles are sketched in Fig. 4.8 (left). Herrmann et al. (2016) further examined the colour and mass surface density trends. They found that, although brighter galaxies tend to have larger R_{Br}, the surface brightness in V, μ_V, at R_{Br} is about 24 ± 1 mag arcsec^{-2}, independent of

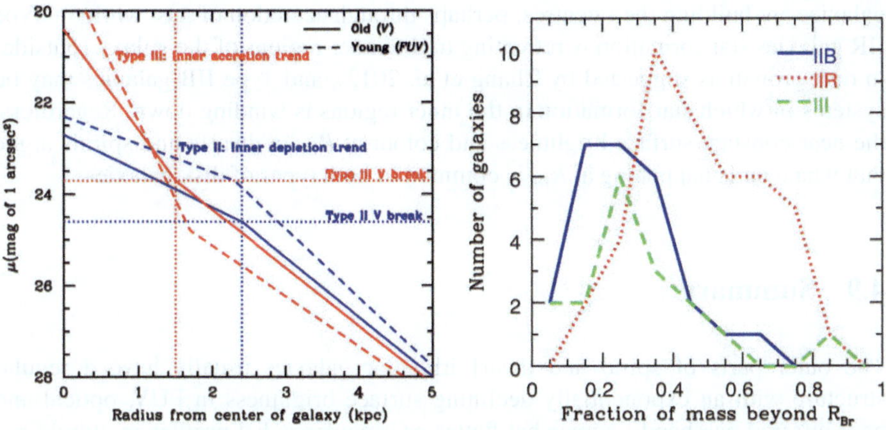

Fig. 4.8 *Left*: representative V and FUV Types II and III surface brightness profiles with parameters for $M_B = -16$ from Herrmann et al. (2013). V highlights older stars, and FUV reveals younger stars. The steep FUV slope of the Type III profile interior to $R_{\rm Br}$ implies an inner accretion trend. The steeper FUV slope in the Type II outskirts is evidence of outside-in shrinking of star formation activity. *Right*: number of galaxies with the given fraction of the stellar mass beyond $R_{\rm Br}$. Type II profiles with *bluing colour* trends with radius (IIB) and with *reddening colour* trends with radius (IIR) are shown separately. Type IIR profiles have a larger fraction of their stellar mass beyond $R_{\rm Br}$ than Type IIB or Type III

M_B and independent of galaxy type. The $B - V$ colour at $R_{\rm Br}$ is also nearly constant. However, when surface photometry is converted to stellar mass surface density for Type II profiles, values for dwarfs are a factor of $\sim$6 lower than those for spirals (Herrmann et al. 2016; Bakos et al. 2008). When separated by radially averaged colour trends, Type II profiles with reddening colour trends (IIR) have a larger fraction of their stellar mass beyond $R_{\rm Br}$ than Type IIs with a bluing colour trend (IIB) or Type IIIs (Herrmann et al. 2016; Fig. 4.8, right).

What is happening at $R_{\rm Br}$? Simulations of spirals by Roškar et al. (2008), Martínez-Serrano et al. (2009), Bakos et al. (2011) and Minchev et al. (2012) suggest that the break radius $R_{\rm Br}$ grows with time and that for Type II profiles stars formed inside $R_{\rm Br}$ migrate outward beyond $R_{\rm Br}$ as a result of secular processes involving bar potentials or spiral arms (see observations by Radburn-Smith et al. 2012). However, scattering of stars from spiral arms is not applicable to dIrrs, and observations of some spirals are inconsistent with this scenario as well (Yoachim et al. 2012). Another possibility is that there is a change in the dominate star formation process or efficiency at $R_{\rm Br}$ (e.g. Schaye 2004; Piontek and Ostriker 2005; Elmegreen and Hunter 2006; Barnes et al. 2012; but for models of star formation without a sharp change with radius, see Ostriker et al. 2010 and Krumholz 2013). Roškar et al. (2008) suggest that, for spirals, it is a combination of a radial star formation cutoff and stellar mass redistribution (see also Zheng et al. 2015). The different radial surface brightness and colour profiles in dwarfs can be understood empirically as the result of different evolutionary histories (Fig. 4.8, left): Type III galaxies are building their centres, perhaps through accretion of gas, while in Type IIR galaxies star formation is retreating to the inner regions of the galaxy (outside-in disk growth as suggested by Zhang et al. 2012), and Type IIB galaxies may be systems in which star formation in the inner regions is winding down. Regardless, the near-constant surface brightness and colour at $R_{\rm Br}$ in dwarfs and spirals argue that whatever is happening at $R_{\rm Br}$ is common to both types of disk galaxies.

4.9 Summary

The outer parts of spiral and dwarf irregular galaxies usually have a regular structure with an exponentially declining surface brightness in FUV, optical and near-infrared passbands, somewhat flatter or irregular radial profiles in atomic gas and frequent evidence for azimuthal asymmetries. Models suggest that these outer parts form by a combination of gas accretion from the halo or beyond, in situ star formation and stellar scattering from the inner disk. The exponential shape is not well understood, but cosmological simulations get close to exponential shapes by approximate angular momentum conservation. Wet mergers also get exponentials after the combined stellar systems relax, and stellar scattering from gas irregularities and spiral arm corotation resonances get exponentials too, all probably for different reasons.

Star formation persists in far-outer disks without any qualitative change in the properties of individual star-forming regions. This happens even though the gas density is very low, Hα is often too weak to see and the dynamical time is long. Gas also tends to dominate stars by mass in the outer parts, but the gas appears to be mostly atomic, making star formation difficult to understand in comparison to inner disks, where it is confined to molecular clouds.

Colour and age gradients suggest that most spiral galaxies have their earliest star formation in the inner disk, with a scale length that increases in time and an outward progress of gas depletion or quenching too. The result is a tendency for spirals to get bluer with increasing radius. Eventually the blue trend stops and spirals get redder after that. This gradient change occurs in all types of spiral galaxies, regardless of the exponential shapes of their radial profiles, and suggests a different process for the formation of inner and outer stellar disks. Most likely stellar scattering from the inner disk to the outer disk is part of the explanation, including stellar scattering from bars and spirals, but there could be other processes at work too, including minor mergers and interactions with other galaxies. Colour gradients in dIrrs are usually the opposite of those in spiral galaxies. Dwarfs tend to get systematically redder with radius in what looks like outside-in star formation. This could reflect an enhanced role for stellar scattering with the first star formation still near the centre, as for spirals, or it could result from radial gas accretion or other truly outside-in processes.

The advent of new surveys that probe galaxies to very faint stellar surface brightnesses, combined with new maps of atomic and molecular emission from the far-outer regions of galaxies, should help us to better understand the origin and evolution of galaxy disks.

Acknowledgements DAH appreciates the support of the Lowell Observatory Research Fund in writing this chapter.

References

Athanassoula, E., Rodionov, S.A., Peschken, N., Lambert, J.C.: Forming disk galaxies in wet major mergers. I. Three fiducial examples. Astrophys. J. **821**, 90 (2016). doi:10.3847/0004-637X/821/2/90, 1602.03189

Aumer, M., White, S.D.M.: Idealized models for galactic disc formation and evolution in 'realistic' ΛCDM haloes. Mon. Not. R. Astron. Soc. **428**, 1055–1076 (2013). doi:10.1093/mnras/sts083, 1203.1190

Aumer, M., White, S.D.M., Naab, T., Scannapieco, C.: Towards a more realistic population of bright spiral galaxies in cosmological simulations. Mon. Not. R. Astron. Soc. **434**, 3142–3164 (2013). doi:10.1093/mnras/stt1230, 1304.1559

Bakos, J., Trujillo, I., Pohlen, M.: Color profiles of spiral galaxies: clues on outer-disk formation scenarios. Astrophys. J. **683**, L103 (2008). doi:10.1086/591671, 0807.2776

Bakos, J., Trujillo, I., Azzollini, R., Beckman, J.E., Pohlen, M.: Outskirts of spiral galaxies: result of a secular evolution process?. Memorie della Societa Astronomica Italiana Supplementi **18**, 113 (2011). 1002.1276

Barker, M.K., Ferguson, A.M.N., Irwin, M.J., Arimoto, N., Jablonka, P.: Quantifying the faint structure of galaxies: the late-type spiral NGC 2403. Mon. Not. R. Astron. Soc. **419**, 1489–1506 (2012). doi:10.1111/j.1365-2966.2011.19814.x, 1109.2625

Barnes, K.L., van Zee, L., Côté, S., Schade, D.: Star formation in the outer disk of spiral galaxies. Astrophys. J. **757**, 64 (2012). doi:10.1088/0004-637X/757/1/64

Bellazzini, M., Beccari, G., Fraternali, F., Oosterloo, T.A., Sollima, A., Testa, V., Galleti, S., Perina, S., Faccini, M., Cusano, F.: The extended structure of the dwarf irregular galaxies Sextans A and Sextans B: signatures of tidal distortion in the outskirts of the local group. Astron. Astrophys. **566**, A44 (2014). doi:10.1051/0004-6361/201423659, 1404.1697

Berrier, J.C., Sellwood, J.A.: Smoothing rotation curves and mass profiles. Astrophys. J. **799**, 213 (2015). doi:10.1088/0004-637X/799/2/213, 1412.0979

Bigiel, F., Blitz, L.: A universal neutral gas profile for nearby disk galaxies. Astrophys. J. **756**, 183 (2012). doi:10.1088/0004-637X/756/2/183, 1208.1505

Bigiel, F., Leroy, A., Walter, F., Blitz, L., Brinks, E., de Blok, W.J.G., Madore, B.: Extremely inefficient star formation in the outer disks of nearby galaxies. Astron. J. **140**, 1194–1213 (2010). doi:10.1088/0004-6256/140/5/1194, 1007.3498

Bird, J.C., Kazantzidis, S., Weinberg, D.H., Guedes, J., Callegari, S., Mayer, L., Madau, P.: Inside out and upside down: tracing the assembly of a simulated disk galaxy using mono-age stellar populations. Astrophys. J. **773**, 43 (2013). doi:10.1088/0004-637X/773/1/43, 1301.0620

Bland-Hawthorn, J., Vlajić, M., Freeman, K.C., Draine, B.T.: NGC 300: an extremely faint, outer stellar disk observed to 10 scale lengths. Astrophys. J. **629**, 239–249 (2005). doi:10.1086/430512, astro-ph/0503488

Borlaff, A., Eliche-Moral, M.C., Rodríguez-Pérez, C., Querejeta, M., Tapia, T., Pérez-González, P.G., Zamorano, J., Gallego, J., Beckman, J.: Formation of S0 galaxies through mergers. Antitruncated stellar discs resulting from major mergers. Astron. Astrophys. **570**, A103 (2014). doi:10.1051/0004-6361/201424299, 1407.5097

Bovy, J., Rix, H.W., Schlafly, E.F., Nidever, D.L., Holtzman, J.A., Shetrone, M., Beers, T.C.: The stellar population structure of the galactic disk. Astrophys. J. **823**, 30 (2016). doi:10.3847/0004-637X/823/1/30, 1509.05796

Bresolin, F., Kennicutt, R.C.: Abundance gradients in low surface brightness spirals: clues on the origin of common gradients in galactic discs. Mon. Not. R. Astron. Soc. **454**, 3664–3673 (2015). doi:10.1093/mnras/stv2245, 1509.07190

Bruzzese, S.M., Meurer, G.R., Lagos, C.D.P., Elson, E.C., Werk, J.K., Blakeslee, J.P., Ford, H.: The initial mass function and star formation law in the outer disc of NGC 2915. Mon. Not. R. Astron. Soc. **447**, 618–635 (2015). doi:10.1093/mnras/stu2461, 1411.6674

Bush, S.J., Cox, T.J., Hernquist, L., Thilker, D., Younger, J.D.: Simulations of XUV disks with a star formation density threshold. Astrophys. J. **683**, L13 (2008). doi:10.1086/591523, 0807.1116

Cooper, A.P., D'Souza, R., Kauffmann, G., Wang, J., Boylan-Kolchin, M., Guo, Q., Frenk, C.S., White, S.D.M.: Galactic accretion and the outer structure of galaxies in the CDM model. Mon. Not. R. Astron. Soc. **434**, 3348–3367 (2013). doi:10.1093/mnras/stt1245, 1303.6283

da Silva, R.L., Fumagalli, M., Krumholz, M.R.: SLUG – Stochastically Lighting Up Galaxies – II. Quantifying the effects of stochasticity on star formation rate indicators. Mon. Not. R. Astron. Soc. **444**, 3275–3287 (2014). doi:10.1093/mnras/stu1688, 1403.4605

Dalcanton, J.J., Williams, B.F., Lang, D., Lauer, T.R., Kalirai, J.S., Seth, A.C., Dolphin, A., Rosenfield, P., Weisz, D.R., Bell, E.F., Bianchi, L.C., Boyer, M.L., Caldwell, N., Dong, H., Dorman, C.E., Gilbert, K.M., Girardi, L., Gogarten, S.M., Gordon, K.D., Guhathakurta, P., Hodge, P.W., Holtzman, J.A., Johnson, L.C., Larsen, S.S., Lewis, A., Melbourne, J.L., Olsen, K.A.G., Rix, H.W., Rosema, K., Saha, A., Sarajedini, A., Skillman, E.D., Stanek, K.Z.: The Panchromatic Hubble Andromeda Treasury. Astrophys. J. Suppl. Ser. **200**, 18 (2012). doi:10.1088/0067-0049/200/2/18, 1204.0010

Dale, D.A., Beltz-Mohrmann, G.D., Egan, A.A., Hatlestad, A.J., Herzog, L.J., Leung, A.S., McLane, J.N., Phenicie, C., Roberts, J.S., Barnes, K.L., Boquien, M., Calzetti, D., Cook, D.O., Kobulnicky, H.A., Staudaher, S.M., van Zee, L.: Radial star formation histories in 15 nearby galaxies. Astron. J. **151**, 4 (2016). doi:10.3847/0004-6256/151/1/4, 1511.03285

de Grijs, R., Kregel, M., Wesson, K.H.: Radially truncated galactic discs. Mon. Not. R. Astron. Soc. **324**, 1074–1086 (2001). doi:10.1046/j.1365-8711.2001.04380.x, astro-ph/0002523

de Vaucouleurs, G.: Photoelectric photometry of messier 33 IN the u, b, v, system. Astrophys. J. **130**, 728 (1959). doi:10.1086/146764

Dopita, M.A., Mathewson, D.S., Ford, V.L.: Shapley constellation III – a region of self-propagating star formation. Astrophys. J. **297**, 599–606 (1985). doi:10.1086/163556

Efstathiou, G.: A model of supernova feedback in galaxy formation. Mon. Not. R. Astron. Soc. **317**, 697–719 (2000). doi:10.1046/j.1365-8711.2000.03665.x, astro-ph/0002245

El-Badry, K., Wetzel, A., Geha, M., Hopkins, P.F., Kereš, D., Chan, T.K., Faucher-Giguère, C.A.: Breathing FIRE: how stellar feedback drives radial migration, rapid size fluctuations, and population gradients in low-mass galaxies. Astrophys. J. **820**, 131 (2016). doi:10.3847/0004-637X/820/2/131, 1512.01235

Eldridge, J.J.: Stochasticity, a variable stellar upper mass limit, binaries and star formation rate indicators. Mon. Not. R. Astron. Soc. **422**, 794–803 (2012). doi:10.1111/j.1365-2966.2012.20662.x, 1106.4311

Elmegreen, B.G.: On the star formation law for spiral and irregular galaxies. Astrophys. J. **814**, L30 (2015). doi:10.1088/2041-8205/814/2/L30, 1511.05633

Elmegreen, B.G., Hunter, D.A.: Radial profiles of star formation in the far outer regions of galaxy disks. Astrophys. J. **636**, 712–720 (2006). doi:10.1086/498082, astro-ph/0509190

Elmegreen, B.G., Hunter, D.A.: A star formation law for dwarf irregular galaxies. Astrophys. J. **805**, 145 (2015). doi:10.1088/0004-637X/805/2/145, 1503.04370

Elmegreen, B.G., Parravano, A.: When star formation stops: galaxy edges and low surface brightness disks. Astrophys. J. **435**, L121–L124 (1994). doi:10.1086/187609

Elmegreen, B.G., Struck, C.: Exponential galaxy disks from stellar scattering. Astrophys. J. **775**, L35 (2013). doi:10.1088/2041-8205/775/2/L35, 1308.5236

Elmegreen, B.G., Struck, C.: Exponential disks from stellar scattering. III. Stochastic models. Astrophys. J. **830**, 115 (2016). doi:10.3847/0004-637X/830/2/115, 1607.07595

Elmegreen, B.G., Kaufman, M., Bournaud, F., Elmegreen, D.M., Struck, C., Brinks, E., Juneau, S.: High star formation rates in turbulent atomic-dominated gas in the interacting galaxies IC 2163 and NGC 2207. Astrophys. J. **823**, 26 (2016). doi:10.3847/0004-637X/823/1/26, 1603.04533

Erwin, P., Beckman, J.E., Pohlen, M.: Antitruncation of disks in early-type barred galaxies. Astrophys. J. **626**, L81–L84 (2005). doi:10.1086/431739, astro-ph/0505216

Erwin, P., Pohlen, M., Beckman, J.E.: The outer disks of early-type galaxies. I. Surface-brightness profiles of barred galaxies. Astron. J. **135**, 20–54 (2008). doi:10.1088/0004-6256/135/1/20, 0709.3505

Erwin, P., Gutiérrez, L., Beckman, J.E.: A strong dichotomy in S0 disk profiles between the Virgo cluster and the field. Astrophys. J. **744**, L11 (2012). doi:10.1088/2041-8205/744/1/L11, 1111.5027

Ferguson, A.M.N., Clarke, C.J.: The evolution of stellar exponential discs. Mon. Not. R. Astron. Soc. **325**, 781–791 (2001). doi:10.1046/j.1365-8711.2001.04501.x, astro-ph/0103205

Ferguson, A.M.N., Wyse, R.F.G., Gallagher, J.S., Hunter, D.A.: Discovery of recent star formation in the extreme outer regions of disk galaxies. Astrophys. J. **506**, L19–L22 (1998). doi:10.1086/311626, astro-ph/9808151

Forbes, J.C., Krumholz, M.R., Goldbaum, N.J., Dekel, A.: Suppression of star formation in dwarf galaxies by photoelectric grain heating feedback. Nature **535**, 523–525 (2016). doi:10.1038/nature18292, 1605.00650

Freeman, K.C.: On the disks of spiral and so galaxies. Astrophys. J. **160**, 811 (1970). doi:10.1086/150474

Fumagalli, M., da Silva, R.L., Krumholz, M.R.: Stochastic star formation and a (nearly) uniform stellar initial mass function. Astrophys. J. **741**, L26 (2011). doi:10.1088/2041-8205/741/2/L26, 1105.6101

Gallart, C., Stetson, P.B., Meschin, I.P., Pont, F., Hardy, E.: Outside-in disk evolution in the large magellanic cloud. Astrophys. J. **682**, L89 (2008). doi:10.1086/590552, 0806.3669

Ghosh, S., Saini, T.D., Jog, C.J.: Effect of dark matter halo on global spiral modes in galaxies. Mon. Not. R. Astron. Soc. **456**, 943–950 (2016). doi:10.1093/mnras/stv2652, 1511.05536

Gil de Paz, A., Madore, B.F., Boissier, S., Swaters, R., Popescu, C.C., Tuffs, R.J., Sheth, K., Kennicutt, R.C. Jr., Bianchi, L., Thilker, D., Martin, D.C.: Discovery of an extended ultraviolet disk in the nearby galaxy NGC 4625. Astrophys. J. **627**, L29–L32 (2005). doi:10.1086/432054, astro-ph/0506357

Glover, S.C.O., Clark, P.C.: Star formation in metal-poor gas clouds. Mon. Not. R. Astron. Soc. **426**, 377–388 (2012). doi:10.1111/j.1365-2966.2012.21737.x, 1203.4251

Grossi, M., Hwang, N., Corbelli, E., Giovanardi, C., Okamoto, S., Arimoto, N.: Stellar structures in the outer regions of M 33. Astron. Astrophys. **533**, A91 (2011). doi:10.1051/0004-6361/201117019, 1106.4704

Gutiérrez, L., Erwin, P., Aladro, R., Beckman, J.E.: The outer disks of early-type galaxies. II. Surface-brightness profiles of unbarred galaxies and trends with hubble type. Astron. J. **142**, 145 (2011). doi:10.1088/0004-6256/142/5/145, 1108.3662

Head, J.T.C.G., Lucey, J.R., Hudson, M.J.: Beyond Sérsic + exponential disc morphologies in the coma cluster. Mon. Not. R. Astron. Soc. **453**, 3729–3753 (2015). doi:10.1093/mnras/stv1662, 1507.07930

Herpich, J., Stinson, G.S., Dutton, A.A., Rix, H.W., Martig, M., Roškar, R., Macciò, A.V., Quinn, T.R., Wadsley, J.: How to bend galaxy disc profiles: the role of halo spin. Mon. Not. R. Astron. Soc. **448**, L99–L103 (2015). doi:10.1093/mnrasl/slv006, 1501.01960

Herrmann, K.A., Hunter, D.A., Elmegreen, B.G.: Surface brightness profiles of dwarf galaxies. I. Profiles and statistics. Astron. J. **146**, 104 (2013). doi:10.1088/0004-6256/146/5/104, 1309.0004

Herrmann, K.A., Hunter, D.A., Elmegreen, B.G.: Surface brightness profiles of dwarf galaxies. II. Color trends and mass profiles. Astron. J. **151**, 145 (2016). doi:10.3847/0004-6256/151/6/145, 1606.00867

Hu, C.Y., Naab, T., Walch, S., Glover, S.C.O., Clark, P.C.: Star formation and molecular hydrogen in dwarf galaxies: a non-equilibrium view. Mon. Not. R. Astron. Soc. **458**, 3528–3553 (2016). doi:10.1093/mnras/stw544, 1510.05644

Hunter, D.A., Elmegreen, B.G., Ludka, B.C.: Galex ultraviolet imaging of dwarf galaxies and star formation rates. Astron. J. **139**, 447–475 (2010). doi:10.1088/0004-6256/139/2/447, 0911.4319

Hunter, D.A., Elmegreen, B.G., Oh, S.H., Anderson, E., Nordgren, T.E., Massey, P., Wilsey, N., Riabokin, M.: The outer disks of dwarf irregular galaxies. Astron. J. **142**, 121 (2011). doi:10.1088/0004-6256/142/4/121, 1107.5587

Hunter, D.A., Ficut-Vicas, D., Ashley, T., Brinks, E., Cigan, P., Elmegreen, B.G., Heesen, V., Herrmann, K.A., Johnson, M., Oh, S.H., Rupen, M.P., Schruba, A., Simpson, C.E., Walter, F., Westpfahl, D.J., Young, L.M., Zhang, H.X.: Little things. Astron. J. **144**, 134 (2012). doi:10.1088/0004-6256/144/5/134, 1208.5834

Hunter, D.A., Elmegreen, B.G., Rubin, V.C., Ashburn, A., Wright, T., Józsa, G.I.G., Struve, C.: Star formation in two luminous spiral galaxies. Astron. J. **146**, 92 (2013). doi:10.1088/0004-6256/146/4/92, 1307.7116

Hunter, D.A., Elmegreen, B.G., Gehret, E.: Young star clusters in the outer disks of LITTLE THINGS dwarf irregular galaxies. Astron. J. **151**, 136 (2016). doi:10.3847/0004-6256/151/6/136, 1603.00495

Hunter, D.A., Ficut-Vicas, D., Ashley, T., Brinks, E., Cigan, P., Elmegreen, B.G., Heesen, V., Herrmann, K.A., Johnson, M., Oh, S.H., Rupen, M.P., Schruba, A., Simpson, C.E., Walter, F., Westpfahl, D.J., Young, L.M., Zhang, H.X.: Erratum: little things. Astron. J. **153**, 47 (2017). doi:10.3847/1538-3881/153/1/47

Izotov, Y.I., Orlitová, I., Schaerer, D., Thuan, T.X., Verhamme, A., Guseva, N.G., Worseck, G.: Eight per cent leakage of Lyman continuum photons from a compact, star-forming dwarf galaxy. Nature **529**, 178–180 (2016a). doi:10.1038/nature16456

Izotov, Y.I., Schaerer, D., Thuan, T.X., Worseck, G., Guseva, N.G., Orlitová, I., Verhamme, A.: Detection of high Lyman continuum leakage from four low-redshift compact star-forming galaxies. Mon. Not. R. Astron. Soc. **461**, 3683–3701 (2016b). doi:10.1093/mnras/stw1205, 1605.05160

Kennicutt, R.C. Jr.: The star formation law in galactic disks. Astrophys. J. **344**, 685–703 (1989). doi:10.1086/167834

Kennicutt, R.C. Jr., Lee, J.C., Funes, S.J.J.G., Sakai, S., Akiyama, S.: An Hα imaging survey of galaxies in the local 11 Mpc volume. Astrophys. J. Suppl. Ser. **178**, 247–279 (2008). doi:10.1086/590058, 0807.2035

Kregel, M., van der Kruit, P.C.: Radial truncations in stellar discs in galaxies. Mon. Not. R. Astron. Soc. **355**, 143–146 (2004). doi:10.1111/j.1365-2966.2004.08307.x, astro-ph/0408125

Kregel, M., van der Kruit, P.C., de Grijs, R.: Flattening and truncation of stellar discs in edge-on spiral galaxies. Mon. Not. R. Astron. Soc. **334**, 646–668 (2002). doi:10.1046/j.1365-8711.2002.05556.x, astro-ph/0204154

Krumholz, M.R.: Star formation in atomic gas. Astrophys. J. **759**, 9 (2012). doi:10.1088/0004-637X/759/1/9, 1208.1504

Krumholz, M.R.: The star formation law in molecule-poor galaxies. Mon. Not. R. Astron. Soc. **436**, 2747–2762 (2013). doi:10.1093/mnras/stt1780, 1309.5100

Krumm, N., Burstein, D.: The extended hydrogen envelope around the dwarf galaxy DDO 154. Astron. J. **89**, 1319–1326 (1984). doi:10.1086/113630

Laine, J., Laurikainen, E., Salo, H., Comerón, S., Buta, R.J., Zaritsky, D., Athanassoula, E., Bosma, A., Muñoz-Mateos, J.C., Gadotti, D.A., Hinz, J.L., Erroz-Ferrer, S., Gil de Paz, A., Kim, T., Menéndez-Delmestre, K., Mizusawa, T., Regan, M.W., Seibert, M., Sheth, K.: Morphology and environment of galaxies with disc breaks in the S^4G and NIRS0S. Mon. Not. R. Astron. Soc. **441**, 1992–2012 (2014). doi:10.1093/mnras/stu628, 1404.0559

Lee, J.C., Gil de Paz, A., Tremonti, C., Kennicutt, R.C. Jr., Salim, S., Bothwell, M., Calzetti, D., Dalcanton, J., Dale, D., Engelbracht, C., Funes, S.J.J.G., Johnson, B., Sakai, S., Skillman, E., van Zee, L., Walter, F., Weisz, D.: Comparison of Hα and UV star formation rates in the local volume: systematic discrepancies for dwarf galaxies. Astrophys. J. **706**, 599–613 (2009). doi:10.1088/0004-637X/706/1/599, 0909.5205

Lee, J.C., Gil de Paz, A., Kennicutt, R.C. Jr., Bothwell, M., Dalcanton, J., Funes, S.J.J.G., Johnson, B.D., Sakai, S., Skillman, E., Tremonti, C., van Zee, L.: A galex ultraviolet imaging survey of galaxies in the local volume. Astrophys. J. Suppl. Ser. **192**, 6 (2011). doi:10.1088/0067-0049/192/1/6, 1009.4705

Lee, J.C., Veilleux, S., McDonald, M., Hilbert, B.: A deeper look at faint Hα emission in nearby dwarf galaxies. Astrophys. J. **817**, 177 (2016). doi:10.3847/0004-637X/817/2/177, 1601.00201

Leitet, E., Bergvall, N., Hayes, M., Linné, S., Zackrisson, E.: Escape of lyman continuum radiation from local galaxies. Detection of leakage from the young starburst Tol 1247-232. Astron. Astrophys. **553**, A106 (2013). doi:10.1051/0004-6361/201118370, 1302.6971

Lemonias, J.J., Schiminovich, D., Thilker, D., Wyder, T.K., Martin, D.C., Seibert, M., Treyer, M.A., Bianchi, L., Heckman, T.M., Madore, B.F., Rich, R.M.: The space density of extended ultraviolet (XUV) disks in the local universe and implications for gas accretion onto galaxies. Astrophys. J. **733**, 74 (2011). doi:10.1088/0004-637X/733/2/74, 1104.4501

Leroy, A.K., Walter, F., Brinks, E., Bigiel, F., de Blok, W.J.G., Madore, B., Thornley, M.D.: The star formation efficiency in nearby galaxies: measuring where gas forms stars effectively. Astron. J. **136**, 2782–2845 (2008). doi:10.1088/0004-6256/136/6/2782, 0810.2556

Mac Low, M.M., Glover, S.C.O.: The abundance of molecular hydrogen and its correlation with midplane pressure in galaxies: non-equilibrium, turbulent, chemical models. Astrophys. J. **746**, 135 (2012). doi:10.1088/0004-637X/746/2/135, 1011.3054

MacArthur, L.A., Courteau, S., Holtzman, J.A.: Structure of disk-dominated galaxies. I. Bulge/disk parameters, simulations, and secular evolution. Astrophys. J. **582**, 689–722 (2003). doi:10.1086/344506, astro-ph/0208404

Maltby, D.T., Gray, M.E., Aragón-Salamanca, A., Wolf, C., Bell, E.F., Jogee, S., Häußler, B., Barazza, F.D., Böhm, A., Jahnke, K.: The environmental dependence of the structure of outer galactic discs in STAGES spiral galaxies. Mon. Not. R. Astron. Soc. **419**, 669–686 (2012). doi:10.1111/j.1365-2966.2011.19727.x, 1108.6206

Maltby, D.T., Aragón-Salamanca, A., Gray, M.E., Hoyos, C., Wolf, C., Jogee, S., Böhm, A.: The environmental dependence of the structure of galactic discs in STAGES S0 galaxies: implications for S0 formation. Mon. Not. R. Astron. Soc. **447**, 1506–1530 (2015). doi:10.1093/mnras/stu2536, 1412.3167

Martig, M., Minchev, I., Flynn, C.: Dissecting simulated disc galaxies – I. The structure of mono-age populations. Mon. Not. R. Astron. Soc. **442**, 2474–2486 (2014). doi:10.1093/mnras/stu1003, 1405.1726

Martin, D.C., Fanson, J., Schiminovich, D., Morrissey, P., Friedman, P.G., Barlow, T.A., Conrow, T., Grange, R., Jelinsky, P.N., Milliard, B., Siegmund, O.H.W., Bianchi, L., Byun, Y.I., Donas, J., Forster, K., Heckman, T.M., Lee, Y.W., Madore, B.F., Malina, R.F., Neff, S.G., Rich, R.M., Small, T., Surber, F., Szalay, A.S., Welsh, B., Wyder, T.K.: The galaxy evolution explorer: a space ultraviolet survey mission. Astrophys. J. **619**, L1–L6 (2005). doi:10.1086/426387, astro-ph/0411302

Martínez-Serrano, F.J., Serna, A., Doménech-Moral, M., Domínguez-Tenreiro, R.: Disk galaxies with broken luminosity profiles from cosmological simulations. Astrophys. J. **705**, L133–L137 (2009). doi:10.1088/0004-637X/705/2/L133, 0906.1118

Melena, N.W., Elmegreen, B.G., Hunter, D.A., Zernow, L.: Bright ultraviolet regions and star formation characteristics in nearby dwarf galaxies. Astron. J. **138**, 1203–1229 (2009). doi:10.1088/0004-6256/138/5/1203, 0908.2837

Meschin, I., Gallart, C., Aparicio, A., Hidalgo, S.L., Monelli, M., Stetson, P.B., Carrera, R.: Spatially resolved LMC star formation history – I. Outside in evolution of the outer LMC disc. Mon. Not. R. Astron. Soc. **438**, 1067–1080 (2014). doi:10.1093/mnras/stt2220, 1312.2584

Mestel, L.: On the galactic law of rotation. Mon. Not. R. Astron. Soc. **126**, 553 (1963). doi:10.1093/mnras/126.6.553

Meurer, G.R., Wong, O.I., Kim, J.H., Hanish, D.J., Heckman, T.M., Werk, J., Bland-Hawthorn, J., Dopita, M.A., Zwaan, M.A., Koribalski, B., Seibert, M., Thilker, D.A., Ferguson, H.C., Webster, R.L., Putman, M.E., Knezek, P.M., Doyle, M.T., Drinkwater, M.J., Hoopes, C.G., Kilborn, V.A., Meyer, M., Ryan-Weber, E.V., Smith, R.C., Staveley-Smith, L.: Evidence for a nonuniform initial mass function in the local universe. Astrophys. J. **695**, 765–780 (2009). doi:10.1088/0004-637X/695/1/765, 0902.0384

Michałowski, M.J., Gentile, G., Hjorth, J., Krumholz, M.R., Tanvir, N.R., Kamphuis, P., Burlon, D., Baes, M., Basa, S., Berta, S., Castro Cerón, J.M., Crosby, D., D'Elia, V., Elliott, J., Greiner, J., Hunt, L.K., Klose, S., Koprowski, M.P., Le Floc'h, E., Malesani, D., Murphy, T., Nicuesa Guelbenzu, A., Palazzi, E., Rasmussen, J., Rossi, A., Savaglio, S., Schady, P., Sollerman, J., de Ugarte Postigo, A., Watson, D., van der Werf, P., Vergani, S.D., Xu, D.: Massive stars formed in atomic hydrogen reservoirs: H I observations of gamma-ray burst host galaxies. Astron. Astrophys. **582**, A78 (2015). doi:10.1051/0004-6361/201526542, 1508.03094

Mihos, J.C., Harding, P., Spengler, C.E., Rudick, C.S., Feldmeier, J.J.: The extended optical disk of M101. Astrophys. J. **762**, 82 (2013). doi:10.1088/0004-637X/762/2/82, 1211.3095

Minchev, I., Famaey, B., Quillen, A.C., Di Matteo, P., Combes, F., Vlajić, M., Erwin, P., Bland-Hawthorn, J.: Evolution of galactic discs: multiple patterns, radial migration, and disc outskirts. Astron. Astrophys. **548**, A126 (2012). doi:10.1051/0004-6361/201219198, 1203.2621

Minchev, I., Martig, M., Streich, D., Scannapieco, C., de Jong, R.S., Steinmetz, M.: On the formation of galactic thick disks. Astrophys. J. **804**, L9 (2015). doi:10.1088/2041-8205/804/1/L9, 1502.06606

Mo, H.J., Mao, S., White, S.D.M.: The formation of galactic discs. Mon. Not. R. Astron. Soc. **295**, 319–336 (1998). doi:10.1046/j.1365-8711.1998.01227.x, astro-ph/9707093

Moffett, A.J., Kannappan, S.J., Baker, A.J., Laine, S.: Extended ultraviolet disks and ultraviolet-bright disks in low-mass E/S0 galaxies. Astrophys. J. **745**, 34 (2012). doi:10.1088/0004-637X/745/1/34, 1111.0959

Muñoz-Mateos, J.C., Gil de Paz, A., Boissier, S., Zamorano, J., Jarrett, T., Gallego, J., Madore, B.F.: Specific star formation rate profiles in nearby spiral galaxies: quantifying the inside-out formation of disks. Astrophys. J. **658**, 1006–1026 (2007). doi:10.1086/511812, astro-ph/0612017

Muñoz-Mateos, J.C., Sheth, K., Gil de Paz, A., Meidt, S., Athanassoula, E., Bosma, A., Comerón, S., Elmegreen, D.M., Elmegreen, B.G., Erroz-Ferrer, S., Gadotti, D.A., Hinz, J.L., Ho, L.C., Holwerda, B., Jarrett, T.H., Kim, T., Knapen, J.H., Laine, J., Laurikainen, E., Madore, B.F., Menendez-Delmestre, K., Mizusawa, T., Regan, M., Salo, H., Schinnerer, E., Seibert, M., Skibba, R., Zaritsky, D.: The impact of bars on disk breaks as probed by S^4G imaging. Astrophys. J. **771**, 59 (2013). doi:10.1088/0004-637X/771/1/59, 1304.6083

Muñoz-Mateos, J.C., Sheth, K., Regan, M., Kim, T., Laine, J., Erroz-Ferrer, S., Gil de Paz, A., Comeron, S., Hinz, J., Laurikainen, E., Salo, H., Athanassoula, E., Bosma, A., Bouquin, A.Y.K., Schinnerer, E., Ho, L., Zaritsky, D., Gadotti, D.A., Madore, B., Holwerda, B., Menéndez-Delmestre, K., Knapen, J.H., Meidt, S., Querejeta, M., Mizusawa, T., Seibert, M., Laine, S., Courtois, H.: The Spitzer survey of stellar structure in galaxies (S^4G): stellar masses, sizes, and radial profiles for 2352 nearby galaxies. Astrophys. J. Suppl. Ser. **219**, 3 (2015). doi:10.1088/0067-0049/219/1/3, 1505.03534

Ostriker, E.C., McKee, C.F., Leroy, A.K.: Regulation of star formation rates in multiphase galactic disks: a thermal/dynamical equilibrium model. Astrophys. J. **721**, 975–994 (2010). doi:10.1088/0004-637X/721/2/975, 1008.0410

Pan, Z., Li, J., Lin, W., Wang, J., Fan, L., Kong, X.: From outside-in to inside-out: galaxy assembly mode depends on stellar mass. Astrophys. J. **804**, L42 (2015). doi:10.1088/2041-8205/804/2/L42, 1504.06821

Pérez, I.: Truncation of stellar disks in galaxies at z 1. Astron. Astrophys. **427**, L17–L20 (2004). doi:10.1051/0004-6361:200400090, astro-ph/0410250

Piontek, R.A., Ostriker, E.C.: Saturated-state turbulence and structure from thermal and magnetorotational instability in the ISM: three-dimensional numerical simulations. Astrophys. J. **629**, 849–864 (2005). doi:10.1086/431549, astro-ph/0504669

Pohlen, M., Trujillo, I.: The structure of galactic disks. Studying late-type spiral galaxies using SDSS. Astron. Astrophys. **454**, 759–772 (2006). doi:10.1051/0004-6361:20064883, astro-ph/0603682

Pohlen, M., Dettmar, R.J., Lütticke, R., Aronica, G.: Outer edges of face-on spiral galaxies. Deep optical imaging of NGC 5923, UGC 9837 and NGC 5434. Astron. Astrophys. **392**, 807–816 (2002). doi:10.1051/0004-6361:20020994

Portas, A.M.P.: From giants to dwarfs: probing the edges of galaxies. Ph.D. thesis, University of Hertfordshire (2010)

Radburn-Smith, D.J., Roškar, R., Debattista, V.P., Dalcanton, J.J., Streich, D., de Jong, R.S., Vlajić, M., Holwerda, B.W., Purcell, C.W., Dolphin, A.E., Zucker, D.B.: Outer-disk populations in NGC 7793: evidence for stellar radial migration. Astrophys. J. **753**, 138 (2012). doi: 10.1088/0004-637X/753/2/138, 1206.1057

Relaño, M., Kennicutt, R.C. Jr., Eldridge, J.J., Lee, J.C., Verley, S.: On how leakage can affect the star formation rate estimation using Hα luminosity. Mon. Not. R. Astron. Soc. **423**, 2933–2940 (2012). doi:10.1111/j.1365-2966.2012.21107.x, 1204.4502

Robertson, B., Yoshida, N., Springel, V., Hernquist, L.: Disk galaxy formation in a Λ cold dark matter universe. Astrophys. J. **606**, 32–45 (2004). doi:10.1086/382871, astro-ph/0401252

Roediger, J.C., Courteau, S., Sánchez-Blázquez, P., McDonald, M.: Stellar populations and radial migrations in virgo disk galaxies. Astrophys. J. **758**, 41 (2012). doi:10.1088/0004-637X/758/1/41, 1201.6361

Rosales-Ortega, F.F., Sánchez, S.F., Iglesias-Páramo, J., Díaz, A.I., Vílchez, J.M., Bland-Hawthorn, J., Husemann, B., Mast, D.: A new scaling relation for H II regions in spiral galaxies: unveiling the true nature of the mass-metallicity relation. Astrophys. J. **756**, L31 (2012). doi:10.1088/2041-8205/756/2/L31, 1207.6216

Roškar, R., Debattista, V.P., Stinson, G.S., Quinn, T.R., Kaufmann, T., Wadsley, J.: Beyond inside-out growth: formation and evolution of disk outskirts. Astrophys. J. **675**, L65 (2008). doi:10.1086/586734, 0710.5523

Ruiz-Lara, T., Pérez, I., Florido, E., Sánchez-Blázquez, P., Méndez-Abreu, J., Lyubenova, M., Falcón-Barroso, J., Sánchez-Menguiano, L., Sánchez, S.F., Galbany, L., García-Benito, R., González Delgado, R.M., Husemann, B., Kehrig, C., López-Sánchez, Á.R., Marino, R.A., Mast, D., Papaderos, P., van de Ven, G., Walcher, C.J., Zibetti, S., CALIFA Team: No direct coupling between bending of galaxy disc stellar age and light profiles. Mon. Not. R. Astron. Soc. **456**, L35–L39 (2016). doi:10.1093/mnrasl/slv174, 1511.03499

Rutkowski, M.J., Scarlata, C., Haardt, F., Siana, B., Henry, A., Rafelski, M., Hayes, M., Salvato, M., Pahl, A.J., Mehta, V., Beck, M., Malkan, M., Teplitz, H.I.: Lyman continuum escape fraction of star-forming dwarf galaxies at z 1. Astrophys. J. **819**, 81 (2016). doi:10.3847/0004-637X/819/1/81, 1511.01998

Safranek-Shrader, C., Krumholz, M.R., Kim, C.G., Ostriker, E.C., Klein, R.I., Li, S., McKee, C.F., Stone, J.M.: Chemistry and radiative shielding in star forming galactic disks. ArXiv e-prints 1605.07618 (2016)

Saha, A., Olszewski, E.W., Brondel, B., Olsen, K., Knezek, P., Harris, J., Smith, C., Subramaniam, A., Claver, J., Rest, A., Seitzer, P., Cook, K.H., Minniti, D., Suntzeff, N.B.: First results from the NOAO survey of the outer limits of the magellanic clouds. Astron. J. **140**, 1719–1738 (2010). doi:10.1088/0004-6256/140/6/1719, 1008.3727

Sánchez-Blázquez, P., Courty, S., Gibson, B.K., Brook, C.B.: The origin of the light distribution in spiral galaxies. Mon. Not. R. Astron. Soc. **398**, 591–606 (2009). doi:10.1111/j.1365-2966.2009.15133.x, 0905.4579

Sandin, C.: The influence of diffuse scattered light. II. Observations of galaxy haloes and thick discs and hosts of blue compact galaxies. Astron. Astrophys. **577**, A106 (2015). doi:10.1051/0004-6361/201425168, 1502.07244

Sanna, N., Bono, G., Stetson, P.B., Ferraro, I., Monelli, M., Nonino, M., Prada Moroni, P.G., Bresolin, R., Buonanno, R., Caputo, F., Cignoni, M., Degl'Innocenti, S., Iannicola, G., Matsunaga, N., Pietrinferni, A., Romaniello, M., Storm, J., Walker, A.R.: On the radial extent of the dwarf irregular galaxy IC10. Astrophys. J. **722**, L244–L249 (2010). doi:10.1088/2041-8205/722/2/L244, 1009.3917

Schaye, J.: Star formation thresholds and galaxy edges: why and where. Astrophys. J. **609**, 667–682 (2004). doi:10.1086/421232, astro-ph/0205125

Schaye, J., Crain, R.A., Bower, R.G., Furlong, M., Schaller, M., Theuns, T., Dalla Vecchia, C., Frenk, C.S., McCarthy, I.G., Helly, J.C., Jenkins, A., Rosas-Guevara, Y.M., White, S.D.M., Baes, M., Booth, C.M., Camps, P., Navarro, J.F., Qu, Y., Rahmati, A., Sawala, T., Thomas, P.A., Trayford, J.: The EAGLE project: simulating the evolution and assembly of galaxies and their environments. Mon. Not. R. Astron. Soc. **446**, 521–554 (2015). doi:10.1093/mnras/stu2058, 1407.7040

Sellwood, J.A., Binney, J.J.: Radial mixing in galactic discs. Mon. Not. R. Astron. Soc. **336**, 785–796 (2002). doi:10.1046/j.1365-8711.2002.05806.x, astro-ph/0203510

Shostak, G.S., van der Kruit, P.C.: Studies of nearly face-on spiral galaxies. II – H I synthesis observations and optical surface photometry of NGC 628. Astron. Astrophys. **132**, 20–32 (1984)

Stinson, G.S., Bovy, J., Rix, H.W., Brook, C., Roškar, R., Dalcanton, J.J., Macciò, A.V., Wadsley, J., Couchman, H.M.P., Quinn, T.R.: MaGICC thick disc – I. Comparing a simulated disc formed with stellar feedback to the Milky Way. Mon. Not. R. Astron. Soc. **436**, 625–634 (2013). doi:10.1093/mnras/stt1600, 1301.5318

Thilker, D.A., Bianchi, L., Meurer, G., Gil de Paz, A., Boissier, S., Madore, B.F., Boselli, A., Ferguson, A.M.N., Muñoz-Mateos, J.C., Madsen, G.J., Hameed, S., Overzier, R.A., Forster, K., Friedman, P.G., Martin, D.C., Morrissey, P., Neff, S.G., Schiminovich, D., Seibert, M., Small, T., Wyder, T.K., Donas, J., Heckman, T.M., Lee, Y.W., Milliard, B., Rich, R.M., Szalay, A.S., Welsh, B.Y., Yi, S.K.: A search for extended ultraviolet disk (XUV-disk) galaxies in the local universe. Astrophys. J. Suppl. Ser. **173**, 538–571 (2007). doi:10.1086/523853, 0712.3555

Toomre, A.: On the gravitational stability of a disk of stars. Astrophys. J. **139**, 1217–1238 (1964). doi:10.1086/147861

Treyer, M., Wyder, T., Neill, J., Seibert, M., Lee, J. (eds.): UV/Hα turmoil. In: Astronomical Society of the Pacific Conference Series, vol. 440 (2011), 1011.2181

van der Kruit, P.C.: Truncations in stellar disks. In: Funes, J.G., Corsini, E.M. (eds.) Galaxy Disks and Disk Galaxies, Astronomical Society of the Pacific Conference Series, vol. 230, pp. 119–126 (2001). astro-ph/0010354

van der Kruit, P.C., Shostak, G.S.: Studies of nearly face-on spiral galaxies. I – the velocity dispersion of the H I gas in NGC 3938. Astron. Astrophys. **105**, 351–358 (1982)

van Dokkum, P.G., Abraham, R., Merritt, A.: First results from the dragonfly telephoto array: the apparent lack of a stellar halo in the massive spiral galaxy M101. Astrophys. J. **782**, L24 (2014). doi:10.1088/2041-8205/782/2/L24, 1401.5467

Vlajić, M., Bland-Hawthorn, J., Freeman, K.C.: The structure and metallicity gradient in the extreme outer disk of NGC 7793. Astrophys. J. **732**, 7 (2011). doi:10.1088/0004-637X/732/1/7, 1101.0607

Vogelsberger, M., Genel, S., Springel, V., Torrey, P., Sijacki, D., Xu, D., Snyder, G., Nelson, D., Hernquist, L.: Introducing the Illustris Project: simulating the coevolution of dark and visible matter in the Universe. Mon. Not. R. Astron. Soc. **444**, 1518–1547 (2014). doi:10.1093/mnras/stu1536, 1405.2921

Walter, F., Brinks, E., de Blok, W.J.G., Bigiel, F., Kennicutt, R.C. Jr., Thornley, M.D., Leroy, A.: THINGS: the H I nearby galaxy survey. Astron. J. **136**, 2563–2647 (2008). doi:10.1088/0004-6256/136/6/2563, 0810.2125

Wang, J., Fu, J., Aumer, M., Kauffmann, G., Józsa, G.I.G., Serra, P., Huang, M.L., Brinchmann, J., van der Hulst, T., Bigiel, F.: An observational and theoretical view of the radial distribution of H I gas in galaxies. Mon. Not. R. Astron. Soc. **441**, 2159–2172 (2014). doi:10.1093/mnras/stu649, 1401.8164

Watkins, A.E., Mihos, J.C., Harding, P.: The red and featureless outer disks of nearby spiral galaxies. Astrophys. J. **826**, 59 (2016). doi:10.3847/0004-637X/826/1/59, 1605.05183

Weiner, B.J., Williams, T.B., van Gorkom, J.H., Sellwood, J.A.: The disk and dark halo mass of the barred galaxy NGC 4123. I. Observations. Astrophys. J. **546**, 916–930 (2001). doi:10.1086/318288, astro-ph/0008204

Weisz, D.R., Johnson, B.D., Johnson, L.C., Skillman, E.D., Lee, J.C., Kennicutt, R.C., Calzetti, D., van Zee, L., Bothwell, M.S., Dalcanton, J.J., Dale, D.A., Williams, B.F.: Modeling the effects of star formation histories on Hα and ultraviolet fluxes in nearby dwarf galaxies. Astrophys. J. **744**, 44 (2012). doi:10.1088/0004-637X/744/1/44, 1109.2905

White, S.D.M., Frenk, C.S.: Galaxy formation through hierarchical clustering. Astrophys. J. **379**, 52–79 (1991). doi:10.1086/170483

Williams, B.F., Dalcanton, J.J., Dolphin, A.E., Holtzman, J., Sarajedini, A.: The detection of inside-out disk growth in M33. Astrophys. J. **695**, L15–L19 (2009). doi:10.1088/0004-637X/695/1/L15, 0902.3460

Yoachim, P., Roškar, R., Debattista, V.P.: Spatially resolved spectroscopic star formation histories of nearby disks: hints of stellar migration. Astrophys. J. **752**, 97 (2012). doi:10.1088/0004-637X/752/2/97, 1204.0026

Younger, J.D., Cox, T.J., Seth, A.C., Hernquist, L.: Antitruncated stellar disks via minor mergers. Astrophys. J. **670**, 269–278 (2007). doi: 10.1086/521976, 0707.4481

Zaritsky, D., Salo, H., Laurikainen, E., Elmegreen, D., Athanassoula, E., Bosma, A., Comerón, S., Erroz-Ferrer, S., Elmegreen, B., Gadotti, D.A., Gil de Paz, A., Hinz, J.L., Ho, L.C., Holwerda, B.W., Kim, T., Knapen, J.H., Laine, J., Laine, S., Madore, B.F., Meidt, S., Menendez-Delmestre, K., Mizusawa, T., Muñoz-Mateos, J.C., Regan, M.W., Seibert, M., Sheth, K.: On the origin of lopsidedness in galaxies as determined from the Spitzer survey of stellar structure in galaxies (S^4G). Astrophys. J. **772**, 135 (2013). doi:10.1088/0004-637X/772/2/135, 1305.2940

Zastrow, J., Oey, M.S., Veilleux, S., McDonald, M.: New constraints on the escape of ionizing photons from starburst galaxies using ionization-parameter mapping. Astrophys. J. **779**, 76 (2013). doi:10.1088/0004-637X/779/1/76, 1311.2227

Zhang, H.X., Hunter, D.A., Elmegreen, B.G., Gao, Y., Schruba, A.: Outside-in shrinking of the star-forming disk of dwarf irregular galaxies. Astron. J. **143**, 47 (2012). doi:10.1088/0004-6256/143/2/47, 1111.3363

Zheng, Z., Thilker, D.A., Heckman, T.M., Meurer, G.R., Burgett, W.S., Chambers, K.C., Huber, M.E., Kaiser, N., Magnier, E.A., Metcalfe, N., Price, P.A., Tonry, J.L., Wainscoat, R.J., Waters, C.: The structure and stellar content of the outer disks of galaxies: a new view from the pan-STARRS1 medium deep survey. Astrophys. J. **800**, 120 (2015). doi:10.1088/0004-637X/800/2/120, 1412.3209

Chapter 5
Metallicities in the Outer Regions of Spiral Galaxies

Fabio Bresolin

Abstract The analysis of the chemical composition of galaxies provides fundamental insights into their evolution. This holds true also in the case of the outer regions of spiral galaxies. This chapter presents the observational data, accumulated in the past few years mostly from the analysis of H II region spectra, concerning the metallicity of the outer disks of spirals that are characterized by extended H I envelopes and low-star formation rates. I present evidence from the literature that the metal radial distribution flattens at large galactocentric distances, with levels of enrichment that exceed those expected given the large gas mass fractions and the weak star formation activity. The interpretation of these results leads to speculations regarding mechanisms of metal mixing in galactic disks and the possibility that metal-enriched gas infall plays a role in determining the chemical evolution of the outskirts of spirals.

5.1 Introduction

The analysis of the chemical abundance composition of galaxies provides essential and unique constraints on their evolutionary status and their star formation properties. Gathering spatially resolved information about the distribution of metals is a well-tested approach to probe not only the metal production in stars across time but also those effects, such as galactic wind outflows, gravitational interactions, secular processes and gas inflows, that can profoundly affect the evolution of galaxies.

This chapter looks at the present-day gas metallicities of outer spiral disks, as derived from the emission-line analysis of H II region spectra, excluding older chemical abundance tracers, such as planetary nebulae and stars, except for a few notable exceptions. The connections between the chemical abundances of the outer disks thus derived and of the circumgalactic medium, probed by resonance absorption lines in the UV, for example, in damped Lyman α systems, are covered elsewhere in this volume (see the review by Chen 2017).

F. Bresolin (✉)
Institute for Astronomy, 2680 Woodlawn Drive, Honolulu, HI 96822, USA
e-mail: bresolin@ifa.hawaii.edu

J.H. Knapen et al. (eds.), *Outskirts of Galaxies*, Astrophysics and Space Science Library 434, DOI 10.1007/978-3-319-56570-5_5

A non-secondary aspect of chemical abundance work in nearby and faraway star-forming galaxies concerns the methodology employed in the measurement of nebular (ionized gas) abundances. Therefore, Sect. 5.2 provides a brief overview of the difficulties and the techniques used. The subsequent sections provide details on the work carried out in a variety of nearby systems (Sects. 5.3 and 5.4), building the framework for interpreting the observed chemical abundance properties (Sect. 5.5). A concise summary concludes the chapter.

5.2 Measuring Nebular Abundances

The measurement of nebular chemical abundances that are free of significant systematic uncertainties remains an unsolved problem in astrophysics, despite decades of observational efforts to investigate the emission-line properties of Galactic and extragalactic H II regions. Many authors (among others, Bresolin et al. 2004; Kewley and Ellison 2008; López-Sánchez et al. 2012) have addressed this issue, showing how the various emission-line diagnostics and the different calibrations proposed in the literature for these diagnostics are afflicted by systematic variations on the derived oxygen abundances that reach up to 0.7 dex.[1] This problem, of course, affects not only the investigation of ionized nebulae in the local Universe, e.g. for the analysis of abundance gradients in spiral galaxies (Vila-Costas and Edmunds 1992; Zaritsky et al. 1994; Kennicutt et al. 2003; Sánchez et al. 2014; Ho et al. 2015; Bresolin and Kennicutt 2015), but also the myriad of studies concerning the chemical composition of star-forming galaxies, notably those at high redshifts (e.g. to investigate the mass-metallicity relation), that rely on the local calibrations and the cosmic evolution of metals (Tremonti et al. 2004; Erb et al. 2006; Maiolino et al. 2008; Mannucci et al. 2010; Zahid et al. 2013; Sanders et al. 2015, to cite only a few).

In order to derive reliable chemical abundances of ionized nebulae, it is necessary to have a good knowledge of the physical conditions of the gas, in particular of the electron temperature T_e, because of the strong temperature sensitivity of the line emissivities of the various ions. An excellent source on the subject of deriving oxygen abundances in ionized nebulae is the monograph published by Stasińska et al. (2012). Nebular electron temperatures can be obtained through the classical, so-called *direct* method (Menzel et al. 1941), utilizing emission lines of the same ions that originate from different excitation levels and in particular from the ratio of the (collisionally excited) auroral [O III] $\lambda 4363$ and nebular [O III] $\lambda\lambda 4959, 5007$ lines. In the case of high-excitation (low-metallicity) extragalactic H II regions

[1]In the literature measuring the *metallicity* of an H II region is equivalent to measuring its *oxygen abundance* O/H, since O constitutes (by number) approximately half of the metals. The standard practice is to report the value of 12 + log(O/H). The Solar value used for reference is taken here to be $12 + \log(\mathrm{O/H})_\odot = 8.69$, from Asplund et al. (2009).

and planetary nebulae, the [O III] λ4363 line is often detected, but it becomes unobservable as the cooling of the gas becomes efficient at high metallicity or whenever the objects are faint, so that properly calibrated *strong-line* abundance diagnostics become necessary in order to infer the oxygen abundances. However, even for nebulae where [O III] λ4363 can be observed, a poorly understood discrepancy exists between the nebular abundances based on the direct method and those obtained from emission-line strengths calibrated via photoionization model grids (McGaugh 1991; Blanc et al. 2015; Vale Asari et al. 2016). A 0.2–0.3 dex discrepancy (T_e-based abundances being lower) is also found when using the weak metal *recombination* lines, in particular the O II lines around 4650 Å, instead of the collisionally excited lines (García-Rojas and Esteban 2007; Esteban et al. 2009; Toribio San Cipriano et al. 2016). On the other hand, comparisons of stellar (B and A supergiants) and nebular chemical compositions in a handful of galaxies (see Bresolin et al. 2009a) provide a generally good agreement when the nebular abundances are calculated from the direct method, at least for subsolar metallicities.

Despite the somewhat unsatisfactory situation illustrated above, we can still derive robust results concerning the metallicities of outer disk H II regions. As will become apparent later on, it is important to highlight two results. Firstly, radial abundance trends in spiral disks are generally found to be qualitatively invariant relative to the selection of nebular abundance diagnostics, although different methods can yield different gradient slopes (see Bresolin et al. 2009a; Arellano-Córdova et al. 2016). Secondly, O/H values that are derived from direct measurements of T_e or from strong-line diagnostics that are calibrated based on [O III] λ 4363 detections lie at the bottom of the possible abundance range, when compared to metallicities derived from other diagnostics, such as those based on theoretical models.

The literature on nebular abundance diagnostics is vast (a recent discussion can be found in Brown et al. 2016), and for the purposes of this review, it is important only to recall a few of the most popular ones and (some of) their respective calibrations:

1. O3N2 $\equiv$ log[([O III] λ5007/Hβ)/([N II] λ6583/Hα)], calibrated empirically (i.e. based on T_e detections in H II regions of nearby galaxies) as given by Pettini and Pagel (2004) and, more recently, by Marino et al. (2013).
2. N2O2 $\equiv$ [N II] λ 6583/[O II] λ3727, calibrated from photoionization models by Kewley and Dopita (2002) and empirically by Bresolin (2007). Bresolin et al. (2009b) showed that these two calibrations yield abundance gradients in spiral disks that have virtually the same slopes, despite a large systematic offset.
3. N2 $\equiv$ [N II] λ6583/Hα, calibrated by Pettini and Pagel (2004) and Marino et al. (2013).
4. R23 $\equiv$ ([O II] λ3727 + [O III] $\lambda\lambda$4959, 5007)/Hβ (Pagel et al. 1979). Many different calibrations have been proposed through the years (e.g. McGaugh 1991; Kobulnicky and Kewley 2004). This diagnostic can be important in order to verify whether the metallicity gradients derived from other strong-line techniques, mostly involving the nitrogen [N II] λ6583 line, are corroborated by

considering only oxygen lines instead. Unfortunately, the use of this indicator for abundance gradient studies is complicated by the non-monotonic behaviour of R23 with oxygen abundance. The simultaneous use of a variety of diagnostics, when the relevant emission lines are available, alleviates this problem. The empirically calibrated P-method (Pilyugin and Thuan 2005) and some related diagnostics (e.g. Pilyugin and Grebel 2016) also make use of both the [O II] and [O III] strong emission lines.

To summarize, a variety of optical emission-line diagnostics are available to derive the metallicities (oxygen abundances) of extragalactic H II regions. Due to poorly known aspects of nebular physics (e.g. temperature fluctuations), absolute metallicities remain a matter of debate, especially at high (near-Solar) values. On the other hand, relative abundances within galaxy disks are quite robust. T_e-based oxygen abundances lie at the bottom of the distribution of values obtained from the set of metallicity diagnostics currently available.

5.3 Chemical Abundances of H II Regions in Outer Disks

While the spectroscopic analysis of H II regions in the inner disks of spiral galaxies is a well-developed activity in extragalactic astronomical research, starting with the pioneering work by Searle (1971) and Shields (1974), who provided the first evidence for the presence of exponential radial abundance gradients in nearby galaxies such as M33 and M101, the investigation of ionized nebulae located beyond the boundary of the main star-forming disk has begun only recently. Observationally, the main difficulty in measuring chemical abundances in these outlying H II regions is represented by their intrinsic faintness. In fact, these nebulae are typically ionized by single hot stars, as deduced from their Hα luminosities (Gil de Paz et al. 2005; Goddard et al. 2010), that are on average about two orders of magnitude fainter than those of the giant H II regions that are routinely observed in the inner disks (Bresolin et al. 2009b; Goddard et al. 2011). This section reviews the investigations of the chemical abundances in the outskirts of nearby spiral galaxies, focussing mostly on oxygen in H II regions. Studies of metals in old stars and nebular nitrogen are also briefly discussed.

5.3.1 Early Work

Spectra of a handful of outlying H II regions in the disks of the late-type spirals NGC 628, NGC 1058 and NGC 6946, known for their extended H I distributions, were first obtained by Ferguson et al. (1998). These authors found that the oxygen abundance gradients measured in the inner disks appear to continue to large galactocentric distances, beyond the isophotal radii[2] R_{25}, and reaching low metallicities, corresponding to 10–15% of the Solar value. Unfortunately, the small sample size concealed the possibility, demonstrated by later work, that the radial metallicity trends could actually be different between inner and outer disks. van Zee et al. (1998) also presented spectra of outlying H II regions in a sample of 13 spirals, but the number of objects observed near R_{25} and beyond was very small.

The oxygen abundance in the outer disks of two iconic representatives of the class of extended UV (XUV) disk galaxies discovered by the *GALEX* satellite, M83 (Thilker et al. 2005) and NGC 4625 (Gil de Paz et al. 2005), was investigated by means of multi-object spectroscopy with the Magellan and Palomar 200-in. telescopes by Gil de Paz et al. (2007). These authors also found a relatively low metal content, around 10–20% of the Solar value, utilizing a combination of photoionization models and the R23 strong-line abundance diagnostic. This study, however, was also limited by the small number of H II regions observed at large galactocentric distances, especially for the galaxy NGC 4625. A feature of the M83 radial abundance gradient that was suggested by Gil de Paz et al. (2007), i.e. a sudden drop in metallicity at a galactocentric distance of 10 kpc, but whose presence was dubious due to the uncertain O/H ratios derived from R23, was also detected later in the larger sample of outlying H II regions studied by Bresolin et al. (2009b). Gil de Paz et al. (2007) remarked that such a sharp decrease, if present, could be the signature of a transition to an outer disk where the star formation efficiency is significantly lower compared to the inner disk.

5.3.2 M83: A Case Study

The first investigation to obtain robust chemical abundances—via a variety of nebular metallicity diagnostics—in the outer disk of a single galaxy was carried out by Bresolin et al. (2009b), who obtained spectra of ionized nebulae in the outer disk of M83 with the ESO Very Large Telescope. Of these H II regions, 32 lie at galactocentric distances larger than the isophotal radius, extending out to 22.3 kpc (2.64 R_{25}) from the galaxy centre.

The principal chemical abundance properties of the outer disks of spiral galaxies, confirmed by subsequent investigations of other targets (as discussed in the next pages), are all showcased in this prototypical XUV disk galaxy (see Fig. 5.1):

- The nearly flat radial gradient beyond R_{25}, contrasting with the steeper exponential decline observed in the inner disk. The gradient slope found for the inner disk

[2] At the isophotal radius R_{25} the surface brightness measures 25 mag arcsec^{-2} and is often reported in the *B* photometric band, as in the Third Reference Catalogue of Bright Galaxies (de Vaucouleurs et al. 1991).

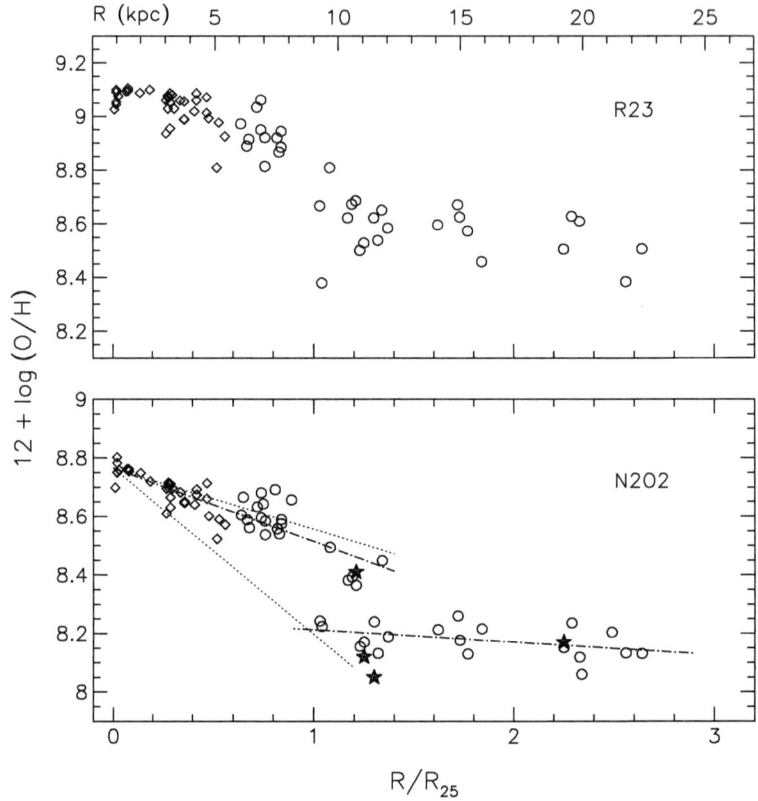

Fig. 5.1 The radial oxygen abundance gradient in M83, determined from two different diagnostics (R23 in the *top panel*, N2O2, as calibrated empirically by Bresolin (2007), in the *bottom panel*), from H II regions located in the inner disk (*open diamond*: Bresolin and Kennicutt 2002; Bresolin et al. 2005) and in the outer disk (*open circle*: Bresolin et al. 2009b). The *star symbol* represents [O III] λ4363-based O/H values. Linear regressions to the radial abundance gradient are shown as separate *dot-dashed lines* for the inner and outer portions of the galactic disks. The *two dotted lines* represent the range of gradient slopes measured by Ho et al. (2015) from a sample of 49 galaxies: $d\log(O/H)/dR = -0.39 \pm 0.18$ dex R_{25}^{-1}. Adapted from the data published by Bresolin et al. (2009b)

using the N2O2 diagnostic

$$\frac{d\log(O/H)}{dR} = -0.25 \pm 0.02 \text{ dex} R_{25}^{-1} \quad (5.1)$$

compares well with the benchmark value of -0.39 ± 0.18 dex R_{25}^{-1} measured by Ho et al. (2015) from a sample of 49 galaxies (the two dotted lines in Fig. 5.1 show the two extreme values of the one-sigma range). On the other hand, in the outer disk, a linear fit to the data displayed in Fig. 5.1 yields a slope

$$\frac{d\log(O/H)}{dR} = -0.04 \pm 0.02 \text{ dex } R_{25}^{-1}, \tag{5.2}$$

i.e. nearly flat. Clearly, the radial behaviour of the gas metallicity differs significantly between the inner, star-forming disk of M83, and the outer disk.

- The relatively high mean O/H value measured at large galactocentric distances. This value depends on the selection of abundance diagnostic, as explained in Sect. 5.2. In the work presented by Bresolin et al. (2009b), it lies in the range $12 + \log(O/H) = 8.2$–8.6, i.e. between $\sim$30% and $\sim$80% of the Solar value, depending on whether the abundances are tied to T_e-based detections (lower value) or whether they are determined from photoionization models (upper value). It is worth pointing out that the few detections of the [O III] λ4363 auroral line lead to O/H ratios (star symbols in Fig. 5.1) that are quite consistent with those determined from the empirically calibrated N2O2 diagnostic.

  This result on the metallicity of the outer disk of M83 is at odds with the expectation that the very outskirts of spiral galaxies are somewhat pristine and chemically unevolved, as would be implied by a simple picture of inside-out galactic formation. I draw attention to the fact that adopting abundance diagnostics that are calibrated via photoionization models, the outer disk of M83 would have a mean metallicity that is nearly Solar. Using a more conservative approach, we can say that the mean metallicity of the outer disk is *at least* 1/3 Solar, based on the summary presented at the end of Sect. 5.2.

  Bresolin et al. (2009b) also pointed out that the extended disk of M83 should be considered chemically over-enriched given its large gas mass fraction (approaching unity) when compared to a closed-box chemical evolution model, the opposite behaviour of what is observed, for example, in dwarf galaxies (Matteucci and Chiosi 1983, see also the explanatory text for Eq. (5.3)). This point will be further discussed in Sect. 5.3.3.

Figure 5.1 also suggests the presence of a $\sim$0.2 dex oxygen abundance discontinuity beyond the isophotal radius. This feature is not confirmed by all abundance diagnostics considered (see also Pilyugin et al. 2012), nor is it detected in other extended disk galaxies, except NGC 4625 (Goddard et al. 2011), but it also appears in the data presented by Gil de Paz et al. (2007). It resembles the break occurring for the α-elements measured for Cepheids in the Milky Way at a galactocentric distance of approximately 9 kpc (Lépine et al. 2014).

The observations in M83 demonstrate that spectroscopy of H II regions located in the extended, gas-rich disks of spiral galaxies allows us to probe the present-day chemical abundances of external galactic disks out to nearly three isophotal radii, equivalent—in the case of M83—to more than 20 kpc. The interesting, and somewhat surprising, result is found that the radial metallicity distribution becomes virtually flat in the extended outskirts, with a value of at least 1/3 of Solar. The extended disk appears to be chemically overabundant for its very large gas mass fraction.

5.3.3 Other Systems

In this section the results obtained from other investigations of single targets or small samples of galaxies, essentially confirming and expanding the general picture outlined in Sect. 5.3.2 for M83, are reviewed. In addition to H II regions as primary probes of the present-day metallicity, information from the older stellar content is included. Some galaxies display a flattening of the gas metallicity already inside the main disk ($R < R_{25}$), perhaps as a result of gas flows induced by the gravitational potential of a stellar bar (Martin and Roy 1995; Zahid and Bresolin 2011; Marino et al. 2012). Here, I will focus on outer disk systems exclusively.

H I-Selected Galaxies The oxygen abundances of outlying H II regions in a sample of 13 H I-selected galaxies were measured using the R23 method by Werk et al. (2011). The sample is dominated by interacting systems and galaxies displaying a disturbed, extended neutral gas morphology. In most cases a flat radial abundance distribution was found across most of the disk of these systems, although the number of H II regions observed per galaxy is sometimes too small to infer variations between the inner and outer disks. Thus, the flattening observed can be of a different nature in this kind of galaxies (see the discussion below) compared with relatively isolated galaxies such as M83, where the flattening is observed to occur only in the outer disk.

Werk et al. (2011) showed that the oxygen abundances in the outskirts of the galaxies included in their sample are considerably higher than expected, given their large gas content compared with the total baryonic (stars + gas) mass. The same result was described by Bresolin et al. (2009b) and Werk et al. (2010) for the extended disks of M83 and the blue compact dwarf galaxy NGC 2915, respectively. While for typical star-forming galaxies the effective oxygen yield, defined by

$$y_{O,\mathrm{eff}} = \frac{Z_O}{\ln(\mu^{-1})}, \tag{5.3}$$

where μ is the gas mass fraction and Z_O is the metallicity equivalent to the O mass fraction,[3] lies below the theoretical oxygen yield for a stellar population, $y_O \simeq 0.007$ (e.g. Kobayashi et al. 2006), in the case of the outer disks, the opposite holds, with $y_{O,\mathrm{eff}} > 0.02$ (Werk et al. 2011; López-Sánchez et al. 2015). This in essence implies that the oxygen abundances measured in the gas located in the outskirts

[3]The gas mass fraction is $\mu = M_{\mathrm{gas}}/(M_{\mathrm{gas}} + M_{\mathrm{stars}})$. The O mass fraction ("metallicity") and the abundance by number are linked by the relation $Z_O = 11.81$ (O/H). The coefficient of proportionality is calculated as $16X$, adopting the hydrogen mass fraction for Solar composition from Asplund et al. (2009).

of these galaxies, including the XUV disk systems, characterized by extended H I envelopes, exceed the values predicted by the closed-box galactic chemical evolution model. How this level of chemical enrichment can be attained, given the low values measured for the star formation rate, will be addressed in Sect. 5.5.

Interacting Systems The majority of the galaxies studied by Werk et al. (2011) are located in interacting systems. Nebular oxygen abundances have been measured in the main star-forming disks and along tidal features of interacting and merging systems by various authors (e.g. Rupke et al. 2010b; Rich et al. 2012; Torres-Flores et al. 2014), finding significantly flatter radial distributions compared to noninteracting systems. These studies are supported by numerical simulations (Rupke et al. 2010a; Torrey et al. 2012), showing that gas flows induced by galaxy interactions redistribute the gas in such a way that the original abundance gradients—present in the galactic disks before the merging process—flatten progressively with merger stage.

This redistribution and radial mixing of metals can take place over very large distances, in extreme cases reaching several tens of kpc. For example, Olave-Rojas et al. (2015) measured the chemical abundances of H II regions located along the main tidal tail of NGC 6845A, part of a compact, interacting group of galaxies, out to almost 70 kpc from the centre (approximately $4\,R_{25}$). The radial oxygen distribution displays a remarkably shallow gradient.

An interesting interacting system is represented by NGC 1512, which is experiencing an encounter with the companion galaxy NGC 1510. The system, an XUV disk galaxy, is embedded in a very extended H I envelope, with a radius of 55 kpc (Koribalski and López-Sánchez 2009), in which low-level star formation is taking place, as shown by the far-UV and Hα emission originating from low-luminosity stellar complexes and associated H II regions. The flat and relatively high O/H abundance values, $12 + \log(\mathrm{O/H}) \simeq 8.3$, out to a galactocentric distance of ~30 kpc, have been studied by Bresolin et al. (2012) and, more recently, by López-Sánchez et al. (2015). The latter authors point out the effect of the interaction on the outlying northern H I spiral arm, where the O/H values have a much larger dispersion than in the opposite side of the galaxy, which remains relatively undisturbed by the ongoing interaction.

Old Stars The data discussed so far refer only to the present-day metallicities, as derived from H II region emission. It is also possible to infer the chemical composition of the outer disks of nearby spirals from stellar photometry of older populations, in particular of red giant branch (RGB) stars. The method requires the photometry of individual stars, and as such has successfully been applied only to nearby systems, out to approximately 3 Mpc. Care must be taken in interpreting these photometric metallicities when discussing disk radial gradients, because of the potential contamination from halo stars.

The metallicity can be derived from a comparison of the observed stellar colours, such as $V-I$, with theoretical stellar tracks, exploiting the fact that these broadband colours are more sensitive to metallicity than age. In this way, Worthey et al. (2005) obtained a flat metallicity gradient in the outer disk of M31, between 20 and

50 kpc from the galaxy centre (1–2.5 R_{25}), with a mean value of [Z/H] $\simeq -0.5$. The published H II region abundances (e.g. Zurita and Bresolin 2012; Sanders et al. 2012) do not extend beyond 25 kpc, and thus whether a similar behaviour is encountered for the younger stellar populations cannot be verified. A flat gradient, with an approximately Solar O/H value, extending out to $\sim$100 kpc, has been reported for planetary nebulae by Balick et al. (2013) and Corradi et al. (2015). These authors attribute this finding to a star formation burst following interactions and merger processes, perhaps related to an encounter with M33, that occurred approximately 3 Gyr ago.

Vlajić et al. (2009, 2011) measured the metallicity of RGB stars from deep g' and i' Gemini photometry in the outer disks of the two Sculptor Group spirals NGC 300 and NGC 7793, out to 15 kpc (2.3 R_{25}) and 11.5 kpc (2.4 R_{25}), respectively. Both galaxies, like M31, display purely exponential surface brightness profiles out to these large galactocentric distances, indicating that the halo contribution is probably negligible. The possible connection between gas-phase metallicity and surface brightness profiles will be briefly discussed in Sect. 5.4.1.

For both NGC 300 and NGC 7793, the stellar metallicity flattens out to an approximately constant value or even slightly increases with radius in the outer disk, in contrast with the exponential decline inferred from H II regions and young stars in the inner disk. The measured metallicity is quite low, [Fe/H] $\simeq -1$ for the outer disk of NGC 300 and [Fe/H] $\simeq -1.5$ (or even lower, depending on the age of the stars) for NGC 7793, but could be compatible with the present-day metallicity of the inner disk if the chemical enrichment due to stellar evolution between the time probed by the RGB stars (8–12 Gyr ago) and the present epoch is taken into account. These results are made somewhat uncertain by the age-metallicity degeneracy, the assumption of a single age for the RGB stars and the potential effects of stellar migration.

Other XUV Disks The chemical abundances of outer H II regions in a few XUV disk galaxies (as defined in the catalogue by Thilker et al. 2007), in addition to M83 and NGC 1512, have been presented by different authors. For convenience, Table 5.1 summarizes these studies. These investigations differ somewhat in spectroscopic depth, abundance diagnostics adopted and radial coverage, but they tend to provide

Table 5.1 Nebular abundance studies in XUV disks

Galaxy	References	Largest radius (R_{25} units)
NGC 628	Rosales-Ortega et al. (2011)[a]	1.7
NGC 1512	Bresolin et al. (2012)	2.2
	López-Sánchez et al. (2012)	2.8
NGC 3621	Bresolin et al. (2012)	2.0
NGC 4625	Goddard et al. (2011)	2.8
NGC 5253 (M83)	Bresolin et al. (2009a,b)	2.6

[a]Data for objects lying beyond R_{25} extracted from Ferguson et al. (1998)

a unified picture regarding the abundance gradients, in particular the presence of a break occurring approximately at the isophotal radius, as a dividing point between the inner disk, characterized by an exponential nebular abundance gradient, and the outer disk, with a shallower or flat abundance gradient.

Not included in Table 5.1 is NGC 3031 (M81), for which a flat outer gradient has been suggested, but this result relies on a very small sample of outlying H II regions (Patterson et al. 2012; Stanghellini et al. 2014). A more recent work by Arellano-Córdova et al. (2016) does not find evidence for a flat gradient out to a galactocentric distance of 33 kpc (2.3 R_{25}). This seems to be consistent with the shallow overall abundance gradient, both in dex kpc^{-1} and normalized to the isophotal radius, which is possibly the consequence of galaxy interactions. It is also worth pointing out that the abundance break observed in NGC 3621 by Bresolin et al. (2012) has been confirmed by the independent spectroscopic analysis of five blue supergiant stars, straddling the isophotal radius, by Kudritzki et al. (2014). The stellar metallicities are intermediate between the nebular metallicities determined from the N2 and R23 diagnostics. Finally, it is important to notice that the sample presented above includes fairly isolated systems (NGC 3621, M83, NGC 628), ruling out the possibility that abundance breaks and significant metal mixing develop only as a consequence of recent galaxy interactions.

The Milky Way Evidence for a flattening of the abundance gradient in the outer disk of the Milky Way comes from observations of various metal tracers, which also sample populations with different ages: Cepheid variables (Korotin et al. 2014), open clusters (Magrini et al. 2009; Yong et al. 2012) and H II regions (Vílchez and Esteban 1996; Esteban et al. 2013). This break appears at a galactocentric distance around 12 kpc, extending outwards to 19–21 kpc, as shown from either Cepheids (Genovali et al. 2015) or open clusters (Carraro et al. 2004). While the flattening in the Cepheid chemical abundances is still somewhat controversial (e.g. Lemasle et al. 2013), the open clusters show a clear bimodal radial gradient in metallicity (Yong et al. 2012), the outer gradient being quite shallow, with a characteristic outer disk metallicity [Fe/H] $\simeq -0.3 \pm 0.1$. Further studies of the behaviour of the radial distribution of the stellar metallicity (and chemical element patterns) in the outer disk of the Galaxy will be important to constrain models of the chemical evolution of the Milky Way and the effects of the corotation resonance and stellar radial migration (Mishurov et al. 2002; Lépine et al. 2011; Korotin et al. 2014).

5.3.4 *Results from Galaxy Surveys*

More recent results about the chemical abundances of the outer disks of spiral galaxies in the nearby Universe have been published as part of relatively large spectroscopic surveys, largely dedicated to the measurement of emission-line abundances of the interstellar medium in the parent galaxies. These surveys, based on 4 m-class telescope observations, do not reach emission-line levels as faint as those probed by some of the single galaxy work illustrated earlier. Therefore, weak lines such as [O III] λ 4363 remain undetected in the low-luminosity H II regions located in the galactic outskirts. In addition, these surveys have provided metallicity information out to $\sim$1–1.5 R_{25}, i.e. to considerably smaller galactocentric distances than possible with 8 m-class facilities (see Table 5.1). On the other hand, the large number of galaxies (hundreds) provides essential statistical information about the properties of the abundance gradients, which are necessary to establish, for instance, how common radial metal distribution breaks are within the general population of spiral galaxies. Furthermore, such surveys also enable the investigation of possible correlations between abundance gradients and galactic attributes, such as mass, star formation rate and structural properties (e.g. the presence or absence of bars).

Integral Field Spectroscopy Sánchez-Menguiano et al. (2016) presented oxygen abundance measurements obtained by the Calar Alto Legacy Integral Field Area (CALIFA) project (Sánchez et al. 2012) in 122 face-on spiral galaxies. Adopting the O3N2 nebular diagnostic, they confirmed earlier results, obtained from the same survey (Sánchez et al. 2014), that a flattening of the gas-phase oxygen abundance taking place around a galactocentric distance corresponding to twice their effective radii[4] (R_e, measured in the r band) is a common occurrence in spiral disks. About 82% of the sample with reliable abundance data in the outer disks show this effect, with no apparent dependence on galactic mass, luminosity and morphological type. The oxygen abundance in the inner disks, on the other hand, follows a gradient having a characteristic slope of approximately $-0.07\,\mathrm{dex}\,R_e^{-1}$, except for the very central parts (Sánchez et al. 2014). This common behaviour is illustrated in Fig. 5.2, which displays data extracted from Sánchez-Menguiano et al. (2016, their Fig. 9).

In order to make these results more easily comparable with those presented earlier, where the radial normalization is done relative to the isophotal radius, we need to define a relation between R_e and R_{25}, which depends on the central surface brightness value (μ_0) for the adopted exponential brightness profile. Taking $\mu_0 = 21.65\,\mathrm{mag\,arcsec^{-2}}$ from Freeman (1970), and using $\mu(R) = \mu_0 + 1.086 R/R_d$, one obtains $R_{25} = 1.84\,R_e$. Thus, the flattening in the abundance gradient observed for the CALIFA sample of galaxies to occur at $R \sim 2\,R_e$ or, equivalently, around $R \sim R_{25}$ is consistent with what is reported in Sects. 5.3.2 and 5.3.3. The O3N2-based oxygen abundances measured in the outer disks, out to $\sim$1.5 R_{25}, are also roughly consistent with those presented earlier; in particular they represent a significant fraction of the Solar value, e.g. approximately 0.5 $(O/H)_\odot$ for the intermediate-mass bin shown in Fig. 5.2.

[4]The effective radius R_e encloses 50% of the light, integrated by adopting an exponential radial profile of the surface brightness (i.e. not including the contribution from the bulge): $I = I_0 \exp[-(R/R_d)]$, with I_0 the central intensity and R_d the disk scale length. The effective radius is given by $R_e = 1.678\,R_d$ (e.g. Graham and Driver 2005).

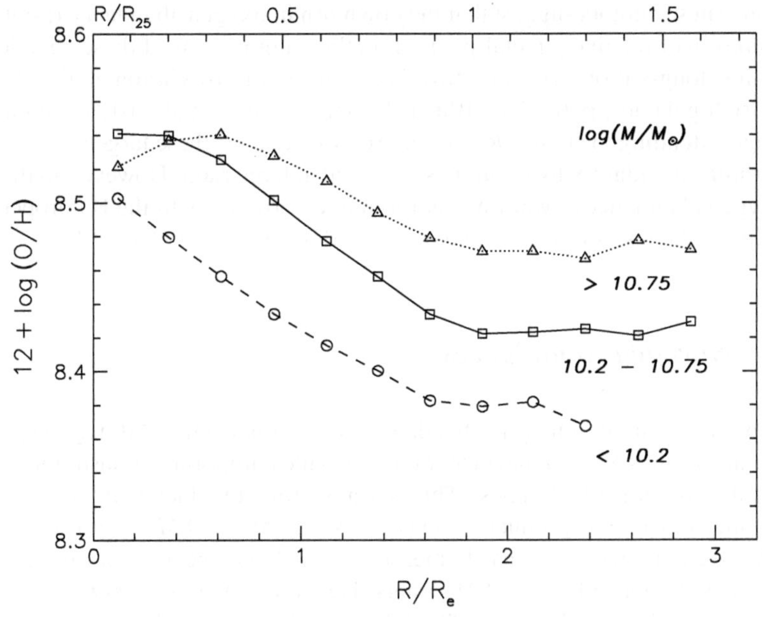

Fig. 5.2 Mean radial oxygen abundance profiles measured for a sample of face-on spirals from the CALIFA survey. The data are plotted in bins of 0.25 R_e, for three different galaxy mass ranges, as indicated in the plot. The upper scale, drawn in units of the isophotal radius, assumes a central surface brightness $\mu_0 = 21.65$ mag arcsec^{-2}. Adapted from Sánchez-Menguiano et al. (2016, Fig. 9)

Long-Slit Spectra For completeness, some additional surveys that obtained spectroscopic observations of the ionized gas in regions close to the edges of spiral galaxies are worth mentioning, even though the radial coverage is not as extended as in the cases discussed so far and is generally limited to regions inside the isophotal radius. Moran et al. (2012) obtained long-slit spectra along the major axis of 174 star-forming galaxies from the *GALEX* Arecibo Sloan Digital Sky Survey (Catinella et al. 2010) with stellar mass $M > 10^{10} M_{\odot}$ and determined O3N2-based gas-phase oxygen abundances in spatial bins for 151 galaxies displaying emission lines. However, their data extend to galactocentric distances of about 1.5 R_{90}[5] or approximately 0.9 R_{25} according to the transformation between the two normalization radii estimated by these authors. Thus, these chemical abundances still refer to the main star-forming disk and should not be compared directly with the outer disk abundance properties presented in the previous sections. Interestingly, however, for about 10% of their galaxies, Moran et al. (2012) measured a significant drop in O/H around $R = R_{90}$, whose magnitude correlates with the total H I mass

[5] R_{90} encloses 90% of the galaxy light, including the bulge.

fraction. These authors suggest that the downturn in oxygen abundance results from the accretion of relatively metal-poor gas in the outer regions of these galaxies.

Similar long-slit observations have been carried out by Carton et al. (2015) for 50 H I-rich galaxies, part of the Bluedisk survey (Wang et al. 2013), with a radial coverage extending to about $2\,R_{90}$ in some cases. Also for these targets, a steepening of the radial abundance distribution is observed at large radii. However, in this work the oxygen abundance downturn is not found to correlate with the H I properties of the parent galaxies as instead found in the work by Moran et al. (2012).

5.3.5 Nitrogen Abundances

The investigation of nitrogen abundances in extragalactic H II regions, and in particular of the N/O abundance ratio, provides important constraints on the chemical evolution of galaxies. This springs from the fact that, while oxygen is the nucleosynthetic product of massive stars ($M > 8M_{\odot}$), nitrogen, whose abundance can in general be easily measured in nebular spectra[6], originates mostly in intermediate-mass ($M = 1-8M_{\odot}$) stars (Henry and Worthey 1999). A look at the N/O ratio variation as a function of metallicity O/H in extragalactic nebulae reveals a bimodal behaviour. The N/O ratio is approximately constant [log(N/O) $\simeq -1.4$, but with a large scatter, see Garnett 1990] below $12+\log(\mathrm{O/H}) = 8.0$ and increases with O/H at larger metallicities. This is interpreted in terms of a primary production of nitrogen, which is what is predominantly being measured at low metallicity, and of a secondary component, proportional to the oxygen abundance, dominating at high O/H (Vila Costas and Edmunds 1993). The N/O ratio measured in outer disk H II regions conforms to this trend, as shown by Bresolin et al. (2012). Since the oxygen abundances measured in outer spiral disks are generally below the level at which secondary nitrogen production becomes predominant, the radial trend of the N/O abundance ratio is virtually flat, with log(N/O) $\simeq -1.3$ to -1.5, in these outer regions (Bresolin et al. 2009b; Berg et al. 2012; López-Sánchez et al. 2015). A similar behaviour has recently been observed by Croxall et al. (2016) in M101 (also an XUV galaxy; Thilker et al. 2007), with the onset of the flattened N/O radial distribution occurring around $0.7\,R_{25}$. These results stress the fact that primary production of nitrogen dominates in the H II regions populating the outskirts of spiral galaxies.

In summary, flat abundance gradients and relatively high oxygen abundances appear to be common features of star-forming outer disks.

[6]The N/O abundance ratio is obtained from the [N II] $\lambda\lambda$6548,6583/[O II] λ3727 line ratio, using the commonly adopted ionization correction scheme $N^+/O^+ = N/O$.

5.4 Additional Considerations

To conclude the discussion of the observational constraints on the metallicity of the outer regions of spiral galaxies, I include two additional topics that have been addressed by recent work. Future investigations of the chemical abundance properties of the outer disks of spiral galaxies will benefit from the study of the spatially resolved gas content (both atomic and molecular), which will help to shed light on the interplay between chemical and secular evolution of the outer disks and accretion events or gas flows taking place in the very outskirts of galaxies.

5.4.1 Relation Between Metallicity and Surface Brightness Breaks

Recent work by Marino et al. (2016) probed into the possible connection between outer disk abundance gradients and surface brightness profile breaks that characterize the disks of spiral galaxies, as discussed elsewhere in this volume. These authors focussed on 131 galaxies, extracted from the larger CALIFA sample, displaying either Type II ("down-bending") or Type III ("up-bending") surface brightness profiles. A correlation was found in the case of Type III galaxies. At lower masses, $\log(M/M_{\odot}) < 10$, a modest flattening in the $g' - r'$ colour, tends to be a common feature, together with a mild flattening of the O/H gradient, while at higher masses, both colour and O/H radial profiles display a pronounced flattening. The different behaviour detected for the Type III galaxies is tentatively attributed by Marino et al. (2016) to a downsizing effect, such that the higher-mass systems have already experienced a phase of inside-out growth, while for the smaller systems, an enhanced disk buildup phase is more recent or still ongoing.

5.4.2 An Analogy with Low Surface Brightness Galaxies?

It has been pointed out by some authors (Thilker et al. 2007; Bresolin et al. 2009b) that the structural parameters and the star-forming properties of outer spiral disks, such as the low mass surface densities and low star formation rates, resemble those observed in low surface brightness (LSB) galaxies. We can then ask the question whether this analogy extends to the chemical abundance properties. In particular, does a low star formation efficiency (Wyder et al. 2009) lead to a flat abundance distribution also in the case of LSB galaxies? The question remained without a clear answer until recently, because very few studies addressed the gas-phase chemical abundance properties of this type of galaxies, and in particular their abundance gradients, in part due to observational challenges. Bresolin and Kennicutt (2015) measured H II region oxygen abundances for a sample of ten LSB spiral galaxies and investigated the presence of radial abundance gradients in this sample. They found that LSB galaxies do display radial abundance gradients that, when normalized by the effective radii, are consistent with those measured for high surface brightness galaxies. Thus, the analogy between LSB galaxies and the outer disks of spiral galaxies does not seem to extend to the chemical abundance properties, despite the similarities outlined above. This result suggests that the chemical evolution of LSB galaxies proceeds in a similar fashion to the high surface brightness galaxies, albeit at a slower pace due to the lower star formation rates, while the outer disks probably follow a different evolutionary path. The latter possibility is addressed in the following section.

5.5 The Evolutionary Status of Outer Disks

A variety of mechanisms can be invoked in order to explain the data presented in the previous section within a coherent picture of galactic chemical evolution. Unfortunately, the theoretical framework is, to a great extent, still lacking, since detailed modelling accounting for the gas-phase chemical abundances measured in the outer disks has not been developed yet. The relatively high metal enrichment observed in these H I-rich, low-star formation rate (SFR) regions of spiral galaxies suggests that some form of mixing mechanism, and perhaps more than one, should be responsible for the observed chemical abundance properties of the outer disks. This section presents some of the possible processes discussed in the literature that could redistribute metals produced in the main star-forming disk of spirals into their very outskirts, tens of kpc from the galactic centres. The effects of galaxy interactions and merging on the radial abundance gradients have already been introduced in Sect. 5.3.3, so that the focus will now be on mixing mechanisms that could affect, in principle, also isolated systems, but we should keep in mind that the chemical abundances in the outer disks could be sensitive to encounters that might have occurred in the distant past, as evidenced, for example, by the presence of warps in the H I envelopes.

The stellar radial migration process (Sellwood and Binney 2002; Roškar et al. 2008) does not affect the present-day abundance gradient of oxygen, since this element, whose abundance we trace with H II region spectroscopy, is produced by massive stars, which do not have sufficient time to migrate radially before ending their lives (Kubryk et al. 2015). Older tracers of ionized gas metallicity, such as planetary nebulae, can have a different behaviour (Magrini et al. 2016).

5.5.1 Flattening the Gradients

While some of the processes discussed below can explain the flattening of the H II region oxygen abundances observed in the outer disks, we start with the remark made by Bresolin et al. (2012), who suggested that the flat O/H distribution could simply be a consequence of relatively flat star formation efficiencies (Bigiel et al. 2010; Espada et al. 2011). Defining the star formation efficiency as SFE $= \Sigma_{\rm SFR}/\Sigma_{\rm HI}$ (e.g. Bigiel et al. 2008), i.e. the ratio between the surface densities of star formation rate (inferred, e.g. from far-UV observations) and H I mass, we can approximate the gas-phase oxygen abundance per unit surface area of the disk, neglecting effects such as gas flows and variable star formation rates, as

$$\frac{\rm O}{\rm H} \sim \frac{y_{\rm O}}{11.81}\, t\, \frac{\Sigma_{\rm SFR}}{\Sigma_{\rm HI}} \propto {\rm SFE}, \tag{5.4}$$

where t is the duration of the star formation activity, since the amount of oxygen produced per unit surface area and unit time is the product of the oxygen yield (by mass) $y_{\rm O}$ and the star formation rate surface density. Then, according to (5.4), the flattened SFE radial profiles traced beyond the isophotal radius, relative to the behaviour in the inner disks, would result in a similarly flattened O/H radial gradient at large galactocentric radii, as observed. This appears to be consistent with the idea that the star formation activity at large galactocentric distances proceeds slowly enough for some metal mixing processes (discussed below) to efficiently erase or reduce chemical abundances, inhomogeneities and large-scale gradients.

Equation (5.4) can also be used to estimate that, given the low-star formation rates measured in the outer disks of spiral galaxies and the large H I content, the timescale necessary to reach the observed metal enrichment can be longer than the star formation timescale within an inside-out scenario for galaxy growth or even longer than a Hubble time, reinforcing the notion that the metallicities measured in outer disks exceed the values attainable by in situ star formation alone (see also Eq. (3) in Kudritzki et al. 2014 for an alternative calculation based on the closed-box model, leading to the same conclusion).

The link between SFE and O/H described above has been shown by recent tailored chemical evolution models to be able to reproduce the flattened gas-phase abundances in the outer disk of the Milky Way beyond 10 kpc from the centre (Esteban et al. 2013) and the flat oxygen abundance gradient in the outer disk of M83 (Bresolin et al. 2016, with an adaptation of the chemical evolution model by Kudritzki et al. 2015). This seems also to be consistent with chemical evolution models of the Milky Way in which the decreasing star formation efficiency with increasing galactocentric distance leads to a flattening of the metallicity gradients in the outer regions (Kubryk et al. 2015).

5.5.2 Bringing Metals to the Outer Disks

The transport of metals produced in the inner disks to large galactocentric radii appears to be necessary in order to explain the relatively high gas-phase metallicities observed in the extended disks of spiral galaxies. A number of different mechanisms have been discussed in the literature. The argumentation contained in the following section is rather speculative since, as already mentioned, a solid theoretical explanation for the chemical properties of extended disks is currently still missing. The mechanisms invoked can be broadly divided into two main categories: mixing and enriched infall. These are succinctly presented below, following in part Bresolin et al. (2009b, 2012) and Werk et al. (2011), to which the reader is referred to for a more in-depth discussion.

5.5.2.1 Mixing

Under this category we can include processes that can be effective in redistributing metals across galactic disks. Some examples are listed below:

- Outward radial flows originating from *viscosity* in the gas layer due, for example, to cloud collisions or gravitational instabilities (Lacey and Fall 1985; Clarke 1989) can produce flat gas-phase abundances in the outer disks (Tsujimoto et al. 1995; Thon and Meusinger 1998).
- Gas flows can also be driven by angular momentum redistribution from *non-axisymmetric structures*, i.e. bars and spiral arms. While stellar bars can affect the chemical evolution in the inner regions of galaxies (Athanassoula 1992; Cavichia et al. 2014), the overlap of spiral and bar resonances (Minchev et al. 2011) can affect the distribution and metallicity of stars at large galactocentric distances. The resonance associated with the corotation of the spiral pattern has been found to correlate with the radial position of breaks in the metallicity gradient by Scarano and Lépine (2013). This suggests that the gas flows occurring in opposite directions, inwards inside corotation and outwards beyond corotation, are connected to the different abundance gradients of the outer disks relative to the inner disks.
- Interstellar *turbulence* plays an important role in homogenizing the metallicity distribution in galactic disks (Scalo and Elmegreen 2004). What constitutes the main driving source, either stellar feedback or gravitational instability, is still uncertain (Krumholz and Burkhart 2016), although in the outer disks, which are characterized by very low star formation rates, it is unlikely that feedback from supernova explosions represents the main source of gas turbulence. The simulations by Yang and Krumholz (2012) indicate that turbulence driven by thermal instability is quite efficient in erasing kpc-scale metallicity gradients, on timescales that are on the order of a few orbital periods (a few 100 Myrs, Petit et al. 2015). Metals can be transported via convective motions of the gas over large distances, which leads to an equilibrium between star formation and turbulent mixing, making this an appealing mechanism for the explanation of the abundance properties of the extended disks of spirals.

5.5.2.2 Enriched Infall

The star formation and chemical enrichment histories of galaxies are profoundly affected by inflows and outflows of gas. Feedback-driven galactic winds eject a large portion of the metals produced in disk stars into the halos, the circumgalactic medium and the intergalactic medium (Kobayashi et al. 2007; Lilly et al. 2013; Côté et al. 2015) out to distances of the order of 100 kpc (Tumlinson et al. 2011; Werk et al. 2013). These outflows are crucial to explain, for example, the existence of the mass-metallicity relation observed for star-forming galaxies (Tremonti et al. 2004; Finlator and Davé 2008).

The gas that has been metal enriched by supernova explosions at early epochs and subsequently ejected from galaxies can later be re-accreted in a wind-recycling process (see, e.g. the models by Oppenheimer and Davé 2008; Davé et al. 2011). This re-accretion on the disk from the halo should take place preferentially in the outskirts, leading to an inside-out growth, on timescales on the order of a few dynamical times, $\sim$1 Gyr (Fu et al. 2013), necessary for the gas to cool down from the hot phase. Some evidence in support of this process has been presented recently by Belfiore et al. (2016) from the spatially resolved metal budget in NGC 628.

Fu et al. (2013) estimated the gas-phase metallicity of the infalling gas to be around 0.4$\times$ Solar for a Milky Way-type galaxy, which is in rough agreement with the observed metallicity of the extended disks. It is also worth pointing out that the metallicity of the circumgalactic medium at $z < 1$, as traced by Lyman limit systems, is bimodal, with a metal-rich branch peaking at a metallicity approximately 0.5$\times$ Solar, as shown by Lehner et al. (2013). According to these authors, this metal-rich branch could be tracing cool, enriched gas originating from galactic outflows and tidally stripped material.

The effects of an enriched infall process on galactic chemical evolution models have already been illustrated before. For example, Tosi (1988) showed how a metal-rich infall would affect the chemical composition of the outer parts of spirals, inducing a flattening of their abundance gradients, and estimated an upper limit for the infalling gas metallicity of 0.4$\times$ Solar from comparisons with the chemical abundances observed in the Milky Way. Figure 5.3 shows a model radial oxygen abundance gradient for the disk of M83, calculated with a galactic wind launched in the inner disk, following Kudritzki et al. (2015), but allowing for an inflow of metal-enriched gas with an oxygen abundance $12 + \log(\mathrm{O/H}) = 8.20$ (equivalent to 0.32$\times$ Solar). Such an enriched infall is required by the model to reproduce the gas metallicity observed in the extended disk of M83, with the flat distribution arising from the assumed constant star formation efficiency.

Minor Mergers Minor merger activity, as a source of cold gas leading to mass growth in galaxies, has also been proposed to be effective at chemically enriching the outer disks (López-Sánchez et al. 2015) and is included in this section because its effects could resemble those described above for enriched infall. Given the low accretion rates of star-forming galaxies due to mergers with low-mass satellites measured in the local Universe (Sancisi et al. 2008; Di Teodoro and Fraternali

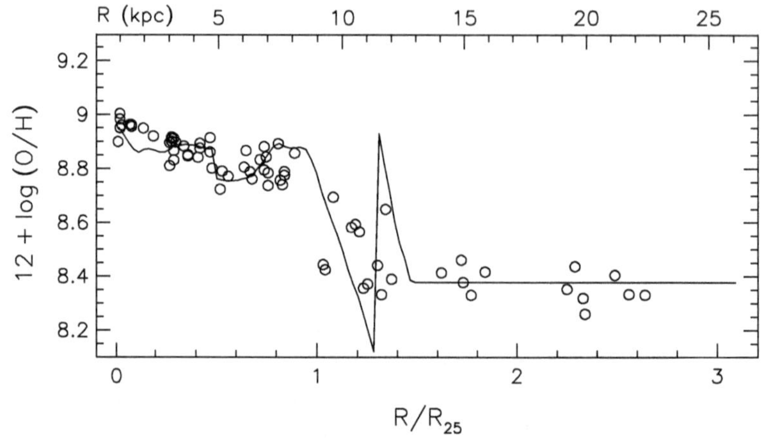

Fig. 5.3 Model radial oxygen abundance gradient for the disk of M83 (*continuous line*), calculated including an enriched gas infall, compared with the H II region metallicities (*open circles*) shown in Fig. 5.1 (Bresolin et al. 2016)

2014), this process is unlikely to be important for the chemical evolution of outer disks at the present time. However, higher merger rates in the past (around a redshift $z \simeq 2$) could have made this process a potential contributor to the accretion of metals in the outskirts of galaxies earlier on during their evolution (Lehnert et al. 2016). Numerical simulations by Zinchenko et al. (2015) also indicate that the stellar migration process induced by minor merging in Milky Way-type spirals cannot generate the flattening of the metallicity observed in the outer disks.

Different mechanisms can be invoked to explain the chemical abundance properties of outer disks. Among these are various mixing processes, including turbulence, and metal-enriched infall of gas from the circumstellar medium.

5.6 Conclusion

The outer disks of spiral galaxies remain a relatively unexplored territory in studies of the evolution of galaxies. This chapter has highlighted the somewhat unexpected attributes of the ionized gas chemical composition discovered in recent years in the outermost parts of galaxy disks, at least for systems with extended H I envelopes and ongoing star formation. At the same time, it is important to stress that metallicity information in the outer disks, gathered until now almost exclusively from H II regions, can provide crucial constraints for models of the chemical evolution of galaxies, considering that these are the most recently assembled regions of the disks according to the inside-out scenario. Future work will more firmly establish how common shallow or flat outer gradients with relatively high oxygen abundances are in spiral disks, which is relevant to ascertain the roles played by enriched galactic inflows and mixing mechanisms, such as turbulence, in regulating the chemical evolution of galaxies. Studies of the evolution of galaxies will benefit from probing the relationship between the gas-phase chemical abundances and the properties of the faint stellar populations present in these low surface brightness structures. A better understanding of the mechanisms leading to the formation of these extended structures and of the importance of the galactic environment in this context is highly desired. From the observational point of view, the combination of deep integral field spectroscopy with spatially resolved radio H I mapping for large samples of spirals will yield a better characterization of the relationship between gas, star formation, environment and metal production in determining the evolutionary status of present-day galaxies out to very large radii.

Acknowledgements The author is grateful to Rob Kennicutt, Emma Ryan-Weber and Rolf Kudritzki for interesting collaborations and stimulating discussions over the years and to the editors of this volume for the invitation to contribute this chapter.

References

Arellano-Córdova, K.Z., Rodríguez, M., Mayya, Y.D., Rosa-González, D.: The oxygen abundance gradient in M81 and the robustness of abundance determinations in H II regions. Mon. Not. R. Astron. Soc. **455**, 2627–2643 (2016). doi:10.1093/mnras/stv2461, 1510.07757

Asplund, M., Grevesse, N., Sauval, A.J., Scott, P.: The chemical composition of the sun. Annu. Rev. Astron. Astrophys. **47**, 481–522 (2009). doi:10.1146/annurev.astro.46.060407.145222, 0909.0948

Athanassoula, E.: The existence and shapes of dust lanes in galactic bars. Mon. Not. R. Astron. Soc. **259**, 345–364 (1992). doi:10.1093/mnras/259.2.345

Balick, B., Kwitter, K.B., Corradi, R.L.M., Henry, R.B.C.: Metal-rich planetary nebulae in the outer reaches of M31. Astrophys. J. **774**, 3 (2013). doi:10.1088/0004-637X/774/1/3, 1307.0472

Belfiore, F., Maiolino, R., Bothwell, M.: Galaxy gas flows inferred from a detailed, spatially resolved metal budget. Mon. Not. R. Astron. Soc. **455**, 1218–1236 (2016). doi:10.1093/mnras/stv2332, 1503.06823

Berg, D.A., Skillman, E.D., Marble, A.R., van Zee, L., Engelbracht, C.W., Lee, J.C., Kennicutt, R.C. Jr., Calzetti, D., Dale, D.A., Johnson, B.D.: Direct oxygen abundances for low-luminosity LVL galaxies. Astrophys. J. **754**, 98 (2012). doi:10.1088/0004-637X/754/2/98, 1205.6782

Bigiel, F., Leroy, A., Walter, F., Brinks, E., de Blok, W.J.G., Madore, B., Thornley, M.D.: The star formation law in nearby galaxies on sub-Kpc scales. Astron. J. **136**, 2846–2871 (2008). doi:10.1088/0004-6256/136/6/2846, 0810.2541

Bigiel, F., Leroy, A., Walter, F., Blitz, L., Brinks, E., de Blok, W.J.G., Madore, B.: Extremely inefficient star formation in the outer disks of nearby galaxies. Astron. J. **140**, 1194–1213 (2010). doi:10.1088/0004-6256/140/5/1194, 1007.3498

Blanc, G.A., Kewley, L., Vogt, F.P.A., Dopita, M.A.: IZI: inferring the gas phase metallicity (Z) and ionization parameter (q) of ionized nebulae using Bayesian statistics. Astrophys. J. **798**, 99 (2015). doi:10.1088/0004-637X/798/2/99, 1410.8146

Bresolin, F.: The oxygen abundance in the inner H II regions of M101: implications for the calibration of strong-line metallicity indicators. Astrophys. J. **656**, 186–197 (2007). doi:10.1086/510380, arXiv:astro-ph/0610690

Bresolin, F., Kennicutt, R.C. Jr.: Optical spectroscopy of metal-rich H II regions and circumnuclear hot spots in M83 and NGC 3351. Astrophys. J. **572**, 838–860 (2002). doi:10.1086/340371, arXiv:astro-ph/0202383

Bresolin, F., Kennicutt, R.C.: Abundance gradients in low surface brightness spirals: clues on the origin of common gradients in galactic discs. Mon. Not. R. Astron. Soc. **454**, 3664–3673 (2015). doi:10.1093/mnras/stv2245, 1509.07190

Bresolin, F., Garnett, D.R., Kennicutt, R.C.: Abundances of metal-rich H II regions in M51. Astrophys. J. **615**, 228–241 (2004). doi:10.1086/424377

Bresolin, F., Schaerer, D., González Delgado, R.M., Stasińska, G.: A VLT study of metal-rich extragalactic H II regions. I. Observations and empirical abundances. Astron. Astrophys. **441**, 981–997 (2005). doi:10.1051/0004-6361:20053369

Bresolin, F., Gieren, W., Kudritzki, R., Pietrzyński, G., Urbaneja, M.A., Carraro, G.: Extragalactic chemical abundances: do H II regions and young stars tell the same story? The case of the spiral galaxy NGC 300. Astrophys. J. **700**, 309–330 (2009a). doi:10.1088/0004-637X/700/1/309, 0905.2791

Bresolin, F., Ryan-Weber, E., Kennicutt, R.C., Goddard, Q.: The flat oxygen abundance gradient in the extended disk of M83. Astrophys. J. **695**, 580–595 (2009b). doi:10.1088/0004-637X/695/1/580, 0901.1127

Bresolin, F., Kennicutt, R.C., Ryan-Weber, E.: Gas metallicities in the extended disks of NGC 1512 and NGC 3621. Chemical signatures of metal mixing or enriched gas accretion? Astrophys. J. **750**, 122 (2012). doi:10.1088/0004-637X/750/2/122, 1203.0956

Bresolin, F., Kudritzki, R.P., Urbaneja, M.A., Gieren, W., Ho, I.T., Pietrzyński, G.: Young stars and ionized nebulae in M83: comparing chemical abundances at high metallicity. Astrophys. J. **830**, 64 (2016). doi:10.3847/0004-637X/830/2/64, 1607.06840

Brown, J.S., Martini, P., Andrews, B.H.: A recalibration of strong-line oxygen abundance diagnostics via the direct method and implications for the high-redshift universe. Mon. Not. R. Astron. Soc. **458**, 1529–1547 (2016). doi:10.1093/mnras/stw392, yy

Carraro, G., Bresolin, F., Villanova, S., Matteucci, F., Patat, F., Romaniello, M.: Metal abundances in extremely distant galactic old open clusters. I. Berkeley 29 and Saurer 1. Astron. J. **128**, 1676–1683 (2004). doi:10.1086/423912, astro-ph/0406679

Carton, D., Brinchmann, J., Wang, J., Bigiel, F., Cormier, D., van der Hulst, T., Józsa, G.I.G., Serra, P., Verheijen, M.A.W.: Gas-phase metallicity profiles of the Bluedisk galaxies: is metallicity in a local star formation regulated equilibrium? Mon. Not. R. Astron. Soc. **451**, 210–235 (2015). doi:10.1093/mnras/stv967, 1505.02797

Catinella, B., Schiminovich, D., Kauffmann, G., Fabello, S., Wang, J., Hummels, C., Lemonias, J., Moran, S.M., Wu, R., Giovanelli, R., Haynes, M.P., Heckman, T.M., Basu-Zych, A.R., Blanton, M.R., Brinchmann, J., Budavári, T., Gonçalves, T., Johnson, B.D., Kennicutt, R.C., Madore, B.F., Martin, C.D., Rich, M.R., Tacconi, L.J., Thilker, D.A., Wild, V., Wyder, T.K.: The GALEX Arecibo SDSS survey - I. Gas fraction scaling relations of massive galaxies and first data release. Mon. Not. R. Astron. Soc. **403**, 683–708 (2010). doi:10.1111/j.1365-2966.2009.16180.x, 0912.1610

Cavichia, O., Mollá, M., Costa, R.D.D., Maciel, W.J.: The role of the galactic bar in the chemical evolution of the Milky Way. Mon. Not. R. Astron. Soc. **437**, 3688–3701 (2014). doi:10.1093/mnras/stt2164, 1311.1518

Chen, H.-W.: Outskirts of distant galaxies in absorption. In: Knapen, J.H., Lee, J.C., Gil de Paz, A. (eds.) Outskirts of Galaxies, vol. 434. Springer, Cham (2017). doi:10.1007/978-3-319-56570-5

Clarke, C.J.: Chemical evolution of viscously evolving galactic discs. Mon. Not. R. Astron. Soc. **238**, 283–292 (1989)

Corradi, R.L.M., Kwitter, K.B., Balick, B., Henry, R.B.C., Hensley, K.: The chemistry of planetary nebulae in the outer regions of M31. Astrophys. J. **807**, 181 (2015). doi:10.1088/0004-637X/807/2/181, 1504.03869

Côté, B., Martel, H., Drissen, L.: Cosmological simulations of the intergalactic medium evolution. II. Galaxy model and feedback. Astrophys. J. **802**, 123 (2015). doi:10.1088/0004-637X/802/2/123

Croxall, K.V., Pogge, R.W., Berg, D.A., Skillman, E.D., Moustakas, J.: CHAOS III: gas-phase abundances in NGC 5457. Astrophys. J. **830**, 4 (2016). doi:10.3847/0004-637X/830/1/4, 1605.01612

Davé, R., Finlator, K., Oppenheimer, B.D.: Galaxy evolution in cosmological simulations with outflows - II. Metallicities and gas fractions. Mon. Not. R. Astron. Soc. **416**, 1354–1376 (2011). doi:10.1111/j.1365-2966.2011.19132.x, 1104.3156

de Vaucouleurs, G., de Vaucouleurs, A., Corwin, H.G. Jr., Buta, R.J., Paturel, G., Fouque, P.: Third Reference Catalogue of Bright Galaxies. Springer, Berlin (1991)

Di Teodoro, E.M., Fraternali, F.: Gas accretion from minor mergers in local spiral galaxies. Astron. Astrophys. **567**, A68 (2014). doi:10.1051/0004-6361/201423596, 1406.0856

Erb, D.K., Shapley, A.E., Pettini, M., Steidel, C.C., Reddy, N.A., Adelberger, K.L.: The mass-metallicity relation at z > 2. Astrophys. J. **644**, 813–828 (2006). doi:10.1086/503623, astro-ph/0602473

Espada, D., et al.: Star formation in the extended gaseous disk of the isolated galaxy CIG 96. Astrophys. J. **736**, 20 (2011). doi:10.1088/0004-637X/736/1/20, 1107.0588

Esteban, C., Bresolin, F., Peimbert, M., García-Rojas, J., Peimbert, A., Mesa-Delgado, A.: Keck HIRES spectroscopy of extragalactic H II regions: C and O abundances from recombination lines. Astrophys. J. **700**, 654–678 (2009). doi:10.1088/0004-637X/700/1/654, 0905.2532

Esteban, C., Carigi, L., Copetti, M.V.F., García-Rojas, J., Mesa-Delgado, A., Castañeda, H.O., Péquignot, D.: NGC 2579 and the carbon and oxygen abundance gradients beyond the solar circle. Mon. Not. R. Astron. Soc. **433**, 382–393 (2013). doi:10.1093/mnras/stt730, 1304.6927

Ferguson, A.M.N., Gallagher, J.S., Wyse, R.F.G.: The extreme outer regions of disk galaxies. I. Chemical abundances of H II regions. Astron. J. **116**, 673–690 (1998). doi:10.1086/300456, y

Finlator, K., Davé, R.: The origin of the galaxy mass-metallicity relation and implications for galactic outflows. Mon. Not. R. Astron. Soc. **385**, 2181–2204 (2008). doi:10.1111/j.1365-2966.2008.12991.x, 0704.3100

Freeman, K.C.: On the disks of spiral and S0 galaxies. Astrophys. J. **160**, 811 (1970). doi:10.1086/150474

Fu, J., Kauffmann, G., Huang, M.l., Yates, R.M., Moran, S., Heckman, T.M., Davé, R., Guo, Q., Henriques, B.M.B.: Star formation and metallicity gradients in semi-analytic models of disc galaxy formation. Mon. Not. R. Astron. Soc. **434**, 1531–1548 (2013). doi:10.1093/mnras/stt1117, 1303.5586

García-Rojas, J., Esteban, C.: On the abundance discrepancy problem in H II regions. Astrophys. J. **670**, 457–470 (2007). doi:10.1086/521871, 0707.3518

Garnett, D.R.: Nitrogen in irregular galaxies. Astrophys. J. **363**, 142–153 (1990). doi:10.1086/169324

Genovali, K., Lemasle, B., da Silva, R., Bono, G., Fabrizio, M., Bergemann, M., Buonanno, R., Ferraro, I., François, P., Iannicola, G., Inno, L., Laney, C.D., Kudritzki, R.P., Matsunaga, N., Nonino, M., Primas, F., Romaniello, M., Urbaneja, M.A., Thévenin, F.: On the α-element gradients of the galactic thin disk using Cepheids. Astron. Astrophys. **580**, A17 (2015). doi:10.1051/0004-6361/201525894, 1503.03758

Gil de Paz, A., Madore, B.F., Boissier, S., Swaters, R., Popescu, C.C., Tuffs, R.J., Sheth, K., Kennicutt, R.C. Jr., Bianchi, L., Thilker, D., Martin, D.C.: Discovery of an extended ultraviolet disk in the nearby galaxy NGC 4625. Astrophys. J. Lett. **627**, L29–L32 (2005). doi:10.1086/432054, arXiv:astro-ph/0506357

Gil de Paz, A., Madore, B.F., Boissier, S., Thilker, D., Bianchi, L., Sánchez Contreras, C., Barlow, T.A., Conrow, T., Forster, K., Friedman, P.G., Martin, D.C., Morrissey, P., Neff, S.G., Rich, R.M., Schiminovich, D., Seibert, M., Small, T., Donas, J., Heckman, T.M., Lee, Y.W., Milliard, B., Szalay, A.S., Wyder, T.K., Yi, S.: Chemical and photometric evolution of extended ultraviolet disks: optical spectroscopy of M83 (NGC 5236) and NGC 4625. Astrophys. J. **661**, 115–134 (2007). doi:10.1086/513730, arXiv:astro-ph/0702302

Goddard, Q.E., Kennicutt, R.C., Ryan-Weber, E.V.: On the nature of star formation at large galactic radii. Mon. Not. R. Astron. Soc. **405**, 2791–2809 (2010). doi:10.1111/j.1365-2966.2010.16661.x, 1003.2520

Goddard, Q.E., Bresolin, F., Kennicutt, R.C., Ryan-Weber, E.V., Rosales-Ortega, F.F.: On the nature of the H II regions in the extended ultraviolet disc of NGC 4625. Mon. Not. R. Astron. Soc. **412**, 1246–1258 (2011). doi:10.1111/j.1365-2966.2010.17990.x, 1011.1967

Graham, A.W., Driver, S.P.: A concise reference to (projected) Sérsic $R^{1/n}$ quantities, including concentration, profile slopes, Petrosian indices, and Kron magnitudes. Publ. Astron. Soc. Aust. **22**, 118–127 (2005). doi:10.1071/AS05001, astro-ph/0503176

Henry, R.B.C., Worthey, G.: The distribution of heavy elements in spiral and elliptical galaxies. Publ. Astron. Soc. Pac. **111**, 919–945 (1999). doi:10.1086/316403, arXiv:astro-ph/9904017

Ho, I.T., Kudritzki, R.P., Kewley, L.J., Zahid, H.J., Dopita, M.A., Bresolin, F., Rupke, D.S.N.: Metallicity gradients in local field star-forming galaxies: insights on inflows, outflows, and the coevolution of gas, stars and metals. Mon. Not. R. Astron. Soc. **448**, 2030–2054 (2015). doi:10.1093/mnras/stv067, 1501.02668

Kennicutt, R.C., Bresolin, F., Garnett, D.R.: The composition gradient in M101 revisited. II. Electron temperatures and implications for the nebular abundance scale. Astrophys. J. **591**, 801–820 (2003). doi:10.1086/375398

Kewley, L.J., Dopita, M.A.: Using strong lines to estimate abundances in extragalactic H II regions and starburst galaxies. Astrophys. J. Suppl. **142**, 35–52 (2002). doi:10.1086/341326

Kewley, L.J., Ellison, S.L.: Metallicity calibrations and the mass-metallicity relation for star-forming galaxies. Astrophys. J. **681**, 1183–1204 (2008). doi:10.1086/587500, arXiv:0801.1849

Kobayashi, C., Umeda, H., Nomoto, K., Tominaga, N., Ohkubo, T.: Galactic chemical evolution: carbon through zinc. Astrophys. J. **653**, 1145–1171 (2006). doi:10.1086/508914, astro-ph/0608688

Kobayashi, C., Springel, V., White, S.D.M.: Simulations of cosmic chemical enrichment. Mon. Not. R. Astron. Soc. **376**, 1465–1479 (2007). doi:10.1111/j.1365-2966.2007.11555.x, astro-ph/0604107

Kobulnicky, H.A., Kewley, L.J.: Metallicities of 0.3<z<1.0 Galaxies in the GOODS-North field. Astrophys. J. **617**, 240–261 (2004). doi:10.1086/425299

Koribalski, B.S., López-Sánchez, Á.R.: Gas dynamics and star formation in the galaxy pair NGC1512/1510. Mon. Not. R. Astron. Soc. **400**, 1749–1767 (2009). doi:10.1111/j.1365-2966.2009.15610.x, 0908.4128

Korotin, S.A., Andrievsky, S.M., Luck, R.E., Lépine, J.R.D., Maciel, W.J., Kovtyukh, V.V.: Oxygen abundance distribution in the galactic disc. Mon. Not. R. Astron. Soc. **444**, 3301–3307 (2014). doi:10.1093/mnras/stu1643, 1408.6103

Krumholz, M.R., Burkhart, B.: Is turbulence in the interstellar medium driven by feedback or gravity? an observational test. Mon. Not. R. Astron. Soc. **458**, 1671–1677 (2016). doi:10.1093/mnras/stw434, 1512.03439

Kubryk, M., Prantzos, N., Athanassoula, E.: Evolution of the Milky Way with radial motions of stars and gas. II. The evolution of abundance profiles from H to Ni. Astron. Astrophys. **580**, A127 (2015). doi:10.1051/0004-6361/201424599, 1412.4859

Kudritzki, R.P., Urbaneja, M.A., Bresolin, F., Hosek, M.W. Jr., Przybilla, N.: Stellar metallicity of the extended disk and distance of the spiral galaxy NGC 3621. Astrophys. J. **788**, 56 (2014). doi:10.1088/0004-637X/788/1/56, 1404.7244

Kudritzki, R.P., Ho, I.T., Schruba, A., Burkert, A., Zahid, H.J., Bresolin, F., Dima, G.I.: The chemical evolution of local star-forming galaxies: radial profiles of ISM metallicity, gas mass, and stellar mass and constraints on galactic accretion and winds. Mon. Not. R. Astron. Soc. **450**, 342–359 (2015). doi:10.1093/mnras/stv522, 1503.01503

Lacey, C.G., Fall, S.M.: Chemical evolution of the galactic disk with radial gas flows. Astrophys. J. **290**, 154–170 (1985). doi:10.1086/162970

Lehner, N., Howk, J.C., Tripp, T.M., Tumlinson, J., Prochaska, J.X., O'Meara, J.M., Thom, C., Werk, J.K., Fox, A.J., Ribaudo, J.: The bimodal metallicity distribution of the cool circum-galactic medium at $z < 1$. Astrophys. J. **770**, 138 (2013). doi:10.1088/0004-637X/770/2/138, 1302.5424

Lehnert, M.D., van Driel, W., Minchin, R.: Can galaxy growth be sustained through HI-rich minor mergers? Astron. Astrophys. **590**, A51 (2016). doi:10.1051/0004-6361/201525743, 1510.01959

Lemasle, B., François, P., Genovali, K., Kovtyukh, V.V., Bono, G., Inno, L., Laney, C.D., Kaper, L., Bergemann, M., Fabrizio, M., Matsunaga, N., Pedicelli, S., Primas, F., Romaniello, M.: Galactic abundance gradients from Cepheids. α and heavy elements in the outer disk. Astron. Astrophys. **558**, A31 (2013). doi:10.1051/0004-6361/201322115, 1308.3249

Lépine, J.R.D., Cruz, P., Scarano, S. Jr., Barros, D.A., Dias, W.S., Pompéia, L., Andrievsky, S.M., Carraro, G., Famaey, B.: Overlapping abundance gradients and azimuthal gradients related to the spiral structure of the galaxy. Mon. Not. R. Astron. Soc. **417**, 698–708 (2011). doi:10.1111/j.1365-2966.2011.19314.x, 1106.3137

Lépine, J.R.D., Andrievky, S., Barros, D.A., Junqueira, T.C., Scarano, S.: Bimodal chemical evolution of the galactic disk and the barium abundance of Cepheids. In: Feltzing, S., Zhao, G., Walton, N.A., Whitelock, P. (eds.) IAU Symposium, IAU Symposium, vol. 298, pp. 86–91 (2014). doi:10.1017/S174392131300625X, 1307.7781

Lilly, S.J., Carollo, C.M., Pipino, A., Renzini, A., Peng, Y.: Gas regulation of galaxies: the evolution of the cosmic specific star formation rate, the metallicity-mass-star-formation rate relation, and the stellar content of Halos. Astrophys. J. **772**, 119 (2013). doi:10.1088/0004-637X/772/2/119, 1303.5059

López-Sánchez, Á.R., Dopita, M.A., Kewley, L.J., Zahid, H.J., Nicholls, D.C., Scharwächter, J.: Eliminating error in the chemical abundance scale for extragalactic H II regions. Mon. Not. R. Astron. Soc. **426**, 2630–2651 (2012). doi:10.1111/j.1365-2966.2012.21145.x, 1203.5021

López-Sánchez, Á.R., Westmeier, T., Esteban, C., Koribalski, B.S.: Ionized gas in the XUV disc of the NGC 1512/1510 system. Mon. Not. R. Astron. Soc. **450**, 3381–3409 (2015). doi:10.1093/mnras/stv703

Magrini, L., Sestito, P., Randich, S., Galli, D.: The evolution of the Galactic metallicity gradient from high-resolution spectroscopy of open clusters. Astron. Astrophys. **494**, 95–108 (2009). doi:10.1051/0004-6361:200810634, 0812.0854

Magrini, L., Coccato, L., Stanghellini, L., Casasola, V., Galli, D.: Metallicity gradients in local Universe galaxies: time evolution and effects of radial migration. Astron. Astrophys. **588**, A91 (2016). doi:10.1051/0004-6361/201527799, 1602.02529

Maiolino, R., Nagao, T., Grazian, A., Cocchia, F., Marconi, A., Mannucci, F., Cimatti, A., Pipino, A., Ballero, S., Calura, F., Chiappini, C., Fontana, A., Granato, G.L., Matteucci, F., Pastorini, G., Pentericci, L., Risaliti, G., Salvati, M., Silva, L.: AMAZE. I. The evolution of the mass-metallicity relation at $z > 3$. Astron. Astrophys. **488**, 463–479 (2008). doi:10.1051/0004-6361:200809678, 0806.2410

Mannucci, F., Cresci, G., Maiolino, R., Marconi, A., Gnerucci, A.: A fundamental relation between mass, star formation rate and metallicity in local and high-redshift galaxies. Mon. Not. R. Astron. Soc. **408**, 2115–2127 (2010). doi:10.1111/j.1365-2966.2010.17291.x, 1005.0006

Marino, R.A., Gil de Paz, A., Castillo-Morales, A., Muñoz-Mateos, J.C., Sánchez, S.F., Pérez-González, P.G., Gallego, J., Zamorano, J., Alonso-Herrero, A., Boissier, S.: Integral field spectroscopy and multi-wavelength imaging of the nearby spiral galaxy NGC 5668: an unusual flattening in metallicity gradient. Astrophys. J. **754**, 61 (2012). doi:10.1088/0004-637X/754/1/61, 1205.5051

Marino, R.A., Rosales-Ortega, F.F., Sánchez, S.F., Gil de Paz, A., Vílchez, J., Miralles-Caballero, D., Kehrig, C., Pérez-Montero, E., Stanishev, V., Iglesias-Páramo, J., Díaz, A.I., Castillo-Morales, A., Kennicutt, R., López-Sánchez, A.R., Galbany, L., García-Benito, R., Mast, D., Mendez-Abreu, J., Monreal-Ibero, A., Husemann, B., Walcher, C.J., García-Lorenzo, B., Masegosa, J., Del Olmo Orozco, A., Mourão, A.M., Ziegler, B., Mollá, M., Papaderos, P., Sánchez-Blázquez, P., González Delgado, R.M., Falcón-Barroso, J., Roth, M.M., van de Ven, G., Califa Team: The O3N2 and N2 abundance indicators revisited: improved calibrations based on CALIFA and T_e-based literature data. Astron. Astrophys. **559**, A114 (2013). doi:10.1051/0004-6361/201321956, 1307.5316

Marino, R.A., Gil de Paz, A., Sánchez, S.F., Sánchez-Blázquez, P., Cardiel, N., Castillo-Morales, A., Pascual, S., Vílchez, J., Kehrig, C., Mollá, M., Mendez-Abreu, J., Catalán-Torrecilla, C., Florido, E., Perez, I., Ruiz-Lara, T., Ellis, S., López-Sánchez, A.R., González Delgado, R.M., de Lorenzo-Cáceres, A., García-Benito, R., Galbany, L., Zibetti, S., Cortijo, C., Kalinova, V., Mast, D., Iglesias-Páramo, J., Papaderos, P., Walcher, C.J., Bland-Hawthorn, J.: Outer-disk reddening and gas-phase metallicities: the CALIFA connection. Astron. Astrophys. **585**, A47 (2016). doi:10.1051/0004-6361/201526986, 1509.07878

Martin, P., Roy, J.R.: The oxygen distribution in NGC 3359 or a disk galaxy in the early phase of bar formation. Astrophys. J. **445**, 161–172 (1995). doi:10.1086/175682

Matteucci, F., Chiosi, C.: Stochastic star formation and chemical evolution of dwarf irregular galaxies. Astron. Astrophys. **123**, 121–134 (1983)

McGaugh, S.S.: H II region abundances - model oxygen line ratios. Astrophys. J. **380**, 140–150 (1991). doi:10.1086/170569

Menzel, D.H., Aller, L.H., Hebb, M.H.: Physical processes in gaseous nebulae. XIII. Astrophys. J. **93**, 230 (1941). doi:10.1086/144259

Minchev, I., Famaey, B., Combes, F., Di Matteo, P., Mouhcine, M., Wozniak, H.: Radial migration in galactic disks caused by resonance overlap of multiple patterns: self-consistent simulations. Astron. Astrophys. **527**, A147 (2011). doi:10.1051/0004-6361/201015139, 1006.0484

Mishurov, Y.N., Lépine, J.R.D., Acharova, I.A.: Corotation: its influence on the chemical abundance pattern of the galaxy. Astrophys. J. Lett. **571**, L113–L115 (2002). doi:10.1086/341360, arXiv:astro-ph/0203458

Moran, S.M., Heckman, T.M., Kauffmann, G., Davé, R., Catinella, B., Brinchmann, J., Wang, J., Schiminovich, D., Saintonge, A., Gracia-Carpio, J., Tacconi, L., Giovanelli, R., Haynes, M., Fabello, S., Hummels, C., Lemonias, J., Wu, R.: The GALEX Arecibo SDSS survey. V. The relation between the H I content of galaxies and metal enrichment at their outskirts. Astrophys. J. **745**, 66 (2012). doi:10.1088/0004-637X/745/1/66, 1112.1084

Olave-Rojas, D., Torres-Flores, S., Carrasco, E.R., Mendes de Oliveira, C., de Mello, D.F., Scarano, S.: NGC 6845: metallicity gradients and star formation in a complex compact group. Mon. Not. R. Astron. Soc. **453**, 2808–2823 (2015). doi:10.1093/mnras/stv1798, 1508.05070

Oppenheimer, B.D., Davé, R.: Mass, metal, and energy feedback in cosmological simulations. Mon. Not. R. Astron. Soc. **387**, 577–600 (2008). doi:10.1111/j.1365-2966.2008.13280.x, 0712.1827

Pagel, B.E.J., Edmunds, M.G., Blackwell, D.E., Chun, M.S., Smith, G.: On the composition of H II regions in southern galaxies. I - NGC 300 and 1365. Mon. Not. R. Astron. Soc. **189**, 95–113 (1979)

Patterson, M.T., Walterbos, R.A.M., Kennicutt, R.C., Chiappini, C., Thilker, D.A.: An oxygen abundance gradient into the outer disc of M81. Mon. Not. R. Astron. Soc. **422**, 401–419 (2012). doi:10.1111/j.1365-2966.2012.20616.x

Petit, A.C., Krumholz, M.R., Goldbaum, N.J., Forbes, J.C.: Mixing and transport of metals by gravitational instability-driven turbulence in galactic discs. Mon. Not. R. Astron. Soc. **449**, 2588–2597 (2015). doi:10.1093/mnras/stv493, 1411.7585

Pettini, M., Pagel, B.E.J.: [OIII]/[NII] as an abundance indicator at high redshift. Mon. Not. R. Astron. Soc. **348**, L59–L63 (2004). doi:10.1111/j.1365-2966.2004.07591.x

Pilyugin, L.S., Grebel, E.K.: New calibrations for abundance determinations in H II regions. Mon. Not. R. Astron. Soc. **457**, 3678–3692 (2016). doi:10.1093/mnras/stw238, 1601.08217

Pilyugin, L.S., Thuan, T.X.: Oxygen abundance determination in H II regions: the strong line intensities-abundance calibration revisited. Astrophys. J. **631**, 231–243 (2005). doi:10.1086/432408

Pilyugin, L.S., Grebel, E.K., Mattsson, L.: 'Counterpart' method for abundance determinations in H II regions. Mon. Not. R. Astron. Soc. **424**, 2316–2329 (2012). doi:10.1111/j.1365-2966.2012.21398.x, 1205.5716

Rich, J.A., Torrey, P., Kewley, L.J., Dopita, M.A., Rupke, D.S.N.: An integral field study of abundance gradients in nearby luminous infrared galaxies. Astrophys. J. **753**, 5 (2012). doi:10.1088/0004-637X/753/1/5, 1204.5520

Rosales-Ortega, F.F., Díaz, A.I., Kennicutt, R.C., Sánchez, S.F.: PPAK wide-field integral field spectroscopy of NGC 628 - II. Emission line abundance analysis. Mon. Not. R. Astron. Soc. **415**, 2439–2474 (2011). doi:10.1111/j.1365-2966.2011.18870.x, 1104.1136

Roškar, R., Debattista, V.P., Quinn, T.R., Stinson, G.S., Wadsley, J.: Riding the spiral waves: implications of stellar migration for the properties of galactic disks. Astrophys. J. Lett. **684**, L79–L82 (2008). doi:10.1086/592231, 0808.0206

Rupke, D.S.N., Kewley, L.J., Barnes, J.E.: Galaxy mergers and the mass-metallicity relation: evidence for nuclear metal dilution and flattened gradients from numerical simulations. Astrophys. J. Lett. **710**, L156–L160 (2010a). doi:10.1088/2041-8205/710/2/L156, 1001.1728

Rupke, D.S.N., Kewley, L.J., Chien, L.H.: Gas-phase oxygen gradients in strongly interacting galaxies. I. Early-stage interactions. Astrophys. J. **723**, 1255–1271 (2010b). doi:10.1088/0004-637X/723/2/1255, 1009.0761

Sánchez, S.F., Rosales-Ortega, F.F., Marino, R.A., Iglesias-Páramo, J., Vílchez, J.M., Kennicutt, R.C., Díaz, A.I., Mast, D., Monreal-Ibero, A., García-Benito, R., Bland-Hawthorn, J., Pérez, E., González Delgado, R., Husemann, B., López-Sánchez, Á.R., Cid Fernandes, R., Kehrig, C., Walcher, C.J., Gil de Paz, A., Ellis, S.: Integral field spectroscopy of a sample of nearby galaxies. II. Properties of the H ii regions. Astron. Astrophys. **546**, A2 (2012). doi:10.1051/0004-6361/201219578, 1208.1117

Sánchez, S.F., Rosales-Ortega, F.F., Iglesias-Páramo, J., Mollá, M., Barrera-Ballesteros, J., Marino, R.A., Pérez, E., Sánchez-Blazquez, P., González Delgado, R., Cid Fernandes, R., de Lorenzo-Cáceres, A., Mendez-Abreu, J., Galbany, L., Falcon-Barroso, J., Miralles-Caballero, D., Husemann, B., García-Benito, R., Mast, D., Walcher, C.J., Gil de Paz, A., García-Lorenzo, B., Jungwiert, B., Vílchez, J.M., Jílková, L., Lyubenova, M., Cortijo-Ferrero, C., Díaz, A.I., Wisotzki, L., Márquez, I., Bland-Hawthorn, J., Ellis, S., van de Ven, G., Jahnke, K., Papaderos, P., Gomes, J.M., Mendoza, M.A., López-Sánchez, Á.R.: A characteristic oxygen abundance gradient in galaxy disks unveiled with CALIFA. Astron. Astrophys. **563**, A49 (2014). doi:10.1051/0004-6361/201322343, 1311.7052

Sánchez-Menguiano, L., Sánchez, S.F., Pérez, I., García-Benito, R., Husemann, B., Mast, D., Mendoza, A., Ruiz-Lara, T., Ascasibar, Y., Bland-Hawthorn, J., Cavichia, O., Díaz, A.I., Florido, E., Galbany, L., Gónzalez Delgado, R.M., Kehrig, C., Marino, R.A., Márquez, I., Masegosa, J., Méndez-Abreu, J., Mollá, M., Del Olmo, A., Pérez, E., Sánchez-Blázquez, P., Stanishev, V., Walcher, C.J., López-Sánchez, Á.R., Califa Collaboration: Shape of the oxygen abundance profiles in CALIFA face-on spiral galaxies. Astron. Astrophys. **587**, A70 (2016). doi:10.1051/0004-6361/201527450, 1601.01542

Sancisi, R., Fraternali, F., Oosterloo, T., van der Hulst, T.: Cold gas accretion in galaxies. Astron. Astrophys. Rev. **15**, 189–223 (2008). doi:10.1007/s00159-008-0010-0, 0803.0109

Sanders, N.E., Caldwell, N., McDowell, J., Harding, P.: The metallicity Profile of M31 from spectroscopy of hundreds of H II regions and PNe. Astrophys. J. **758**, 133 (2012). doi:10.1088/0004-637X/758/2/133, 1209.2251

Sanders, R.L., Shapley, A.E., Kriek, M., Reddy, N.A., Freeman, W.R., Coil, A.L., Siana, B., Mobasher, B., Shivaei, I., Price, S.H., de Groot, L.: The MOSDEF survey: mass, metallicity, and star-formation rate at $z \sim 2.3$. Astrophys. J. **799**, 138 (2015). doi:10.1088/0004-637X/799/2/138, 1408.2521

Scalo, J., Elmegreen, B.G.: Interstellar turbulence II: implications and effects. Annu. Rev. Astron. Astrophys. **42**, 275–316 (2004). doi:10.1146/annurev.astro.42.120403.143327, arXiv:astro-ph/0404452

Scarano, S., Lépine, J.R.D.: Radial metallicity distribution breaks at corotation radius in spiral galaxies. Mon. Not. R. Astron. Soc. **428**, 625–640 (2013). doi:10.1093/mnras/sts048, 1209.5031

Searle, L.: Evidence for composition gradients across the disks of spiral galaxies. Astrophys. J. **168**, 327 (1971). doi:10.1086/151090

Sellwood, J.A., Binney, J.J.: Radial mixing in galactic discs. Mon. Not. R. Astron. Soc. **336**, 785–796 (2002). doi:10.1046/j.1365-8711.2002.05806.x, arXiv:astro-ph/0203510

Shields, G.A.: Composition gradients across spiral galaxies. Astrophys. J. **193**, 335–341 (1974). doi:10.1086/153167

Stanghellini, L., Magrini, L., Casasola, V., Villaver, E.: The radial metallicity gradient and the history of elemental enrichment in M 81 through emission-line probes. Astron. Astrophys. **567**, A88 (2014). doi:10.1051/0004-6361/201423423, 1403.5547

Stasińska, G., Prantzos, N., Meynet, G., Simón-Díaz, S., Chiappini, C., Dessauges-Zavadsky, M., Charbonnel, C., Ludwig, H.G., Mendoza, C., Grevesse, N., Arnould, M., Barbuy, B., Lebreton, Y., Decourchelle, A., Hill, V., Ferrando, P., Hébrard, G., Durret, F., Katsuma, M., Zeippen, C.J. (eds.): Oxygen in the Universe. EAS Publications Series, vol. 54. EDP Sciences, Paris (2012)

Thilker, D.A., Bianchi, L., Boissier, S., Gil de Paz, A., Madore, B.F., Martin, D.C., Meurer, G.R., Neff, S.G., Rich, R.M., Schiminovich, D., Seibert, M., Wyder, T.K., Barlow, T.A., Byun, Y.I., Donas, J., Forster, K., Friedman, P.G., Heckman, T.M., Jelinsky, P.N., Lee, Y.W., Malina, R.F., Milliard, B., Morrissey, P., Siegmund, O.H.W., Small, T., Szalay, A.S., Welsh, B.Y.: Recent star formation in the extreme outer disk of M83. Astrophys. J. Lett. **619**, L79–L82 (2005), doi:10.1086/425251, arXiv:astro-ph/0411306

Thilker, D.A., Bianchi, L., Meurer, G., Gil de Paz, A., Boissier, S., Madore, B.F., Boselli, A., Ferguson, A.M.N., Muñoz-Mateos, J.C., Madsen, G.J., Hameed, S., Overzier, R.A., Forster, K., Friedman, P.G., Martin, D.C., Morrissey, P., Neff, S.G., Schiminovich, D., Seibert, M., Small, T., Wyder, T.K., Donas, J., Heckman, T.M., Lee, Y.W., Milliard, B., Rich, R.M., Szalay, A.S., Welsh, B.Y., Yi, S.K.: A search for extended ultraviolet disk (XUV-Disk) galaxies in the local universe. Astrophys. J. Suppl. **173**, 538–571 (2007). doi:10.1086/523853, arXiv:0712.3555

Thon, R., Meusinger, H.: Models of the long-term evolution of the Galactic disk with viscous flows and gas infall. Astron. Astrophys. **338**, 413–434 (1998)

Toribio San Cipriano, L., García-Rojas, J., Esteban, C., Bresolin, F., Peimbert, M.: Carbon and oxygen abundance gradients in NGC 300 and M33 from optical recombination lines. Mon. Not. R. Astron. Soc. **458**, 1866–1890 (2016). doi:10.1093/mnras/stw397, 1602.05587

Torres-Flores, S., Scarano, S., Mendes de Oliveira, C., de Mello, D.F., Amram, P., Plana, H.: Star-forming regions and the metallicity gradients in the tidal tails: the case of NGC 92. Mon. Not. R. Astron. Soc. **438**, 1894–1908 (2014). doi:10.1093/mnras/stt2340, 1312.0812

Torrey, P., Cox, T.J., Kewley, L., Hernquist, L.: The metallicity evolution of interacting galaxies. Astrophys. J. **746**, 108 (2012). doi:10.1088/0004-637X/746/1/108, 1107.0001

Tosi, M.: The effect of metal-rich infall on galactic chemical evolution. Astron. Astrophys. **197**, 47–51 (1988)

Tremonti, C.A., Heckman, T.M., Kauffmann, G., Brinchmann, J., Charlot, S., White, S.D.M., Seibert, M., Peng, E.W., Schlegel, D.J., Uomoto, A., Fukugita, M., Brinkmann, J.: The origin of the mass-metallicity relation: insights from 53,000 star-forming galaxies in the Sloan Digital Sky Survey. Astrophys. J. **613**, 898–913 (2004). doi:10.1086/423264

Tsujimoto, T., Yoshii, Y., Nomoto, K., Shigeyama, T.: Abundance gradients in the star-forming viscous disk and chemical properties of the bulge. Astron. Astrophys. **302**, 704 (1995)

Tumlinson, J., et al.: The large, oxygen-rich halos of star-forming galaxies are a major reservoir of galactic metals. Science **334**, 948 (2011). doi:10.1126/science.1209840, 1111.3980

Vale Asari, N., Stasińska, G., Morisset, C., Cid Fernandes, R.: BOND: Bayesian oxygen and nitrogen abundance determinations in giant H II regions using strong and semi-strong lines. Mon. Not. R. Astron. Soc. (2016) doi:10.1093/mnras/stw971, 1605.01057

van Zee, L., Salzer, J.J., Haynes, M.P., O'Donoghue, A.A., Balonek, T.J.: Spectroscopy of outlying H II regions in spiral galaxies: abundances and radial gradients. Astron. J. **116**, 2805–2833 (1998). doi:10.1086/300647, arXiv:astro-ph/9808315

Vila-Costas, M.B., Edmunds, M.G.: The relation between abundance gradients and the physical properties of spiral galaxies. Mon. Not. R. Astron. Soc. **259**, 121–145 (1992)

Vila Costas, M.B., Edmunds, M.G.: The nitrogen-to ratio in galaxies and its implications for the origin of nitrogen. Mon. Not. R. Astron. Soc. **265**, 199–212 (1993)

Vílchez, J.M., Esteban, C.: The chemical composition of HII regions in the outer Galaxy. Mon. Not. R. Astron. Soc. **280**, 720–734 (1996)

Vlajić, M., Bland-Hawthorn, J., Freeman, K.C.: The abundance gradient in the extremely faint outer disk of NGC 300. Astrophys. J. **697**, 361–372 (2009). doi:10.1088/0004-637X/697/1/361, 0903.1855

Vlajić, M., Bland-Hawthorn, J., Freeman, K.C.: The structure and metallicity gradient in the extreme outer disk of NGC 7793. Astrophys. J. **732**, 7 (2011). doi:10.1088/0004-637X/732/1/7, 1101.0607

Wang, J., Kauffmann, G., Józsa, G.I.G., Serra, P., van der Hulst, T., Bigiel, F., Brinchmann, J., Verheijen, M.A.W., Oosterloo, T., Wang, E., Li, C., den Heijer, M., Kerp, J.: The Bluedisks project, a study of unusually H I-rich galaxies - I. H I sizes and morphology. Mon. Not. R. Astron. Soc. **433**, 270–294 (2013). doi:10.1093/mnras/stt722, 1303.3538

Werk, J.K., Putman, M.E., Meurer, G.R., Ryan-Weber, E.V., Kehrig, C., Thilker, D.A., Bland-Hawthorn, J., Drinkwater, M.J., Kennicutt, R.C., Wong, O.I., Freeman, K.C., Oey, M.S., Dopita, M.A., Doyle, M.T., Ferguson, H.C., Hanish, D.J., Heckman, T.M., Kilborn, V.A., Kim, J.H., Knezek, P.M., Koribalski, B., Meyer, M., Smith, R.C., Zwaan, M.A.: Outlying H II regions in H I-selected galaxies. Astron. J. **139**, 279–295 (2010). doi:10.1088/0004-6256/139/1/279, 0911.1791

Werk, J.K., Putman, M.E., Meurer, G.R., Santiago-Figueroa, N.: Metal transport to the gaseous outskirts of galaxies. Astrophys. J. **735**, 71 (2011). doi:10.1088/0004-637X/735/2/71, 1104.3897

Werk, J.K., Prochaska, J.X., Thom, C., Tumlinson, J., Tripp, T.M., O'Meara, J.M., Peeples, M.S.: The COS-halos survey: an empirical description of metal-line absorption in the low-redshift circumgalactic medium. Astrophys. J. Suppl. **204**, 17 (2013). doi:10.1088/0067-0049/204/2/17, 1212.0558

Worthey, G., España, A., MacArthur, L.A., Courteau, S.: M31's heavy-element distribution and outer disk. Astrophys. J. **631**, 820–831 (2005). doi:10.1086/432785, arXiv:astro-ph/0410454

Wyder, T.K., Martin, D.C., Barlow, T.A., Foster, K., Friedman, P.G., Morrissey, P., Neff, S.G., Neill, J.D., Schiminovich, D., Seibert, M., Bianchi, L., Donas, J., Heckman, T.M., Lee, Y., Madore, B.F., Milliard, B., Rich, R.M., Szalay, A.S., Yi, S.K.: The star formation law at low surface density. Astrophys. J. **696**, 1834–1853 (2009). doi:10.1088/0004-637X/696/2/1834, 0903.3015

Yang, C.C., Krumholz, M.: Thermal-instability-driven turbulent mixing in galactic disks. I. Effective mixing of metals. Astrophys. J. **758**, 48 (2012). doi:10.1088/0004-637X/758/1/48, 1208.4625

Yong, D., Carney, B.W., Friel, E.D.: Elemental abundance ratios in stars of the outer galactic disk. IV. A new sample of open clusters. Astron. J. **144**, 95 (2012). doi:10.1088/0004-6256/144/4/95, 1206.6931

Zahid, H.J., Bresolin, F.: Reexamination of the radial abundance gradient break in NGC 3359. Astron. J. **141**, 192 (2011). doi:10.1088/0004-6256/141/6/192

Zahid, H.J., Geller, M.J., Kewley, L.J., Hwang, H.S., Fabricant, D.G., Kurtz, M.J.: The chemical evolution of star-forming galaxies over the last 11 billion years. Astrophys. J. Lett. **771**, L19 (2013). doi:10.1088/2041-8205/771/2/L19, 1303.5987

Zaritsky, D., Kennicutt, R.C. Jr., Huchra, J.P.: H II regions and the abundance properties of spiral galaxies. Astrophys. J. **420**, 87–109 (1994). doi:10.1086/173544

Zinchenko, I.A., Berczik, P., Grebel, E.K., Pilyugin, L.S., Just, A.: On the influence of minor mergers on the radial abundance gradient in disks of Milky-Way-like galaxies. Astrophys. J. **806**, 267 (2015). doi:10.1088/0004-637X/806/2/267, 1504.07483

Zurita, A., Bresolin, F.: The chemical abundance in M31 from H II regions. Mon. Not. R. Astron. Soc. **427**, 1463–1481 (2012). doi:10.1111/j.1365-2966.2012.22075.x, 1209.1505

Chapter 6
Molecular Gas in the Outskirts

Linda C. Watson and Jin Koda

Abstract The outskirts of galaxies offer extreme environments where we can test our understanding of the formation, evolution and destruction of molecules and their relationship with star formation and galaxy evolution. We review the basic equations that are used in normal environments to estimate physical parameters like the molecular gas mass from CO line emission and dust continuum emission. Then we discuss how those estimates may be affected when applied to the outskirts, where the average gas density, metallicity, stellar radiation field and temperature may be lower. We focus on observations of molecular gas in the outskirts of the Milky Way, extragalactic disk galaxies, early-type galaxies, groups and clusters. The scientific results show the versatility of molecular gas, as it has been used to trace Milky Way spiral arms out to a galactocentric radius of 15 kpc, to study star formation in extended ultraviolet disk galaxies, to probe galaxy interactions in polar-ring S0 galaxies and to investigate ram pressure stripping in clusters. Throughout the chapter, we highlight the physical stimuli that accelerate the formation of molecular gas, including internal processes such as spiral arm compression and external processes such as interactions.

6.1 Introduction

Despite early discoveries of OB stars and molecular gas in the outer Milky Way (MW; e.g. Fich and Blitz 1984; Brand and Wouterloot 1988), not much attention had been paid to molecular gas in galaxy outskirts primarily because there was

L.C. Watson (✉)
European Southern Observatory, Alonso de Córdova 3107, Vitacura, Casilla 19001, Santiago, Chile
e-mail: lwatson@eso.org

J. Koda
NAOJ Chile Observatory, Joaquín Montero 3000, Oficina 702, Vitacura, Santiago, Chile

Joint ALMA Office, Alonso de Córdova 3107, Casilla 19001, Vitacura, Santiago, Chile

Department of Physics and Astronomy, Stony Brook University, Stony Brook, NY 11794, USA
e-mail: jin.koda@stonybrook.edu

J.H. Knapen et al. (eds.), *Outskirts of Galaxies*, Astrophysics and Space Science Library 434, DOI 10.1007/978-3-319-56570-5_6

a notion that virtually no star formation occurs there. This notion was altered entirely by the *Galaxy Evolution Explorer* (*GALEX*), which revealed that ultraviolet emission often extends far beyond the edges of optical disks (namely, extended ultraviolet disks or XUV disks; Thilker et al. 2005; Gil de Paz et al. 2007b). The UV emission suggests the presence of massive stars, at least B stars, and hence that there was recent star formation within the lifetime of B stars ($\sim$100 Myr). These young stars must have been born nearby, perhaps requiring unnoticed molecular gas and clouds somewhere in the extended galaxy outskirts. Average gas densities there are extremely low compared to typical star-forming regions within the MW. Understanding the conditions of parental molecular gas in such an extreme condition is vital to expand our knowledge of the physics of star formation. We need to understand the internal properties of molecular clouds, including the atomic-to-molecular gas-phase transition, the distribution of molecular clouds and the external environment in galaxy outskirts.

A blind search for molecular gas has been difficult for the large outskirts of nearby galaxies due to the limited capability of existing facilities. The Atacama Large Millimeter/submillimeter Array (ALMA) improved the sensitivity remarkably, but even ALMA would need to invest hours to days to carry out a large areal search for molecular gas over extended disks. This review summarizes the current knowledge on molecular gas and star formation in the outskirts, but this research field is still in a phase of discovery. The space to explore is large, and more systematic understanding will become possible with future observations.

Studies of molecular gas in the outskirts will also reveal the yet unknown physical properties of the interstellar medium (ISM) in the outskirts. Most observational tools were developed and calibrated in the inner parts of galactic disks and may not be applicable as they are to the outskirts. Many studies are subject to *systematic biases*, especially when molecular gas in the outskirts is compared with inner disks. For example, the rotational transition of carbon monoxide (CO) is often used to measure the mass of molecular gas in normal galaxies; however, its presence and excitation conditions depend on the metal abundance, stellar radiation field, internal volume and column densities and kinetic temperature, all of which may change in the outskirts.

In this review, we start from a summary of how the ISM evolves in the inner parts of the MW and nearby galaxies with an emphasis on molecular gas (Sect. 6.2). We then discuss the observational methods, including the equations needed to plan for a future observational search of molecular gas with a radio telescope (Sect. 6.3). We explain the potential effects of applying these equations under the extreme conditions in galaxy outskirts, which may cause systematic biases when the ISM is compared between galaxies' inner parts and outskirts (Sect. 6.3.4). Although not many observations have been carried out in galaxy outskirts, we summarize the current state of molecular gas observations in spiral (Sect. 6.4) and elliptical galaxies (Sect. 6.5) and in galaxy groups and clusters (Sect. 6.6). We finish the review with possible future directions (Sect. 6.7). The term "outskirts" is abstract and has been used differently in different contexts. In this review we use this term for the area beyond the optical radius of galaxy, e.g. beyond r_{25}, which is the radius where the

B-band surface brightness of a galaxy falls to 25 mag arcsec^{-2}. We should, however, note that in some circumstances r_{25} is not defined well, and we have to rely on a loose definition of "outskirts".

The measurements of gas properties, such as molecular mass, often depend on some assumptions of the gas properties themselves. However, galaxy outskirts are an extreme environment, and the assumptions based on previous measurements in inner disks may not be appropriate. This problem needs to be resolved iteratively by adjusting the assumptions to match future observations. We therefore spend a number of pages on the methods of basic measurements (Sect. 6.3), so that the equations and assumptions can be revisited easily in future studies. Readers who already understand the basic methods and assumptions may skip Sect. 6.3 entirely and move from Sect. 6.2 to Sect.6.4.

6.2 Molecular Gas from the Inner to the Outer Regions of Galaxies

The most abundant molecule H_2 does not have significant emission at the cold temperatures that are typical in molecular clouds (<30 K). Hence, the emission from CO, the second-most abundant molecule, is commonly used to trace molecular gas. Molecular gas is typically concentrated towards the centres of galaxies and its surface density decreases with galactic radius (Young and Scoville 1991; Wong and Blitz 2002). The gas phase changes from mostly molecular in the central regions to more atomic in the outer regions (Sofue et al. 1995; Koda et al. 2016; Sofue and Nakanishi 2016). These trends apparently continue into the outskirts, as HI disks often extend beyond the edges of optical disks (Bosma 1981).

We may infer the properties of gas in the outskirts by extending our knowledge from the inner disks. Recently, Koda et al. (2016) concluded that the HI-H_2 gas-phase transition between spiral arm and interarm regions changes as a function of radius in the MW and other nearby galaxies. In the molecule-dominant inner parts, the gas remains highly molecular as it moves from an interarm region into a spiral arm and back into the next interarm region. Stellar feedback does not dissociate molecules much, and perhaps the coagulation and fragmentation of molecular clouds dominate the evolution of the ISM at these radii. The trend differs in the outer regions where the gas phase is atomic on average. The HI gas is converted to H_2 in spiral arm compression and goes back into the HI phase after passing spiral arms. These different regimes of ISM evolution are also seen in the LMC, M33 and M51, depending on the dominant gas phase there (Heyer and Terebey 1998; Engargiola et al. 2003; Koda et al. 2009; Fukui et al. 2009; Tosaki et al. 2011; Colombo et al. 2014).

Even in regions of relatively low gas densities, a natural fluctuation may occasionally lead to gravitational collapse into molecular gas and clouds. For example, many low-density dwarf galaxies show some molecular gas and star formation.

However, some stimulus, such as spiral arm compression, seems necessary to accelerate the HI to H_2 phase transition. In addition to such internal stimuli, there are external stimuli, such as interactions with satellite galaxies, which may also trigger the phase transition into molecular gas in the outskirts.

6.3 Molecular ISM Masses: Basic Equations

The molecular ISM is typically cold and is observed at radio wavelengths. To search for the molecular ISM in galaxy outskirts, one needs to be familiar with conventional notations in radio astronomy. Here we summarize the basic equations and assumptions that have been used in studies of the molecular ISM in traditional environments, such as in the MW's inner disk. In particular, we focus on the $J = 1 - 0, 2 - 1$ rotational transitions of CO molecules and dust continuum emission at millimetre/submillimetre wavelengths. The molecular ISM in galaxy outskirts may have different properties from those in the inner disks. We discuss how expected differences could affect the measurements with CO $J = 1 - 0, 2 - 1$, and dust continuum emission.

6.3.1 Brightness Temperature, Flux Density and Luminosity

The definitions of brightness temperature T_ν, brightness I_ν, flux density S_ν, and luminosity L_ν are often confusing. It is useful to go back to the amount of energy ($\mathrm{d}E$) that passes through an aperture (e.g. detector or sometimes the 4π sky area):

$$\mathrm{d}E = I_\nu \mathrm{d}\Omega_\mathrm{B} \mathrm{d}A \mathrm{d}t \mathrm{d}\nu = \{[I_\nu \mathrm{d}\Omega_\mathrm{B}]\, \mathrm{d}A\}\, \mathrm{d}t \mathrm{d}\nu = \{S_\nu \mathrm{d}A\}\, \mathrm{d}t \mathrm{d}\nu = L_\nu \mathrm{d}t \mathrm{d}\nu, \tag{6.1}$$

where $S_\nu = \int I_\nu \mathrm{d}\Omega_\mathrm{B}$ and $L_\nu = \int\int I_\nu \mathrm{d}\Omega_\mathrm{B} \mathrm{d}A$ (see Fig. 6.1). The $\mathrm{d}t$ and $\mathrm{d}\nu$ denote unit time and frequency, respectively. The $\mathrm{d}\Omega_\mathrm{B}$ is the solid angle of the source and

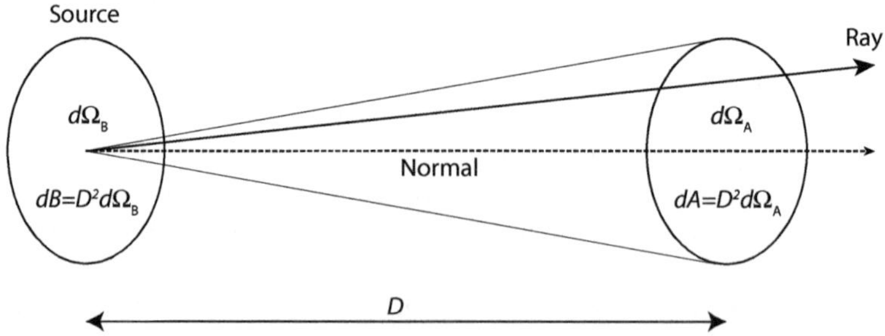

Fig. 6.1 Definitions of parameters. The rays emitted from the source with the area $\mathrm{d}B = D^2 \mathrm{d}\Omega_\mathrm{B}$ pass through the solid angle $\mathrm{d}\Omega_\mathrm{A}$ (or the area $D^2 \mathrm{d}\Omega_\mathrm{A}$) at the distance of D

has the relation with the physical area $dB = D^2 d\Omega_B$ with the distance D. Similarly, $dA = D^2 d\Omega_A$ using the solid angle of the aperture area seen from the source $d\Omega_A$. The aperture dA can be a portion of the 4π sky sphere as it is seen from the source and is $4\pi D^2$ when integrated over the entire sphere to calculate luminosity. The dA could also represent an area of a detector (or a pixel of a detector).

The flux density S_ν is often expressed in the unit of "Jansky (Jy)", which is equivalent to "$10^{-23}\,\mathrm{erg\,s^{-1}\,cm^{-2}\,Hz^{-1}}$". An integration of I_ν over a solid angle $d\Omega_B$ (e.g. telescope beam area or synthesized beam area) provides S_ν. In reverse, I_ν is S_ν divided by the solid angle Ω_B [$= \int d\Omega_B$]. Therefore, the brightness I_ν [$= S_\nu/\Omega_B$] is expressed in the unit of "Jy/beam".

The brightness temperature T_ν is the temperature that makes the black body function $B_\nu(T_\nu)$ have the same brightness as the observed I_ν at a frequency ν (i.e. $I_\nu = B_\nu(T_\nu)$), even when I_ν does not follow the black body law! In the Rayleigh-Jeans regime ($h\nu \ll kT$),

$$T_\nu = \frac{c^2}{2\nu^2 k} I_\nu = \frac{c^2}{2\nu^2 k}\left(\frac{S_\nu}{\Omega_B}\right). \tag{6.2}$$

The T_ν characterizes radiation and is *not necessarily* a physical temperature of an emitting body. However, if the emitting body is an optically thick black body and is filling the beam Ω_B, T_ν is equivalent to the physical temperature of the emitting body when the Rayleigh-Jeans criterion is satisfied.

The T_ν is measured in "Kelvin". This unit is convenient in radio astronomy since radio single-dish observations calibrate a flux scale in the Kelvin unit using hot and cold loads of known temperatures. Giant molecular clouds (GMCs) in the MW have a typical temperature of $\sim$10 K (Scoville and Sanders 1987), and the black body radiation $B_\nu(T)$ at this temperature peaks at $\nu \sim 588$ GHz ($\sim$510 μm). Therefore, most radio observations of molecular gas are in the Rayleigh-Jeans range.

A numerical expression of Eq. (6.2) is useful in practice,

$$\left(\frac{T_\nu}{\mathrm{K}}\right) = 13.6\left(\frac{\lambda}{\mathrm{mm}}\right)^2\left(\frac{S_\nu}{\mathrm{Jy}}\right)\left(\frac{b_{\mathrm{maj}} \times b_{\mathrm{min}}}{1'' \times 1''}\right)^{-1}. \tag{6.3}$$

The last term corresponds to Ω_B in Eq. (6.2) and is calculated as

$$\Omega_B = \frac{\pi b_{\mathrm{maj}} b_{\mathrm{min}}}{4\ln 2} \sim 1.133 b_{\mathrm{maj}} b_{\mathrm{min}}, \tag{6.4}$$

which represents the area of interest (e.g. source size, telescope beam) as a 2-d Gaussian with the major and minor axis FWHM diameters of b_{maj} and b_{min}, respectively. Equation (6.3) is sometimes written with brightness as

$$\left(\frac{T_\nu}{\mathrm{K}}\right) = 13.6\left(\frac{\lambda}{\mathrm{mm}}\right)^2\left(\frac{I_\nu}{\mathrm{Jy/beam}}\right)\left(\frac{b_{\mathrm{maj}} \times b_{\mathrm{min}}}{1'' \times 1''}\right)^{-1}, \tag{6.5}$$

where in this case the last term is for the unit conversion from "beam" into arcsec^2 and b_{maj} and b_{min} must refer to the telescope beam or synthesized beam.

6.3.2 *Observations of the Molecular ISM Using CO Line Emission*

Molecular hydrogen (H_2) is the principal component of the ISM at a high density, $>100\,\mathrm{cm}^{-3}$. This molecule has virtually no emission at cold temperatures. Hence, CO emission is typically used to trace the molecular ISM. Conventionally, the molecular ISM mass M_{mol} includes the masses of helium and other elements. $M_{\mathrm{mol}} = 1.36\,M_{\mathrm{H_2}}$ is used to convert the H_2 mass into M_{mol}.

6.3.2.1 CO($J = 1 - 0$) Line Emission

The fundamental CO rotational transition $J = 1 - 0$ at $\nu_{\mathrm{CO}}(1 - 0) = 115.271208\,\mathrm{GHz}$ has been used to measure the molecular ISM mass since the 1980s. For simplicity we omit "CO(1 − 0)" in subscript and instead write "10". Hence, $\nu_{\mathrm{CO}}(1-0) = \nu_{10}$.

The dynamical masses of GMCs and their CO(1 − 0) luminosities are linearly correlated in the MW's inner disk (Scoville et al. 1987; Solomon et al. 1987). If a great majority of molecules reside in GMCs, the CO(1 − 0) luminosity L'_{10} integrated over an area (i.e. an ensemble of GMCs in the area) can be linearly translated to the molecular mass M_{mol},

$$M_{\mathrm{mol}} = \alpha_{10} L'_{10}, \tag{6.6}$$

where α_{10} (or X_{CO}; see below) is a mass-to-light ratio and is called the CO-to-H_2 conversion factor (Bolatto et al. 2013).

By convention we define L'_{10}, instead of L_{10} (Eq. (6.1)). With the CO(1 − 0) brightness temperature T_{10} (instead of I_ν or I_{10}), velocity width $\mathrm{d}v$ (instead of frequency width $\mathrm{d}\nu$) and beam area in physical scale $\mathrm{d}B = D^2 \mathrm{d}\Omega_{\mathrm{B}}$, it is defined as

$$L'_{10} \equiv \int\int T_{10} \mathrm{d}v \mathrm{d}B = \frac{c^2}{2\nu_{10}^2 k} \left[\int S_{10} \mathrm{d}v \right] D^2, \tag{6.7}$$

where we used Eq. (6.2) for T_{10}. The molecular mass is

$$M_{\mathrm{mol}} = \alpha_{10} \frac{c^2}{2\nu_{10}^2 k} \left[\int S_{10} \mathrm{d}v \right] D^2. \tag{6.8}$$

Numerically, this can be expressed as

$$\left(\frac{M_{\rm mol}}{M_\odot}\right) = 1.1\times10^4 \left(\frac{\alpha_{10}}{4.3\,M_\odot\,{\rm pc}^{-2}[{\rm K\cdot km/s}]^{-1}}\right)\left(\frac{\int S_{10}{\rm d}v}{\rm Jy\cdot km/s}\right)\left(\frac{D}{\rm Mpc}\right)^2. \tag{6.9}$$

Note that S_{10} [$= \int I_{10}{\rm d}\Omega_{\rm B}$] is an integration over an area of interest (or summation over all pixels within the area). The $\alpha_{10} = 4.3\,M_\odot\,{\rm pc}^{-2}$ corresponds to the conversion factor of $X_{\rm CO} = 2.0\times10^{20}\,{\rm cm}^{-2}\,[{\rm K\cdot km/s}]^{-1}$ multiplied by the factor of 1.36 to account for the masses of helium and other elements. α_{10} includes helium, while $X_{\rm CO}$ does not. The calibration of α_{10} (or $X_{\rm CO}$) is discussed in Bolatto et al. (2013).

A typical GMC in the MW has a mass of $4\times10^5\,M_\odot$ and $\mathrm{d}v = 8.9\,{\rm km/s}$ (FWHM) (Scoville and Sanders 1987), which is $\int S_{10}{\rm d}v \sim 1.5\,{\rm Jy\,km/s}$ or $S_{10} \sim 170\,{\rm mJy}$ at $D = 5\,{\rm Mpc}$.

6.3.2.2 CO($J = 2-1$) Line Emission

The CO($J = 2-1$) emission (230.538 GHz) is also useful for a rough estimation of molecular mass though an excitation condition may play a role (see below). We can redefine Eq. (6.8) for CO(2−1) by replacing the subscripts from 10 to 21 and using a new CO(2 − 1)-to-H_2 conversion factor $\alpha_{21} \equiv \alpha_{10}/R_{21/10}$, where $R_{21/10}[\equiv T_{21}/T_{10}]$ is the CO $J = 2-1/1-0$ line ratio in brightness temperature.

In practice, α_{10} and $R_{21/10}$ are carried over in use of CO($J = 2-1$) as these are the parameters that have been measured. Equation (6.8) is now

$$M_{\rm mol} = \left(\frac{\alpha_{10}}{R_{21/10}}\right)\frac{c^2}{2\nu_{21}^2 k}\left[\int S_{21}{\rm d}v\right]D^2. \tag{6.10}$$

A numerical evaluation gives

$$\left(\frac{M_{\rm H_2}}{M_\odot}\right) = 3.8\times10^3 \left(\frac{\alpha_{10}}{4.3\,M_\odot\,{\rm pc}^{-2}[{\rm K\cdot km/s}]^{-1}}\right)\left(\frac{R_{21/10}}{0.7}\right)^{-1}\left(\frac{\int S_{21}{\rm d}v}{\rm Jy\cdot km/s}\right)\left(\frac{D}{\rm Mpc}\right)^2. \tag{6.11}$$

The typical GMC with $4\times10^5\,M_\odot$ and ${\rm d}v = 8.9\,{\rm km/s}$ has $\int S_{21}{\rm d}v \sim 4.2\,{\rm Jy\,km/s}$ or $S_{21} \sim 470\,{\rm mJy}$ at $D = 5\,{\rm Mpc}$. Note $S_{21} > S_{10}$ for the same GMC because $S_{21}/S_{10} = (\nu_{21}/\nu_{10})^2 T_{21}/T_{10} = (\nu_{21}/\nu_{10})^2 R_{21/10} \sim 2.8$ from Eq. (6.2), where the $(\nu_{21}/\nu_{10})^2$ term arises from two facts: at the higher frequency, (a) each photon carries twice the energy, and (b) there are two times more photons in each frequency interval ${\rm d}\nu$, which is in the denominator of the definition of flux density S. Empirically, $R_{21/10} \sim 0.7$ on average in the MW (Sakamoto et al. 1997; Hasegawa 1997), which is consistent with a theoretical explanation under the conditions of the MW disk (Scoville and Solomon 1974; Goldreich and Kwan 1974; see Sect. 6.3.4).

6.3.3 Observations of the Molecular ISM Using Dust Continuum Emission

Continuum emission from dust provides an alternative means for ISM mass measurement. Dust is mixed in the gas-phase ISM, and its emission at millimetre/submillimetre waves correlates well with the fluxes of both atomic gas (HI 21 cm emission) and molecular gas (CO emission). Scoville et al. (2016) discussed the usage and calibration of dust emission for ISM mass measurement. We briefly summarize the basic equations, whose normalization will be adjusted with an empirical fitting in the end.

The radiative transfer equation gives the brightness of dust emission

$$I_\nu = (1 - e^{-\tau_\nu}) B_\nu(T_\mathrm{d}) \tag{6.12}$$

with the black body radiation $B_\nu(T_\mathrm{d})$ at the dust temperature T_d and the optical depth τ_ν. The flux density of dust is an integration:

$$S_\nu = \int (1 - e^{-\tau_\nu}) B_\nu(T_\mathrm{d}) \mathrm{d}\Omega_\mathrm{B} = (1 - e^{-\tau_\nu}) B_\nu(T_\mathrm{d}) \Omega_\mathrm{B}, \tag{6.13}$$

where B_ν and τ_ν are assumed constant within Ω_B $[= \int \mathrm{d}\Omega_\mathrm{B}]$. When the integration is over the beam area, S_ν is the flux density within the beam, and $(S_\nu/\Omega_\mathrm{B})$, from Eq. (6.13), is in Jy/beam.

An integration of S_ν over the entire sky area at the distance of D (i.e. $\int dA = D^2 \int_{4\pi} d\Omega_\mathrm{A} = 4\pi D^2$) gives the luminosity

$$L_\nu = \int (1 - e^{-\tau_\nu}) B_\nu(T_\mathrm{d}) \Omega_\mathrm{B} \mathrm{d}A = (1 - e^{-\tau_\nu}) B_\nu(T_\mathrm{d}) \Omega_\mathrm{B} 4\pi D^2 \tag{6.14}$$

$$\approx 4\pi \tau_\nu B_\nu(T_\mathrm{d}) D^2 \Omega_\mathrm{B} = 4\pi \kappa_\nu \Sigma_\mathrm{d} B_\nu(T_\mathrm{d}) D^2 \Omega_\mathrm{B} = 4\pi \kappa_\nu M_\mathrm{d} B_\nu(T_\mathrm{d}). \tag{6.15}$$

The dust is optically thin at mm/sub-mm wavelengths, and we used $(1 - e^{-\tau_\nu}) \sim \tau_\nu = \kappa_\nu \Sigma_\mathrm{d}$, where κ_ν and Σ_d are the absorption coefficient and surface density of dust. The dust mass within the beam is $M_\mathrm{d} = \Sigma_\mathrm{d} D^2 \Omega_\mathrm{B}$. Obviously, the dust continuum luminosity depends on the dust properties (e.g. compositions and size distribution; via κ_ν), amount (M_d) and temperature (T_d).

Equation (6.15) gives the mass-to-light ratio for dust

$$\frac{M_\mathrm{d}}{L_\nu} = \frac{1}{4\pi \kappa_\nu B_\nu(T_\mathrm{d})}. \tag{6.16}$$

We convert M_d into gas mass, $M_\mathrm{gas} = \delta_\mathrm{GDR} M_\mathrm{d}$, with the gas-to-dust ratio δ_GDR. By redefining the dust absorption coefficient $\kappa'_\nu \equiv \kappa_\nu/\delta_\mathrm{GDR}$ (the absorption coefficient per unit total mass of gas), the gas mass-to-dust continuum flux ratio γ_ν at the frequency ν becomes

$$\gamma_\nu \equiv \frac{M_{\rm gas}}{L_\nu} = \frac{1}{4\pi\kappa'_\nu B_\nu(T_{\rm d})}. \tag{6.17}$$

Once γ_ν is obtained, the gas mass is estimated as $M_{\rm gas} = \gamma_\nu L_\nu$. Here, we use the character γ, instead of α that Scoville et al. (2016) used, to avoid a confusion with the CO-to-H_2 conversion factor. Dust continuum emission is associated with HI and H_2, and $M_{\rm gas} \sim M_{\rm mol}$ in dense, molecule-dominated regions ($\gtrsim 100\,{\rm cm}^{-3}$).

The κ'_ν can be approximated as a power-law $\kappa'_\nu = \kappa'_{850\,\mu{\rm m}}(\lambda/850\,\mu{\rm m})^{-\beta}$ with the spectral index $\beta \sim 1.8$ (Planck Collaboration et al. 2011) and coefficient $\kappa'_{850\,\mu{\rm m}}$ at $\lambda = 850\,\mu{\rm m}$ (352 GHz). In order to show the frequency dependence explicitly, we separate $B_\nu(T_{\rm d})$ into the Rayleigh-Jeans term and the correction term $\Gamma_\nu(T_{\rm d})$ as $B_\nu(T_{\rm d}) = (2\nu^2 kT_{\rm d}/c^2)\Gamma_\nu(T_{\rm d})$, where

$$\Gamma_\nu(T_{\rm d}) = \frac{x}{e^x - 1} \quad \text{with} \quad x = \frac{h\nu}{kT_{\rm d}}. \tag{6.18}$$

Equation (6.17) has the dependence $\gamma_\nu \propto \nu^{-(\beta+2)} T_{\rm d}^{-1} \Gamma_\nu(T_{\rm d})^{-1}$, and the proportionality coefficient, including $\kappa'_{850\,\mu{\rm m}}$ and $\delta_{\rm GDR}$, is evaluated empirically.

Scoville et al. (2016) cautioned that $T_{\rm d}$ should not be derived from a spectral energy distribution fit (which gives a luminosity-weighted average $T_{\rm d}$ biased towards hot dust with a peak in the infrared). Instead, they suggested to use a mass-weighted $T_{\rm d}$ for the bulk dust component where the most mass resides. Scoville et al. (2016) adopted $T_{\rm d} = 25$ K and calibrated $\gamma_{\nu 850\,\mu{\rm m}}$ from an empirical comparison of $M_{\rm mol}$ (from CO measurements) and L_ν,

$$\left(\frac{\gamma_\nu}{M_\odot [{\rm Jy\,cm^2}]^{-1}}\right) = 1.5 \pm 0.4 \times 10^3 \left(\frac{\nu}{352\,{\rm GHz}}\right)^{-3.8} \left(\frac{T_{\rm d}}{25\,{\rm K}}\right)^{-1} \left(\frac{\Gamma_\nu(T_{\rm d})}{\Gamma_{\nu 850\,\mu{\rm m}}(25\,{\rm K})}\right)^{-1}. \tag{6.19}$$

The luminosity is calculated from the observed S_ν in Jy and distance D in centimetre as $L_\nu = 4\pi D^2 S_\nu$ [Jy cm^2]. The gas mass is then $M_{\rm mol} = \gamma_\nu L_\nu$.

6.3.4 The ISM in Extreme Environments Such as the Outskirts

The methods for molecular ISM mass measurement that we discussed above were developed and calibrated mainly for the inner parts of galaxies. However, it is not guaranteed that these calibrations are valid in extreme environments such as galaxy outskirts. In fact, metallicities appear to be lower in the outskirts than in the inner part (see Bresolin 2017). On a 1 kpc scale average, gas and stellar surface densities, and hence stellar radiation fields, are also lower, although it is not clear if these trends persist at smaller scales, e.g. cloud scales, where the molecular ISM typically exists. Empirically, α_{10} could be larger when metallicities are lower, and $R_{21/10}$ could be smaller when gas density and/or temperature are lower.

In order to search for the molecular ISM and to understand star formation in the outskirts, it is important to take into account the properties and conditions of the ISM there. Here we explain some aspects that may bias measurements if the above equations are applied naively as they are. These potential biases should not discourage future research and, instead, should be adjusted continuously as we learn more about the ISM in the extreme environment.

6.3.4.1 Variations of α_{10} (or X_{CO})

The CO-to-H_2 conversion factor α_{10} (or X_{CO}) is a mass-to-light ratio between the CO($1-0$) luminosity and the molecular ISM mass (Bolatto et al. 2013). Empirically, this factor increases with decreasing metallicity (Arimoto et al. 1996; Leroy et al. 2011) due to the decreasing abundance of CO over H_2. At the low metallicity of the Small Magellanic Cloud ($\sim$1/10$Z_{\odot}$), α_{10} appears $\sim$10–20 times larger (Arimoto et al. 1996; Leroy et al. 2011).

This trend can be understood based on the self-shielding nature of molecular clouds. Molecules on cloud surfaces are constantly photodissociated by stellar UV radiation. At high densities within clouds, the formation rate of molecules can be as fast as the dissociation rate, and hence molecules are maintained in molecular clouds. The depth where molecules are maintained depends on the strength of the ambient UV radiation field and its attenuation by line absorptions by the molecules themselves as well as by continuum absorption by dust (van Dishoeck and Black 1988).

H_2 is $\sim 10^4$ times more abundant than CO. It can easily become optically thick on the skin of cloud surfaces and be self-shielded (Fig. 6.2). On the other hand, UV photons for CO dissociation penetrate deeper into the cloud due to its lower abundance. This process generates the CO-dark H_2 layer around molecular clouds (Fig. 6.2b; Wolfire et al. 2010). Shielding by dust is more important for CO than H_2. Therefore, if the metallicity or dust abundance is low, the UV photons for CO dissociation reach deeper and deeper and eventually destroy all CO molecules, while H_2 still remains (Fig. 6.2c). As the CO-dark H_2 layer becomes thicker, L_{10} decreases, while M_{H_2} stays high, resulting in a larger α_{10} in a low-metallicity environment, such as galaxy outskirts. Since this process depends on the depth that photons can penetrate (through dust attenuation as well as line absorption), the visual extinction A_V is often used as a parameter to characterize α_{10} (or X_{CO}).

6.3.4.2 Variations of $R_{21/10}$

The CO($2-1$) line emission is useful to locate the molecular ISM and to derive a rough estimation of its mass. However, the higher transitions inevitably suffer from excitation conditions. Indeed, $R_{21/10}$ ($\equiv T_{21}/T_{10}$) has been observed to vary by a

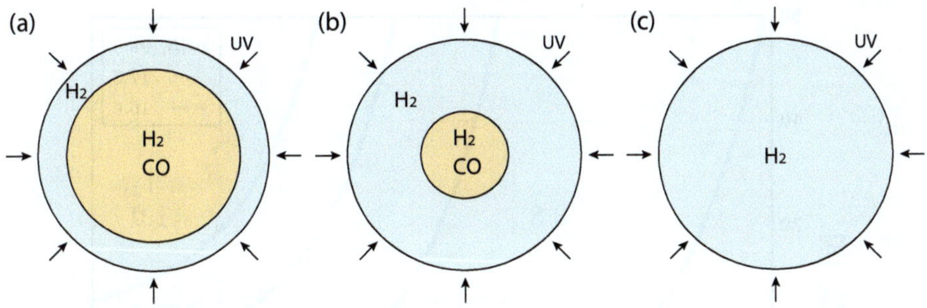

Fig. 6.2 Self-shielding nature of molecules in molecular clouds. The abundance of molecules is maintained in clouds, since the destruction (photodissociation by UV radiation) and formation rates are in balance. The shielding from ambient UV radiation is mainly due to line absorption by molecules themselves. Therefore, the abundant H_2 molecules become optically thick at the absorption-line wavelengths on the skin of clouds, while UV photons for CO dissociation can get deeper into clouds. This mechanism generates the CO-dark H_2 layer on the surface of molecular clouds. This layer can become thicker (*panels* (**a**), (**b**), (**c**)) under several conditions, e.g. lower metallicity or stronger local radiation field. The CO-to-H_2 conversion factor α_{10} (or X_{CO}) increases with the increasing thickness of the CO-dark H_2 layer and, therefore, with lower metallicity or stronger local radiation field

factor of 2–3 in the MW and in other nearby galaxies, e.g. between star-forming molecular clouds (typically $R_{21/10} \sim 0.7$–1.0 and occasionally up to 1.2) and dormant clouds (∼0.4–0.7) and between spiral arms (>0.7) and interarm regions (<0.7; Sakamoto et al. 1997; Koda et al. 2012). The variation may be negligible for finding molecular gas but may cause a systematic bias, for example, in comparing galaxy outskirts with inner disks. It is noteworthy that $R_{21/10}$ changes systematically with star formation activity and varies along the direction of the Kennicutt-Schmidt relation, which can introduce a bias.

Theoretically, $R_{21/10}$ is controlled by three parameters: the volume density n_{H_2} and kinetic temperature T_k—which determine the CO excitation condition due to collisions—and the column density N_{CO}, which controls radiative transfer and photon trapping (Scoville and Solomon 1974; Goldreich and Kwan 1974). Figure 6.3 shows the variation of $R_{21/10}$ with respect to n_{H_2} and T_k under the large velocity gradient (LVG) approximation. In this approximation, the Doppler shift due to a cloud's internal velocity gradient is assumed to be large enough such that any two parcels along the line of sight do not overlap in velocity space. The front parcel does not block emission from the back parcel, and the optical depth is determined only locally within the parcel (or in small dv). Therefore, the column density is expressed per velocity N_{CO}/dv. A typical velocity range in molecular clouds is adopted for this figure. An average GMC in the MW has $n_{H_2} \sim 300\,\mathrm{cm}^{-3}$ and $T_k \sim 10\,\mathrm{K}$ (Scoville and Sanders 1987), which results in $R_{21/10}$ of ∼0.6–0.7. If the density and/or temperature is a factor of 2–3 higher due to a contraction before star formation or feedback from young stars, the ratio increases to $R_{21/10} > 0.7$.

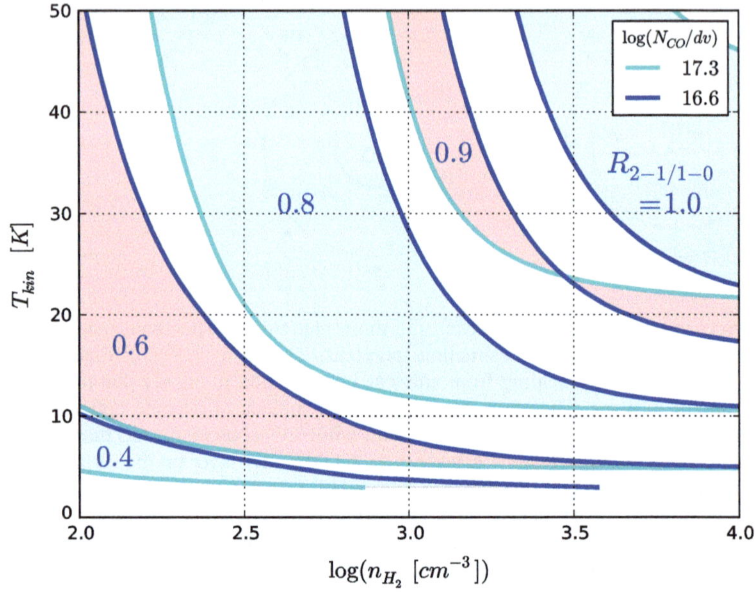

Fig. 6.3 The CO $J = 2 - 1/1 - 0$ line ratios as function of the gas kinetic temperature $T_{\rm kin}$ and H_2 density n_{H2} under the LVG approximation (from Koda et al. 2012). Most GMCs in the MW have CO column density in the range of $\log(N_{\rm CO}/dv) \sim$ 16.6–17.3, assuming the CO fractional abundance to H_2 of 8×10^{-5}. An average GMC in the MW has $n_{H_2} \sim 300\,{\rm cm}^{-3}$ and $T_{\rm k} \sim 10\,{\rm K}$ and therefore shows $R_{21/10}$ ~0.6–0.7. $R_{21/10}$ is <0.7 if the density and/or temperature decrease by a factor of 2–3, and $R_{21/10}$ is >0.7 if the density and/or temperature increase by a factor of 2–3. Observationally, dormant clouds typically have $R_{21/10} = 0.4$–0.7, while actively star-forming clouds have $R_{21/10} = 0.7$–1.0 (and occasionally up to ~1.2; Sakamoto et al. 1997; Hasegawa 1997). There is also a systematic variation between spiral arms ($R_{21/10} > 0.7$) and interarm regions ($R_{21/10} < 0.7$; Koda et al. 2012)

On the contrary, if a cloud is dormant compared to the average, the ratio is lower $R_{21/10} < 0.7$.

In the MW, cloud properties appear to change with the galactocentric radius (Heyer and Dame 2015). If their densities or temperatures are lower in the outskirts, it would result in a lower $R_{21/10}$ and, hence, a higher H_2 mass at a given CO(2 − 1) luminosity. If the $R_{21/10}$ variation is not accounted for, it could result in a bias when clouds within the inner disk and in the outskirts are compared.

6.3.4.3 Variations of Dust Properties and Temperature

The gas mass-to-dust luminosity $M_{\rm gas}/L_{\nu}$ depends on the dust properties/emissivity (κ_{ν}), dust temperature ($T_{\rm d}$) and gas-to-dust ratio ($\delta_{\rm GDR}$)—see Eqs. (6.16) and (6.17). All of these parameters could change in galaxy outskirts, which have low average metallicity, density and stellar radiation field. Of course, the assumption of a single

T_d casts a limitation to the measurement as the ISM is multiphase in reality, although the key idea of using Eqs. (6.16) and (6.17) is to target regions where the cold, molecular ISM is dominant (Scoville et al. 2016). The δ_{GDR} may increase with decreasing metallicity by about an order of magnitude ($\delta_{GDR} \sim 40 \rightarrow 400$) for the change of metallicity $12 + \log(O/H)$ from $\sim 9.0 \rightarrow 8.0$ (their Fig. 6; Leroy et al. 2011). If this trend applies to the outskirts, Eq. (6.17) would tend to underestimate the gas mass by up to an order of magnitude.

Excess dust emission at millimetre/submillimetre wavelengths has been reported in the Small and Large Magellanic Clouds (SMC and LMC) and other dwarfs (Bot et al. 2010; Dale et al. 2012; see also Kirkpatrick et al. 2013). This excess emission appears significant when spectral energy distribution fits to infrared data are extrapolated to millimetre/submillimetre wavelengths. Among the possible explanations are the presence of very cold dust, a change of the dust spectral index and spinning dust emission (e.g. Bot et al. 2010). Gordon et al. (2014) suggested that variations in the dust emissivity are the most probable cause in the LMC and SMC from their analysis of infrared data from the *Herschel Space Observatory*. The environment of galaxy outskirts may be similar to those of the LMC/SMC. The excess emission (27 and 43% for the LMC and SMC, respectively; Gordon et al. 2014) can be ignored if one only needs to locate dust in the vast outskirts, but could cause a systematic bias when the ISM is compared between inner disks and outskirts.

6.4 Molecular Gas Observations in the Outskirts of Disk Galaxies

A primary motivation for molecular gas observations in the outskirts of disk galaxies has been to study molecular clouds and star formation in an extreme environment with lower average density and metallicity. Many researchers highlight that these studies may teach us about the early Universe, where these conditions were more prevalent.

6.4.1 The Milky Way

The MW is the disk galaxy with the most molecular gas detections in the outskirts, with pioneering studies of the outer disk molecular gas and star formation properties beginning in the 1980s (e.g. Fich and Blitz 1984; Brand and Wouterloot 1988). The MW can serve as a model for the types of studies that can be done in nearby galaxies with larger and more sensitive facilities. We will use "outer" MW to refer to galactocentric radii between the Solar circle ($R_{Gal} > R_{\odot} = 8.5$ kpc) and the edge of the optical disk, which is estimated to be at $R_{Gal} \sim 13$–19 kpc (Ruffle et al.

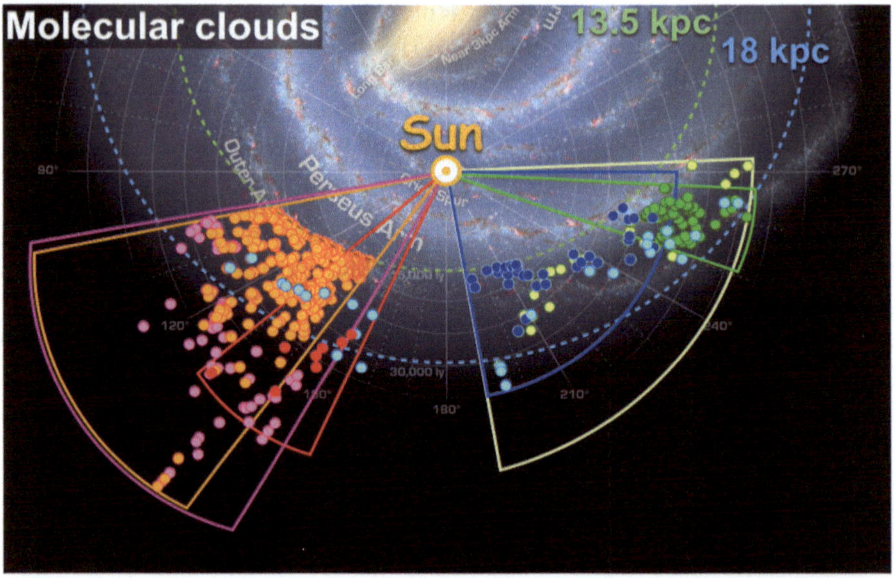

Fig. 6.4 Figure from N. Izumi (personal communication) showing the known molecular clouds at $R_{\rm Gal} > 13.5$ kpc in the second and third quadrants overlaid on an artist's conception of the MW (R. Hurt: NASA/JPL-Caltech/SSC). The *colours* correspond to the following surveys: *orange*, Brunt et al. (2003); *magenta*, Sun et al. (2015); *red*, Digel et al. (1994); *cyan*, Brand and Wouterloot (1994); *blue*, May et al. (1997); *green*, Nakagawa et al. (2005); and *yellow*, Vázquez et al. (2008). The *points* represent molecular clouds and the *fan-shaped regions* represent the survey area. The distances were derived assuming $R_{\odot} = 8.5$ kpc and a Solar orbital speed of $V_{\odot} = 220\,{\rm km\,s^{-1}}$

2007; Sale et al. 2010 and references therein). We will use "outskirts" to refer to galactocentric radii beyond the edge of the optical disk.

Only about 2% of the molecular mass of the MW is at $R_{\rm Gal} > 14.5$ kpc (Nakagawa et al. 2005 estimated the molecular mass at $R_{\rm Gal} > 14.5$ kpc to be $2\times10^7\,M_{\odot}$, while Heyer and Dame (2015) estimated the total molecular mass of the Galaxy to be $(1\pm0.3)\times10^9\,M_{\odot}$). N. Izumi (personal communication) collected the known molecular clouds with $R_{\rm Gal} > 13.5$ kpc in the second and third quadrants (Fig. 6.4). The molecular cloud with the largest known galactocentric radius is probably Digel Cloud 1 with a kinematic galactocentric radius of $R_{\rm Gal} = 22$ kpc, dynamical mass of $\sim 6\times10^4\,M_{\odot}$ and radius of 36 pc (Digel Cloud 2 has a larger kinematic distance of $R_{\rm Gal} = 24$ kpc, but the photometric distance is $R_{\rm Gal} = 15$–19 kpc based on optical spectroscopy of an associated B star; Digel et al. 1994; Yasui et al. 2006, 2008; Izumi et al. 2014). Digel Cloud 1 is beyond the edge of the optical disk but well within the HI disk, which extends to $R_{\rm Gal} \sim 30$ kpc (Digel et al. 1994; Ruffle et al. 2007 and references therein).

Extremely tenuous H_2 gas is mixed with the HI gas in the Galactic halo with a fraction of H_2 over HI of only $10^{-4\,{\rm to}\,-5}$ (Lehner 2002). Such tenuous H_2 is observed via UV absorption, e.g. towards the Magellanic Stream (Lehner 2002) and high velocity clouds (HVCs; Bluhm et al. 2001). This component is important for understanding the complex physics of the ISM, but is not a major molecular component in galaxy outskirts. We therefore do not discuss this component further in this review.

6.4.1.1 Properties of Molecular Clouds in the Outer Milky Way

In this section we highlight studies that have compared the mass, size and mass surface density of molecular clouds in the outer MW to clouds in the inner MW. Molecular clouds are the site of star formation, and hence, comparisons of their properties between the inner and outer MW are important. In general, molecular clouds in the outer MW have lower mass and mass surface density than clouds in the inner disk. We also describe how molecular clouds have been used to trace spiral arms into the outskirts and to study relatively high-mass star formation.

Heyer and Dame (2015) combined published data on the CO surface brightness out to $R_{\rm Gal} \sim 20\,{\rm kpc}$. The clouds in the outer MW and outskirts are ~7 times fainter than clouds in the inner MW (and even fainter relative to the Galactic centre). Assuming a constant $X_{\rm CO}$, this corresponds to a factor of ~7 decrease in the mass surface density of molecular clouds. Heyer and Dame (2015) argued that there is a real decrease in the mass surface density of the molecular clouds, perhaps caused by the lower midplane pressure or stronger local FUV radiation field in the outer Galaxy. However, there is also evidence that the outer MW requires a larger $X_{\rm CO}$ to convert the CO surface brightness into the mass surface density (see Sect.6.3.4). Therefore the mass surface density likely decreases by somewhat less than a factor of ~7.

The mass function of molecular clouds in the outer MW ($9.5\,{\rm kpc} \lesssim R_{\rm Gal} \lesssim 13.5\,{\rm kpc}$ in this study) has a steeper power-law index than that in the inner MW, such that the outer disk hosts more of its molecular mass in lower-mass clouds (Rosolowsky 2005, based on the 330 deg^2 Heyer et al. 1998 catalogue and analysis in Heyer et al. 2001 and Brunt et al. 2003), although this conclusion may at some level be a result of variable angular resolution (Heyer and Dame 2015). The mass function of the outer MW shows no clear evidence for a truncation at the high-mass end, but under some assumptions, Rosolowsky (2005) estimated that the maximum molecular cloud mass is $\sim$2–3 $\times 10^5\,M_{\odot}$. In contrast, Rosolowsky (2005) concluded that the inner MW shows a clear truncation with maximum molecular cloud mass of $\sim 3\times 10^6\,M_{\odot}$. Because of the small number of known clouds, the apparent lack of massive clouds in the outer MW might be due to a sampling effect. This possibility should be addressed in future studies, as a truncation, if it exists, would be an important clue to understanding cloud physics in the outskirts.

Heyer et al. (2001) concluded that the size distribution of molecular clouds in the outer MW is similar to the distribution in the inner MW from Solomon et al. (1987), but note that surveys with fewer clouds and different galactocentric distance ranges reached different conclusions. May et al. (1997) concluded that outer MW clouds have smaller sizes than the inner MW, while Brand and Wouterloot (1995) concluded that the outer MW clouds have larger sizes than inner MW clouds at the same mass. While there are conflicting results in the literature, it seems natural to conclude that an outer MW cloud must have a larger radius than an inner MW cloud at the same mass because it appears that the mass surface density of clouds is lower in the outer MW (see above and Heyer and Dame 2015).

Molecular gas observations in the outskirts of the MW have been used to identify spiral arms. Dame and Thaddeus (2011) discovered a spiral arm in the first quadrant at $R_{\rm Gal} \sim 15$ kpc, based on HI and CO data. Their new arm is consistent with being an extension of the Scutum-Centaurus Arm. Sun et al. (2015) also used HI and CO data to discover an arm in the second quadrant at $R_{\rm Gal} = 15$–19 kpc. This arm could be a further continuation of the Scutum-Centaurus Arm and the Dame and Thaddeus (2011) Arm. These kinds of studies are important not only to map the spiral structure of the MW but also to help understand the observation that star formation in the outskirts of other galaxies often follows spiral arms.

Another important goal of molecular gas studies in the outskirts of the MW has been to understand the connection with star formation under low density and metallicity conditions. For example, Brand and Wouterloot (2007) studied an IRAS-selected molecular cloud with a mass of 4.5–6.6 $\times$ $10^3\,M_\odot$ at $R_{\rm Gal} \sim 20.2$ kpc. They discovered an embedded cluster of 60 stars, and the lack of radio continuum emission limits the most massive star to be later than B0.5. In addition, Kobayashi et al. (2008) studied Digel Cloud 2, which is really two clouds each with a mass of $\sim 5 \times 10^3\,M_\odot$. They discovered embedded clusters in each of the clouds. One cluster likely contains a Herbig Ae/Be star, and there are also several Herbig Ae/Be star candidates, a B0-B1 star and an HII region nearby. Therefore, high-mass star formation has occurred near this low-mass molecular cloud. We encourage more study on the relationship between cloud mass and the most massive star present, as extragalactic studies can trace O and B stars relatively easily, but have difficulty detecting the parent molecular clouds (see Sect. 6.4.2.1).

In the outskirts of the MW and other galaxies, it is important to ask what triggers molecular cloud and star formation. In Digel Cloud 2, star formation may have been triggered by the expanding HI shell of a nearby supernova remnant (Kobayashi and Tokunaga 2000; Yasui et al. 2006; Kobayashi et al. 2008), while Izumi et al. (2014) hypothesized that the star formation in Digel Cloud 1 may have been triggered by interaction with a nearby HVC.

6.4.2 Extragalactic Disk Galaxies

We can study molecular gas in more varied environments by moving from the MW to extragalactic disk galaxies. In this section, we use “outskirts” to refer to galactocentric radii greater than the optical radius ($R_{\rm Gal} > r_{25}$).

6.4.2.1 Molecular Gas Detections

Numerous attempts to detect CO beyond the optical radius in the disks of spiral galaxies have failed, although many of the non-detections are unpublished (Watson et al. 2016; Morokuma-Matsui et al. 2016; J. Braine, F. Combes, J. Donovan Meyer and A. Gil de Paz, personal communications). To our knowledge, there are only four isolated spiral galaxies with published CO detections beyond the optical radius (Braine and Herpin 2004; Braine et al. 2007, 2010, 2012; Dessauges-Zavadsky et al. 2014). Table 6.1 summarizes the number of detected regions and their range of galactocentric radii and molecular gas masses. Extragalactic studies have not yet reached the molecular gas masses that are typical in the outskirts of the MW (2–20 × $10^3\,M_\odot$ for the eleven Digel clouds at $R_{\rm Gal}$ = 18–22 kpc; Digel et al. 1994; Kobayashi et al. 2008; see also Braine et al. 2007).

It would be useful to be able to predict where CO will be detected in the outskirts of disk galaxies, both as a test of our understanding of the physics of CO formation and destruction in extreme conditions (see Sect. 6.3.4), and to help us efficiently collect more detections. Most of the published CO studies selected high HI column density regions or regions near young stars traced by Hα, FUV or FIR emission. None of these selection methods is completely reliable. Braine et al. (2010) concluded that CO is often associated with large HI and FIR structures, but it is not necessarily located at HI, FIR or Hα peaks. Many factors might affect the association between HI, CO and star formation tracers. For example, the star-forming regions may drift away from their birthplaces over the 10–100 Myr timescales traced by Hα, FUV and FIR emission. In addition, feedback from massive stars might destroy molecular clouds more easily in the low-density outskirt environment. Finally, higher-resolution HI maps may show better correlation with CO emission. Sensitive, large-scale (>kpc) maps of the outskirts of disk galaxies may allow for a more impartial study of the conditions that maximize the CO detection rate.

Table 6.1 Extragalactic disk galaxies in relative isolation with CO detections beyond the optical radius (Braine and Herpin 2004; Braine et al. 2007, 2010, 2012; Dessauges-Zavadsky et al. 2014)

Galaxy	Detected regions (#)	Galactocentric radius (r_{25})	Molecular gas mass ($10^5\,M_\odot$)	Method used for mass
NGC 4414	4	1.1–1.5	10–20	Within 21" IRAM 30 m beam
NGC 6946	4	1.0–1.4	1.7–3.3	Within 21" IRAM 30 m beam
M33	6	1.0–1.1	0.43	Virial mass using resolved PdBI data
M63	2	1.36	7.1	Sum of 12 IRAM 30 m pointings

For M33, the molecular gas mass is for one of the detected clouds. For M63, the molecular gas mass is based on a sum of the CO line intensities in 12 pointings, two of which are detections. The NGC 4414, NGC 6946 and M63 masses were computed assuming $X_{\rm CO}$=2 × $10^{20}\,{\rm cm^{-2}}({\rm K\,km\,s^{-1}})^{-1}$

6.4.2.2 Star Formation in Extragalactic Disk Galaxies

It is generally accepted that stars form from molecular gas (e.g. Fukui and Kawamura 2010) and that an important stage before star formation is the conversion of HI to H_2 (e.g. Leroy et al. 2008). A main tool to study the connection between gas and star formation is the Kennicutt-Schmidt law (Schmidt 1959; Kennicutt 1998), which is an empirical relationship between the star formation rate (SFR) surface density (Σ_{SFR}) and the gas surface density. Within the optical disk of spiral galaxies, there is an approximately linear correlation between Σ_{SFR} and the molecular-hydrogen surface density (Σ_{H_2}) but no correlation between Σ_{SFR} and the atomic hydrogen surface density (Σ_{HI}; e.g. Bigiel et al. 2008; Schruba et al. 2011).

The majority of the published work connecting the SFR and gas density in the outskirts of disk galaxies has focussed on the atomic gas because molecular gas is difficult to detect (Sect. 6.4.2.1) and because the ISM is dominantly atomic in the outskirts, at least on $\gtrsim$ kpc scales. Bigiel et al. (2010) concluded that there is a correlation between the FUV-based Σ_{SFR} and Σ_{HI} in the outskirts of 17 disk galaxies and 5 dwarf galaxies. They measured a longer depletion time in the outskirts, such that it will take on average 10^{11} years to deplete the HI gas reservoir in the outskirts versus 10^9 years to deplete the H_2 gas reservoir within the optical disk. Roychowdhury et al. (2015) reached a similar conclusion using HI-dominated regions in disks and dwarfs, including some regions in the outskirts, although they concluded that the depletion time is somewhat shorter than in the outskirts of the Bigiel et al. (2010) sample (see also Boissier et al. 2007; Dong et al. 2008; Barnes et al. 2012). The correlation between Σ_{SFR} and Σ_{HI} is surprising because there is no correlation within the optical disk. Bigiel et al. (2010) suggested that high HI column density is important for determining where stars will form in the outskirts.

The study of the connection between molecular gas and star formation in the outskirts has been limited by the few molecular gas detections. Figures 6.5 and 6.6 show the relationship between Σ_{SFR} and Σ_{H_2} for the molecular gas detections from Table 6.1 plus a number of deep CO upper limits. In both panels the SFR was computed based on FUV and 24 μm data to account for the star formation that is unobscured and obscured by dust.

Dessauges-Zavadsky et al. (2014) studied a UV-bright region at $r = 1.36\, r_{25}$ in the XUV disk of M63 (Fig. 6.5). They detected CO in 2 out of 12 pointings and concluded that the molecular gas has a low star formation efficiency (or, equivalently, the molecular gas has a long depletion time) compared to regions within the optical disk. They suggested that the low star formation efficiency may be caused by a warp or by high turbulence. Watson et al. (2016) measured a deep CO upper limit in a region at $r = 3.4\, r_{25}$ in the XUV disk of NGC 4625 and compiled published CO measurements and upper limits for 15 regions in the XUV disk or outskirts of NGC 4414, NGC 6946 and M33 from Braine and Herpin (2004) and Braine et al. (2007, 2010) (see Table 6.1 and Fig. 6.6). They concluded that star-forming regions in the outskirts are in general consistent with the same Σ_{SFR}–Σ_{H_2} relationship that exists in the optical disk. However, some points are offset to high star formation efficiency (short depletion time), which may be because the authors

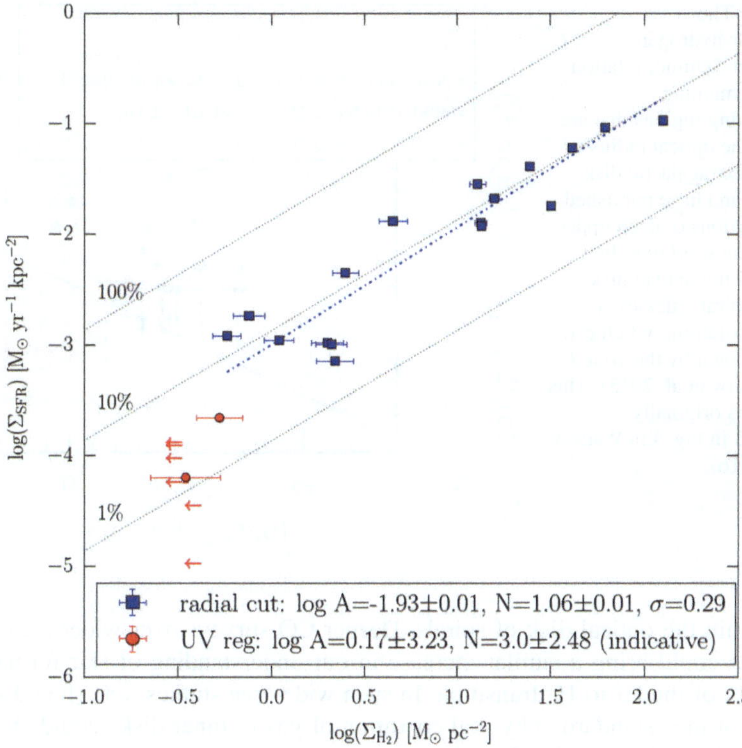

Fig. 6.5 Figure 7 from Dessauges-Zavadsky et al. (2014) showing the molecular-hydrogen Kennicutt-Schmidt relation for the star-forming regions in the UV complex at $r = 1.36\,r_{25}$ in M63 (*red points*) compared to regions within the optical disk (*blue points*). The *blue line* shows the fit for the optical disk. The *black lines* represent constant star formation efficiency, assuming a timescale of 10^8 years. Credit: Dessauges-Zavadsky et al., A&A, 566, A147, 2014, reproduced with permission © ESO

selected Hα- or FUV-bright regions that could have already exhausted some of the molecular gas supply (as in Schruba et al. 2010; Kruijssen and Longmore 2014).

We should ask what stimulates the formation of molecular gas and stars in the outskirts of disk galaxies. Thilker et al. (2007) suggested that interactions may trigger the extended star formation in XUV disks, while Holwerda et al. (2012) suggested that cold accretion may be more important. Bush et al. (2008, 2010) carried out hydrodynamic simulations and concluded that spiral density waves can raise the density in an extended gas disk to induce star formation (see also Sect. 6.4.1.1. of Debattista et al. 2017).

The state-of-the-art data from SINGS (Kennicutt et al. 2003), the *GALEX* Nearby Galaxy Survey (Gil de Paz et al. 2007a), THINGS (Walter et al. 2008) and HERACLES (Leroy et al. 2009) brought new insight into the Kennicutt-Schmidt

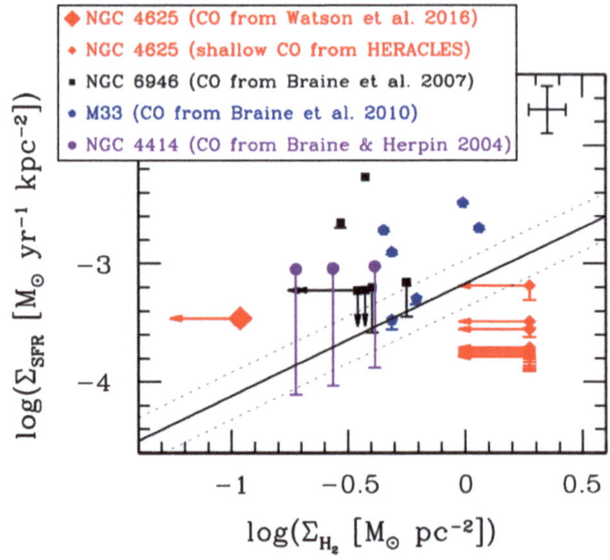

Fig. 6.6 The molecular-hydrogen Kennicutt-Schmidt relation for the remaining star-forming regions that are beyond the optical radius in isolated extragalactic disk galaxies and have published CO detections or deep upper limits. The *solid line* shows the fit for the optical disk of normal spiral galaxies at ∼kpc resolution, with the 1σ scatter shown by the *dotted lines* (Leroy et al. 2013). This figure was originally presented in Fig. 4 in Watson et al. (2016)

law within the optical disk of spirals. Deeper CO surveys over wider areas in the outskirts could bring a similar increase in our understanding of star formation at the onset of the HI-to-H_2 transition. In such wide-area studies, one should keep in mind that the "standard" physical condition of gas in inner disks could change in the outskirts, which could affect the measurements (Sect. 6.3.4).

6.4.2.3 Theory

This chapter focusses on observations, but here we briefly highlight theoretical works that are related to molecular gas in the outskirts. The majority of the relevant theoretical studies have concentrated on the origin of gas in the outskirts (e.g. Dekel and Birnboim 2006; Sancisi et al. 2008; Sánchez Almeida et al. 2014; Mitra et al. 2015) and star formation in the outskirts (Bush et al. 2008, 2010; Ostriker et al. 2010; Krumholz 2013; Sánchez Almeida et al. 2014; see also Roškar et al. 2010; Khoperskov and Bertin 2015). Krumholz (2013) is particularly relevant because he extended earlier work to develop an analytic model for the atomic and molecular ISM and star formation in outer disks. Krumholz assumed that hydrostatic equilibrium sets the density of cold neutral gas in the outskirts and was able to match the Bigiel et al. (2010) observations that show a correlation between $\Sigma_{\rm SFR}$ and $\Sigma_{\rm HI}$ (see also Sect. 6.7 of Elmegreen and Hunter 2017).

6.5 Molecular Gas Observations in the Outskirts of Early-Type Galaxies

Early-type galaxies were historically viewed as "red and dead", with little gas to form new stars. However, more recent surveys have found reservoirs of cold gas both at galaxy centres and in the outskirts. Molecular gas in the centres of early-type galaxies can have an internal and/or external origin, while the molecular gas in the outskirts often originated in a gas-rich companion that has interacted or merged with the early-type. As in all of the environments we have explored, stimuli can also trigger new molecule formation in the outskirts of early-types.

We start with a review of HI in the inner and outer regions of early-type galaxies to put the molecular gas observations in context. The ATLAS^{3D} survey detected HI in 32% of 166 early-type galaxies in a volume-limited sample, down to a 3σ upper limit of $M_{HI} = 5 \times 10^6$–$5 \times 10^7\,M_\odot$. Atomic gas in the outskirts of early-type galaxies is even relatively common, as 14% of the ATLAS^{3D} sample have HI that extends out to more than 3.5 times the optical effective radius (Serra et al. 2012).

Most surveys of molecular gas in early-type galaxies have focussed on the inner regions. Twenty-two percent of 260 early-type galaxies in the ATLAS^{3D} sample were detected in CO, down to a 3σ upper limit of $M_{H_2} \sim 10^7$–$10^8\,M_\odot$ (Young et al. 2011; see also Sage and Wrobel 1989; Knapp and Rupen 1996; Welch and Sage 2003; Combes et al. 2007; Welch et al. 2010). Within the areas searched, the molecular gas is generally confined to the central few kpc and is distributed in disks, bars plus rings and spiral arms or with a disrupted morphology (Young 2002; Welch and Sage 2003; Young et al. 2008; Davis et al. 2013; Alatalo et al. 2013).

One important motivation for studies of molecular gas in early-type galaxies has been to determine whether the gas is of internal or external origin. Some of the molecular gas has likely either been present since the galaxies transitioned to being early-type or has accumulated from stellar mass loss (Faber and Gallagher 1976; Young 2002; Young et al. 2008; Mathews and Brighenti 2003; Ciotti et al. 2010). In contrast, some molecular gas has likely been accreted more recently through minor mergers and/or cold accretion. This external origin is most clearly exhibited by galaxies that display a misalignment between the kinematic axes of the molecular/ionized gas and the stars (Young et al. 2008; Crocker et al. 2008; Davis et al. 2011; Alatalo et al. 2013). In particular, Alatalo et al. (2013) concluded that 15 galaxies out of a sample of 40 show a kinematic misalignment of at least 30 degrees, which is consistent with gas accretion via minor mergers.

The majority of accreting gas is perhaps in the atomic form, but the outskirts of early-type galaxies also offer the opportunity to study recently accreted molecular gas, which has mainly been detected in polar rings of elliptical and S0 galaxies (see Fig. 6.7 for an example). These polar rings are present in about 0.5% of nearby S0 galaxies (Whitmore et al. 1990). CO has been detected in polar rings at galactocentric radii of 12 kpc in NGC 660 (Combes et al. 1992) and 2 kpc in NGC 2685 (Schinnerer and Scoville 2002; see also Watson et al. 1994; Galletta et al. 1997; Combes et al. 2013). Published values for the mass of molecular hydrogen in

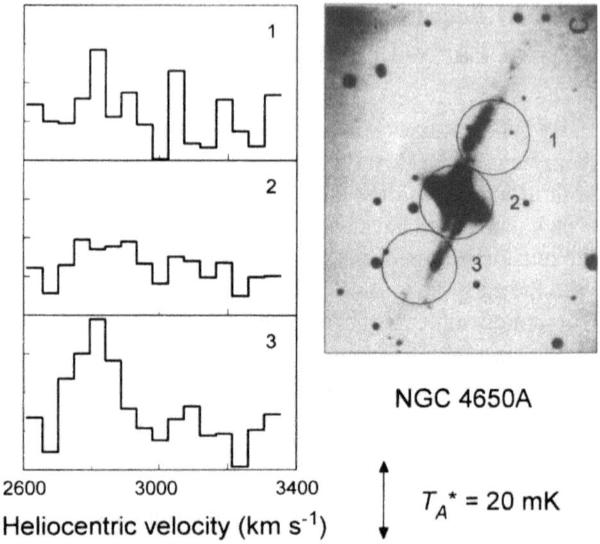

Fig. 6.7 Figure 2 from Watson et al. (1994) showing the Caltech Submillimeter Observatory CO(2 − 1) spectra (*left*) at three pointings, which are indicated by *circles* in the B-band image of the polar-ring galaxy NGC 4650A (Whitmore et al. 1987) on the *right*. Watson et al. (1994) estimated the mass of molecular hydrogen in the polar ring of NGC 4650A to be $M_{H_2} = 8$–$16 \times 10^8\, M_\odot$. © AAS. Reproduced with permission

the polar rings range from 8–11 × $10^6\, M_\odot$ in NGC 2685 (Schinnerer and Scoville 2002) to $10^9\, M_\odot$ in NGC 660 (Combes et al. 1992), although the handful of polar rings with CO detections are likely biased towards high M_{H_2}.

Polar rings are likely caused by tidal accretion from or, a merger with, a gas-rich companion and are stable on timescales of a few Gyr as a result of self-gravity (Bournaud and Combes 2003). The molecular gas observations generally support this hypothesis because the molecular gas masses are consistent with those of a dwarf or spiral galaxy (Watson et al. 1994; Galletta et al. 1997; Schinnerer and Scoville 2002).

Mergers between an early-type galaxy and a gas-rich companion can manifest in non-polar-ring systems as well. Buta et al. (1995) studied the spheroid-dominated spiral galaxy NGC 7217 and concluded that most of the molecular mass is in an outer star-forming ring at $R_{\rm Gal} \sim 0.6\, r_{25}$ that could have an H_2 mass that is equal to or greater than the HI mass. More recent work by Sil'chenko et al. (2011) indicates that minor mergers may be responsible for the outer ring structures.

Molecular gas has also been detected in shells at a galactocentric radius of 15 kpc ($1.16\, r_{25}$) in the elliptical galaxy Centaurus A (Charmandaris et al. 2000). Charmandaris et al. (2000) calculated the mass of molecular hydrogen in the CenA shells to be $M_{H_2} = 4.3 \times 10^7\, M_\odot$. Like polar rings, shells are likely caused by galaxy interactions, and Charmandaris et al. (2000) concluded that CenA interacted with a massive spiral galaxy rather than a low-mass dwarf galaxy because of the large total gas mass and large ratio of molecular to atomic gas in CenA. Additional molecular cloud formation may have been triggered by the interaction between the shells and the CenA radio jet (see also Salomé et al. 2016).

6.6 Molecular Gas Observations in Galaxy Groups and Clusters

Extended HI gas disks beyond optical edges are common around spiral galaxies, and as already discussed, some stimulus seems necessary to accelerate molecule formation there. In the group/cluster environment, galaxy interactions and interactions with the intergalactic medium (IGM) are triggers for the HI to H_2 phase transition. In the nearby M81 triplet (M82, M81 and NGC 3077), tidal interactions stretch the atomic gas in the outskirts into tidal spiral arms, leading to gravitational collapse to form molecular gas and stars (Brouillet et al. 1992; Walter et al. 2006). Even an interaction with a minor partner can be a trigger, e.g. in the M51 system, CO emission is detected along the tidal arm/bridge between the main galaxy NGC 5194 and its companion NGC 5195 (Koda et al. 2009).

Interaction with the IGM in clusters is also important for the gas-phase transition. Most HI gas in galaxy outskirts is stripped away by the ram pressure from the IGM (van Gorkom 2004), while the molecular gas, which resides mostly in inner disks, remains less affected (Kenney and Young 1989; Boselli et al. 1997). Some compression acts on the molecular gas near the transition from the molecular-dominant inner disks to the atomic-dominant outer disks, as the extents of molecular disks are smaller when the HI in the outskirts is stripped away (Boselli et al. 2014).

The stripped gas in the outskirts is seen as multiphase and has been detected in HI (e.g. Chung et al. 2009), Hα (e.g. Yagi et al. 2010) and X-rays (e.g. Wang et al. 2004; Sun et al. 2010). Stripped molecular gas is found in NGC 4438 and NGC 4435, which are interacting galaxies in the Virgo cluster (Vollmer et al. 2005). CO emission has also been discovered in the trailing tails of the stripped gas from the disk galaxies ESO137-001 and NGC 4388 in the Norma and Virgo clusters, respectively (Jáchym et al. 2014; Verdugo et al. 2015).

The ram pressure from the IGM can also heat up and excite H_2 molecules, and H_2 rotational emission lines are detected in the mid-infrared in spiral galaxies in the Virgo cluster (Wong et al. 2014). The emission from warm H_2 is also detected over large scales in the intergalactic space of Stephan's Quintet galaxy group with the *Spitzer Space Telescope* (Appleton et al. 2006). An analysis of the rotational transition ladder of its ground vibrational state suggests that the molecular gas has temperatures of 185 ± 30 and 675 ± 80 K. This H_2 emission coincides with and extends along the X-ray-emitting shock front that is generated by the galaxy NGC 7318b passing through the IGM at a high velocity.

A final example of the cluster environment affecting molecular gas formation is that CO has been detected in cooling flows in the outskirts of galaxies in cluster cores (e.g. Salomé et al. 2006). Clearly, the group and cluster environments produce some triggers for the formation of molecular gas in galaxy outskirts and therefore represent another extreme environment where we can test our understanding of the physics of the ISM and star formation.

6.7 Conclusions and Future Directions

Throughout the chapter, we have highlighted that some stimuli seem necessary to accelerate the formation of molecular gas in galaxy outskirts. In the outskirts of the MW, stimuli include spiral arm compression, expanding shells from supernova remnants and interactions with HVCs (Yasui et al. 2006; Izumi et al. 2014; Koda et al. 2016). These same processes are likely at play in the outskirts of extragalactic disk galaxies. In particular, spiral density waves, interactions and/or cold accretion may stimulate molecule formation and the subsequent star formation activity in XUV disks (Thilker et al. 2007; Bush et al. 2008; Holwerda et al. 2012). Interactions and mergers likely cause the polar rings in the outskirts of S0 galaxies, although it may be more likely that the molecules form in the gas-rich companion before the merger (Bournaud and Combes 2003). Finally, in groups and clusters, interactions and ram pressure stripping may accelerate molecular gas formation in some localized areas of galaxies even as the overall effect is to remove the star-forming fuel from the galaxies (Vollmer et al. 2005; Jáchym et al. 2014). Galaxy outskirts offer opportunities to study the formation of molecular gas over a variety of conditions and will be the key to understanding if there are different modes of star formation.

Fundamental questions remain about the physical conditions of the ISM in the outskirts. Where is the molecular gas? What are the basic properties of the molecular clouds, e.g. the H_2 volume density, H_2 column density, temperature, mass and size? How do these properties differ from the properties of molecular clouds in the inner regions of galaxies? Is the transition from HI to H_2 and the transition from H_2 to stars more or less efficient in the outskirts? Are these phase transitions affected by different large-scale processes, stimuli or environmental conditions compared to inner regions? Measurements of molecular gas properties often depend on assumptions about the gas properties themselves. Right now, those assumptions are based on our knowledge of molecular gas in inner disks. Those assumptions need to be revisited and adjusted continuously as we learn more about molecular gas in the outskirts. This iterative improvement of our knowledge is now starting in the field of galaxy outskirts.

Building on the research that has already been done, we have identified a number of specific studies that would begin to address the fundamental questions above. In the outskirts of the MW, we can study whether the relationship between the mass of the molecular cloud and the most massive associated star is different than in the inner MW. In the outskirts of extragalactic disk galaxies, we need to measure the mass and size functions of molecular clouds and compare to the MW results. In addition, theoretical studies can work towards predicting where and how molecular gas will form in the outskirts. To test these predictions, we encourage sensitive and wide-area mapping of CO and/or dust continuum emission. Higher-resolution (cloud-scale) maps of HI may also be required to accurately locate potential sites of molecular gas formation. After each discovery of molecular gas, subsequent multiwavelength studies including excitation ladders of molecular line emission are necessary to refine our knowledge of the physical conditions of molecular gas there. In early-type galaxies, we should search for molecular gas in XUV disks, as XUV emission could be even more common in early-type galaxies than late-type galaxies (Moffett et al. 2012). We hope those researchers will take note and learn from the high failure rate of previous (published and unpublished) searches for molecular gas in the outskirts of disk galaxies.

Acknowledgements We are grateful to Françoise Combes, Jennifer Donovan Meyer, Natsuko Izumi, and Hiroyuki Nakanishi for their advice, reading suggestions, and comments. We also thank Natsuko Izumi for allowing us to use her figure for the distribution of molecular clouds in the outer MW (Fig. 6.4). JK thanks the NAOJ Chile observatory, a branch of the National Astronomical Observatory of Japan, and the Joint ALMA Observatory for hospitality during his sabbatical visit. JK acknowledges the support from NASA through grant NNX14AF74G.

References

Alatalo, K., Davis, T.A., Bureau, M., Young, L.M., Blitz, L., Crocker, A.F., Bayet, E., Bois, M., Bournaud, F., Cappellari, M., Davies, R.L., de Zeeuw, P.T., Duc, P.A., Emsellem, E., Khochfar, S., Krajnović, D., Kuntschner, H., Lablanche, P.Y., Morganti, R., McDermid, R.M., Naab, T., Oosterloo, T., Sarzi, M., Scott, N., Serra, P., Weijmans, A.M.: The ATLAS^{3D} project - XVIII. CARMA CO imaging survey of early-type galaxies. Mon. Not. R. Astron. Soc. **432**, 1796–1844 (2013). doi:10.1093/mnras/sts299, 1210.5524

Appleton, P.N., Xu, K.C., Reach, W., Dopita, M.A., Gao, Y., Lu, N., Popescu, C.C., Sulentic, J.W., Tuffs, R.J., Yun, M.S.: Powerful high-velocity dispersion molecular hydrogen associated with an intergalactic shock wave in Stephan's quintet. Astrophys. J. **639**, L51–L54 (2006). doi:10.1086/502646, astro-ph/0602554

Arimoto, N., Sofue, Y., Tsujimoto, T.: CO-to-H_2 conversion factor in galaxies. Publ. Astron. Soc. Jpn. **48**, 275–284 (1996). doi:10.1093/pasj/48.2.275

Barnes, K.L., van Zee, L., Côté, S., Schade, D.: Star formation in the outer disk of spiral galaxies. Astrophys. J. **757**, 64 (2012). doi:10.1088/0004-637X/757/1/64

Bigiel, F., Leroy, A., Walter, F., Brinks, E., de Blok, W.J.G., Madore, B., Thornley, M.D.: The star formation law in nearby galaxies on sub-Kpc scales. Astron. J. **136**, 2846–2871 (2008). doi:10.1088/0004-6256/136/6/2846, 0810.2541

Bigiel, F., Leroy, A., Walter, F., Blitz, L., Brinks, E., de Blok, W.J.G., Madore, B.: Extremely inefficient star formation in the outer disks of nearby galaxies. Astron. J. **140**, 1194–1213 (2010). doi:10.1088/0004-6256/140/5/1194, 1007.3498

Bluhm, H., de Boer, K.S., Marggraf, O., Richter, P.: ORFEUS echelle spectra: molecular hydrogen in disk, IVC, and HVC gas in front of the LMC. Astron. Astrophys. **367**, 299–310 (2001). doi:10.1051/0004-6361:20000346, astro-ph/0012238

Boissier, S., Gil de Paz, A., Boselli, A., Madore, B.F., Buat, V., Cortese, L., Burgarella, D., Muñoz-Mateos, J.C., Barlow, T.A., Forster, K., Friedman, P.G., Martin, D.C., Morrissey, P., Neff, S.G., Schiminovich, D., Seibert, M., Small, T., Wyder, T.K., Bianchi. L., Donas. J., Heckman, T.M., Lee, Y.W., Milliard, B., Rich, R.M., Szalay, A.S., Welsh, B.Y., Yi, S.K.: Radial variation of attenuation and star formation in the largest late-type disks observed with GALEX. Astrophys. J. Suppl. Ser. **173**, 524–537 (2007) doi:10.1086/516642, astro-ph/0609071

Bolatto, A.D., Wolfire, M., Leroy, A.K.: The CO-to-H_2 conversion factor. Annu. Rev. Astron. Astrophys. **51**, 207–268 (2013). doi:10.1146/annurev-astro-082812-140944, 1301.3498

Boselli, A., Gavazzi, G., Lequeux, J., Buat, V., Casoli, F., Dickey, J., Donas, J.: The molecular gas content of spiral galaxies in the Coma/A1367 supercluster. Astron. Astrophys. **327**, 522–538 (1997)

Boselli, A., Cortese, L., Boquien, M., Boissier, S., Catinella, B., Gavazzi, G., Lagos, C., Saintonge, A.: Cold gas properties of the Herschel reference survey. III. Molecular gas stripping in cluster galaxies. Astron. Astrophys. **564**, A67 (2014). doi:10.1051/0004-6361/201322313, 1402.0326

Bosma, A.: 21-cm line studies of spiral galaxies. I - Observations of the galaxies NGC 5033, 3198, 5055, 2841, and 7331. II - The distribution and kinematics of neutral hydrogen in spiral galaxies of various morphological types. Astron. J. **86**, 1791–1846 (1981). doi:10.1086/113062

Bot, C., Ysard, N., Paradis, D., Bernard, J.P., Lagache, G., Israel, F.P., Wall, W.F.: Submillimeter to centimeter excess emission from the magellanic clouds. II. On the nature of the excess. Astron. Astrophys. **523**, A20 (2010). doi:10.1051/0004-6361/201014986, 1008.2875

Bournaud, F., Combes, F.: Formation of polar ring galaxies. Astron. Astrophys. **401**, 817–833. (2003). doi:10.1051/0004-6361:20030150, astro-ph/0301391

Braine, J., Herpin, F.: Molecular hydrogen beyond the optical edge of an isolated spiral galaxy. Nature **432**, 369–371 (2004). doi:10.1038/nature03054, astro-ph/0412283

Braine, J., Ferguson, A.M.N., Bertoldi, F., Wilson, C.D.: The detection of molecular gas in the outskirts of NGC 6946. Astrophys. J. **669**, L73–L76 (2007). doi:10.1086/524135, 0710.1863

Braine, J., Gratier, P., Kramer, C., Schuster, K.F., Tabatabaei, F., Gardan, E.: Molecular cloud formation and the star formation efficiency in M 33. Molecule and star formation in M 33. Astron. Astrophys. **520**, A107 (2010). doi:10.1051/0004-6361/201014166, 1007.0702

Braine, J., Gratier, P., Contreras, Y., Schuster, K.F., Brouillet, N.: A detailed view of a molecular cloud in the far outer disk of M 33. Molecular cloud formation in M 33. Astron. Astrophys. **548**, A52 (2012). doi:10.1051/0004-6361/201220093, 1210.6470

Brand, J., Wouterloot, J.G.A.: The velocity field of the outer galaxy in the southern hemisphere. III - determination of distances to O, B, and A type stars in the Walraven photometric system. Astron. Astrophys. Suppl. Ser. **75**, 117–137 (1988)

Brand, J., Wouterloot, J.G.A.: IRAS sources beyond the solar circle. IV. Maps of far-outer galaxy molecular clouds. Astron. Astrophys. Suppl. Ser. **103**, 503–540 (1994)

Brand, J., Wouterloot, J.G.A.: IRAS sources beyond the solar circle. V. Properties of far-outer galaxy molecular clouds. Astron. Astrophys. **303**, 851 (1995)

Brand, J., Wouterloot, J.G.A.: A star cluster at the edge of the galaxy. Astron. Astrophys. **464**, 909–920 (2007). doi:10.1051/0004-6361:20065437, astro-ph/0702541

Bresolin, F.: Metallicities in the outer regions of spiral galaxies. In: Knapen, J.H., Lee, J.C., Gil de Paz, A. (eds.) Outskirts of Galaxies, vol. 434. Springer, Cham (2017). doi:10.1007/978-3-319-56570-5

Brouillet, N., Henkel, C., Baudry, A.: Detection of an intergalactic molecular complex? Astron. Astrophys. **262**, L5–L8 (1992)

Brunt, C.M., Kerton, C.R., Pomerleau, C.: An outer galaxy molecular cloud catalog. Astrophys. J. Suppl. Ser. **144**, 47–70 (2003). doi:10.1086/344245

Bush, S.J., Cox, T.J., Hernquist, L., Thilker, D., Younger, J.D.: Simulations of XUV disks with a star formation density threshold. Astrophys. J. **683**, L13 (2008). doi:10.1086/591523, 0807.1116

Bush, S.J., Cox, T.J., Hayward, C.C., Thilker, D., Hernquist, L., Besla, G.: Spiral-induced star formation in the outer disks of galaxies. Astrophys. J. **713**, 780–799 (2010). doi:10.1088/0004-637X/713/2/780, 1003.5672

Buta, R., van Driel, W., Braine, J., Combes, F., Wakamatsu, K., Sofue, Y., Tomita, A.: NGC 7217: a spheroid-dominated, early-type resonance ring spiral galaxy. Astrophys. J. **450**, 593 (1995). doi:10.1086/176169

Charmandaris, V., Combes, F., van der Hulst, J.M.: First detection of molecular gas in the shells of CenA. Astron. Astrophys. **356**, L1–L4 (2000). astro-ph/0003175

Chung, A., van Gorkom, J.H., Kenney, J.D.P., Crowl, H., Vollmer, B.: VLA imaging of Virgo Spirals in Atomic Gas (VIVA). I. The atlas and the H I properties. Astron. J. **138**, 1741–1816 (2009). doi:10.1088/0004-6256/138/6/1741

Ciotti, L., Ostriker, J.P., Proga, D.: Feedback from central black holes in elliptical galaxies. III. Models with both radiative and mechanical feedback. Astrophys. J. **717**, 708–723 (2010). doi:10.1088/0004-637X/717/2/708, 1003.0578

Colombo, D., Hughes, A., Schinnerer, E., Meidt, S.E., Leroy, A.K., Pety, J., Dobbs, C.L., García-Burillo, S., Dumas, G., Thompson, T.A., Schuster, K,F., Kramer, C.: The PdBI arcsecond whirlpool survey (PAWS): environmental dependence of giant molecular cloud properties in M51. Astrophys. J. **784**, 3 (2014). doi:10.1088/0004-637X/784/1/3, 1401.1505

Combes, F., Braine, J., Casoli, F., Gerin, M., van Driel, W.: Molecular clouds in a polar ring. Astron. Astrophys. **259**, L65–L68 (1992)

Combes, F., Young, L.M., Bureau, M.: Molecular gas and star formation in the SAURON early-type galaxies. Mon. Not. R. Astron. Soc. **377**, 1795–1807 (2007). doi:10.1111/j.1365-2966.2007.11759.x, astro-ph/0703557

Combes, F., Moiseev, A., Reshetnikov, V.: Molecular content of polar-ring galaxies. Astron. Astrophys. **554**, A11 (2013). doi:10.1051/0004-6361/201321385, 1302.7273

Crocker, A.F., Bureau, M., Young, L.M., Combes, F.: The molecular polar disc in NGC 2768. Mon. Not. R. Astron. Soc. **386**, 1811–1820 (2008). doi:10.1111/j.1365-2966.2008.13177.x, 0803.0426

Dale, D.A., Aniano, G., Engelbracht, C.W., Hinz, J.L., Krause, O., Montiel, E.J., Roussel, H., Appleton, P.N., Armus, L., Beirão, P., Bolatto, A.D., Brandl, B.R., Calzetti, D., Crocker, A.F., Croxall, K.V., Draine, B.T., Galametz, M., Gordon, K.D., Groves, B.A., Hao, C.N., Helou, G., Hunt, L.K., Johnson, B.D., Kennicutt, R.C., Koda, J., Leroy, A.K., Li, Y., Meidt, S.E., Miller, A.E., Murphy, E.J., Rahman, N., Rix, H.W., Sandstrom, K.M., Sauvage, M., Schinnerer, E., Skibba, R.A., Smith, J.D.T., Tabatabaei, F.S., Walter, F., Wilson, C.D., Wolfire, M.G., Zibetti, S.: Herschel far-infrared and submillimeter photometry for the KINGFISH sample of nearby galaxies. Astrophys. J. **745**, 95 (2012). doi:10.1088/0004-637X/745/1/95, 1112.1093

Dame, T.M., Thaddeus, P.: A molecular spiral arm in the far outer galaxy. Astrophys. J. **734**, L24 (2011). doi:10.1088/2041-8205/734/1/L24, 1105.2523

Davis, T.A., Alatalo, K., Sarzi, M., Bureau, M., Young, L.M., Blitz, L., Serra, P., Crocker, A.F., Krajnović, D., McDermid, R.M., Bois, M., Bournaud, F., Cappellari, M., Davies, R.L., Duc, P.A., de Zeeuw, P.T., Emsellem, E., Khochfar, S., Kuntschner, H., Lablanche, P.Y., Morganti, R., Naab, T., Oosterloo, T., Scott, N., Weijmans, A.M.: The ATLAS^{3D} project - X. On the origin of the molecular and ionized gas in early-type galaxies. Mon. Not. R. Astron. Soc. **417**, 882–899 (2011). doi:10.1111/j.1365-2966.2011.19355.x, 1107.0002

Davis, T.A., Alatalo, K., Bureau, M., Cappellari, M., Scott, N., Young, L.M., Blitz, L., Crocker, A., Bayet, E., Bois, M., Bournaud, F., Davies, R.L., de Zeeuw, P.T., Duc, P.A., Emsellem, E., Khochfar, S., Krajnović, D., Kuntschner, H., Lablanche, P.Y., McDermid, R.M., Morganti, R., Naab, T., Oosterloo, T., Sarzi, M., Serra, P., Weijmans, A.M.: The ATLAS^{3D} Project - XIV. The extent and kinematics of the molecular gas in early-type galaxies. Mon. Not. R. Astron. Soc. **429**, 534–555 (2013). doi:10.1093/mnras/sts353, 1211.1011

Debattista, V.P., Roškar, R., Loebman, S.R.: The impact of stellar migration on disk outskirts. In: Knapen, J.H., Lee, J.C., Gil de Paz, A. (eds.) Outskirts of Galaxies, vol. 434. Springer, Cham (2017). doi: 10.1007/978-3-319-56570-5

Dekel, A., Birnboim, Y.: Galaxy bimodality due to cold flows and shock heating. Mon. Not. R. Astron. Soc. **368**, 2–20 (2006). doi:10.1111/j.1365-2966.2006.10145.x, astro-ph/0412300

Dessauges-Zavadsky, M., Verdugo, C., Combes, F., Pfenniger, D.: CO map and steep Kennicutt–Schmidt relation in the extended UV disk of M 63. Astron. Astrophys. **566**, A147 (2014). doi:10.1051/0004-6361/201323330, 1406.0310

Digel, S., de Geus, E., Thaddeus, P.: Molecular clouds in the extreme outer galaxy. Astrophys. J. **422**, 92–101 (1994). doi:10.1086/173706

Dong, H., Calzetti, D., Regan, M., Thilker, D., Bianchi, L., Meurer, G.R., Walter, F.: Spitzer observations of star formation in the extreme outer disk of M83 (NGC5236). Astron. J. **136**, 479–497 (2008). doi:10.1088/0004-6256/136/1/479, 0804.3632

Elmegreen, B.G., Hunter, D.A.: Outskirts of nearby disk galaxies: star formation and stellar populations. In: Knapen, J.H., Lee, J.C., Gil de Paz, A. (eds.) Outskirts of Galaxies, vol. 434. Springer, Cham (2017). doi: 10.1007/978-3-319-56570-5

Engargiola, G., Plambeck, R.L., Rosolowsky, E., Blitz, L.: Giant molecular clouds in M33. I. BIMA all-disk survey. Astrophys. J. Suppl. Ser. **149**, 343–363 (2003). doi:10.1086/379165, arXiv:astro-ph/0308388

Faber, S.M., Gallagher, J.S.: H I in early-type galaxies. II - mass loss and galactic winds. Astrophys. J. **204**, 365–375 (1976). doi:10.1086/154180

Fich, M., Blitz, L.: Optical H II regions in the outer galaxy. Astrophys. J. **279**, 125–135 (1984). doi:10.1086/161872

Fukui, Y., Kawamura, A.: Molecular clouds in nearby galaxies. Annu. Rev. Astron. Astrophys. **48**, 547–580 (2010). doi:10.1146/annurev-astro-081309-130854

Fukui, Y., Kawamura, A., Wong, T., Murai, M., Iritani, H., Mizuno, N., Mizuno, Y., Onishi, T., Hughes, A., Ott, J., Muller, E., Staveley-Smith, L., Kim, S.: Molecular and atomic gas in the large magellanic cloud. II. Three-dimensional correlation between CO and H I. Astrophys. J. **705**, 144–155 (2009). doi:10.1088/0004-637X/705/1/144, 0909.0382

Galletta, G., Sage, L.J., Sparke, L.S.: Molecular gas in polar-ring galaxies. Mon. Not. R. Astron. Soc. **284**, 773–784 (1997). doi:10.1093/mnras/284.3.773

Gil de Paz, A., Boissier, S., Madore, B.F., Seibert, M., Joe, Y.H., Boselli, A., Wyder, T.K., Thilker, D., Bianchi, L., Rey, S.C., Rich, R.M., Barlow, T.A., Conrow, T., Forster, K., Friedman, P.G., Martin, D.C., Morrissey, P., Neff, S.G., Schiminovich, D., Small, T., Donas, J., Heckman, T.M., Lee, Y.W., Milliard, B., Szalay, A.S., Yi, S.: The GALEX ultraviolet atlas of nearby galaxies. Astrophys. J. Suppl. Ser. **173**, 185–255 (2007a) . doi:10.1086/516636, astro-ph/0606440

Gil de Paz, A., Madore, B.F., Boissier, S., Thilker, D., Bianchi, L., Sánchez Contreras, C., Barlow, T.A., Conrow, T., Forster. K., Friedman, P.G., Martin, D.C., Morrissey, P., Neff, S.G., Rich, R.M., Schiminovich, D., Seibert, M., Small, T., Donas, J., Heckman, T.M., Lee, Y.W., Milliard, B., Szalay, A.S., Wyder, T.K., Yi, S.: Chemical and photometric evolution of extended ultraviolet disks: optical spectroscopy of M83 (NGC 5236) and NGC 4625. Astrophys. J. **661**, 115–134 (2007b) . doi:10.1086/513730, arXiv:astro-ph/0702302

Goldreich, P., Kwan, J.: Molecular clouds. Astrophys. J. **189**, 441–454 (1974). doi:10.1086/152821

Gordon, K.D., Roman-Duval, J., Bot, C., Meixner, M., Babler, B., Bernard, J.P., Bolatto, A., Boyer, M.L., Clayton, G.C., Engelbracht, C., Fukui, Y., Galametz, M., Galliano, F., Hony, S., Hughes, A., Indebetouw, R., Israel, F.P., Jameson, K., Kawamura, A., Lebouteiller, V., Li, A., Madden, S.C., Matsuura, M., Misselt, K., Montiel, E., Okumura, K., Onishi, T., Panuzzo, P., Paradis, D., Rubio, M., Sandstrom, K., Sauvage, M., Seale, J., Sewiło, M., Tchernyshyov, K., Skibba, R.: Dust and gas in the magellanic clouds from the HERITAGE Herschel key project. I. Dust properties and insights into the origin of the submillimeter excess emission. Astrophys. J. **797**, 85 (2014). doi:10.1088/0004-637X/797/2/85, 1406.6066

Hasegawa, T.: The CO 2-1/1-0 ratio. In: Latter, W.B., Radford, S.J.E. Jewell, P.R., Mangum, J.G., Bally, J. (eds.) IAU Symposium, IAU Symposium, vol. 170, pp. 39–46 (1997)

Heyer, M., Dame, T.M.: Molecular clouds in the milky way. Annu. Rev. Astron. Astrophys. **53**, 583–629 (2015). doi:10.1146/annurev-astro-082214-122324

Heyer, M.H., Terebey, S.: The anatomy of the Perseus spiral arm: ^{12}CO and IRAS imaging observations of the W3-W4-W5 cloud complex. Astrophys. J. **502**, 265–277 (1998). doi:10.1086/305881

Heyer, M.H., Brunt, C., Snell, R.L., Howe, J.E., Schloerb, F.P., Carpenter, J.M.: The five college radio astronomy observatory CO survey of the outer galaxy. Astrophys. J. Suppl. Ser. **115**, 241–258 (1998). doi:10.1086/313086

Heyer, M.H., Carpenter, J.M., Snell, R.L.: The equilibrium state of molecular regions in the outer galaxy. Astrophys. J. **551**, 852–866 (2001). doi:10.1086/320218, astro-ph/0101133

Holwerda, B.W., Pirzkal, N., Heiner, J.S.: Quantified H I morphology - VI. The morphology of extended discs in UV and H I. Mon. Not. R. Astron. Soc. **427**, 3159–3175 (2012). doi:10.1111/j.1365-2966.2012.21975.x, 1207.4916

Izumi, N., Kobayashi, N., Yasui, C., Tokunaga, A.T., Saito, M., Hamano, S.: Discovery of star formation in the extreme outer galaxy possibly induced by a high-velocity cloud impact. Astrophys. J. **795**, 66 (2014). doi:10.1088/0004-637X/795/1/66, 1411.7290

Jáchym, P., Combes, F., Cortese, L., Sun, M., Kenney, J.D.P.: Abundant molecular gas and inefficient star formation in intracluster regions: ram pressure stripped tail of the normal galaxy ESO137-001. Astrophys. J. **792**, 11 (2014) doi:10.1088/0004-637X/792/1/11, 1403.2328

Kenney, J.D.P., Young, J.S.: The effects of environment on the molecular and atomic gas properties of large Virgo cluster spirals. Astrophys. J. **344**, 171–199 (1989). doi:10.1086/167787

Kennicutt, R.C. Jr.: The global Schmidt law in star-forming galaxies. Astrophys. J. **498**, 541–552 (1998). doi:10.1086/305588, astro-ph/9712213

Kennicutt, R.C. Jr., Armus, L., Bendo, G., Calzetti, D., Dale, D.A., Draine, B.T., Engelbracht, C.W., Gordon, K.D., Grauer, A.D., Helou, G., Hollenbach, D.J., Jarrett, T.H., Kewley, L.J., Leitherer, C., Li, A., Malhotra, S., Regan, M.W., Rieke, G.H., Rieke, M.J., Roussel, H., Smith, J.D.T., Thornley, M.D., Walter, F.: SINGS: the SIRTF nearby galaxies survey. Publ. Astron. Soc. Pac. **115**, 928–952 (2003). doi:10.1086/376941, astro-ph/0305437

Khoperskov, S.A., Bertin, G.: Spiral density waves in the outer galactic gaseous discs. Mon. Not. R. Astron. Soc. **451**, 2889–2899 (2015). doi:10.1093/mnras/stv1145, 1505.04598

Kirkpatrick, A., Calzetti, D., Galametz, M., Kennicutt, R. Jr., Dale, D., Aniano, G., Sandstrom, K., Armus, L., Crocker, A., Hinz, J., Hunt, L., Koda, J., Walter, F.: Investigating the presence of 500 μm submillimeter excess emission in local star forming galaxies. Astrophys. J. **778**, 51 (2013). doi:10.1088/0004-637X/778/1/51, 1310.0456

Knapp, G.R., Rupen, M.P.: Molecular gas in elliptical galaxies: CO observations of an IRAS flux-limited sample. Astrophys. J. **460**, 271 (1996). doi:10.1086/176967

Kobayashi, N., Tokunaga, A.T.: Discovery of young stellar objects at the edge of the optical disk of our galaxy. Astrophys. J. **532**, 423–429 (2000) doi:10.1086/308564, astro-ph/9909327

Kobayashi, N., Yasui, C., Tokunaga, A.T., Saito, M.: Star formation in the most distant molecular cloud in the extreme outer galaxy: a laboratory of star formation in an early epoch of the galaxy's formation. Astrophys. J. **683**, 178–188 (2008). doi:10.1086/588421, 0803.3369

Koda, J., Scoville, N., Sawada, T., La Vigne, M.A., Vogel, S.N., Potts, A.E., Carpenter, J.M., Corder, S.A., Wright, M.C.H., White, S.M., Zauderer, B.A., Patience, J., Sargent, A.I., Bock, D.C.J., Hawkins, D., Hodges, M., Kemball, A., Lamb, J.W., Plambeck, R.L., Pound, M.W., Scott, S.L., Teuben, P., Woody, D.P.: Dynamically driven evolution of the interstellar medium in M51. Astrophys. J. **700**, L132–L136 (2009). doi:10.1088/0004-637X/700/2/L132, 0907.1656

Koda, J., Scoville, N., Hasegawa, T., Calzetti, D., Donovan Meyer, J., Egusa, F., Kennicutt, R., Kuno, N., Louie, M., Momose, R., Sawada, T., Sorai, K., Umei, M.: Physical conditions in molecular clouds in the arm and interarm regions of M51. Astrophys. J. **761**, 41 (2012). doi:10.1088/0004-637X/761/1/41, 1210.6349

Koda, J., Scoville, N., Heyer, M.: Evolution of molecular and atomic gas phases in the milky way. Astrophys. J. **823**, 76 (2016). doi:10.3847/0004-637X/823/2/76, 1604.01053

Kruijssen, J.M.D., Longmore, S.N.: An uncertainty principle for star formation - I. Why galactic star formation relations break down below a certain spatial scale. Mon. Not. R. Astron. Soc. **439**, 3239–3252 (2014). doi:10.1093/mnras/stu098, 1401.4459

Krumholz, M.R.: The star formation law in molecule-poor galaxies. Mon. Not. R. Astron. Soc. **436**, 2747–2762 (2013). doi:10.1093/mnras/stt1780, 1309.5100

Lehner, N.: Far-ultraviolet spectroscopic explorer observations of the magellanic bridge gas toward two early-type stars: molecules, physical conditions, and relative abundances. Astrophys. J. **578**, 126–143 (2002). doi:10.1086/342349, astro-ph/0206250

Leroy, A.K., Walter, F., Brinks, E., Bigiel, F., de Blok, W.J.G., Madore, B., Thornley, M.D.: The star formation efficiency in nearby galaxies: measuring where gas forms stars effectively. Astron. J. **136**, 2782–2845 (2008). doi:10.1088/0004-6256/136/6/2782, 0810.2556

Leroy, A.K., Walter, F., Bigiel, F., Usero, A., Weiss, A., Brinks, E., de Blok, W.J.G., Kennicutt, R.C., Schuster, K.F., Kramer, C., Wiesemeyer, H.W., Roussel, H.: Heracles: The HERA co line extragalactic survey. Astron. J. **137**, 4670–4696 (2009). doi:10.1088/0004-6256/137/6/4670, 0905.4742

Leroy, A.K., Bolatto, A., Gordon, K., Sandstrom, K., Gratier, P., Rosolowsky, E., Engelbracht, C.W., Mizuno, N., Corbelli, E., Fukui, Y., Kawamura, A.: The CO-to-H_2 conversion factor from infrared dust emission across the local group. Astrophys. J. **737**, 12 (2011). doi:10.1088/0004-637X/737/1/12, 1102.4618

Leroy, A.K., Walter, F., Sandstrom, K., Schruba, A., Munoz-Mateos, J.C., Bigiel, F., Bolatto, A., Brinks, E., de Blok, W.J.G., Meidt, S., Rix, H.W., Rosolowsky, E., Schinnerer, E., Schuster, K.F., Usero, A.: Molecular gas and star formation in nearby disk galaxies. Astron. J. **146**, 19 (2013). doi:10.1088/0004-6256/146/2/19, 1301.2328

Mathews, W.G., Brighenti, F.: Hot gas in and around elliptical galaxies. Annu. Rev. Astron. Astrophys. **41**, 191–239 (2003). doi:10.1146/annurev.astro.41.090401.094542, astro-ph/0309553

May, J., Alvarez, H., Bronfman, L.: Physical properties of molecular clouds in the southern outer galaxy. Astron. Astrophys. **327**, 325–332 (1997)

Mitra, S., Davé, R., Finlator, K.: Equilibrium model constraints on baryon cycling across cosmic time. Mon. Not. R. Astron. Soc. **452**, 1184–1200 (2015). doi:10.1093/mnras/stv1387, 1411.1157

Moffett, A.J., Kannappan, S.J., Baker, A.J., Laine, S.: Extended ultraviolet disks and ultraviolet-bright disks in low-mass E/S0 galaxies. Astrophys. J. **745**, 34 (2012). doi:10.1088/0004-637X/745/1/34, 1111.0959

Morokuma-Matsui, K., Koda, J., Takekoshi, T., Saito, M., Nakanishi, H., Boissier, S., Madore, B., Boselli, A., Gil de Paz, A., Thilker, D., Yagi, M., Sorai, K., Kuno, N.: Search for molecular gas in XUV disk of M83. In: Gil de Paz, A., Lee, J.C., Knapen, J.H. (eds.) Proceedings of IAU Symposium 321, "Formation and Evolution of Galaxy Outskirts". Cambridge University Press, Cambridge (2016)

Nakagawa, M., Onishi, T., Mizuno, A., Fukui, Y.: An unbiased search for molecular clouds in the southern galactic warp. Publ. Astron. Soc. Jpn. **57**, 917–931 (2005). doi:10.1093/pasj/57.6.917, astro-ph/0510473

Ostriker, E.C., McKee, C.F., Leroy, A.K.: Regulation of star formation rates in multiphase galactic disks: a thermal/dynamical equilibrium model. Astrophys. J. **721**, 975–994 (2010). doi:10.1088/0004-637X/721/2/975, 1008.0410

Planck Collaboration, Ade, P.A.R., Aghanim, N., Arnaud, M., Ashdown, M., Aumont, J., Baccigalupi, C., Balbi, A., Banday, A.J., Barreiro, R.B., et al.: Planck early results. XIX. All-sky temperature and dust optical depth from Planck and IRAS. Constraints on the "dark gas" in our galaxy. Astron. Astrophys. **536**, A19 (2011). doi:10.1051/0004-6361/201116479, 1101.2029

Rosolowsky, E.: The mass spectra of giant molecular clouds in the local group. Publ. Astron. Soc. Pac. **117**, 1403–1410 (2005). doi:10.1086/497582, astro-ph/0508679

Roškar, R., Debattista, V.P., Brooks, A.M., Quinn, T.R., Brook, C.B., Governato, F., Dalcanton, J.J., Wadsley, J.: Misaligned angular momentum in hydrodynamic cosmological simulations: warps, outer discs and thick discs. Mon. Not. R. Astron. Soc. **408**, 783–796 (2010). doi:10.1111/j.1365-2966.2010.17178.x, 1006.1659

Roychowdhury, S., Huang, M.L., Kauffmann, G., Wang, J., Chengalur, J.N.: The spatially resolved Kennicutt–Schmidt relation in the H I-dominated regions of spiral and dwarf irregular galaxies. Mon. Not. R. Astron. Soc. **449**, 3700–3709 (2015). doi:10.1093/mnras/stv515, 1503.02667

Ruffle, P.M.E., Millar, T.J., Roberts, H., Lubowich, D.A., Henkel, C., Pasachoff, J.M., Brammer, G.: Galactic edge clouds. I. Molecular line observations and chemical modeling of edge cloud 2. Astrophys. J. **671**, 1766–1783 (2007) doi:10.1086/522775, 0708.2740

Sage, L.J., Wrobel, J.M.: Detection of CO emission from S0 galaxies. Astrophys. J. **344**, 204–209 (1989). doi:10.1086/167789

Sakamoto, S., Hasegawa, T., Handa, T., Hayashi, M., Oka, T.: An out-of-plane CO (J = 2–1) survey of the milky way. II. Physical conditions of molecular gas. Astrophys. J. **486**, 276 (1997). doi:10.1086/304479

Sale, S.E., Drew, J.E., Knigge, C., Zijlstra, A.A., Irwin, M.J., Morris, R.A.H., Phillipps, S., Drake, J.J., Greimel, R., Unruh, Y.C., Groot, P.J., Mampaso, A., Walton, N.A.: The structure of the outer galactic disc as revealed by IPHAS early A stars. Mon. Not. R. Astron. Soc. **402**, 713–723 (2010). doi:10.1111/j.1365-2966.2009.15746.x, 0909.3857

Salomé, P., Combes, F., Edge, A.C., Crawford, C., Erlund, M., Fabian, A.C., Hatch, N.A., Johnstone, R.M., Sanders, J.S., Wilman, R.J.: Cold molecular gas in the Perseus cluster core. Association with X-ray cavity, Hα filaments and cooling flow. Astron. Astrophys. **454**, 437–445 (2006). doi:10.1051/0004-6361:20054745, astro-ph/0603350

Salomé, Q., Salomé, P., Combes, F., Hamer, S., Heywood, I.: Star formation efficiency along the radio jet in Centaurus A. Astron. Astrophys. **586**, A45 (2016). doi:10.1051/0004-6361/201526409, 1511.04310

Sánchez Almeida, J., Elmegreen, B.G., Muñoz-Tuñón, C., Elmegreen, D.M.: Star formation sustained by gas accretion. Astron. Astrophys. Rev. **22**, 71 (2014) doi:10.1007/s00159-014-0071-1, 1405.3178

Sancisi, R., Fraternali, F., Oosterloo, T., van der Hulst, T.: Cold gas accretion in galaxies. Astron. Astrophys. Rev. **15**, 189–223 (2008). doi:10.1007/s00159-008-0010-0, 0803.0109

Schinnerer, E., Scoville, N.: First interferometric observations of molecular gas in a polar ring: the helix galaxy NGC 2685. Astrophys. J. **577**, L103–L106 (2002). doi:10.1086/344242, astro-ph/0209004

Schmidt, M.: The rate of star formation. Astrophys. J. **129**, 243 (1959). doi:10.1086/146614

Schruba, A., Leroy, A.K., Walter, F., Sandstrom, K., Rosolowsky, E.: The scale dependence of the molecular gas depletion time in M33. Astrophys. J. **722**, 1699–1706 (2010). doi:10.1088/0004-637X/722/2/1699, 1009.1651

Schruba, A., Leroy, A.K., Walter, F., Bigiel, F., Brinks, E., de Blok, W.J.G., Dumas, G., Kramer, C., Rosolowsky, E., Sandstrom, K., Schuster, K., Usero, A., Weiss, A., Wiesemeyer, H.: A molecular star formation law in the atomic-gas-dominated regime in nearby galaxies. Astron. J. **142**, 37 (2011). doi:10.1088/0004-6256/142/2/37, 1105.4605

Scoville, N.Z., Sanders, D.B.: H2 in the galaxy. In: Hollenbach, D.J., Thronson, H.A. Jr. (eds.) Interstellar Processes, Astrophysics and Space Science Library, vol. 134, pp. 21–50 (1987)

Scoville, N.Z., Solomon, P.M.: Radiative transfer, excitation, and cooling of molecular emission lines (co and Cs). Astrophys. J. **187**, L67 (1974). doi:10.1086/181398

Scoville, N.Z., Yun, M.S., Sanders, D.B., Clemens, D.P., Waller, W.H.: Molecular clouds and cloud cores in the inner galaxy. Astrophys. J. Suppl. Ser. **63**, 821–915 (1987). doi:10.1086/191185

Scoville, N., Sheth, K., Aussel, H., Vanden Bout, P., Capak, P., Bongiorno, A., Casey, C.M., Murchikova, L., Koda, J., Álvarez-Márquez, J., Lee, N., Laigle, C., McCracken, H.J., Ilbert, O., Pope, A., Sanders, D., Chu, J., Toft, S., Ivison, R.J., Manohar, S.: ISM masses and the star formation law at Z = 1 to 6: ALMA observations of dust continuum in 145 galaxies in the COSMOS survey field. Astrophys. J. **820**, 83 (2016). doi:10.3847/0004-637X/820/2/83, 1511.05149

Serra, P., Oosterloo, T., Morganti, R., Alatalo, K., Blitz, L., Bois, M., Bournaud, F., Bureau, M., Cappellari, M., Crocker, A.F., Davies, R.L., Davis, T.A., de Zeeuw, P.T., Duc, P.A., Emsellem, E., Khochfar, S., Krajnović. D., Kuntschner, H., Lablanche, P.Y., McDermid, R.M., Naab, T., Sarzi, M., Scott, N., Trager, S.C., Weijmans, A.M., Young, L.M.: The ATLAS^{3D} project - XIII. Mass and morphology of H I in early-type galaxies as a function of environment. Mon. Not. R. Astron. Soc. **422**, 1835–1862 (2012). doi:10.1111/j.1365-2966.2012.20219.x, 1111.4241

Sil'chenko, O.K., Chilingarian. I.V., Sotnikova, N.Y., Afanasiev, V.L.: Large-scale nested stellar discs in NGC 7217. Mon. Not. R. Astron. Soc. **414**, 3645–3655 (2011). doi:10.1111/j.1365-2966.2011.18665.x, 1103.1692

Sofue, Y., Nakanishi, H.: Three-dimensional distribution of the ISM in the Milky Way galaxy. IV. 3D molecular fraction and Galactic-scale H I-to-H_2 transition. Publ. Astron. Soc. Jpn. (2016). doi:10.1093/pasj/psw062, 1604.05794

Sofue, Y., Honma, M., Arimoto, N.: The molecular front in galaxies. I. CO VS HI in position-velocity diagrams. Astron. Astrophys. **296**, 33 (1995)

Solomon, P.M., Rivolo, A.R., Barrett, J., Yahil, A.: Mass, luminosity, and line width relations of Galactic molecular clouds. Astrophys. J. **319**, 730–741 (1987). doi:10.1086/165493

Sun, M., Donahue, M., Roediger, E., Nulsen, P.E.J., Voit, G.M., Sarazin, C., Forman, W., Jones, C.: Spectacular X-ray tails, intracluster star formation, and ULXs in A3627. Astrophys. J. **708**, 946–964 (2010). doi:10.1088/0004-637X/708/2/946, 0910.0853

Sun, Y., Xu, Y., Yang, J., Li, F.C., Du, X.Y., Zhang, S.B., Zhou, X.: A possible extension of the Scutum–Centaurus arm into the outer second quadrant. Astrophys. J. **798**, L27 (2015). doi:10.1088/2041-8205/798/2/L27, 1412.2425

Thilker, D.A., Bianchi, L., Boissier, S., Gil de Paz, A., Madore, B.F., Martin, D.C., Meurer, G.R., Neff, S.G., Rich, R.M., Schiminovich, D., Seibert, M., Wyder, T.K., Barlow, T.A., Byun, Y.I., Donas, J., Forster, K., Friedman, P.G., Heckman, T.M., Jelinsky, P.N., Lee, Y.W., Malina, R.F., Milliard, B., Morrissey, P., Siegmund, O.H.W., Small, T., Szalay, A.S., Welsh, B.Y.: Recent star formation in the extreme outer disk of M83. Astrophys. J. **619**, L79–L82 (2005). doi:10.1086/425251, arXiv:astro-ph/0411306

Thilker, D.A., Bianchi, L., Meurer, G., Gil de Paz, A., Boissier, S., Madore, B.F., Boselli, A., Ferguson, A.M.N., Muñoz-Mateos, J.C., Madsen, G.J., Hameed, S., Overzier, R.A., Forster, K., Friedman, P.G., Martin, D.C., Morrissey, P., Neff, S.G., Schiminovich, D., Seibert, M., Small, T., Wyder, T.K., Donas, J., Heckman, T.M., Lee, Y.W., Milliard, B., Rich, R.M., Szalay, A.S., Welsh, B.Y., Yi, S.K.: A search for extended ultraviolet disk (XUV-Disk) galaxies in the local universe. Astrophys. J. Suppl. Ser. **173**, 538–571 (2007). doi:10.1086/523853, 0712.3555

Tosaki, T., Kuno, N., Onodera, S. Rie, M., Sawada, T., Muraoka, K., Nakanishi, K., Komugi, S., Nakanishi, H., Kaneko, H., Hirota, A., Kohno, K., Kawabe, R.: NRO M33 all-disk survey of giant molecular clouds (NRO MAGiC). I. H I to H_2 transition. Publ. Astron. Soc. Jpn. **63**, 1171–1179 (2011). 1106.4115

van Dishoeck, E.F., Black, J.H.: The photodissociation and chemistry of interstellar CO. Astrophys. J. **334**, 771–802 (1988). doi:10.1086/166877

van Gorkom, J.H.: Interaction of galaxies with the intracluster medium. clusters of galaxies: probes of cosmological structure and galaxy evolution (2004), p. 305. astro-ph/0308209

Vázquez, R.A., May, J., Carraro, G., Bronfman, L., Moitinho, A., Baume, G.: Spiral structure in the outer galactic disk. I. The third galactic quadrant. Astrophys. J. **672**, 930–939 (2008). doi:10.1086/524003, 0709.3973

Verdugo, C., Combes, F., Dasyra, K., Salomé, P., Braine, J.: Ram pressure stripping in the Virgo Cluster. Astron. Astrophys. **582**, A6 (2015). doi:10.1051/0004-6361/201526551, 1507.04388

Vollmer, B., Braine, J., Combes, F., Sofue, Y.: New CO observations and simulations of the NGC 4438/NGC 4435 system. Interaction diagnostics of the Virgo cluster galaxy NGC 4438. Astron. Astrophys. **441**, 473–489 (2005). doi:10.1051/0004-6361:20041389

Walter, F., Martin, C.L., Ott, J.: Extended star formation and molecular gas in the tidal arms near NGC 3077. Astron. J. **132**, 2289–2295 (2006). doi:10.1086/508273, astro-ph/0608169

Walter, F., Brinks, E., de Blok, W.J.G., Bigiel, F., Kennicutt, R.C. Jr., Thornley, M.D., Leroy, A.: THINGS: The H I nearby galaxy survey. Astron. J. **136**, 2563–2647 (2008) doi:10.1088/0004-6256/136/6/2563, 0810.2125

Wang, Q.D., Owen, F., Ledlow, M.: X-Raying A2125: A large-scale hierarchical complex of galaxies and hot gas. Astrophys. J. **611**, 821–834 (2004). doi:10.1086/422332, astro-ph/0404602

Watson, D.M., Guptill, M.T., Buchholz, L.M.: Detection of CO J = 2 goes to 1 emission from the polar rings of NGC 2685 and NGC 4650A. Astrophys. J. **420**, L21–L24 (1994). doi:10.1086/187153

Watson, L.C., Martini, P., Lisenfeld, U., Böker, T., Schinnerer, E.: Testing the molecular-hydrogen Kennicutt–Schmidt law in the low-density environments of extended ultraviolet disc galaxies. Mon. Not. R. Astron. Soc. **455**, 1807–1818 (2016). doi:10.1093/mnras/stv2412

Welch, G.A., Sage, L.J.: The cool interstellar medium in S0 galaxies. I. A survey of molecular gas. Astrophys. J. **584**, 260–277 (2003). doi:10.1086/345537, astro-ph/0210337

Welch, G.A., Sage, L.J., Young, L.M.: The cool interstellar medium in elliptical galaxies. II. Gas content in the volume-limited sample and results from the combined elliptical and lenticular surveys. Astrophys. J. **725**, 100–114 (2010). doi:10.1088/0004-637X/725/1/100, 1009.5259

Whitmore, B.C., McElroy, D.B., Schweizer, F.: The shape of the dark halo in polar-ring galaxies. Astrophys. J. **314**, 439–456 (1987) doi:10.1086/165077

Whitmore, B.C., Lucas, R.A., McElroy, D.B., Steiman-Cameron, T.Y., Sackett, P.D., Olling, R.P.: New observations and a photographic atlas of polar-ring galaxies. Astron. J. **100**, 1489–1522 (1990). doi:10.1086/115614

Wolfire, M.G., Hollenbach, D., McKee, C.F.: The dark molecular gas. Astrophys. J. **716**, 1191–1207 (2010). doi:10.1088/0004-637X/716/2/1191, 1004.5401

Wong, T., Blitz, L.: The relationship between gas content and star formation in molecule-rich spiral galaxies. Astrophys. J. **569**, 157–183 (2002). doi:10.1086/339287, astro-ph/0112204

Wong, O.I., Kenney, J.D.P, Murphy, E.J., Helou, G.: The search for shock-excited H_2 in Virgo spirals experiencing ram pressure stripping. Astrophys. J. **783**, 109 (2014). doi:10.1088/0004-637X/783/2/109, 1401.6223

Yagi, M., Yoshida, M., Komiyama, Y., Kashikawa, N., Furusawa, H., Okamura, S., Graham, A.W., Miller, N.A., Carter, D., Mobasher, B., Jogee, S.: A dozen new galaxies caught in the act: gas stripping and extended emission line regions in the coma cluster. Astron. J. **140**, 1814–1829 (2010). doi:10.1088/0004-6256/140/6/1814, 1005.3874

Yasui, C., Kobayashi, N., Tokunaga, A.T., Terada, H., Saito, M.: Deep near-infrared imaging of an embedded cluster in the extreme outer galaxy: census of supernova-triggered star formation. Astrophys. J. **649**, 753–758 (2006). doi:10.1086/506382, astro-ph/0606023

Yasui, C., Kobayashi, N., Tokunaga, A.T., Terada, H., Saito, M.: Star formation in the extreme outer galaxy: Digel cloud 2 clusters. Astrophys. J. **675**, 443–453 (2008). doi:10.1086/524356, 0711.0257

Young, L.M.: Molecular gas in elliptical galaxies: distribution and kinematics. Astron. J. **124**, 788–810 (2002). doi:10.1086/341648, astro-ph/0205162

Young, J.S., Scoville, N.Z.: Molecular gas in galaxies. Annu. Rev. Astron. Astrophys. **29**, 581–625 (1991). doi:10.1146/annurev.aa.29.090191.003053

Young, L.M., Bureau, M., Cappellari, M.: Structure and kinematics of molecular disks in fast-rotator early-type galaxies. Astrophys. J. **676**, 317–334 (2008). doi:10.1086/529019, 0712.4189

Young, L.M., Bureau, M., Davis, T.A., Combes, F., McDermid, R.M., Alatalo, K., Blitz, L., Bois, M., Bournaud, F., Cappellari, M., Davies, R.L., de Zeeuw, P.T., Emsellem, E., Khochfar, S., Krajnović, D., Kuntschner, H., Lablanche, P.Y., Morganti, R., Naab, T., Oosterloo, T., Sarzi, M., Scott, N., Serra, P., Weijmans, A.M.: The ATLAS^{3D} project - IV. The molecular gas content of early-type galaxies. Mon. Not. R. Astron. Soc. **414**, 940–967 (2011). doi:10.1111/j.1365-2966.2011.18561.x, 1102.4633

Chapter 7
HI in the Outskirts of Nearby Galaxies

Albert Bosma

Abstract The HI in disk galaxies frequently extends beyond the optical image and can trace the dark matter there. I briefly highlight the history of high spatial resolution HI imaging, the contribution it made to the dark matter problem, and the current tension between several dynamical methods to break the disk-halo degeneracy. I then turn to the flaring problem, which could in principle probe the shape of the dark halo. Instead, however, a lot of attention is now devoted to understanding the role of gas accretion via galactic fountains. The current Λ cold dark matter theory has problems on galactic scales, such as the core-cusp problem, which can be addressed with HI observations of dwarf galaxies. For a similar range in rotation velocities, galaxies of Type Sd have thin disks, while those of Type Im are much thicker. After a few comments on Modified Newtonian Dynamics and on irregular galaxies, I close with statistics on the HI extent of galaxies.

7.1 Introduction

In this review, I will discuss the development of HI imaging in nearby galaxies, with emphasis on the galaxy outskirts, and take stock of the subject just before the start of the new surveys using novel instrumentation enabled by the developments in the framework of the Square Kilometer Array (SKA), which was originally partly inspired by HI imaging (Wilkinson 1991). I refer to other reviews on more specific subjects when appropriate. Issues related to star formation are dealt with by Elmegreen and Hunter (2017) and Koda and Watson (2017).

In the late 1950s, it became clear that there was more to a galaxy than just its optical image. This was principally due to the prediction of the 21 cm HI hyperfine structure line by Van de Hulst in 1944 and its detection in the Milky Way (Ewen and Purcell 1951; Muller and Oort 1951). A first rotation curve of the Milky Way was determined by Kwee et al. (1954). The first observations of M31 were done using the Dwingeloo 25 m telescope (van de Hulst et al. 1957), and a mass model for this

A. Bosma (✉)
Laboratoire d'Astrophysique de Marseille, Aix Marseille University, CNRS, LAM, Marseille, France
e-mail: bosma@lam.fr

J.H. Knapen et al. (eds.), *Outskirts of Galaxies*, Astrophysics and Space Science Library 434, DOI 10.1007/978-3-319-56570-5_7

galaxy was considered by Schmidt (1957). A possible increase in the mass-to-light (M/L) ratio in the outermost parts of M31 prompted the photoelectric study of its light distribution by de Vaucouleurs (1958), who confirmed that the local M/L ratio at the last point measured by Van de Hulst et al. exceeds the expectation from a model with constant M/L ratio throughout this galaxy's disk, in contrast with an earlier model by Schwarzschild (1954). The outer HI layer of the Milky Way was found to be warped (Burke 1957; Kerr 1957) and flaring (Gum et al. 1960). Dark matter in the Local Group was inferred by Kahn and Woltjer (1959), who attributed the high mass of the Local Group to intergalactic gas and explained the warp as due to our Milky Way moving through it.

7.2 HI in Galaxies and the Dark Matter Problem: Early Work

Early HI work on external galaxies was done with single-dish telescopes, often prone to side lobe effects (cf. discussion after Salpeter 1978). Progress was slow and hampered by low spatial resolution—a small ratio of HI radius to beam size (Bosma 1978, his Chap. 3.4)—of the observations. With the Dwingeloo telescope, work on M31 was followed by major axis measurements of M33 and M101 (Volders 1959). The Parkes telescope was used to image the LMC (McGee and Milton 1966), the SMC (Hindman 1967), NGC 300 (Shobbrook and Robinson 1967) and M83 (Lewis 1968). In the northern hemisphere, data were reported for several galaxies by Roberts (1966, for M31), Roberts (1972) and Davies (1974). In an Appendix, Freeman (1970) discusses rotation data for the LMC, SMC, NGC 300 and M33 and, for the latter two galaxies, found that the turnover velocity indicated by the low-resolution HI data is larger than the one indicated by fitting an exponential disk to the optical surface photometry, suggesting the presence of matter in the outer parts with a different distribution than that in the inner parts.

The quest for higher spatial resolution was first achieved routinely with the Caltech interferometer (e.g. for M101, Rogstad and Shostak 1971; Rogstad 1971), and results for five Scd galaxies were summarized in Rogstad and Shostak (1972), who found high M/L material in the outer parts of these galaxies. Work began on M31 with the Cambridge half-mile telescope with a first report (Emerson and Baldwin 1973) showing "normal" M/L values. Their data were in disagreement with data by Roberts on the same galaxy, as debated in a meeting in Besançon (Roberts 1975; Baldwin 1975). This problem was settled by new data reported by Newton and Emerson (1977), who by and large confirmed the data of Roberts and Whitehurst (1975). Meanwhile, HI work with the Westerbork Synthesis Radio Telescope (WSRT) had started, with as first target M81. The resulting rotation data, extending beyond the optical image, were discussed in Roberts and Rots (1973), with curves for M31 (300 ft data), M81 (WSRT and 300 ft data), M101 (from Rogstad's Caltech data) and the Milky Way. These data, in particular for M31, also

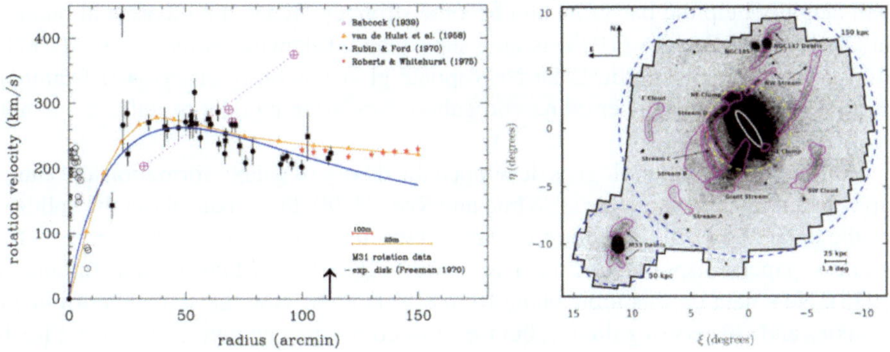

Fig. 7.1 *Left panel*: rotation curves of M31, as determined by Babcock (1939, *purple points)*, van de Hulst et al. (1957, *orange points)*, Rubin and Ford (1970, *black points)* and Roberts and Whitehurst (1975, *red points)*. The *blue line* indicates the expected maximum disk rotation curve based on an exponential disk with the scale length given in Freeman (1970) based on the study of de Vaucouleurs (1958). The *arrow* indicates the optical radius. *Right panel*: modern picture of the outskirts of M31, as determined from the PAndAS survey (reproduced with permission from Ferguson and Mackey 2016), where the outline of the optical image is in *white*. Note the change in scale

suggested that there could be more than meets the eye in the outer parts of galaxies, i.e. material with high M/L ratio (cf. Fig. 7.1, left panel).

A second line of argument for material with high M/L ratio in the outer parts of galaxies came from theory. Early numerical experiments, e.g. by Hohl (1971), showed that simulating a galaxy disk with N-body particles and letting it evolve generated the formation of a bar-like structure with relatively high velocity dispersion. Since strong bars are present only in about 30% of the disk galaxies, Ostriker and Peebles (1973) proposed to stabilize the disk by immersing it in a halo of "dark matter", in such a way that the gravitational forces of the halo material acting on the disk could prohibit the bar to form. This requires that out to the radius of the "optical" disk, the mass of the halo is 1–2.5 times the disk mass. Once hypothesized, it followed that the masses of galaxies exterior to this radius could be extremely large. This was subsequently investigated by Einasto et al. (1974), as well as Ostriker et al. (1974). While the former considered five galaxies, among which IC 342, and data on binary galaxies, the latter put forward at least half a dozen probes (the rotation curve data discussed above, the Local Group timing argument already considered in Kahn and Woltjer (1959), binary galaxy samples, etc.). Both papers concluded that there must be additional, dark, matter beyond the optical radius of a galaxy, since the mass increases almost linearly with radius, and the light converges asymptotically.

Athanassoula (2002, 2003), using much-improved N-body simulations with live haloes (i.e. haloes responding to gravitational forces), confirmed the above results for the initial phases of the evolution. She found, however, that at later times, when the bar evolves and increases in strength, the halo material at resonance with the bar will actually help the bar grow and become stronger. Hence the theoretical picture in Ostriker and Peebles (1973) is now superseded. Likewise, Athanassoula (2008) showed that the Efstathiou-Lake-Negroponte global stability criterion (Efstathiou et al. 1982), popular in semi-analytic galaxy formation models of galaxies, is not really valid.

Although some cosmologists developed the theory of galaxy formation and came up with a two-stage model (e.g. White and Rees 1978), the debate about the validity of the data, and even the notion of dark matter itself, continued for several years, witness papers disputing the idea (e.g. Burbidge 1975; Materne and Tammann 1975). New data on this topic came from efforts improving the statistics of binary galaxies and clusters of galaxies, but the most convincing evidence came from fresh data on the rotation curves of spiral galaxies, discussed in Sect. 7.4.

7.3 Warps

One observational problem hindering the acceptance of the existence of the dark matter indicated by the HI data concerned the presence of large-scale non-circular motions in the outer parts, as frequently emphasized by sceptical observers (e.g. Baldwin 1974, 1975). Einasto et al. (1974) simply rejected this hypothesis—without stating a reason—even though the one galaxy they show data for, IC 342, was found to be clearly warped (Newton 1980).

Rogstad et al. (1974) came up with a novel approach to model the velocity field of M83, using new data obtained with the two-element Owens Valley interferometer. They introduced the notion of what later was called the "tilted ring model", by modelling a galaxy with a set of concentric annuli each having a different spatial orientation. Although they were forced to use a model rotation curve on account of M83's low inclination, later on rotation curves were derived by determining the rotation velocity for each annulus separately (Bosma 1978, 1981a). It is of interest to display the results of Rogstad et al. (1974) side by side with more modern results, as has been arranged in Fig. 7.2. Since the outer contour in the 1974 HI image is $1.37 \times 10^{20}\,\mathrm{cm}^{-2}$, the presence of extended HI disks in M83-like systems is detectable with the upcoming large HI surveys, even for those with shorter integration times such as planned for the APERTIF-SNS and WALLABY surveys.

The first study of HI warps in spiral galaxies seen edge-on was done by Sancisi (1976), whose clearest case was NGC 5907. This galaxy has subsequently been studied extensively in the optical regime, in order to find the presence of extraplanar light. A "faint glow" was found by Sackett et al. (1994a), whose report made the cover of Nature, but later work by Zheng et al. (1999) indicated the presence of an arc. A deeper picture was published by Martínez-Delgado et al. (2008) and is reproduced in Fig. 7.3. Further modelling of the warp has been done in Allaert et al. (2015). The surmise in Sancisi (1976) that warps are occurring in galaxies seemingly free of signs of interactions thus turns out to be not justified, but the stellar mass

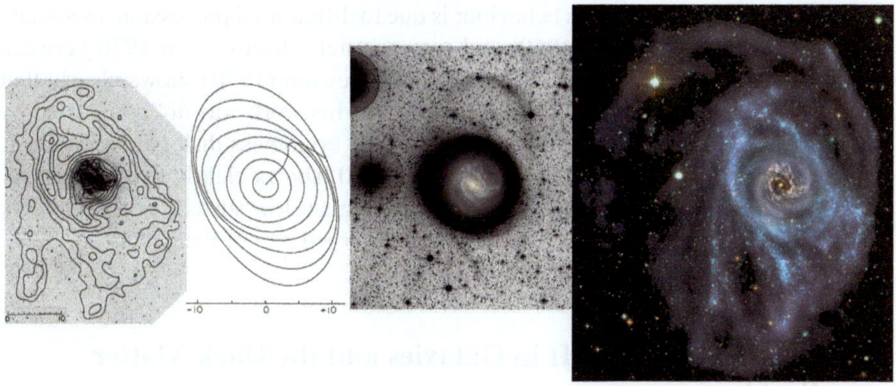

Fig. 7.2 *Left two panels*: HI distribution and tilted ring model of M83 (reproduced with permission from Rogstad et al. 1974); next to it a deep optical image obtained by Malin and Hadley (1997) with a shallower colour image superimposed (credit: D. Malin). The *rightmost image* was constructed from *GALEX* UV data, optical images and the HI distribution obtained in the Local Volume HI Survey conducted with CSIRO's Australia Telescope Compact Array (ATCA; credit, B. Koribalski and A.R. López-Sánchez). All images are on the same scale

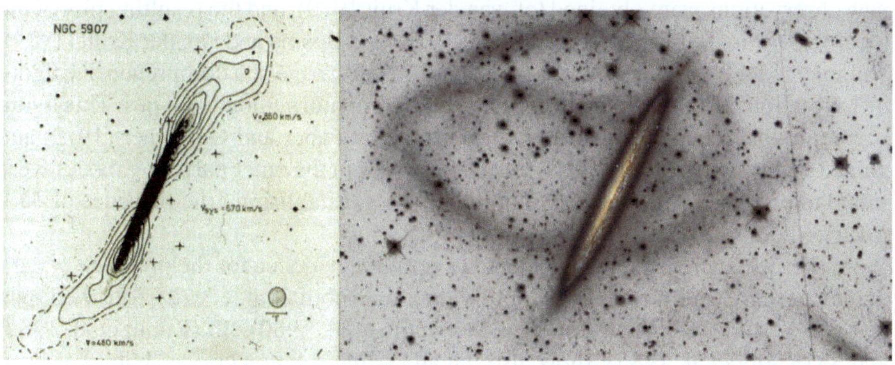

Fig. 7.3 *Left*: two HI channel maps at either side of the systemic velocity of NGC 5907, superimposed on an optical image of the galaxy, indicate clearly the presence of a warp in the HI distribution (reproduced with permission from Sancisi 1976). *Right*: at the same scale, a deep optical image showing the presence of a diffuse tidal stellar stream around this galaxy (reproduced with permission from Martínez-Delgado et al. 2008)

involved in the stream appears minor compared to that in the main galaxy. As is the case for M83, the streamer indicates a past interaction or merging event, a process which could have caused the warp as well.

Rogstad et al. (1976, 1979) found warps in M33 and NGC 300, and I found several more during my thesis work, as discussed below. Briggs (1990) outlined a set of "rules of behaviour" for warps by studying HI images of a number of warped galaxies studied by the end of the 1980s. They start in the region beyond the optical radius (r_{opt}) and become prominent at the Holmberg radius (r_{Ho}), while the line of nodes turns in such a manner that the direction in which it is turning is leading with respect to the spiral arms. This behaviour is due to differential precession, as already noted in Kahn and Woltjer (1959) and was privately discussed in 1976 between Rogstad and Bosma for five cases then known. Newton (1980) shows clearly that the data of IC 342 do not support this picture, but his work was not considered by Briggs (1990). Note that for the giant low surface brightness disk galaxy Malin 1, the HI velocity field observed by Lelli et al. (2010) shows clearly that this galaxy also violates Briggs's rule no. 3 and that the faint giant low surface brightness disk imaged recently by Galaz et al. (2015) is in the warped part of the galaxy.

7.4 Further Data on HI in Galaxies and the Dark Matter Problem

In my thesis work (Bosma 1978, 1981a,b), I collected extensive HI data on a number of galaxies using the WSRT and its 80-channel receiver. I produced warp models using tilted ring models and derived mass models from rotation curves ignoring possible vertical motions associated with the warps. For the photometry, new observations were obtained (cf. van der Kruit 1979), and the resulting plot of the local M/L ratios was reported in Bosma (1978) and Bosma and van der Kruit (1979). I augmented the sample with literature data so as to arrive at a comprehensive figure of 25 rotation curves split over 6 panels of different morphological types. This figure was reproduced in an influential review article by Faber and Gallagher (1979) and gained widespread attention. The local M/L ratios in the outer parts of galaxies were found to be very large in quite a number of cases, establishing the presence of dark matter.

At present, such data are considered as primary evidence for the presence of dark matter in spiral galaxies. Contemporary data by Rubin et al. (1978), followed by a systematic study of galaxies of Type Sc (Rubin et al. 1980), Sb (Rubin et al. 1982) and Sa (Rubin et al. 1985), show also flat and rising rotation curves, but it has been convincingly shown first by Kalnajs (1983) and later by Kent (1986, 1987, 1988), as well as by Athanassoula et al. (1987), that for most of the galaxies in Rubin's survey, the data do not go out far enough in radius to unambiguously demonstrate the need for dark matter (see also Bertone and Hooper 2016).

For this review, I will concentrate on the most extended HI disks found in Bosma (1978, 1981a,b) and show as example the galaxy NGC 2841 (Fig. 7.4). The HI extends out to ~2.5 times the Holmberg radius, i.e. $\sim 3.5 \times r_{\rm opt}$. The tilted ring model describing the warp leaves out the northernmost feature, an asymmetry further emphasized in Baldwin et al. (1980), although of much smaller amplitude than that in the HI distribution in M101. A similar extended HI disk was found for NGC 5055 (Fig. 7.5), which became later a prototype of a Type I extended UV disk (Thilker et al. 2007), and for which Martínez-Delgado et al. (2010) found extensive streamers, some of which are not associated with the HI gas or the young stars. In Bosma (1978, 1981b), I summarize data on the extent of the HI disks in my sample

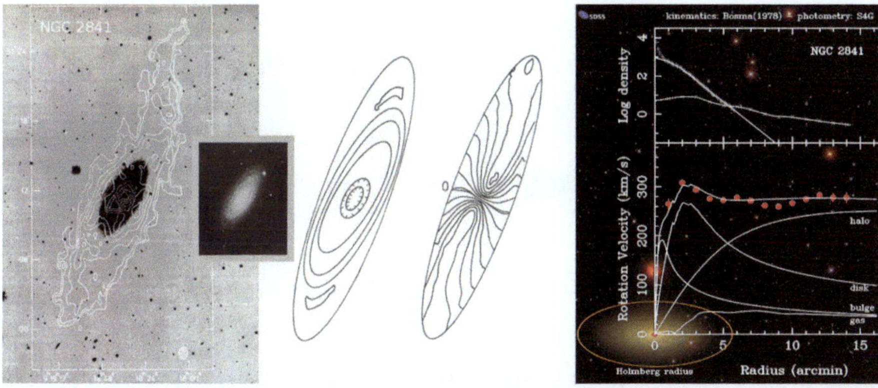

Fig. 7.4 At *left*, the HI distribution in the galaxy NGC 2841 observed by Bosma (1978), overlaid on a deep IIIaJ image provided by H.C. Arp; the *inset* shows the Hubble Atlas image (Sandage 1961). The *middle panels* show the warp model. At *right* a mass model of the galaxy adjusted to the HI rotation data in Bosma (1978), calculated as described in Athanassoula et al. (1987), using surface photometry data from the *Spitzer* Survey of Stellar Structure in Galaxies (S^4G; Muñoz-Mateos et al. 2015), and overlaid on scale on a Sloan Digital Sky Survey (SDSS) colour image. The *orange ellipse* around the galaxy outlines the Holmberg (1958) dimensions. Montage as in Roberts (1988), suggested in this form by Bernard Jones (private communication)

and found a wide variety of values of the ratio $r_{\rm HI}/r_{\rm opt}$, where $r_{\rm HI}$ is defined as the isophote where the HI column density is $1.82 \times 10^{20}\,{\rm cm}^{-2}$, with a mean of 2.2 ± 1.1.

7.5 The Disk-Halo Degeneracy in the Dark Matter Problem

Van Albada and Sancisi (1986) pointed out that mass modelling based on the assumption that the M/L ratio is constant as function of radius in the disk contains a degeneracy: it is a priori not clear whether a maximum disk, i.e. a disk so massive that its rotation curve fits the inner parts of the observed one without overshooting it, is the correct answer. Already it is not entirely justified to assume that the M/L ratio is constant throughout the disk, even though this is customary, since if there is a colour gradient in bright spirals, it is usually in the sense that the outer parts are bluer. Yet another problem is that the assumption of maximum disk generally leads to haloes which are cored and thus do not follow a Navarro-Frenk-White (NFW) model (e.g. Navarro 1998).

This disk-halo degeneracy is a serious problem which is even now under debate. Relating a value of the M/L ratio to a disk colour, and working out whether "reasonable" stellar populations can then be assumed, does not appear entirely satisfactory, due to possible variations in the initial mass function (IMF). Various groups have thus tried to marshall dynamical arguments to break the degeneracy, but the answers are mixed. Mechanisms of spiral structure generation, in particular

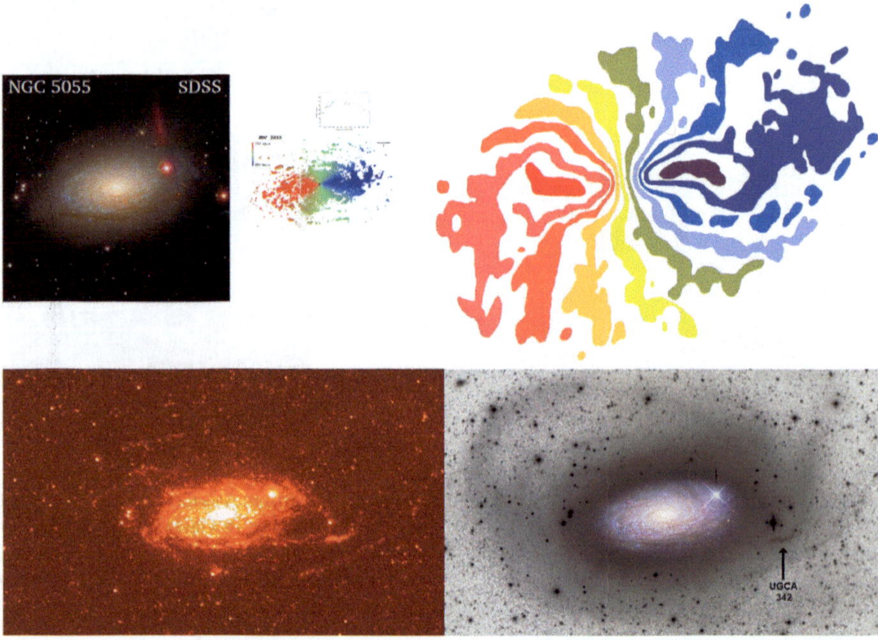

Fig. 7.5 *Top left*, a SDSS colour image of NGC 5055, with next to it a representation of the Hα velocity field from the GHASP survey (reproduced with permission from Blais-Ouellette et al. 2004) and just above it the Hα rotation curve of Burbidge et al. (1960). The *top-right panel* shows the velocity field of NGC 5055 shown on the front cover of Bosma (1978). The colour lookup table used there is such that the boundaries of each coloured patch correspond to equispaced isovelocity contours. *Bottom left* is the *GALEX* near-UV image (data taken from the NASA/IPAC Extragalactic Database, NED) and *bottom right* a deep optical image from Martínez-Delgado et al. (2010) (credit: D. Martinez-Delgado and T. Chonis). All images are on the same scale. The star-forming patch west of the centre of NGC 5055, UGCA 342, can already be seen on the deep IIIaJ image (van der Kruit 1979) also shown as Fig. 4.3.1b in Bosma (1978)

the swing amplifier mechanism (Toomre 1981), depend on the ratio of disk mass to halo mass. Athanassoula et al. (1987) used this in detail to set a range of allowed values of the disk-halo mass ratio, by remarking that most spirals are dominated by an $m = 2$ component and thus requiring that an $m = 2$ spiral be allowed to amplify, while at the upper end suppressing an $m = 1$ component. This is further discussed in Bosma (1999) and illustrated in Fig. 7.6 (top-right panel).

Fuller dynamical modelling has been done for a number of galaxies. Kranz et al. (2003) calculated spiral structure models based on potentials derived from K'-band photometry and compared the gas flow in these with the observed velocity fields for a sample of five galaxies. Similarly, for bars, the gas flow can be calculated in a potential derived from imaging in the near- or mid-infrared. Seeking to fit the amplitude of the jump in radial velocity across a dust lane, which outlines the location of a shock in the flow, will constrain the M/L ratio of the disk (Lindblad et al. 1996; Weiner et al. 2001; Weiner 2004; Zánmar Sánchez et al. 2008). For

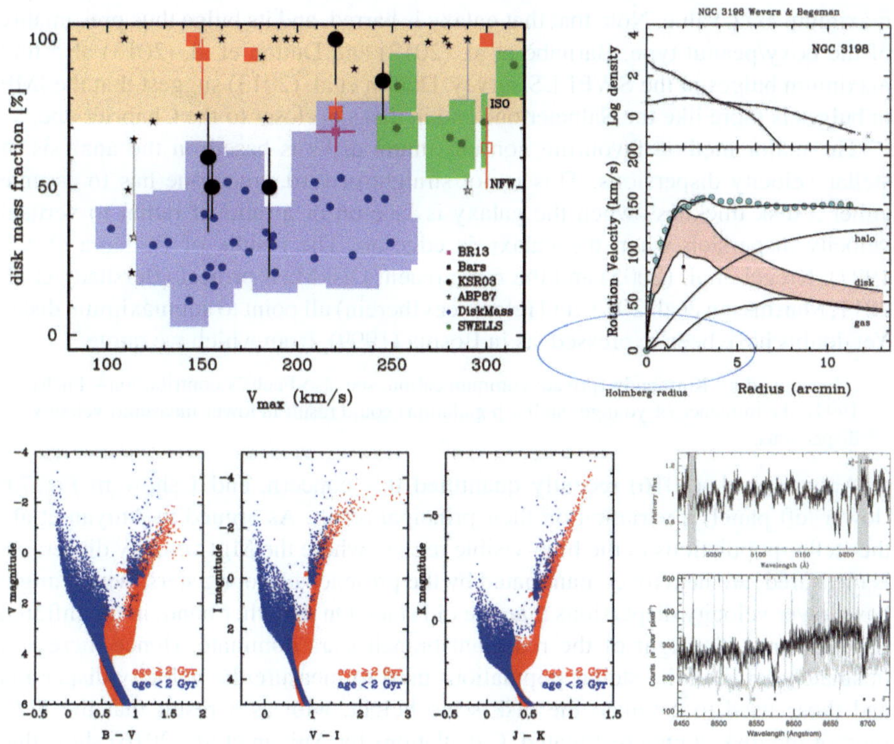

Fig. 7.6 *Top left*: disk mass fraction as function of the maximum velocity of the rotation curve, determined with several methods. "BR13" (Bovy and Rix 2013) indicates the value for our Galaxy. "Bars" concern the determination using gas flow models in barred spirals (see text). "KSR03" concern five galaxies studied by Kranz et al. (2003). *Black filled stars* concern the results from Athanassoula et al. (1987) for their "maximum disk with no $m = 1$" models, except for *two vertical lines* at $V_{\max} = 114.0$ and 280.0 km/s which indicate also the "no $m = 2$" models. For the DiskMass project, the results are taken as in Courteau and Dutton (2015), but the error bars are replaced by the area spanned by them. A similar representation has been done for the SWELLS survey (Barnabè et al. 2012; Dutton et al. 2013). *Top right*: mass models for NGC 3198 according to the method described in Athanassoula et al. (1987) and shown in detail in Bosma (1999). *Bottom left*: colour-magnitude diagrams calculated with IAC-STAR (Aparicio and Gallart 2004) in a manner similar to that used in Aniyan et al. (2016). *Bottom right*: spectra obtained by Westfall et al. (2011) in the MgI region and by Bershady et al. (2005) in the CaII region (reproduced by permission)

the face-on barred spiral NGC 1291, Fragkoudi et al. (2016) calculated a range of models trying to fit the shape of the dust lane. Most of the barred spiral models require close to maximum disk, while for the spiral models, a range of values depending on the maximum rotational velocity has been found, as shown in Fig. 7.6 (top-left panel).

The study of the lensing galaxy associated with the quad-lens Q2237+0305 done by Trott et al. (2010) shows that at least the central part of that galaxy needs a maximum M/L value. Note that that galaxy is barred, and its bulge thus presumably of the boxy/peanut type. Barnabè et al. (2012) and Dutton et al. (2013) also find maximum bulges in the SWELLS survey. Dutton et al. (2013) suggest that the IMF in bulges is more like the Salpeter one and in disks is closer to the Chabrier one.

The major method favouring non-maximum disks is based on the analysis of stellar velocity dispersions. This is not straightforward, since one has to assume either a disk thickness when the galaxy is face-on or a ratio of radial to vertical velocity dispersion when the galaxy is edge-on. The results of Bottema (1993, 1997), Kregel et al. (2005) and the more recent DiskMass project (Bershady et al. 2011; Martinsson et al. 2013, and references therein) all point to submaximum disks. Yet doubts have been expressed, as in Bosma (1999), from which we quote

> ... as argued by Kormendy (private communication, see also Fuchs's contribution—Fuchs 1999), the influence of younger stellar populations could result in lower measured velocity dispersions.

Aniyan et al. (2016) recently quantified this concern, and I show in Fig. 7.6 (lower left panel) a variation on their principal result. As argued in Aniyan et al., the stellar populations in the blue-visible region, where the MgI velocity dispersion is measured, are heavily contaminated by the presence of young stars, which might have lower velocity dispersions than the old stars. On the other hand, in the infrared (e.g. K-band), the light of the red giant branch stars dominate. Hence there is a mismatch between the stellar populations used to measure the velocity dispersion and those used to estimate the disk scale height, with as a result that the M/L ratio of the disk is underestimated. Calculations by Aniyan et al. (2016) show that this can roughly explain the difference between the results of the DiskMass project (submaximum disks) and those from several other dynamical estimators (maximum disks).

Although the authors of the DiskMass project were aware of a stellar population effect, witness the extensive discussions of this in, e.g. Bershady et al. (2005, 2010a,b) and Westfall et al. (2011), they argue clearly for the use of the MgI region as the preferred region for doing velocity dispersion work, since the CaII region has (1) larger intrinsic line widths, (2) a higher background and (3) more scattered light (cf. Bershady et al. 2005 and the spectra shown in Fig. 7.6, lower right panel). Most galaxies they discuss have been observed only in the MgI region, and, in the few cases where data at both wavelength regions were available, no difference was noticed.

It should be possible, however, to investigate a possible systematic effect of stellar population differences on the velocity dispersions further, since even in the i-band the older stellar populations are substantially more prominent than in the g-band (see Fig. 7.6, lower row, second panel from the left). There are now several spectrographs being built which will have a setup allowing the simultaneous measurement of the velocity dispersions of the same galaxy in both the MgI region and the CaII region. In particular, the WEAVE spectrograph (Dalton 2016) has high spectral resolution and can thus suitably be used to test whether CaII-derived velocity dispersion measurements are systematically larger than MgI-derived ones in face-on spiral galaxies.

7.6 Flaring of the Outer HI Layer: Probing the Shape of the Dark Matter Halo

As mentioned in Sect. 7.1, a flaring HI disk was found in the Milky Way (Gum et al. 1960). An early study of the flaring HI disk in the edge-on galaxy NGC 891 was made by van der Kruit (1981), who found that the HI thickness as function of radius observed by Sancisi and Allen (1979) could best be modelled by assuming that the disk potential was furnished mainly by the old stellar disk. For the face-on galaxies NGC 3938, NGC 628 and NGC 1058, a constant gas velocity dispersion as function of radius was found, of order 10 km/s (van der Kruit and Shostak 1982; Shostak and van der Kruit 1984; van der Kruit and Shostak 1984). It was thus thought that flaring HI disks could be used to probe the shape of the dark halo.

7.6.1 Early Work on Case Studies

With the advent of the high-sensitivity, high-resolution capabilities of the Very Large Array (VLA), NGC 4565 and NGC 891 were reobserved by Rupen (1991). These observations indicated clearly the warps in these systems, as well as asymmetries in the HI distribution. A further effort was made by Olling (1996a,b) for the small galaxy NGC 4244, who found that the best fit was a very flattened halo, with an axial ratio c/a of $\sim$0.2. Becquaert and Combes (1997) found a similar result for NGC 891, but if the vertical velocity dispersion was smaller than the radial one, a value of $\sim$0.4–0.5 could be found. Applying the same technique to observations of our Galaxy, Olling and Merrifield (2000) found a relatively round halo ($c/a \sim 0.8$) but needed to adopt rather low values for some Galactic constants. Using a completely different method (lensing and optical spectroscopy) for the massive galaxy SDSS J2141-0001, Barnabè et al. (2012) recently found an oblate dark halo, with $c/a = 0.75\pm^{0.27}_{0.16}$.

Also in the 1980s–1990s, efforts were made to use another probe of the shape of the dark matter halo, i.e. polar-ring galaxies. Here, there was a remarkable shift of the "most likely" halo shape: the early studies of the polar-ring galaxy A0136-0801 by Schweizer et al. (1983) and a few others (Whitmore et al. 1987) indicated a nearly round halo, while a later study of Sackett et al. (1994b) showed the halo around NGC 4650A as flat as E6. However, Arnaboldi et al. (1997) showed that the polar ring in that galaxy is very massive and has spiral structure. Khoperskov et al. (2014) worked on a larger sample and showed that a variety of halo shapes describe the data.

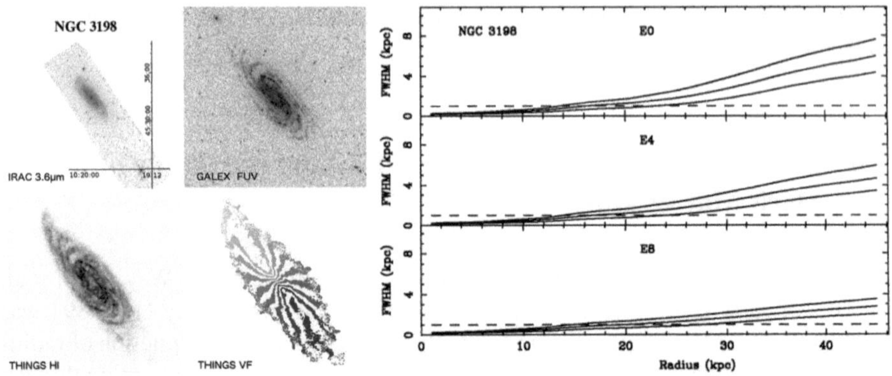

Fig. 7.7 *Spitzer* 3.6 μm, *GALEX* far-UV, and HI data from the THINGS survey for NGC 3198 (data taken from NED). At *right* is the flaring calculation from Bosma (1994), using the mass model in the *top-right panel* of Fig. 7.6. The *three curves* are for constant velocity dispersions of 6, 8 and 10 km/s, and the halo flattening is expressed as for elliptical galaxies

The expected HI disk flaring was calculated for NGC 3198 and DDO 154 by Bosma (1994), and the result for NGC 3198, based on the maximum disk model in Fig. 7.6 (top-right panel), is shown in Fig. 7.7. This calculation assumes equilibrium and shows that the flaring depends on the velocity dispersion of the HI disk, as well as the axial ratio of the dark halo. For a dwarf galaxy, assumed also to be flat, the flaring starts well inside the optical radius, since the dark matter dominates also in the inner parts of the disk. Thus the study of small, thin edge-on galaxies seems ideal for this problem.

7.6.2 Recent Results for Small, Flat Galaxies

"Super thin" galaxies were studied initially by Goad and Roberts (1981) and later investigated by, e.g. Uson and Matthews (2003), Matthews and Wood (2003) and Matthews and Uson (2008) and in thesis work by O'Brien et al. (2010a,b,c,d) and Peters et al. (2013, 2016a,b,c,d). A lot of work hides behind these results, since HI disks are frequently warped, and candidate galaxies have to be observed first before judging whether they are suitable for further study for the flaring problem. Even if the obviously warped galaxies are excluded from further study in this respect, subtle variations can influence the result. Matthews and Wood (2003) thus considered for their best case, UGC 7321, several models: a smooth distribution, a warped one, a flaring one and a model with an HI halo, which either corotates or lags, and combinations of these. None of them give a very satisfactory fit if attention is paid to detail. O'Brien et al. (2010d) found $c/a = 1.0 \pm 0.1$ for their best case, again UGC 7321. Peters et al. (2013) extended the database of O'Brien et al., but Peters et al. (2016d) found in the end only good flaring models for two galaxies, ESO

274-G001 and UGC 7321. For ESO 274-G001, they found an oblate halo with shape $c/a = 0.7 \pm 0.1$, while for UGC 7321 they found a prolate halo, with $c/a = 1.9\pm^{0.1}_{0.3}$ in case the HI is treated as optically thin and 2.0 ± 0.1 in case the optical thickness of the HI is taken into account. They point out that O'Brien et al. (2010d) did not consider prolate halo shapes, hence the difference.

Most of these late-type Sd galaxies have small maximum rotation velocities and low star formation activity. Systematic studies by Karachentsev et al. (1999) have resulted in an extensive catalogue of these flat galaxies. Such data can be explored further in the era of large galaxy surveys.

Peters et al. (2016a) emphasize that it cannot be assumed that the HI in galaxies is optically thin and constructed a modelling procedure. Braun (1997) already emphasized this point 20 years ago in a study of seven nearby spiral galaxies at high spatial and spectral resolution with the VLA and argued that there is significant opacity in the high surface brightness HI gas. Braun et al. (2009) show this in their M31 data with specific examples and estimate that the total HI gas mass in that galaxy is about 30% higher than the value inferred using the assumption of optically thin gas.

7.6.3 Large Galaxies with a High Star Formation Rate: Accretion

Deep imaging (for that time) of M101 (van der Hulst and Sancisi 1988) showed that there are numerous holes in the HI distribution in the main disk. In addition, HI gas was detected at velocities not corresponding to the disk rotation, indicating the possible presence of the equivalence with the Galactic high-velocity clouds (HVCs), either due to accretion of fresh gas or a collision with remnant gas clouds related to a tidal interaction event between this galaxy and a smaller neighbour. Deeper HI images of edge-on galaxies also became available, in particular for the galaxy NGC 891 (cf. Fig. 7.8). The observations of this very actively star-forming galaxy show a thick HI disk (Swaters et al. 1997), and yet more sensitive data show a more extensive thick HI disk and a streamer (Oosterloo et al. 2007).

The main result for the thick HI disk in NGC 891 is that its rotation rate is lagging with respect to the thin disk, as already shown by Swaters et al. (1997) and more extensively in Fraternali et al. (2005). In Oosterloo et al. (2007) more modelling of NGC 891 was done, with a catalogue of possibilities: a thin disk, a strong warp along the line of sight, a flaring disk, a disk and corotating halo, a lagging halo with constant gradient, a lagging halo with high velocity dispersion, a lagging halo with a radial inflow motion and a lagging halo with velocity gradient increasing in the inner parts.

It was realized that there could also be an effect of lagging HI halo gas seen in the observations of more face-on galaxies. Indeed, a position-velocity diagram for the galaxy NGC 2403 (Fraternali et al. 2001, 2002; Fig. 7.9) shows that there is an

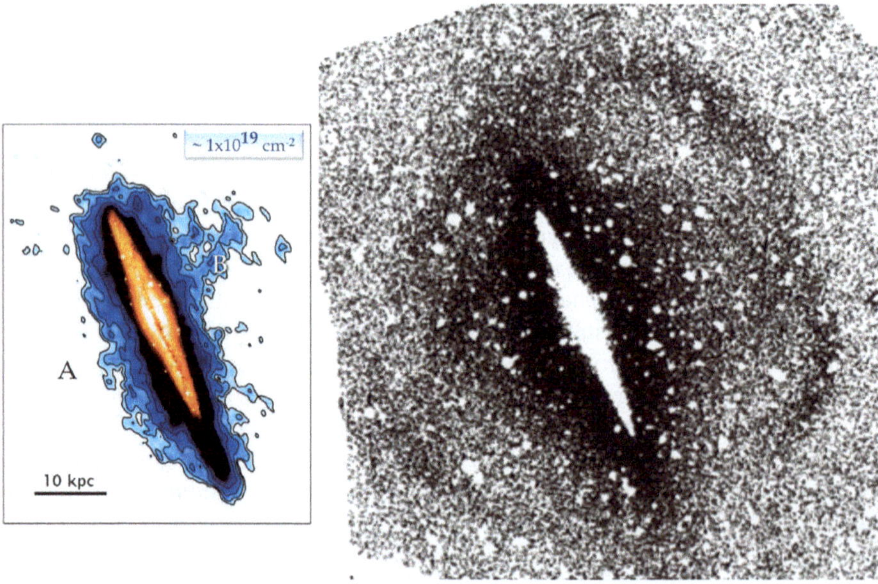

Fig. 7.8 At *left* the HI image of NGC 891 (reproduced with permission from Oosterloo et al. 2007), with an outer contour column density of $1.0 \times 10^{19}\,\mathrm{cm}^{-2}$, and at *right*, on the same scale, the image of the RGB stars (reproduced with permission from Mouhcine et al. 2010)

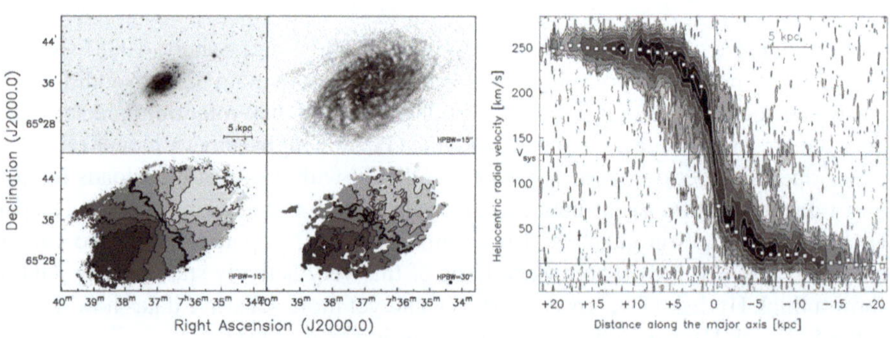

Fig. 7.9 VLA observations of the galaxy NGC 2403 (reproduced with permission from Fraternali et al. 2001). *Left panel of four*: *top*: Optical image from the Digital Sky Survey (*left*) and HI distribution (*right*) of the main disk component; *bottom*: velocity field of the main disk component (*left*) and of the anomalous gas component (*right*). *Right panel*: position-velocity diagram along the major axis, clearly showing the anomalous velocities (the "beard"). The derived *rotation curve* of the main disk component is superimposed, and the *lines at the bottom of the plot* show the range of velocities where contamination by galactic emission was filtered out

anomalous component with ∼10% of the total HI mass. The rotation velocity of the anomalous gas is 25–50 km/s lower than that of the disk. Its velocity field has non-orthogonal major and minor axes implying an overall inflow motion of $10-20$ km/s towards the centre of the galaxy. A similar phenomenon is also seen in the THINGS observations of the galaxy NGC 3198 shown in Fig. 7.11 (upper panel) and is better brought out by the deeper imaging of this galaxy by Gentile et al. (2013), who find that ∼15% of the total HI gas mass is in a thick disk with a scale height of about 3 ± 1 kpc.

Fraternali and Binney (2006) explored an initial model, where galactic fountain activity leads to clouds being sent up in the halo and then falling back to the disk again, but this did not explain the lagging of the rotation of the extraplanar gas, nor the inflow towards the disk. Sancisi et al. (2008), in a review, discuss the signatures of HI gas accretion around nearby galaxies and found an average visible HI accretion rate of $\sim 0.2\, M_\odot$ year^{-1}. This is an order of magnitude too low to sustain the current star formation rates in some of the galaxies studied. Fraternali and Binney (2008) improved their galactic fountain model by considering the interaction of the fountain clouds with the hot coronal gas in the halo, the presence of which is expected theoretically and ought to be observable in X-ray observations (see Hodges-Kluck and Bregman 2013 for a detection of that gas in NGC 891). A higher accretion rate can then be obtained, which is more of order of the star formation rate, so that the latter could be sustained from infalling gas over a longer timescale (e.g. Fraternali 2014, and references therein). For a recent in-depth review of this accretion model, see Fraternali (2016).

To follow up on this, the HALOGAS project was undertaken with the WSRT, but while first results have been published, a conclusive paper has still to be completed. From presentations at conferences, it appears that NGC 891 is a relatively rare case, given its active star formation and extensive extraplanar gas distribution, which has been detected at other wavelengths as well (e.g. radio continuum: Allen et al. 1978, Hα: Rand et al. 1990, X-ray: Hodges-Kluck and Bregman 2013). Dahlem et al. (2006) argue on the basis of several examples that the presence of radio continuum thick disks is correlated with active star formation.

Several HALOGAS case studies have been published. New data on NGC 4414 (de Blok et al. 2014a) show that the outer warped HI layer is dynamically somewhat offset from the central parts, while on a deep optical image, a shell structure is seen in the western part. Both the warp and the thickening of the main galaxy (the inclination is higher in the central parts of the galaxy than the outline of the main galaxy body seen in Fig. 7.10) could be due to an interaction which caused also the shell structure itself.

Concurrently, the HALOGAS galaxies were observed with the Green Bank Telescope (GBT), to look for further HI gas at lower column densities. de Blok et al. (2014b) found a cloud near NGC 2403 in the GBT data, which has an HI mass of $4 \times 10^6\, M_\odot$ which is 0.15% of the total HI mass of the galaxy. It seems to link up with a $\sim 10^7\, M_\odot$ filament in the anomalous HI gas described in Fraternali et al. (2002). Likewise, Heald (2015) reports about a few HVC-like clouds with a total HI mass of $4 \times 10^6\, M_\odot$ around NGC 1003. The very low column density HI filament between M31 and M33 discovered by Braun and Thilker (2004) was reobserved by Wolfe et al. (2013) and shown to be composed for ∼50% of discrete clouds, embedded in more diffuse gas. To conclude, it can be said that there is HI around

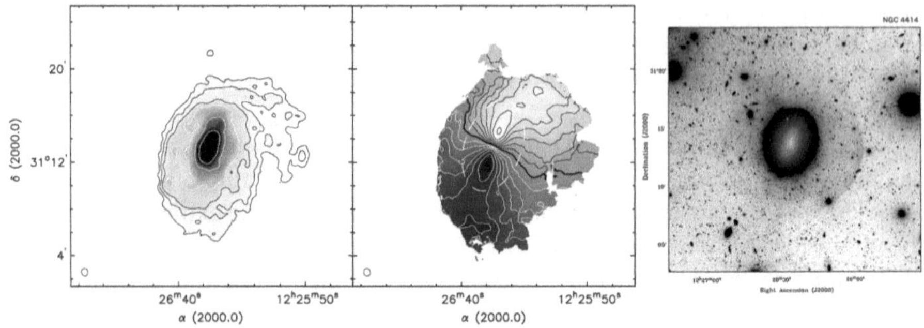

Fig. 7.10 Deep HI imaging of the galaxy NGC 4414 (reproduced with permission from de Blok et al. 2014a). From *left* to *right*, the HI image, with the outer column density contour being $2 \times 10^{19}\,\text{cm}^{-2}$, the velocity field indicating a strong warp in the northwestern part of the galaxy, and a deep optical image showing an outer shell feature, with a colour SDSS image overlaid in the inner parts. The outer isophotes of the SDSS image have a lower axial ratio than those of the outer isophotes of the main galaxy body in the deeper image

disk galaxies which could be accreted, but the amount of mass is not enough to sustain the star formation rate in the main disk (cf. Heald 2015).

This gas accretion problem has almost completely superseded the attention given to the flaring of the HI gas layer. Allaert et al. (2015) have taken the latter problem up again and find that the HI disk in NGC 5907 is flaring, while they do not exclude a moderate flaring for other edge-on galaxies. Kalberla and Dedes (2008) show a significant flaring for the gas disk of the Milky Way, in agreement with earlier studies. Note that although a lot of modelling of the stellar disk of edge-on galaxies has been done assuming no flaring, the work by Saha et al. (2009) and Streich et al. (2016) shows evidence for a mild flaring of the older stellar disk in a number of edge-on galaxies.

7.6.4 Velocity Dispersions in the Outer HI Layers of Spiral Galaxies

As discussed at the start of this section, the flaring problem depends in part on the determination of the vertical velocity dispersion of the HI gas, and early work with the WSRT indicated a value around 10 km/s, with a slight enhancement in star-forming regions as found for NGC 628 by Shostak and van der Kruit (1984). Dickey et al. (1990) determined that for NGC 1058 the velocity dispersion in the gas outside the optical image is remarkably constant at 6 km/s. Further work on this was done by Kamphuis (1993) and Kamphuis and Sancisi (1993), in particular for the relatively face-on galaxy NGC 6946, where the velocity dispersion was determined after derotating the data cube by resetting the intensity-weighted mean velocities of the individual profiles to the same central velocity before adding them, as suggested already by Boulanger and Viallefond (1992). It was shown more clearly that the higher velocity dispersions are related to star formation activity.

In the framework of the THINGS project, Tamburro et al. (2009) determine HI gas velocity dispersions for 11 galaxies and found again evidence for lower velocity dispersions in the extended HI envelopes beyond the optical image and high velocity dispersions in the inner parts of actively star-forming galaxies. If correct, the model calculation in Fig. 7.7 (right panels) suggests that a gradient in the velocity dispersion from, e.g. 10 to 6 km/s, as seen in several cases, can lead to the absence of significant flaring.

There are various ways to go about the determination of the HI gas velocity dispersion, and Tamburro et al. (2009) took the simplest way, i.e. considering the second moment map derived from the profiles in the data cube. However, various geometric effects need to be taken into account. One is that the HI might not be a single layer but in a combination of a low velocity dispersion thin disk and a higher velocity dispersion thick disk. This has been taken up recently for a number of galaxies in the THINGS sample (Ianjamasimanana et al. 2012, 2015; Mogotsi et al. 2016). The results for NGC 3198 are shown in Fig. 7.11, side by side with the position-velocity diagram along the major axis. This galaxy does not have the most favourable orientation for this work (perhaps too edge-on), but the results are illustrative of the difficulties involved. Ianjamasimanana et al. (2015) use "super-profiles", i.e. profiles corrected for the effect of rotation and stacked in annuli 0.2 R_{25} wide. They fit both a double Gaussian to low- and high-velocity components and a single Gauss fit which allows comparison with older data in the literature. The low-velocity component is thought to represent the cold neutral medium in the thin HI disk, and the higher-velocity component the warm neutral medium in a thicker disk. Mogotsi et al. (2016) present in an Appendix the results from fitting a single Gaussian to the profiles as well as taking the second moment. The comparison in Fig. 7.11 shows that the results differ widely. In the very central parts, the data are influenced by the broadening of the profiles inside a spatial resolution element (i.e. beam smearing). Outside, there is a large difference between the second moment and a Gaussian fit, on account of the skewness of the profiles. As discussed already in Sect. 7.6.3, for NGC 3198, this skewness is due to the presence of a thick disk with lagging rotation (Gentile et al. 2013).

For the more face-on galaxy NGC 6946, the results from both studies also show again the presence of a low- and high-velocity dispersion component. Older WSRT data by Boomsma et al. (2008) show an HI velocity dispersion profile similar to the one found in Kamphuis (1993), and the profile determined by Gaussian fitting of Mogotsi et al. (2016) agrees reasonably well with this. However, the single Gauss fit by Ianjamasimanana et al. (2015) is definitely lower, presumably due to the selection of only profiles with signal-to-noise ratio of three or more to be included into the stacked profiles. What is most striking, however, is that the velocity dispersions in NGC 6946 are lower than those in NGC 3198, despite the fact that NGC 6946 is forming stars more actively. The difference is perhaps due to the presence of more extraplanar gas in NGC 3198. Moreover, NGC 3198 is seen relatively edge-on, so there are line-of-sight integration effects. However, analysis of the publicly available

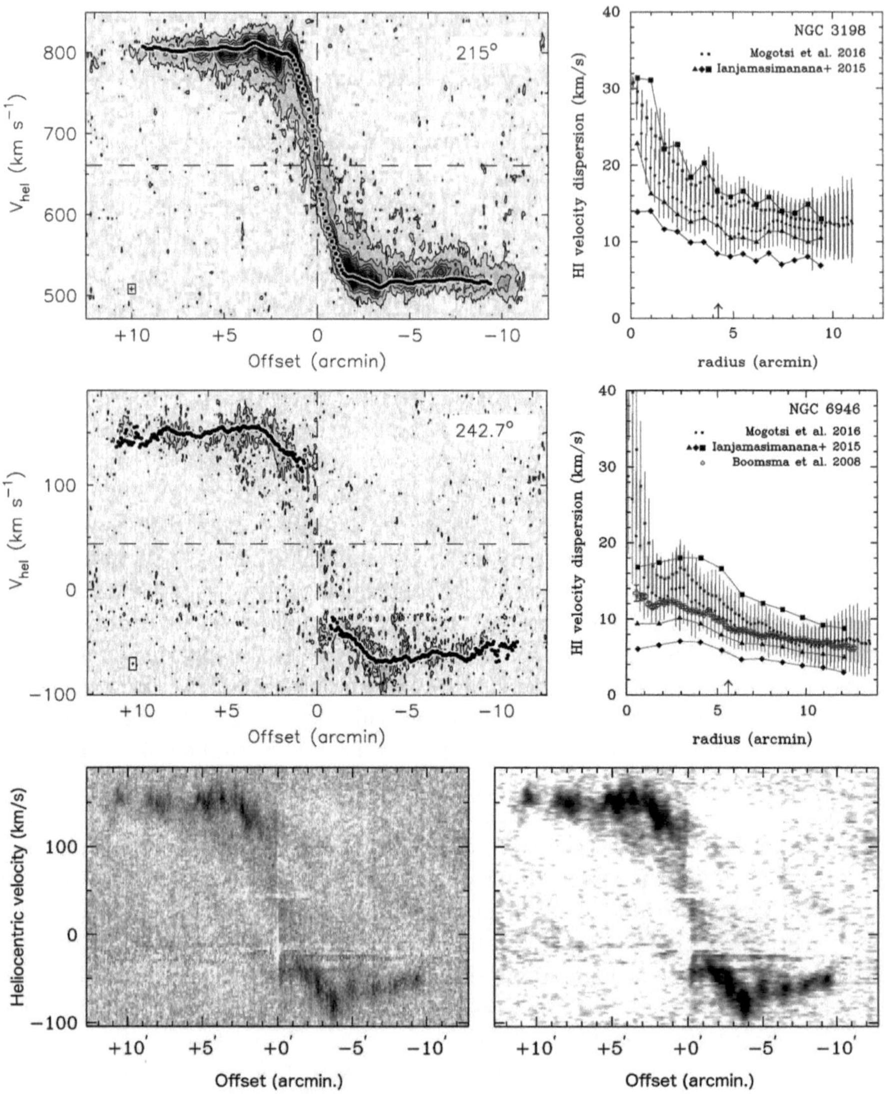

Fig. 7.11 *Top*: position-velocity diagram along the major axis of NGC 3198 (reproduced with permission from de Blok et al. 2008, *left*) and various radial profiles of the HI gas velocity dispersion (*right*). *Middle*: idem for NGC 6946. *Bottom*: Position-velocity diagrams along the major axis of NGC 6946, using the available data cube from NED, at the *left* the data with beam size of 6.′0 × 5.′6 and at the *right* the 18.′0 × 18.′0 data. The "beard" shows up rather well in the smoothed data. The *horizontal stripes* are due to incomplete subtraction of galactic foreground emission

data cube of NGC 6946 suggests another cause: if I smooth the cube from $6''.0 \times 5''.6$ to $18''.0 \times 18''.0$, the "beard" shows up in the smoothed version of the position-velocity diagram (cf. Fig. 7.11, lower panels). Thus considerations of spatial resolution and sensitivity as well as the spatial orientation of the galaxy with respect to the line of sight all play a role in the outcome of the studies of this problem. And, of course, if the question is asked, "what about the magnetic field?", it is strong in NGC 6946 and much weaker in NGC 3198. Beck (2007) shows that in the outer parts of NGC 6946, the magnetic field energy density is higher than the energy in turbulent motions, so this cannot be neglected in the calculations of the thickness of the gas layer, at least for that galaxy.

7.6.5 Star Formation in Warped HI Layers

Occasionally, star formation is seen in the warped HI layers, as in, e.g. NGC 3642 (Verdes-Montenegro et al. 2002), where the outer disk of low surface brightness has a different spatial orientation than the inner parts. This galaxy was later taken as a prototype of a Type III radial surface brightness profile in Laine et al. (2014), i.e. having an exponential disk with in the outer parts an up-bending profile. This situation is noted also for Malin 1 (see also Sect. 7.3). van der Kruit (2007) notes that for the survey of warps in edge-on galaxies by García-Ruiz et al. (2002), the onset of the warp is rather abrupt and occurs, for most of the galaxies considered, just outside the truncation radius if there is one. However, faint star-forming regions can be found in a number of XUV disks, as shown in Fig. 7.5 (lower left) for NGC 5055 and Fig. 7.7 (top, second panel from the left) for NGC 3198. Koribalski and López-Sánchez (2009) present HI observations of the NGC 1510/1512 system and state that the faint disk of NGC 1512, which contains strong spiral features, is warped with respect to the inner disk. Radburn-Smith et al. (2014) find that in the warped HI disk of NGC 4565, the young stellar populations participate in the warp, but not the older ones. Detection of molecular gas in some of these galaxies is discussed in Koda and Watson (2017).

7.7 The Core-Cusp Problem

The ΛCDM theory of galaxy formation and evolution predicts that in dark matter-dominated galaxies, the tracers of the potential should indicate a cuspy NFW profile (Navarro et al. 1996, 1997). However, this has not been observed, as already remarked by Moore (1994) and Flores and Primack (1994). Observations of late-type low surface brightness disk galaxies were done by a number of groups, first in HI, and later augmented with long-slit Hα observations in the inner parts (e.g. de Blok et al. 2001, 2003; de Blok and Bosma 2002). These confirmed the presence of cores in such galaxies or, at the most, mildly cusp slopes with $\alpha = -0.2 \pm 0.2$, not

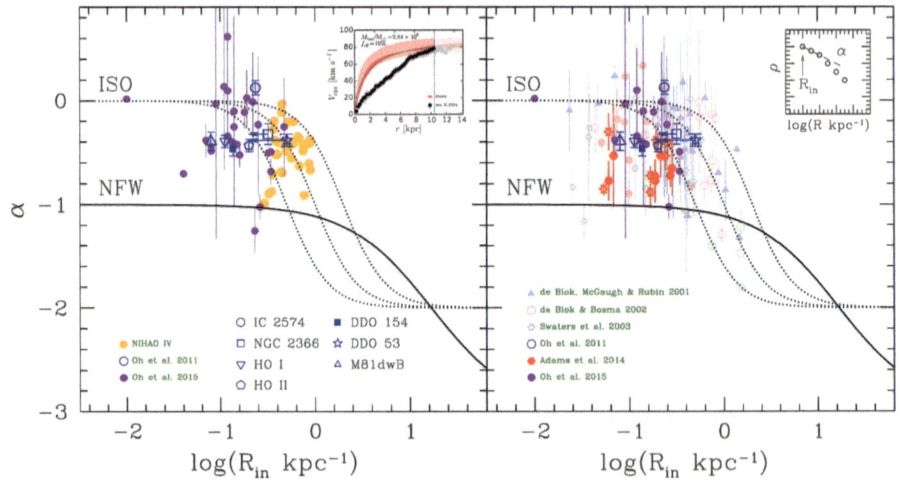

Fig. 7.12 *Left*: results from Oh et al. (2011, 2015) for the slope of the dark matter density profile as function of the radius of the innermost point on the rotation curve. The *orange points* are from hydrodynamical zoom simulations from the NIHAO project (Tollet et al. 2016, their Fig. 13). The *inset* shows the current data for IC 2574 and the APOSTLE simulation result for it (reproduced with permission from Oman et al. 2016). *Right*: a similar plot which collects older data (de Blok et al. 2001; de Blok and Bosma 2002; Swaters et al. 2003), the data from Oh et al. (2011), data based on stellar and gas velocity dispersions (Adams et al. 2014) and the selected data for 15 galaxies from Oh et al. (2015) as described in the text. The *dotted lines* converging on $\alpha = 0$ represent the theoretical variations in slope for ISO haloes with $R_C = 0.5$, 1 and 2 kpc. The *line* converging on $\alpha = -1$ shows the variation in slope for $c/V_{200} = 8.0/100$

compatible with the NFW profiles. These results were (e.g. Swaters et al. 2003) and still are the subject of intense debate. CO and three-dimensional Hα observations (Simon et al. 2003, 2005; Blais-Ouellette et al. 2004; Kuzio de Naray et al. 2006, 2008) were also brought to bear on this problem. A review has been given in de Blok (2010).

Projects such as THINGS, and LITTLE THINGS, using HI observations at higher resolution and sensitivity, keep finding cores (Oh et al. 2008, 2011, 2015). These observations can now be compared with modified predictions of the ΛCDM theory of galaxy formation and evolution, where star formation and feedback have been added to the ingredients ("subgrid physics") of the numerical simulations. The results are given in Fig. 7.12, together with the results from cosmological zoom simulations of the NIHAO project (Tollet et al. 2016). The addition of baryonic physics changes the prediction for the dark matter profile towards a slope shallower than the NFW profile. However, the current resolution achieved in the NIHAO project does not yet probe the full range in inner radii comparable to those of the observations. The debate is ongoing about whether the ΛCDM theory should be further revised to accommodate the new observational results. I select here a couple of issues which can be addressed with HI data in the context of galaxy outskirts.

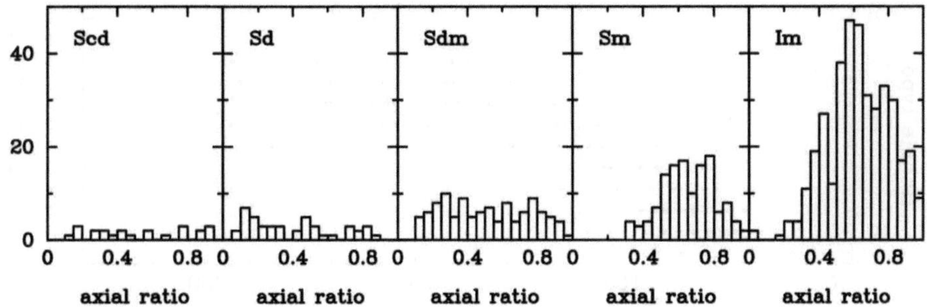

Fig. 7.13 Statistics of axial ratios for late-type galaxies (numerical Hubble type ≥ 6) in the Local Volume Galaxy catalogue (Karachentsev et al. 2013), split out by morphological type

Oman et al. (2015, 2016) emphasize that there is a large variety in the data of late-type low surface brightness dwarfs and find that the observations of the rotation properties of some dwarfs are closer to their models than others. They suggest that observers somehow do not have the systematics of galaxy inclinations under control. Read et al. (2016) claim good agreement with ΛCDM by discarding half their sample of four galaxies: IC 1613 is thought to be in disequilibrium due to starburst activity, and for DDO 101 the distance is too uncertain.

To examine the problem further, I looked at some of the underlying assumptions in the modelling of the observational data. One is that galaxy disks are thin, with a typical vertical axial ratio of 0.2. Such a value can be checked statistically, as has been done in the exemplary work of Sandage et al. (1970). For data in the Second Reference Catalog of Bright Galaxies (de Vaucouleurs et al. 1976), Binney and de Vaucouleurs (1981) point out that for Hubble types close to the end of the sequence (Type 10), the distribution of apparent axial ratios does not indicate flattened disks. This was discussed further in Bosma (1994) and is illustrated here again in Fig. 7.13, now using data from the Local Volume Galaxy catalogue (Karachentsev et al. 2013), split by morphological type for the later types. While Scd, Sd and Sdm galaxies have histograms that seem compatible with those for flattened disks viewed from different orientation angles, the data for Sm and in particular Im galaxies show an apparent axial ratio distribution not compatible with this. It is thus likely that some of the dwarf galaxies used in the core-cusp debate are not modelled correctly when it is assumed that they are thin disks: instead, models with a considerable thickness should be explored. The results from the FIGGS sample (Begum et al. 2008; Roychowdhury et al. 2010) confirm this and show that the thickness of the HI maps peaks around 0.5. Roychowdhury et al. (2013) also analyse the Local Volume catalogue and find a similar thickening of the galaxies of later type, but their analysis considers only a few bins. As discussed further in Sect. 7.9.2, thick HI disks have been recognized as such by Puche et al. (1992) and even earlier by Bottema et al. (1986).

In Fig. 7.14 I use data from the recent collection of rotation curves studied by Lelli et al. (2016) concerning the mean velocity in the outer parts, $V_{\rm flat}$, if defined.

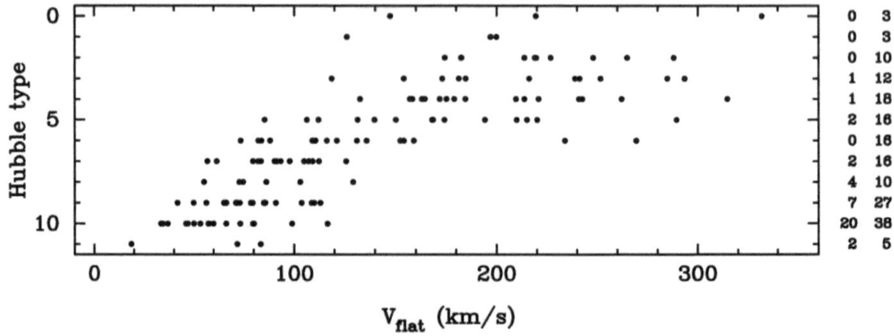

Fig. 7.14 Statistics of the $V_{\rm flat}$ value from the SPARC sample (Lelli et al. 2016) split out by morphological type. The number of galaxies not reaching a $V_{\rm flat}$ value and the total number per type bin are indicated at the *right*. Note the overlap in $V_{\rm flat}$ for Types 7–10

For the thin disk galaxies of Types Sd and Sdm (numerical Hubble Types 7 and 8), the range in $V_{\rm flat}$ overlaps the one of the thicker disk galaxies of Types Sm, Im and BCD (Types 9, 10 and 11, respectively). There is thus more to a galaxy than just its rotation curve, since the three-dimensional morphology of a galaxy does not follow automatically from the one-dimensional rotation curve. Therefore, for galaxies with $V_{\rm flat}$ between 70 and 110 km/s, corresponding to a stellar mass range of $\sim 5\times10^8$–$4\times10^9\,M_\odot$, numerical simulations of galaxy formation and evolution ought to reproduce a variety of shapes in the stellar mass distribution, rather than a single one, and for smaller galaxies, a thicker stellar disk should be produced.

I further consider the data on the LITTLE THINGS project discussed by Oh et al. (2015). In that paper, the HI layer is assumed to be infinitesimally thin, and the thickness of the stellar component is computed from the ratio of disk scale length to disk scale height (h/z_0) of 5.0, which holds for large spiral galaxies seen edge-on, but not necessarily for dwarfs. To minimize problems, I selected only galaxies for which the difference in position angle and inclination derived from the HI data and those from the axial ratio of the optical images, as given in Hunter et al. (2012), is less than 25° (as $\sqrt{\Delta(\mathrm{PA})^2 + \Delta(\mathrm{Inc})^2}$). This leaves 15 galaxies, which are shown in Fig. 7.12 (right panel). Selecting "better behaving" dwarf galaxies, with little warping of the HI disk, thus does not alleviate the core-cusp problem. A more elaborate analysis is beyond the scope of this review.

7.8 Alternative Gravity Theories

A recurrent issue with the interpretation of HI rotation curves is the argument that the dark matter interpretation is not necessarily correct, starting with a paper by Milgrom (1983) proposing MOND ("MOdified Newtonian Dynamics"). Begeman et al. (1991) made models for ten galaxies, both for Newtonian gravity and for

MOND, and pointed out that a common critical acceleration parameter can be found once some leeway is allowed for galaxy distances. Reviews about MOND and its relative success in reproducing rotation curves have been produced by Sanders and McGaugh (2002) and Famaey and McGaugh (2012). However, the applicability of MOND to larger scales, such as clusters of galaxies, is problematic, and Angus (2009) argues for the presence of 11 eV neutrinos to cure this. Such particles have not been found yet, but then, the nature of the dark matter remains unknown. The fate of the Vulcan hypothesis by Le Verrier (cf. Fontenrose 1973) is frequently quoted to illustrate the idea behind MOND (modifying the theory of gravity rather than searching for additional planets), but the history of dark matter detection resembles now more the discovery story of Pluto, rather than of Neptune.

My own contribution to this has been limited to refereeing some of the papers, and I will neither treat the debate here nor discuss other alternative theories. What is of interest are the attempts to disprove MOND, by emphasizing the discrepancies with HI imaging data. There are manifest problems with some galaxies if their Cepheid distance is adopted, the clearest case being NGC 3198 (cf. Bottema et al. 2002; Gentile et al. 2011, 2013). For late-type dwarf galaxies, the recent results of Carignan et al. (2013) for NGC 3109 and Randriamampandry and Carignan (2014) for DDO 154, IC 2574, NGC 925 and NGC 7793—in addition to NGC 3109 and NGC 3198—clearly bring out discrepancies as well. However, Angus et al. (2012) could reduce the discrepancy for DDO 154 by considering a thick gaseous disk, up to the point of getting a reasonable fit for most of the radial extent of this galaxy. Such thick HI gas disks are not unreasonable for DDO 154 and IC 2574, in view of the discussion above, in Sect. 7.7 and further in Sect. 7.9.2, but less so for the Sd galaxies NGC 925 and NGC 7793. Finally, the often cited "dip" in the rotation curve of NGC 1560 (Broeils 1992; Gentile et al. 2010) poses a problem, since the curve determined from the data for the northern half of this galaxy does not have a dip while the one for the southern part does, due to a local absence of gas and stars there; the mean thus has "half a dip"!?

MOND does not go away easily, and the recent result from 153 galaxies in the SPARC sample (12 galaxies too face-on and 10 galaxies too asymmetric were rejected, cf. McGaugh et al. 2016), i.e. a tight relation, with a scatter of $\sim$30% on a log-log plot, of the acceleration determined from rotation curves compared to the one based on 3.6 μm radial surface brightness profiles assuming a constant M/L ratio for galactic disks, revived some interest in it. Ludlow et al. (2016) argue that the small scatter in this mass discrepancy-acceleration relation is due to different feedback processes moving data points along this line, rather than deviating strongly from it, so that the ΛCDM theory does not have much difficulty explaining it. However, this leaves the core-cusp problem unsolved (cf. Fig. 7.12, inset of left panel).

The diversity of the 3.6 μm radial surface brightness profiles (e.g. Types I, II and III profiles—Martín-Navarro et al. 2012; Muñoz-Mateos et al. 2013; Kim et al. 2014; Laine et al. 2014), and the various physical processes invoked to explain these down- or up-bending profiles, cannot be ignored. Furthermore, in Famaey and McGaugh (2012), the two galaxies presented as examples for good MOND fits, NGC 6946 and NGC 1560, have a different K-band M/L ratio (0.37 vs. 0.18), i.e. a factor of two difference, which is just the order of magnitude difference discussed in Sect. 7.5. Moreover, data from the high angular resolution THINGS and LITTLE THINGS samples are not included in the SPARC sample, since, according to Lelli et al. (2016), their rotation curves are characterized by many small-scale bumps and wiggles thought to be due to non-circular components such as streaming motions along spiral arms. This exclusion is ironic, seen that when MOND is discussed, the capacity of reproducing such bumps is deemed very important. Anyway, these "details" seem to operate on a different level of complexity than the overall mass discrepancy-acceleration relation. Perusal of the individual mass models of each of the SPARC galaxies confirms this: the limiting surface brightness of the 3.6 μm profiles used in those mass models is 23.8 ± 1.8 mag arcsec^{-2}, i.e. faint surface brightness levels are not considered for every galaxy.

7.9 Irregular Galaxies

*7.9.1 Very Large H*I *Envelopes*

Irregular galaxies can be intriguing, as is shown by observations of the giant Magellanic irregular galaxy NGC 4449, whose stellar mass is $1.8 \times 10^9\, M_\odot$ (Muñoz-Mateos et al. 2015), i.e. 89% of that of NGC 300. A large HI envelope around this galaxy had already been discovered by van Woerden et al. (1975), using the 100 m Effelsberg telescope. A 3×3-point mosaic with the VLA of this galaxy has been made by Hunter et al. (1998). The small irregular galaxy DDO 125 is present as the southernmost blob in the images in Fig. 7.15. A further faint dwarf, or rather an extended stellar stream, discussed more recently by Rich et al. (2012), Martínez-Delgado et al. (2012) and Toloba et al. (2016), is indicated by a cross in the HI image: there is no immediate connection between this object and the HI features. The outer envelope shows a regular velocity gradient from north to south, while the body of the main galaxy seen in the visible light shows a regular velocity gradient at a position angle of ~60° in the opposite sense. Lelli et al. (2014) remark that if for NGC 4449 the rotation velocity of the outer envelope is used, this galaxy falls on the baryonic Tully-Fisher relation. This suggests that it is the outer envelope which traces the dark halo and that the inner parts, including the visible galaxy, are not yet settled.

Galaxies with such large HI envelopes (more than five times larger than the optical radius) are relatively rare, and only a handful of other cases have been discussed. Blue compact dwarfs can sometimes be surprisingly large in HI, as is the case for NGC 2915, which has a relatively regular HI disk (cf. Meurer et al. 1996; Elson et al. 2010, 2011a,b, see Fig. 7.16, upper panels), UGC 5288 (van Zee 2004), NGC 3741 (Begum et al. 2005), ADBS 113845+2008 (Cannon et al. 2009), IZw18 (Lelli et al. 2014) and IIZw40 (Brinks and Klein 1988). Note that DDO154 also has a very large HI size compared to its optical size (Carignan and Freeman 1988; de Blok et al. 2008).

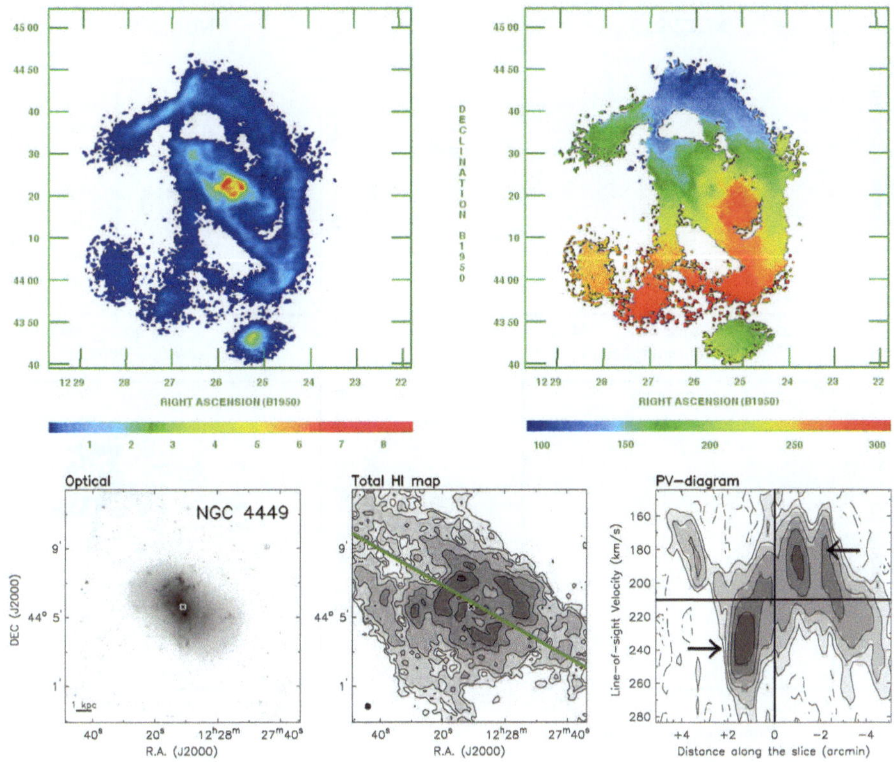

Fig. 7.15 *Top*: HI image (*left*) and velocity field (*right*) of the Magellanic irregular galaxy NGC 4449 (adapted with permission from Hunter et al. 1998). The southernmost blob is the companion galaxy DDO 125. The *cross* is the position of the stellar stream as given in Rich et al. (2012). *Bottom*: details of the central parts of this galaxy (reproduced with permission from Lelli et al. 2014), a *V*-band optical image (*left*), the HI image (*centre*) and a position-velocity diagram (*right*) at a position angle of 60° along the *green line* indicated in the HI image showing the inversion of the line-of-sight velocities in the inner parts

Some galaxies look a bit surprising, such as NGC 4214, shown in Fig. 7.16 (middle panels). If the HI is in a circular disk seen in projection, there is a large misalignment of $\sim$55° between the position angle of its major axis and the one derived from the velocity field. Lelli et al. (2014) fit a tilted ring model to the observations of this galaxy and derive a variable inclination, which tends to 0° in the outer parts. This seems hard to square with the apparent axial ratio of the HI distribution.

Also shown in Fig. 7.16 (lower panels) is the galaxy Holmberg II, imaged already by Puche et al. (1992) and later in the THINGS and VLA-ANGST projects. From the velocity field, a clear warp can be inferred, and the HI image also shows a tail towards the west side. Nevertheless, several analyses have been performed on this galaxy. One of the striking aspects in the HI distribution is the presence of holes, also discussed for bright galaxies, such as M101 by van der Hulst and Sancisi (1988) and

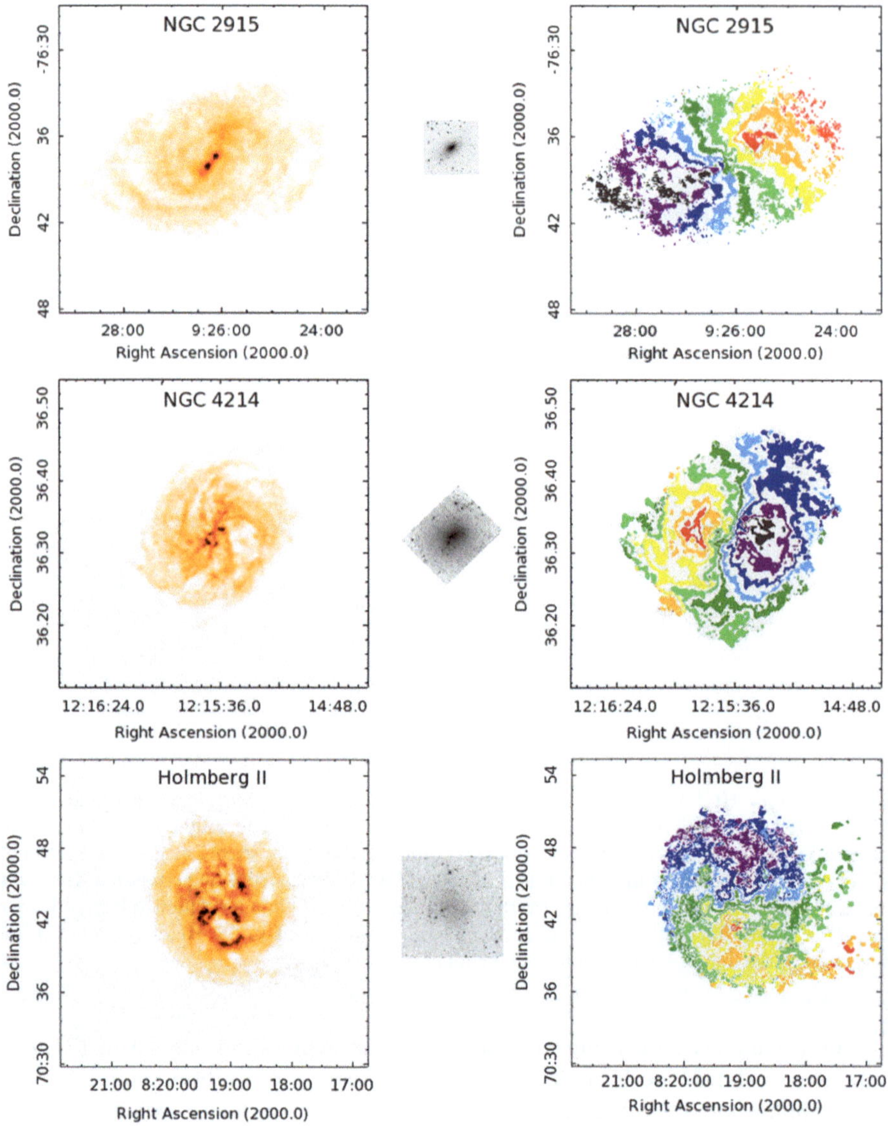

Fig. 7.16 *Top*: HI image (*left*) and velocity field (*right*) of the blue compact dwarf NGC 2915, made from data kindly supplied by Elson, as reported in Elson et al. (2010). The 3.6 μm image in the *middle* is on the same scale. *Middle*: HI image (*left*) and velocity field (*right*) of NGC 4214, from the THINGS project, and a 3.6 μm image. *Bottom*: HI image (*left*) and velocity field (*right*) of Holmberg II, from the ANGST project, and a 3.6 μm image. The THINGS, ANGST and 3.6 μm data were downloaded from the NED

NGC 6946 by Boomsma et al. (2008). Puche et al. (1992) remark that HI holes in late-type dwarf galaxies are larger than the HI holes in large spiral galaxies.

7.9.2 *Velocity Dispersions in Dwarf Irregular Galaxies*

A number of recent surveys concern almost uniquely dwarf irregular galaxies. In particular, the ANGST survey (Ott et al. 2012) has observed or reobserved a number of dwarfs with the VLA at high spectral resolution. An extensive discussion of HI gas velocity dispersions based on these observations is given in Stilp et al. (2013a,b). The dispersions have again been calculated on the basis of "super-profiles", derived after derotating the data cube and stacking the individual profiles. The general shape of the profiles can be described by a double Gaussian, with a narrow centre of order 7.2–8.5 km/s and a wider wing of order 20–25 km/s. The latter is presumably due to the influence of star formation, while the former is attributed to turbulence. Stilp et al. (2013a) do not think it is possible to discriminate between the "cold neutral medium" (CNM) and the "warm neutral medium" (WNM), as done by Ianjamasimanana et al. (2012) for the THINGS data, since the typical velocity dispersions expected for the typical temperatures associated with these do not match. They also argue that the data cubes with robust weighting should be used and that the signal-to-noise threshold matters.

Interestingly, thicknesses of the gas layer have been derived, based on a hydrostatic equilibrium used already in Ott et al. (2001) and even before that by Puche et al. (1992). Banerjee et al. (2011) studied the flaring of the gas layer in four of the THINGS dwarf galaxies, again using the hydrostatic equilibrium approach. For DDO 154, the disk flares to $\sim$1 kpc at a radius of $\sim$5 kpc. This is to be compared with an overall thickness derived by Stilp et al. (2013b) of 708 $\pm$ 139 pc based on the method discussed by Ott et al. (2001) and a thickness of 650 pc calculated by Angus et al. (2012) to get a more or less acceptable MOND fit (cf. Sect. 7.8). Very recently, Johnson et al. (2017) also argue that the HI disks in galaxies in the LITTLE THINGS survey are thick, rather than thin.

7.10 The Relation Between HI Extent and the Optical Radius

A specific search for galaxies large in HI has been executed by Broeils and van Woerden (1994) and Broeils and Rhee (1997), using short observations with the WSRT. Broeils and Rhee reported an interesting correlation between the HI mass and the HI size, the latter defined as an isophotal radius at the level of 1 $M_{\odot}$ pc^{-2}. More recently, a number of surveys have been executed on various telescopes, typically with sample sizes of order 10–60 galaxies. Wang et al. (2016) collected HI sizes as defined by Broeils and Rhee (1997) for 437 galaxies—spread over 14 projects—although not for every galaxy an HI size was determined, and there might

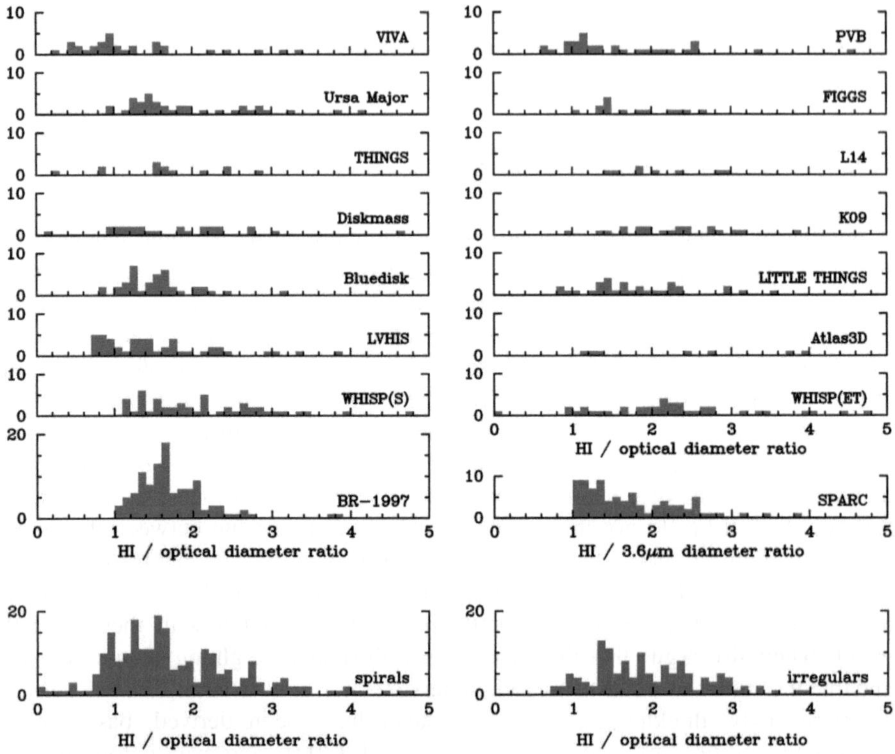

Fig. 7.17 Statistics of the ratio of HI diameter, evaluated at $\Sigma_{HI} = 1\,M_\odot\,pc^{-2}$, to the optical diameter, following Wang et al. (2016), except for the Bluedisk sample, for which Wang et al. (2013) data are used. The samples are BR-1997, Broeils and Rhee (1997); WHISP(S), Swaters et al. (2002); LVHIS, Koribalski (2008), Westmeier et al. (2011, 2013); Bluedisk, Wang et al. (2013); Diskmass, Martinsson et al. (2016); THINGS, Walter et al. (2008); Ursa Major, Verheijen and Sancisi (2001); VIVA, Chung et al. (2009); WHISP(ET), Noordermeer et al. (2005); Atlas3D, Serra et al. (2012, 2014); LITTLE THINGS, Hunter et al. (2012); K09, Kovač et al. (2009); L14, Lelli et al. (2014); FIGGS, Begum et al. (2008); and PVB, Ponomareva et al. (2016). Note that several large galaxies are missing from the LVHIS, THINGS, LITTLE THINGS, FIGGS and VIVA samples, while for the WHISP samples, a maximum diameter of 400″ is imposed, on account of missing flux when comparing the interferometric data with single-dish observations or too large in extent compared to the primary beam. For the SPARC sample (Lelli et al. 2016), I determined the 3.6 μm radius from the tables associated with their publication, by interpolating those radial luminosity profiles which reached a depth of $1.0\,L_\odot pc^{-2}$, which could be done for 84 of the 174 galaxies in that sample. The *bottom row* refers to the total number of spirals (*left*) and irregulars (numerical Hubble type ≥ 8.5, *right*) in the samples studied by Wang et al. (2016)

be a slight overlap in the sense that several galaxies are in more than one sample. They found again a very tight relationship between the HI mass and the HI size. It is instructive to examine for each survey they considered the distribution of the ratio of HI to optical size, which is shown in Fig. 7.17. The survey done by Broeils and Rhee (1997) shows a peak at $R_{HI}/R_{opt} \sim 1.5$, and they found relatively few galaxies with extended HI disks. The statistics based on the data by Wang et al. (2016) show that the proportion of extended HI disks is not too different for spirals compared to irregulars (even though statistically there are more irregulars with large $R_{\rm HI}/R_{\rm opt}$), and the fraction of very extended disks with an HI size larger than, e.g. three times the optical size, is about 10%.

Even though Broeils and Rhee (1997) found a very good relation between the HI mass and the size of the HI disk, indicating a roughly constant mean HI surface density, there is still no clear indication about when we can expect an HI disk which is much larger than the optical disk. This is due to the fact that the amount of HI in the extended HI disk is usually only a modest fraction of the total HI mass, so that differences are drowned in the "bigger things are bigger in many quantities" effect.

Not all surveys peak at the same HI-to-optical size ratio. This is to be expected for the Virgo cluster survey, VIVA (Chung et al. 2009), since in the centre of that cluster, the HI gas in galaxy outskirts is subject to ram pressure stripping against the hot intergalactic medium (IGM) probed by the X-ray emission. Indeed, several galaxies have been reported undergoing stripping, such as NGC 4522 (Kenney et al. 2004); NGC 4388, where a long HI tail has been observed (Oosterloo and van Gorkom 2005); NGC 4254 and its HI tail, including an HI dwarf almost devoid of stars (Minchin et al. 2007); and NGC 4569, where Boselli et al. (2016) report very extended Hα emission.

Studies have shown that HI sizes of galaxies in a group environment could also be affected by the presence of the IGM. Obviously this is the case for compact groups (Verdes-Montenegro et al. 2001), where an evolutionary scenario is suggested depending on the state of the merging.

For loose groups, the effects are rather more subtle. In a recent extensive study, Wolfinger et al. (2016) analysed data for the Ursa Major region and constructed a complete sample of 1209 galaxies limited in systemic velocity ($300 \leq V_{\rm LG} \leq 3000\,{\rm km/s}$), absolute magnitude ($M_{\rm r} \leq -15.3$) and stellar mass ($M_* \geq 10^8\,M_\odot$). They identified six groups with more than three members, centred around NGCs 4449, 4258, 4278, 4026, 3938 and 5033, while 74% of the galaxies in their sample reside outside those groups. The high-resolution interferometer data on the "Ursa Major cluster" (Verheijen and Sancisi 2001) and the CVn region (Kovač et al. 2009) are part of this sample, but for other parts of the region, single-dish data from Jodrell Bank and Arecibo were used. There are HI-deficient galaxies associated with the densest parts of some of the groups, but not all, as, e.g. NGC 4449 (shown in Fig. 7.15), which does not seem HI deficient. There are also galaxies with excess HI mass, usually in regions of low galactic density.

For the Sculptor group, Westmeier et al. (2011) find that the deep HI image of NGC 300 is asymmetric in the very outer parts and attribute this to a possible interaction with the IGM in that group. The process of the formation of warps due to the interaction with the IGM has been explored further by Haan and Braun (2014). Deep HI imaging data on M83 with the KAT-7 telescope (Heald et al. 2016) shows sharp edges in the outer HI there as well, which can be either due to a ram pressure effect with respect to the surrounding IGM, or photoionisation, as discussed below.

The Void Galaxy Survey (Kreckel et al. 2011, 2012) investigated the properties of HI disks in galaxies in voids and found that on average the statistics of the size ratio R_{HI}/R_{opt} are roughly similar to those observed by Swaters et al. (2002) for the late-type WHISP sample. The galaxy with the largest ratio, VGS_12, has an HI disk in the polar direction of a small S0-like galaxy (cf. Stanonik et al. 2009). No star formation is associated with this HI gas. However, apart from this, there is nothing outstanding about the properties of void galaxies with respect to the field galaxies studied in most other surveys.

The outermost HI is prone to be ionized by the ultraviolet background radiation, as already suggested by Sunyaev (1969). An early attempt to get limits on this has been done for NGC 3198 by Maloney (1993), using an unpublished deep VLA HI image. The realization that there ought to be detectable ionized gas emission around HI envelopes, which could be used to probe the dark matter at larger radii, has led to some further studies of this gas using Fabry-Pérot Hα observations, but these are technically very challenging (cf. Bland-Hawthorn et al. 1997).

Hlavacek-Larrondo et al. (2011a,b) report deep Hα observations of three Sculptor group galaxies with the wide-field Fabry-Pérot system on the 36 cm Marseille Telescope in La Silla. For NGC 247, the Hα and HI data extend out to $\sim$13.5 arcmin, barely beyond the Holmberg radius of 12.2 arcmin. For NGC 300, the field of view was limited, and the region of the outer, warped HI disk not imaged in Hα. For NGC 253, the Hα disk goes out to similar radii (11.5 arcmin) as the HI disk seen with the VLA, but faint [NII] emission has been detected out to 19.0 arcmin at the southwest part of the galaxy. The kinematic data seem to indicate a declining rotation curve, but the galaxy is heavily perturbed there as seen on recent deep optical images, and recent KAT-7 HI data show the presence of extraplanar cold gas (Lucero et al. 2015). Earlier Hα observations of the Sculptor group galaxy NGC 7793 with the same instrument (Dicaire et al. 2008) led to the conclusion that the ionizing sources of the Hα disk of this galaxy are likely to be internal, rather than the UV background.

For more on the interface between the cold HI gas and the hotter circumgalactic medium, see the review by Chen (2017).

7.11 Concluding Remarks

In this review, I have discussed a number of issues related to detailed HI imaging of the outskirts of nearby spiral and irregular galaxies done with current interferometers, such as the WSRT, the VLA, the ATCA and the GMRT. I have been selective in my topics and avoided to discuss the issue of star formation in extended HI envelopes, since this is covered elsewhere in this volume. Most topics are related to the dark matter problem in one way or another. A number of results have been obtained by making case studies with the emphasis on improved sensitivity, rather than by simply observing more objects from a list.

The Square Kilometer Array project is advancing at great strides, and the associated approach of setting a challenging imaging target requiring the development of new instrumentation to reach it is soon going to bear fruit. Indeed, extensive HI imaging surveys of relatively nearby galaxies will start next year with SKA pathfinders, such as the WSRT Apertif survey, and the SKA precursors ASKAP (the WALLABY survey) and MeerKAT (the deep, targeted MHONGOOSE survey and also the MALS survey). Some of the projects for the first phase of the SKA telescope, SKA1, are described in the SKA science book (e.g. Blyth et al. 2015; de Blok et al. 2015). All these new surveys will bring fresh data of high quality, which can be brought to bear on the scientific problems discussed above and on other related subjects, and will most likely lead to new insights and discoveries.

Acknowledgements I thank Lia Athanassoula for fruitful discussions and a careful reading of the manuscript. I also thank the editors for inviting me to write this review and in particular Johan Knapen for his careful editing. I thank David Malin for sending me the deep image of M83 with the shallower image superimposed and David Martínez-Delgado and Taylor Chonis for providing the image in the lower right panel of Fig. 7.5. Erwin de Blok supplied the script and Se-Heon Oh and Andrea Macciò supplied data for the construction of Fig. 7.12. Deidre Hunter helped out with Fig. 7.15. Ed Elson supplied the HI data on NGC 2915 from which I could make the top panel of Fig. 7.16. Jing Wang clarified an issue with the Bluedisk sample.

I acknowledge financial support from the People Programme (Marie Curie Actions) of the European Union's Seventh Framework Programme FP7/2007–2013/ under REA grant agreement number PITN-GA-2011-289313 to the DAGAL network, from the CNES (Centre National d'Etudes Spatiales—France) and from the "Programme National de Cosmologie et Galaxies" (PNCG) of CNRS/INSU, France. This research has made use of NASA's Astrophysics Data System. Use was made of the NASA/IPAC Extragalactic Database (NED) which is operated by the Jet Propulsion Laboratory, California Institute of Technology, under contract with the National Aeronautics and Space Administration to produce parts of Figs. 7.5, 7.7, 7.11 and 7.16.

References

Adams, J.J., Simon, J.D., Fabricius, M.H., van den Bosch, R.C.E., Barentine, J.C., Bender, R., Gebhardt, K., Hill, G.J., Murphy, J.D., Swaters, R.A., Thomas, J., van de Ven, G.: Dwarf galaxy dark matter density profiles inferred from stellar and gas kinematics. Astrophys. J. **789**, 63 (2014). doi:10.1088/0004-637X/789/1/63, 1405.4854

Allaert, F., Gentile, G., Baes, M., De Geyter, G., Hughes, T.M., Lewis, F., Bianchi, S., De Looze, I., Fritz, J., Holwerda, B.W., Verstappen, J., Viaene, S.: HERschel observations of edge-on spirals (HEROES). II. Tilted-ring modelling of the atomic gas disks. Astron. Astrophys. **582**, A18 (2015). doi:10.1051/0004-6361/201526667, 1507.03095

Allen, R.J., Sancisi, R., Baldwin, J.E.: Radio continuum observations of the edge-on disc galaxy NGC 891. Astron. Astrophys. **62**, 397–409 (1978)

Angus, G.W.: Is an 11eV sterile neutrino consistent with clusters, the cosmic microwave background and modified Newtonian dynamics? Mon. Not. R. Astron. Soc. **394**, 527–532 (2009). doi:10.1111/j.1365-2966.2008.14341.x, 0805.4014

Angus, G.W., van der Heyden, K.J., Famaey, B., Gentile, G., McGaugh, S.S., de Blok, W.J.G.: A QUMOND galactic N-body code - I. Poisson solver and rotation curve fitting. Mon. Not. R. Astron. Soc. **421**, 2598–2609 (2012). doi:10.1111/j.1365-2966.2012.20532.x, 1201.3185

Aniyan, S., Freeman, K.C., Gerhard, O.E., Arnaboldi, M., Flynn, C.: The influence of a kinematically cold young component on disc-halo decompositions in spiral galaxies: insights from solar neighbourhood K-giants. Mon. Not. R. Astron. Soc. **456**, 1484–1494 (2016). doi:10.1093/mnras/stv2730, 1511.06047

Aparicio, A., Gallart, C.: IAC-STAR: a code for synthetic color-magnitude diagram computation. Astron. J. **128**, 1465–1477 (2004). doi:10.1086/382836, astro-ph/0407589

Arnaboldi, M., Oosterloo, T., Combes, F., Freeman, K.C., Koribalski, B.: New H I observations of the prototype polar ring galaxy NGC 4650A. Astron. J. **113**, 585–598 (1997). doi:10.1086/118278

Athanassoula, E.: Bar-halo interaction and bar growth. Astrophys. J. Lett. **569**, L83–L86 (2002). doi:10.1086/340784, astro-ph/0203368

Athanassoula, E.: What determines the strength and the slowdown rate of bars? Mon. Not. R. Astron. Soc. **341**, 1179–1198 (2003). doi:10.1046/j.1365-8711.2003.06473.x, astro-ph/0302519

Athanassoula, E.: Disc instabilities and semi-analytic modelling of galaxy formation. Mon. Not. R. Astron. Soc. **390**, L69–L72 (2008). doi:10.1111/j.1745-3933.2008.00541.x, 0808.0016

Athanassoula, E., Bosma, A., Papaioannou, S.: Halo parameters of spiral galaxies. Astron. Astrophys. **179**, 23–40 (1987)

Babcock, H.W.: The rotation of the Andromeda Nebula. Lick Observatory Bull. **19**, 41–51 (1939). doi:10.5479/ADS/bib/1939LicOB.19.41B

Baldwin, J.E.: The distribution of mass-to ratio with radius in M31 and M33. In: Shakeshaft, J.R. (ed.) The Formation and Dynamics of Galaxies, IAU Symposium, vol. 58, p. 139 (1974)

Baldwin, J.E.: M/L ratios in galactic disks. In: Hayli, A. (ed.) Dynamics of the Solar Systems. IAU Symposium, vol. 69, p. 341 (1975)

Baldwin, J.E., Lynden-Bell, D., Sancisi, R.: Lopsided galaxies. Mon. Not. R. Astron. Soc. 193, 313–319 (1980). doi:10.1093/mnras/193.2.313

Banerjee, A., Jog, C.J., Brinks, E., Bagetakos, I.: Theoretical determination of H I vertical scale heights in the dwarf galaxies DDO 154, Ho II, IC 2574 and NGC 2366. Mon. Not. R. Astron. Soc. **415**, 687–694 (2011). doi:10.1111/j.1365-2966.2011.18745.x, 1103.4494

Barnabè, M., Dutton, A.A., Marshall, P.J., Auger, M.W., Brewer, B.J., Treu, T., Bolton, A.S., Koo, D.C., Koopmans, L.V.E.: The SWELLS survey - IV. Precision measurements of the stellar and dark matter distributions in a spiral lens galaxy. Mon. Not. R. Astron. Soc. **423**, 1073–1088 (2012). doi:10.1111/j.1365-2966.2012.20934.x, 1201.1692

Beck, R.: Magnetism in the spiral galaxy NGC 6946: magnetic arms, depolarization rings, dynamo modes, and helical fields. Astron. Astrophys. **470**, 539–556 (2007). doi:10.1051/0004-6361:20066988, 0705.4163

Becquaert, J.F., Combes, F.: The 3D geometry of dark matter halos. Astron. Astrophys. **325**, 41–56 (1997). astro-ph/9704088

Begeman, K.G., Broeils, A.H., Sanders, R.H.: Extended rotation curves of spiral galaxies - dark haloes and modified dynamics. Mon. Not. R. Astron. Soc. **249**, 523–537 (1991). doi:10.1093/mnras/249.3.523

Begum, A., Chengalur, J.N., Karachentsev, I.D.: A dwarf galaxy with a giant HI disk. Astron. Astrophys. **433**, L1–L4 (2005). doi:10.1051/0004-6361:200500026, astro-ph/0502307

Begum, A., Chengalur, J.N., Karachentsev, I.D., Sharina, M.E., Kaisin, S.S.: FIGGS: faint irregular galaxies GMRT survey - overview, observations and first results. Mon. Not. R. Astron. Soc. **386**, 1667–1682 (2008). doi:10.1111/j.1365-2966.2008.13150.x, 0802.3982

Bershady, M.A., Andersen, D.R., Verheijen, M.A.W., Westfall, K.B., Crawford, S.M., Swaters, R.A.: SparsePak: a formatted fiber field unit for the WIYN telescope bench spectrograph. II. On-Sky performance. Astrophys. J. Suppl. Ser. **156**, 311–344 (2005). doi:10.1086/426479, astro-ph/0410404

Bershady, M.A., Verheijen, M.A.W., Swaters, R.A., Andersen, D.R., Westfall, K.B., Martinsson, T.: The DiskMass survey. I. Overview. Astrophys. J. **716**, 198–233 (2010a). doi:10.1088/0004-637X/716/1/198, 1004.4816

Bershady, M.A., Verheijen, M.A.W., Westfall, K.B., Andersen, D.R., Swaters, R.A., Martinsson, T.: The DiskMass survey. II. Error budget. Astrophys. J. **716**, 234–268 (2010b). doi:10.1088/0004-637X/716/1/234, 1004.5043

Bershady, M.A., Martinsson, T.P.K., Verheijen, M.A.W., Westfall, K.B., Andersen, D.R., Swaters, R.A.: Galaxy disks are submaximal. Astrophys. J. Lett. **739**, L47 (2011). doi:10.1088/2041-8205/739/2/L47, 1108.4314

Bertone, G., Hooper, D.: A history of dark matter (2016). ArXiv e-prints 1605.04909

Binney, J., de Vaucouleurs, G.: The apparent and true ellipticities of galaxies of different Hubble types in the Second Reference Catalogue. Mon. Not. R. Astron. Soc. **194**, 679–691 (1981). doi:10.1093/mnras/194.3.679

Blais-Ouellette, S., Amram, P., Carignan, C., Swaters, R.: Accurate determination of the mass distribution in spiral galaxies. III. Fabry-Perot imaging spectroscopy of 6 spiral galaxies. Astron. Astrophys. **420**, 147–161 (2004). doi:10.1051/0004-6361:20034263, astro-ph/0411001

Bland-Hawthorn, J., Freeman, K.C., Quinn, P.J.: Where do the disks of spiral galaxies end? Astrophys. J. **490**, 143–155 (1997). astro-ph/9706210

Blyth, S., van der Hulst, T.M., Verheijen, M.A.W., Catinella, B., Fraternali, F., Haynes, M.P., Hess, K.M., Koribalski, B., Lagos, C., Meyer, M., Obreschkow, D., Popping, A., Power, C., Verdes-Montenegro, L.L., Zwaan, M.: Exploring neutral hydrogen and galaxy evolution with the SKA. In: Advancing Astrophysics with the Square Kilometre Array (AASKA14), vol. 128 (2015). 1501.01295

Boomsma, R., Oosterloo, T.A., Fraternali, F., van der Hulst, J.M., Sancisi, R.: HI holes and high-velocity clouds in the spiral galaxy NGC 6946. Astron. Astrophys. **490**, 555–570 (2008). doi:10.1051/0004-6361:200810120, 0807.3339

Boselli, A., Cuillandre, J.C., Fossati, M., Boissier, S., Bomans, D., Consolandi, G., Anselmi, G., Cortese, L., Côté, P., Durrell, P., Ferrarese, L., Fumagalli, M., Gavazzi, G., Gwyn, S., Hensler, G., Sun, M., Toloba, E.: Spectacular tails of ionized gas in the Virgo cluster galaxy NGC 4569. Astron. Astrophys. **587**, A68 (2016). doi:10.1051/0004-6361/201527795, 1601.04978

Bosma, A.: The distribution and kinematics of neutral hydrogen in spiral galaxies of various morphological types. Ph.D. thesis, Groningen University (1978)

Bosma, A.: 21-cm line studies of spiral galaxies. I - Observations of the galaxies NGC 5033, 3198, 5055, 2841, and 7331. II - The distribution and kinematics of neutral hydrogen in spiral galaxies of various morphological types. Astron. J. **86**, 1791–1824 (1981a). doi:10.1086/113062

Bosma, A.: 21-cm line studies of spiral galaxies. II. The distribution and kinematics of neutral hydrogen in spiral galaxies of various morphological types. Astron. J. **86**, 1825–1846 (1981b). doi:10.1086/113063

Bosma, A.: Kinematics and dynamics of dwarf spirals and irregulars. In: Meylan, G., Prugniel, P. (eds.) European Southern Observatory Conference and Workshop Proceedings, European Southern Observatory Conference and Workshop Proceedings, vol. 49, pp. 187–196 (1994)

Bosma, A.: Dark matter in disk galaxies. In: Merritt, D.R., Valluri, M., Sellwood, J.A. (eds.) Galaxy Dynamics - A Rutgers Symposium, Astronomical Society of the Pacific Conference Series, vol. 182 (1999)

Bosma, A., van der Kruit, P.C.: The local mass-to-light ratio in spiral galaxies. Astron. Astrophys. **79**, 281–286 (1979)

Bottema, R.: The stellar kinematics of galactic disks. Astron. Astrophys. **275**, 16 (1993)

Bottema, R.: The maximum rotation of a galactic disc. Astron. Astrophys. **328**, 517–525 (1997). astro-ph/9706230

Bottema, R., Shostak, G.S., van der Kruit, P.C.: The thickness of the hydrogen layer of the edge-on dwarf galaxy NGC 5023. Astron. Astrophys. **167**, 34–40 (1986)

Bottema, R., Pestaña, J.L.G., Rothberg, B., Sanders, R.H.: MOND rotation curves for spiral galaxies with Cepheid-based distances. Astron. Astrophys. **393**, 453–460 (2002). doi:10.1051/0004-6361:20021021, astro-ph/0207469

Boulanger, F., Viallefond, F.: Observational study of the spiral galaxy NGC 6946. I - H I and radio-continuum observations. Astron. Astrophys. **266**, 37–56 (1992)

Bovy, J., Rix, H.W.: A direct dynamical measurement of the milky way's disk surface density profile, disk scale length, and dark matter profile at 4 kpc $\leq$ R $\leq$ 9 kpc. Astrophys. J. **779**, 115 (2013). doi:10.1088/0004-637X/779/2/115, 1309.0809

Braun, R.: The temperature and opacity of atomic hydrogen in spiral galaxies. Astrophys. J. **484**, 637–655 (1997). astro-ph/9702111

Braun, R., Thilker, D.A.: The WSRT wide-field H I survey. II. Local group features. Astron. Astrophys. **417**, 421–435 (2004). doi:10.1051/0004-6361:20034423, astro-ph/0312323

Braun, R., Thilker, D.A., Walterbos, R.A.M., Corbelli, E.: A wide-field high-resolution H I mosaic of Messier 31. I. Opaque atomic gas and star formation rate density. Astrophys. J. **695**, 937–953 (2009). doi:10.1088/0004-637X/695/2/937, 0901.4154

Briggs, F.H.: Rules of behavior for galactic WARPS. Astrophys. J. **352**, 15–29 (1990). doi:10.1086/168512

Brinks, E., Klein, U.: Dark matter in the dwarf galaxy II Zwicky 40. Mon. Not. R. Astron. Soc. **231**, 63–67 (1988). doi:10.1093/mnras/231.1.63P

Broeils, A.H.: The mass distribution of the dwarf spiral NGC 1560. Astron. Astrophys. **256**, 19–32 (1992)

Broeils, A.H., Rhee, M.H.: Short 21-cm WSRT observations of spiral and irregular galaxies. HI properties. Astron. Astrophys. **324**, 877–887 (1997)

Broeils, A.H., van Woerden, H.: A search for spiral galaxies with extended HI disks. Astron. Astrophys. Suppl. Ser. **107**, 129–176 (1994)

Burbidge, G.: On the masses and relative velocities of galaxies. Astrophys. J. Lett. **196**, L7–L10 (1975). doi:10.1086/181731

Burbidge, E.M., Burbidge, G.R., Prendergast, K.H.: The rotation, mass distribution, and mass of NGC 5055. Astrophys. J. **131**, 282 (1960). doi:10.1086/146832

Burke, B.F.: Systematic distortion of the outer regions of the galaxy. Astron. J. **62**, 90 (1957). doi:10.1086/107463

Cannon, J.M., Salzer, J.J., Rosenberg, J.L.: Quiescent isolation: the extremely extended H I halo of the optically compact dwarf galaxy ADBS 113845+2008. Astrophys. J. **696**, 2104–2114 (2009). doi:10.1088/0004-637X/696/2/2104, 0902.3988

Carignan, C., Freeman, K.C.: DDO 154 - A 'dark' galaxy? Astrophys. J. Lett. **332**, L33–L36 (1988). doi:10.1086/185260

Carignan, C., Frank, B.S., Hess, K.M., Lucero, D.M., Randriamampandry, T.H., Goedhart, S., Passmoor, S.S.: KAT-7 science verification: using H I observations of NGC 3109 to understand its kinematics and mass distribution. Astron. J. **146**, 48 (2013). doi:10.1088/0004-6256/146/3/48, 1306.3227

Chen, H.-W.: Outskirts of distant galaxies in absorption. In: Knapen, J.H., Lee, J.C., Gil de Paz, A. (eds.) Outskirts of Galaxies, vol. 434. Springer, Cham (2017). doi:10.1007/978-3-319-56570-5

Chung, A., van Gorkom, J.H., Kenney, J.D.P., Crowl, H., Vollmer, B.: VLA imaging of virgo spirals in atomic gas (VIVA). I. The ttlas and the H I properties. Astron. J. **138**, 1741–1816 (2009). doi:10.1088/0004-6256/138/6/1741

Courteau, S., Dutton, A.A.: On the global mass distribution in disk galaxies. Astrophys. J. Lett. **801**, L20 (2015). doi:10.1088/2041-8205/801/2/L20, 1502.04709

Dahlem, M., Lisenfeld, U., Rossa, J.: Dependence of radio halo properties on star formation activity and galaxy mass. Astron. Astrophys. **457**, 121–131 (2006). doi:10.1051/0004-6361:20054787, astro-ph/0607102

Dalton, G.: WEAVE: The Next Generation Spectroscopy Facility for the WHT. In: Skillen, I., Barcells, M., Trager, S. (eds.) Multi-Object Spectroscopy in the Next Decade: Big Questions, Large Surveys, and Wide Fields, Astronomical Society of the Pacific Conference Series, vol. 507, p. 97 (2016)

Davies, R.D.: Neutral hydrogen at large distances from parent galaxies. In: Shakeshaft, J.R. (ed.) The Formation and Dynamics of Galaxies. IAU Symposium, vol. 58, pp. 119–128 (1974)

de Blok, W.J.G.: The core-cusp problem. Adv. Astron. **2010**, 789293 (2010). doi:10.1155/2010/789293, 0910.3538 de Blok, W.J.G., Bosma, A.: High-resolution rotation curves of low surface brightness galaxies. Astron. Astrophys. **385**, 816–846 (2002). doi:10.1051/0004-6361: 20020080, astro-ph/0201276

de Blok, W.J.G., McGaugh, S.S., Bosma, A., Rubin, V.C.: Mass density profiles of low surface brightness galaxies. Astrophys. J. Lett. **552**, L23–L26 (2001). doi:10.1086/320262, astro-ph/0103102

de Blok, W.J.G., Bosma, A., McGaugh, S.: Simulating observations of dark matter dominated galaxies: towards the optimal halo profile. Mon. Not. R. Astron. Soc. **340**, 657–678 (2003). doi:10.1046/j.1365-8711.2003.06330.x, astro-ph/0212102

de Blok, W.J.G., Walter, F., Brinks, E., Trachternach, C., Oh, S.H., Kennicutt, R.C. Jr.: High-resolution rotation curves and galaxy mass models from THINGS. Astron. J. **136**, 2648–2719 (2008). doi:10.1088/0004-6256/136/6/2648, 0810.2100

de Blok, W.J.G., Józsa, G.I.G., Patterson, M., Gentile, G., Heald, G.H., Jütte, E., Kamphuis, P., Rand, R.J., Serra, P., Walterbos, R.A.M.: HALOGAS observations of NGC 4414: fountains, interaction, and ram pressure. Astron. Astrophys. **566**, A80 (2014a). doi:10.1051/0004-6361/201322517, 1405.2160

de Blok, W.J.G., Keating, K.M., Pisano, D.J., Fraternali, F., Walter, F., Oosterloo, T., Brinks, E., Bigiel, F., Leroy, A.: A low H I column density filament in NGC 2403: signature of interaction or accretion. Astron. Astrophys. **569**, A68 (2014b). doi:10.1051/0004-6361/201423880, 1407.3648

de Blok, E., Fraternali, F., Heald, G., Adams, B., Bosma, A., Koribalski. B.: The SKA view of the neutral interstellar medium in galaxies. In: Advancing Astrophysics with the Square Kilometre Array (AASKA14), vol. 129 (2015)

de Vaucouleurs, G.: Photoelectric photometry of the Andromeda nebula in the UBV system. Astrophys. J. **128**, 465 (1958). doi:10.1086/146564

de Vaucouleurs, G., de Vaucouleurs, A., Corwin, J.R.: Second reference catalogue of bright galaxies. In: Second Reference Catalogue of Bright Galaxies, vol. 1976. University of Texas Press, Austin (1976)

Dicaire, I., Carignan, C., Amram, P., Marcelin, M., Hlavacek-Larrondo, J., de Denus-Baillargeon, M.M., Daigle, O., Hernandez, O.: Deep Fabry-Perot Hα observations of NGC 7793: a very extended Hα disk and a truly declining rotation curve. Astron. J. **135**, 2038–2047 (2008). doi:10.1088/0004-6256/135/6/2038, 0802.2545

Dickey, J.M., Hanson, M.M., Helou, G.: NGC 1058 - Gas motions in an extended, quiescent spiral disk. Astrophys. J. **352**, 522–531 (1990). doi:10.1086/168555

Dutton, A.A., Treu, T., Brewer, B.J., Marshall, P.J., Auger, M.W., Barnabè, M., Koo, D.C., Bolton, A.S., Koopmans, L.V.E.: The SWELLS survey - V. A Salpeter stellar initial mass function in the bulges of massive spiral galaxies. Mon. Not. R. Astron. Soc. **428**, 3183–3195 (2013). doi:10.1093/mnras/sts262, 1206.4310

Efstathiou, G., Lake, G., Negroponte, J.: The stability and masses of disc galaxies. Mon. Not. R. Astron. Soc. **199**, 1069–1088 (1982). doi:10.1093/mnras/199.4.1069

Einasto, J., Kaasik, A., Saar, E.: Dynamic evidence on massive coronas of galaxies. Nature **250**, 309–310 (1974). doi:10.1038/250309a0

Elmegreen, B.G., Hunter, D.A.: Outskirts of nearby disk galaxies: star formation and stellar populations. In: Knapen, J.H., Lee, J.C., Gil de Paz, A. (eds.) Outskirts of Galaxies, vol. 434. Springer, Cham (2017). doi: 10.1007/978-3-319-56570-5

Elson, E.C., de Blok, W.J.G., Kraan-Korteweg, R.C.: The dark matter content of the blue compact dwarf NGC 2915. Mon. Not. R. Astron. Soc. **404**, 2061–2076 (2010). doi:10.1111/j.1365-2966.2010.16422.x, 1002.0403

Elson, E.C., de Blok, W.J.G., Kraan-Korteweg, R.C.: A search for non-circular flows in the extended H I disc of NGC 2915. Mon. Not. R. Astron. Soc. **411**, 200–210 (2011a). doi:10.1111/j.1365-2966.2010.17672.x, 1009.1753

Elson, E.C., de Blok, W.J.G., Kraan-Korteweg, R.C.: Three-dimensional modelling of the H I kinematics of NGC 2915. Mon. Not. R. Astron. Soc. **415**, 323–332 (2011b). doi:10.1111/j.1365-2966.2011.18701.x, 1103.4427

Emerson, D.T., Baldwin, J.E.: The rotation curve and mass distribution in M31. Mon. Not. R. Astron. Soc. **165**, 9P–13P (1973). doi:10.1093/mnras/165.1.9P

Ewen, H.I., Purcell, E.M.: Observation of a line in the Galactic radio spectrum: radiation from galactic hydrogen at 1,420 Mc./sec. Nature **168**, 356 (1951). doi:10.1038/168356a0

Faber, S.M., Gallagher, J.S.: Masses and mass-to-light ratios of galaxies. Annu. Rev. Astron. Astrophys. **17**, 135–187 (1979). doi:10.1146/annurev.aa.17.090179.001031

Famaey, B., McGaugh, S.S.: Modified Newtonian Dynamics (MOND): observational phenomenology and relativistic extensions. Living Rev. Relat. **15** (2012). doi:10.12942/lrr-2012-10, 1112.3960

Ferguson, A.M.N., Mackey, A.D.: Substructure and tidal streams in the andromeda galaxy and its satellites. In: Newberg, H.J., Carlin, J.L. (eds.) Astrophysics and Space Science Library, Astrophysics and Space Science Library, vol. 420, p. 191 (2016). doi:10.1007/978-3-319-19336-6_8, 1603.01993

Flores, R.A., Primack, J.R.: Observational and theoretical constraints on singular dark matter halos. Astrophys. J. Lett. **427**, L1–L4 (1994). doi:10.1086/187350, astro-ph/9402004

Fontenrose, R.: In search of vulcan. J. Hist. Astron. **4**, 145–158 (1973)

Fragkoudi, F., Athanassoula, E., Bosma, A.: Breaking the disc-halo degeneracy in NGC 1291 using hydrodynamic simulations (2016). ArXiv e-prints 1605.05754

Fraternali, F.: How can star formation be sustained? In: Feltzing, S., Zhao, G., Walton, N.A., Whitelock, P. (eds.) Setting the Scene for Gaia and LAMOST. IAU Symposium, vol. 298, pp. 228–239 (2014). doi:10.1017/S1743921313006418, 1310.2956

Fraternali, F.: Gas accretion via condensation and fountains (2016). ArXiv e-prints 1612.00477

Fraternali, F., Binney, J.J.: A dynamical model for the extraplanar gas in spiral galaxies. Mon. Not. R. Astron. Soc. **366**, 449–466 (2006). doi:10.1111/j.1365-2966.2005.09816.x, astro-ph/0511334

Fraternali, F., Binney, J.J.: Accretion of gas on to nearby spiral galaxies. Mon. Not. R. Astron. Soc. **386**, 935–944 (2008). doi:10.1111/j.1365-2966.2008.13071.x, 0802.0496

Fraternali, F., Oosterloo, T., Sancisi, R., van Moorsel, G.: A new, kinematically anomalous H I component in the spiral galaxy NGC 2403. Astrophys. J. Lett. **562**, L47–L50 (2001). doi:10.1086/338102, astro-ph/0110369

Fraternali, F., van Moorsel, G., Sancisi, R., Oosterloo, T.: Deep H I survey of the spiral galaxy NGC 2403. Astron. J. **123**, 3124–3140 (2002). doi:10.1086/340358, astro-ph/0203405

Fraternali, F., Oosterloo, T.A., Sancisi, R., Swaters, R.: The extra-planar neutral gas in the edge-on spiral galaxy NGC 891. In: Braun, R. (ed.) Extra-Planar Gas, Astronomical Society of the Pacific Conference Series, vol. 331, p. 239 (2005). astro-ph/0410375

Freeman, K.C.: On the disks of spiral and S0 galaxies. Astrophys. J. **160**, 811 (1970). doi:10.1086/150474

Fuchs, B.: NGC 2613, 3198, 6503, 7184: Case studies against 'Maximum' Disks. In: Merritt, D.R., Valluri, M., Sellwood, J.A. (eds.) Galaxy Dynamics - A Rutgers Symposium, Astronomical Society of the Pacific Conference Series, vol. 182 (1999). astro-ph/9812048

Galaz, G., Milovic, C., Suc, V., Busta, L., Lizana, G., Infante, L., Royo, S.: Deep optical images of Malin 1 reveal new features. Astrophys. J. Lett. **815**, L29 (2015). doi:10.1088/2041-8205/815/2/L29, 1512.01095

García-Ruiz, I., Sancisi, R., Kuijken, K.: Neutral hydrogen and optical observations of edge-on galaxies: hunting for warps. Astron. Astrophys. **394**, 769–789 (2002). doi:10.1051/0004-6361:20020976, astro-ph/0207112

Gentile, G., Baes, M., Famaey, B., van Acoleyen, K.: Mass models from high-resolution HI data of the dwarf galaxy NGC 1560. Mon. Not. R. Astron. Soc. **406**, 2493–2503 (2010). doi:10.1111/j.1365-2966.2010.16838.x, 1004.3421

Gentile, G., Famaey, B., de Blok, W.J.G.: THINGS about MOND. Astron. Astrophys. **527**, A76 (2011). doi:10.1051/0004-6361/201015283, 1011.4148

Gentile, G., Józsa, G.I.G., Serra, P., Heald, G.H., de Blok, W.J.G., Fraternali, F., Patterson, M.T., Walterbos, R.A.M., Oosterloo, T.: HALOGAS: extraplanar gas in NGC 3198. Astron. Astrophys. **554**, A125 (2013). doi:10.1051/0004-6361/201321116, 1304.4232

Goad, J.W., Roberts, M.S.: Spectroscopic observations of superthin galaxies. Astrophys. J. **250**, 79–86 (1981). doi:10.1086/159349

Gum, C.S., Kerr, F.J., Westerhout, G.: A 21-cm determination of the principal plane of the Galaxy. Mon. Not. R. Astron. Soc. **121**, 132 (1960). doi:10.1093/mnras/121.2.132

Haan, S., Braun, R.: On the formation of warped gas discs in galaxies. Mon. Not. R. Astron. Soc. **440**, L21–L25 (2014). doi:10.1093/mnrasl/slu008

Heald, G.: The Wsrt Halogas Survey. In: Ziegler, B.L., Combes, F., Dannerbauer, H., Verdugo, M. (eds.) Galaxies in 3D across the Universe. IAU Symposium, vol. 309, pp. 69–72 (2015). doi:10.1017/S1743921314009338, 1409.7599

Heald, G., de Blok, W.J.G., Lucero, D., Carignan, C., Jarrett, T., Elson, E., Oozeer, N., Randriamampandry, T.H., van Zee, L.: Neutral hydrogen and magnetic fields in M83 observed with the SKA Pathfinder KAT-7. Mon. Not. R. Astron. Soc. **462**, 1238–1255 (2016). doi:10.1093/mnras/stw1698, 1607.03365

Hindman, J.V.: A high resolution study of the distribution and motions of neutral hydrogen in the Small Cloud of Magellan. Aust. J. Phys. **20**, 147 (1967)

Hlavacek-Larrondo, J., Carignan, C., Daigle, O., de Denus-Baillargeon, M.M., Marcelin, M., Epinat, B., Hernandez, O.: Deep Hα observations of NGC 253: a very extended and possibly declining rotation curve? Mon. Not. R. Astron. Soc. **411**, 71–84 (2011a). doi:10.1111/j.1365-2966.2010.17662.x, 1009.1177

Hlavacek-Larrondo, J., Marcelin, M., Epinat, B., Carignan, C., de Denus-Baillargeon, M.M., Daigle, O., Hernandez, O.: Deep Fabry-Perot Hα observations of two Sculptor group galaxies, NGC 247 and 300. Mon. Not. R. Astron. Soc. **416**, 509–521 (2011b). doi:10.1111/j.1365-2966.2011.19063.x, 1105.2456

Hodges-Kluck, E.J., Bregman, J.N.: A deep X-ray view of the hot halo in the edge-on spiral galaxy NGC 891. Astrophys. J. **762**, 12 (2013). doi:10.1088/0004-637X/762/1/12, 1211.1669

Hohl, F.: Numerical experiments with a disk of stars. Astrophys. J. **168**, 343 (1971). doi:10.1086/151091

Holmberg, E.: A photographic photometry of extragalactic nebulae. Meddelanden fran Lunds Astronomiska Observatorium Serie II **136**, 1 (1958)

Hunter, D.A., Wilcots, E.M., van Woerden, H., Gallagher, J.S., Kohle, S.: The nature of the extended H I Gas around NGC 4449: The Dr. Jekyll/Mr. Hyde of irregular galaxies. Astrophys. J. Lett. **495**, L47–L50 (1998). doi:10.1086/311213

Hunter, D.A., Ficut-Vicas, D., Ashley, T., Brinks, E., Cigan, P., Elmegreen, B.G., Heesen, V., Herrmann, K.A., Johnson, M., Oh, S.H., Rupen, M.P., Schruba, A., Simpson, C.E., Walter, F., Westpfahl, D.J., Young, L.M., Zhang, H.X.: Little things. Astron. J. **144**, 134 (2012). doi:10.1088/0004-6256/144/5/134, 1208.5834

Ianjamasimanana, R., de Blok, W.J.G., Walter, F., Heald, G.H.: The shapes of the H I velocity profiles of the THINGS galaxies. Astron. J. **144**, 96 (2012). doi:10.1088/0004-6256/144/4/96, 1207.5041

Ianjamasimanana, R., de Blok, W.J.G., Walter, F., Heald, G.H., Caldú-Primo, A., Jarrett, T.H.: The radial variation of H I velocity dispersions in dwarfs and spirals. Astron. J. **150**, 47 (2015). doi:10.1088/0004-6256/150/2/47, 1506.04156

Johnson, M.C., Hunter, D.A., Kamphuis, P., Wang, J.: A constant intrinsic thickness for dwarf irregular galaxies? Mon. Not. R. Astron. Soc. **465**, L49–L53 (2017). doi:10.1093/mnrasl/slw203

Kahn, F.D., Woltjer, L.: Intergalactic matter and the galaxy. Astrophys. J. **130**, 705 (1959). doi:10.1086/146762

Kalberla, P.M.W., Dedes, L.: Global properties of the H I distribution in the outer Milky Way. Planar and extra-planar gas. Astron. Astrophys. **487**, 951–963 (2008). doi:10.1051/0004-6361:20079240, 0804.4831

Kalnajs, A.: Mass distribution and dark halos: discussion. In: Athanassoula, E. (ed.) Internal Kinematics and Dynamics of Galaxies. IAU Symposium, vol. 100, pp. 87–88 (1983)

Kamphuis, J.J.: Neutral hydrogen in nearby spiral galaxies: holes and high velocity clouds. Ph.D. thesis, University of Groningen (1993)

Kamphuis, J., Sancisi, R.: Widespread high velocity gas in the spiral galaxy NGC6946. Astron. Astrophys. **273**, L31 (1993)

Karachentsev, I.D., Karachentseva, V.E., Kudrya, Y.N., Sharina, M.E., Parnovskij, S.L.: The revised flat galaxy catalogue. Bull. Special Astrophys. Obs. **47** (1999). astro-ph/0305566

Karachentsev, I.D., Makarov, D.I., Kaisina, E.I.: Updated nearby galaxy catalog. Astron. J. **145**, 101 (2013). doi:10.1088/0004-6256/145/4/101, 1303.5328

Kenney, J.D.P., van Gorkom, J.H., Vollmer, B.: VLA H I observations of gas stripping in the virgo cluster spiral NGC 4522. Astron. J. **127**, 3361–3374 (2004), doi:10.1086/420805, astro-ph/0403103

Kent, S.M.: Dark matter in spiral galaxies. I - galaxies with optical rotation curves. Astron. J. **91**, 1301–1327 (1986). doi:10.1086/114106

Kent, S.M.: Dark matter in spiral galaxies. II - galaxies with H I rotation curves. Astron. J. **93**, 816–832 (1987). doi:10.1086/114366

Kent, S.M.: Dark matter in spiral galaxies. III - the Sa galaxies. Astron. J. **96**, 514–527 (1988). doi:10.1086/114829

Kerr, F.J.: A magellanic effect on the galaxy. Astron. J. **62**, 93–93 (1957). doi:10.1086/107466

Khoperskov, S.A., Moiseev, A.V., Khoperskov, A.V., Saburova, A.S.: To be or not to be oblate: the shape of the dark matter halo in polar ring galaxies (2014). Mon. Not. R. Astron. Soc. **441**, 2650–2662. doi:10.1093/mnras/stu692

Kim, T., Gadotti, D.A., Sheth, K., Athanassoula, E., Bosma, A., Lee, M.G., Madore, B.F., Elmegreen, B., Knapen, J.H., Zaritsky, D., Ho, L.C., Comerón, S., Holwerda, B., Hinz, J.L., Muñoz-Mateos, J.C., Cisternas, M., Erroz-Ferrer, S., Buta, R., Laurikainen, E., Salo, H., Laine, J., Menéndez-Delmestre, K., Regan, M.W., de Swardt, B., Gil de Paz, A., Seibert, M., Mizusawa, T.: Unveiling the structure of barred galaxies at 3.6 μm with the Spitzer survey of stellar structure in galaxies (S^4G). I. Disk breaks. Astrophys. J. **782**, 64 (2014). doi:10.1088/0004-637X/782/2/64, 1312.3384

Koda, L.C., Watson, J.: Molecular gas in the outskirts. In: Knapen, J.H., Lee, J.C., Gil de Paz, A. (eds.) Outskirts of Galaxies, vol. 434. Springer, Cham (2017). doi: 10.1007/978-3-319-56570-5

Koribalski, B.S.: The Local Volume HI Survey (LVHIS). Astrophys. Space Sci. Proc. **5**, 41 (2008). doi:10.1007/978-1-4020-6933-8-7

Koribalski, B.S., López-Sánchez, Á.R.: Gas dynamics and star formation in the galaxy pair NGC1512/1510. Mon. Not. R. Astron. Soc. **400**, 1749–1767 (2009). doi:10.1111/j.1365-2966.2009.15610.x, 0908.4128

Kovač, K., Oosterloo, T.A., van der Hulst, J.M.: A blind HI survey in the Canes Venatici region. Mon. Not. R. Astron. Soc. **400**, 743–765 (2009). doi:10.1111/j.1365-2966.2009.14662.x, 0904.2775

Kranz, T., Slyz, A., Rix, H.W.: Dark matter within high surface brightness spiral galaxies. Astrophys. J. **586**, 143–151 (2003). doi:10.1086/367551, astro-ph/0212290

Kreckel, K., Platen, E., Aragón-Calvo, M.A., van Gorkom, J.H., van de Weygaert, R., van der Hulst, J.M., Kovač, K., Yip, C.W., Peebles, P.J.E.: Only the lonely: H I imaging of void galaxies. Astron. J. **141**, 4 (2011). doi:10.1088/0004-6256/141/1/4, 1008.4616

Kreckel, K., Platen, E., Aragón-Calvo, M.A., van Gorkom, J.H., van de Weygaert, R., van der Hulst, J.M., Beygu, B.: The void galaxy survey: optical properties and H I morphology and kinematics. Astron. J. **144**, 16 (2012). doi:10.1088/0004-6256/144/1/16, 1204.5185

Kregel, M., van der Kruit, P.C., Freeman, K.C.: Structure and kinematics of edge-on galaxy discs - V. The dynamics of stellar discs. Mon. Not. R. Astron. Soc. **358**, 503–520 (2005). doi:10.1111/j.1365-2966.2005.08855.x, astro-ph/0501503

Kuzio de Naray, R., McGaugh, S.S., de Blok, W.J.G., Bosma, A.: High-resolution optical velocity fields of 11 low surface brightness galaxies. Astrophys. J. Suppl. Ser. **165**, 461–479 (2006). doi:10.1086/505345, astro-ph/0604576

Kuzio de Naray, R., McGaugh, S.S., de Blok, W.J.G.: Mass models for low surface brightness galaxies with high-resolution optical velocity fields. Astrophys. J. **676**, 920–943 (2008). doi:10.1086/527543, 0712.0860

Kwee, K.K., Muller, C.A., Westerhout, G.: The rotation of the inner parts of the Galactic System. Bull. Astron. Inst. Neth. **12**, 211 (1954)

Laine, J., Laurikainen, E., Salo, H., Comerón, S., Buta, R.J., Zaritsky, D., Athanassoula, E., Bosma, A., Muñoz-Mateos, J.C., Gadotti, D.A., Hinz, J.L., Erroz-Ferrer, S., Gil de Paz, A., Kim, T., Menéndez-Delmestre, K., Mizusawa, T., Regan, M.W., Seibert, M., Sheth, K.: Morphology and environment of galaxies with disc breaks in the S^{4}G and NIRSOS. Mon. Not. R. Astron. Soc. **441**, 1992–2012 (2014). doi:10.1093/mnras/stu628, 1404.0559

Lelli, F., Fraternali, F., Sancisi, R.: Structure and dynamics of giant low surface brightness galaxies. Astron. Astrophys. **516**, A11 (2010). doi:10.1051/0004-6361/200913808, 1003.1312

Lelli, F., Verheijen, M., Fraternali, F.: Dynamics of starbursting dwarf galaxies. III. A H I study of 18 nearby objects. Astron. Astrophys. **566**, A71 (2014). doi:10.1051/0004-6361/201322657, 1404.6252

Lelli, F., McGaugh, S.S., Schombert, J.M.: SPARC: mass Models for 175 disk galaxies with Spitzer photometry and accurate rotation curves (2016). ArXiv e-prints 1606.09251

Lewis, B.M.: Neutral hydrogen observations of NGC 5236. Proc. Astron. Soc. Aust. **1**, 104 (1968)

Lindblad, P.A.B., Lindblad, P.O., Athanassoula, E.: Hydrodynamical simulations of the barred spiral galaxy NGC 1365. Dynamical interpretation of observations. Astron. Astrophys. **313**, 65–90 (1996)

Lucero, D.M., Carignan, C., Elson, E.C., Randriamampandry, T.H., Jarrett, T.H., Oosterloo, T.A., Heald, G.H.: H I observations of the nearest starburst galaxy NGC 253 with the SKA precursor KAT-7. Mon. Not. R. Astron. Soc. **450**, 3935–3951 (2015). doi:10.1093/mnras/stv856

Ludlow, A.D., Benitez-Llambay, A., Schaller, M., Theuns, T., Frenk, C.S., Bower R, Schaye, J., Crain, R.A., Navarro, J.F., Fattahi, A., Oman, K.A.: The mass-discrepancy acceleration relation: a natural outcome of galaxy formation in CDM halos (2016). ArXiv e-prints 1610.07663

Malin, D., Hadley, B.: HI in shell galaxies and other merger remnants. Publ. Astron. Soc. Aust. **14**, 52–58 (1997). doi:10.1071/AS97052

Maloney, P.: Sharp edges to neutral hydrogen disks in galaxies and the extragalactic radiation field. Astrophys. J. **414**, 41–56 (1993). doi:10.1086/173055

Martín-Navarro, I., Bakos, J., Trujillo, I., Knapen, J.H., Athanassoula E, Bosma, A., Comerón, S., Elmegreen, B.G., Erroz-Ferrer, S., Gadotti, D.A., Gil de Paz, A., Hinz, J.L., Ho, L.C., Holwerda, B.W., Kim, T., Laine, J., Laurikainen, E., Menéndez-Delmestre, K., Mizusawa, T., Muñoz-Mateos, J.C., Regan, M.W., Salo, H., Seibert, M., Sheth, K.: A unified picture of breaks and truncations in spiral galaxies from SDSS and S^{4}G imaging. Mon. Not. R. Astron. Soc. **427**, 1102–1134 (2012). doi:10.1111/j.1365-2966.2012.21929.x, 1208.2893

Martínez-Delgado, D., Peñarrubia, J., Gabany, R.J., Trujillo, I., Majewski, S.R., Pohlen, M.: The ghost of a dwarf galaxy: fossils of the hierarchical formation of the nearby spiral galaxy NGC 5907. Astrophys. J. **689**, 184–193 (2008). doi:10.1086/592555, 0805.1137

Martínez-Delgado, D., Gabany, R.J., Crawford, K., Zibetti, S., Majewski, S.R., Rix, H.W., Fliri, J., Carballo-Bello, J.A., Bardalez-Gagliuffi, D.C., Peñarrubia, J., Chonis, T.S., Madore, B., Trujillo, I., Schirmer, M., McDavid, D.A.: Stellar tidal streams in spiral galaxies of the local volume: a pilot survey with modest aperture telescopes. Astron. J. **140**, 962–967 (2010). doi:10.1088/0004-6256/140/4/962, 1003.4860

Martínez-Delgado, D., Romanowsky, A.J., Gabany, R.J., Annibali, F., Arnold, J.A., Fliri, J., Zibetti, S., van der Marel, R.P., Rix, H.W., Chonis, T.S., Carballo-Bello, J.A., Aloisi, A., Macciò, A.V., Gallego-Laborda, J., Brodie, J.P., Merrifield, M.R.: Dwarfs gobbling dwarfs: a stellar tidal stream around NGC 4449 and hierarchical galaxy formation on small scales. Astrophys. J. Lett. **748**, L24 (2012). doi:10.1088/2041-8205/748/2/L24, 1112.2154

Martinsson, T.P.K., Verheijen, M.A.W., Westfall, K.B., Bershady, M.A., Andersen, D.R., Swaters, R.A.: The DiskMass survey. VII. The distribution of luminous and dark matter in spiral galaxies. Astron. Astrophys. **557**, A131 (2013). doi:10.1051/0004-6361/201321390, 1308.0336

Martinsson, T.P.K., Verheijen, M.A.W., Bershady, M.A., Westfall, K.B., Andersen, D.R., Swaters, R.A.: The DiskMass survey. X. Radio synthesis imaging of spiral galaxies. Astron. Astrophys. **585**, A99 (2016). doi:10.1051/0004-6361/201527067, 1510.07666

Materne, J., Tammann, G.A.: On the stability of groups of galaxies and the question of hidden matter. In: Kharadze, E.K. (ed.) Proceedings of the Third European Astronomical Meeting, Tbilisi. Springer (1975)

Matthews, L.D., Uson, J.M.: H I imaging observations of superthin galaxies. II. IC 2233 and the blue compact dwarf NGC 2537. Astron. J. **135**, 291–318 (2008). doi:10.1088/0004-6256/135/1/291, 0709.4249

Matthews, L.D., Wood, K.: High-latitude H I in the low surface brightness galaxy UGC 7321. Astrophys. J. **593**, 721–732 (2003). doi:10.1086/376602, astro-ph/0305402

McGaugh, S., Lelli, F., Schombert, J.: The radial acceleration relation in rotationally supported galaxies (2016). ArXiv e-prints 1609.05917

McGee, R.X., Milton, J.A.: 21 cm hydrogen-line survey of the Large Magellanic Cloud. II. Distribution and motions of neutral hydrogen. Aust. J. Phys. **19**, 343 (1966)

Meurer, G.R., Carignan, C., Beaulieu, S.F., Freeman, K.C.: NGC 2915.II.A dark spiral galaxy with a blue compact dwarf core. Astron. J. **111**, 1551 (1996). doi:10.1086/117895, astro-ph/9601191

Milgrom, M.: A modification of the Newtonian dynamics as a possible alternative to the hidden mass hypothesis. Astrophys. J. **270**, 365–370 (1983). doi:10.1086/161130

Minchin, R., Davies, J., Disney, M., Grossi, M., Sabatini, S., Boyce, P., Garcia, D., Impey, C., Jordan, C., Lang, R., Marble, A., Roberts, S., van Driel, W.: 21 cm Synthesis Observations of VIRGOHI 21-A possible dark galaxy in the virgo cluster. Astrophys. J. **670**, 1056–1064 (2007). doi:10.1086/520620, 0706.1586

Mogotsi, K.M., de Blok, W.J.G., Caldú-Primo, A., Walter, F., Ianjamasimanana R, Leroy, A.K.: H I and CO velocity dispersions in nearby galaxies. Astron. J. **151**, 15 (2016). doi:10.3847/0004-6256/151/1/15, 1511.06006

Moore, B.: Evidence against dissipation-less dark matter from observations of galaxy haloes. Nature **370**, 629–631 (1994). doi:10.1038/370629a0

Mouhcine, M., Ibata, R., Rejkuba, M.: A panoramic view of the milky way analog NGC 891. Astrophys. J. Lett. **714**, L12–L15 (2010). doi:10.1088/2041-8205/714/1/L12, 1002.0461

Muller, C.A., Oort, J.H.: Observation of a line in the galactic radio spectrum: the interstellar hydrogen line at 1,420 Mc./sec., and an estimate of galactic rotation. Nature **168**, 357–358 (1951). doi:10.1038/168357a0

Muñoz-Mateos, J.C., Sheth, K., Gil de Paz, A., Meidt, S., Athanassoula, E., Bosma, A., Comerón, S., Elmegreen, D.M., Elmegreen, B.G., Erroz-Ferrer, S., Gadotti, D.A., Hinz, J.L., Ho, L.C., Holwerda, B., Jarrett, T.H., Kim, T., Knapen, J.H., Laine, J., Laurikainen, E., Madore, B.F., Menendez-Delmestre, K., Mizusawa, T., Regan, M., Salo, H., Schinnerer, E., Seibert, M., Skibba, R., Zaritsky, D.: The impact of bars on disk breaks as probed by S^4G imaging. Astrophys. J. **771**, 59 (2013). doi:10.1088/0004-637X/771/1/59, 1304.6083

Muñoz-Mateos, J.C., Sheth, K., Regan, M., Kim, T., Laine, J., Erroz-Ferrer S, Gil de Paz, A., Comeron, S., Hinz, J., Laurikainen, E., Salo, H., Athanassoula, E., Bosma, A., Bouquin, A.Y.K., Schinnerer, E., Ho, L., Zaritsky, D., Gadotti, D.A., Madore, B., Holwerda, B., Menéndez-Delmestre, K., Knapen, J.H., Meidt, S., Querejeta, M., Mizusawa T, Seibert, M., Laine, S., Courtois, H.: The Spitzer survey of stellar structure in galaxies (S^4G): stellar masses, sizes, and radial profiles for 2352 nearby galaxies. Astrophys. J. Suppl. Ser. **219**, 3 (2015). doi:10.1088/0067-0049/219/1/3, 1505.03534

Navarro, J.F.: The cosmological significance of disk galaxy rotation curves (1998). ArXiv e-prints astro-ph/9807084

Navarro, J.F., Frenk, C.S., White, S.D.M.: The structure of cold dark matter halos. Astrophys. J. **462**, 563 (1996). doi:10.1086/177173, astro-ph/9508025

Navarro, J.F., Frenk, C.S., White, S.D.M.: A universal density profile from hierarchical clustering. Astrophys. J. **490**, 493–508 (1997). astro-ph/9611107

Newton, K.: Neutral hydrogen in IC 342. I - The large-scale structure. Mon. Not. R. Astron. Soc. **191**, 169–184 (1980). doi:10.1093/mnras/191.2.169

Newton, K., Emerson, D.T.: Neutral hydrogen in the outer regions of M31. Mon. Not. R. Astron. Soc. **181**, 573–590 (1977). doi:10.1093/mnras/181.3.573

Noordermeer, E., van der Hulst, J.M., Sancisi, R., Swaters, R.A., van Albada, T.S.: (2005) The Westerbork HI survey of spiral and irregular galaxies. III. HI observations of early-type disk galaxies. Astron. Astrophys. **442**, 137–157. doi:10.1051/0004-6361: 20053172, astro-ph/0508319

O'Brien, J.C., Freeman, K.C., van der Kruit, P.C.: The dark matter halo shape of edge-on disk galaxies. III. Modelling the HI observations: results. Astron. Astrophys. **515**, A62 (2010a). doi:10.1051/0004-6361/200912567, 1003.3109

O'Brien, J.C., Freeman, K.C., van der Kruit, P.C.: The dark matter halo shape of edge-on disk galaxies. II. Modelling the HI observations: methods. Astron. Astrophys. **515**, A61 (2010b). doi:10.1051/0004-6361/200912566, 1003.3108

O'Brien, J.C., Freeman, K.C., van der Kruit, P.C.: The dark matter halo shape of edge-on disk galaxies. IV. UGC 7321. Astron. Astrophys. **515**, A63 (2010c). doi:10.1051/0004-6361/200912568, 1002.3098

O'Brien, J.C., Freeman, K.C., van der Kruit, P.C., Bosma, A.: The dark matter halo shape of edge-on disk galaxies. I. HI observations. Astron. Astrophys. **515**, A60 (2010d). doi:10.1051/0004-6361/200912565, 1003.3110

Oh, S.H., de Blok, W.J.G., Walter, F., Brinks, E., Kennicutt, R.C. Jr.: High-resolution dark matter density profiles of THINGS dwarf galaxies: correcting for noncircular motions. Astron. J. **136**, 2761–2781 (2008). doi:10.1088/0004-6256/136/6/2761, 0810.2119

Oh, S.H., Brook, C., Governato, F., Brinks, E., Mayer, L., de Blok, W.J.G., Brooks, A., Walter, F.: The central slope of dark matter cores in dwarf galaxies: simulations versus THINGS. Astron. J. **142**, 24 (2011). doi:10.1088/0004-6256/142/1/24, 1011.2777

Oh, S.H., Hunter, D.A., Brinks, E., Elmegreen, B.G., Schruba, A., Walter, F., Rupen, M.P., Young, L.M., Simpson, C.E., Johnson, M.C., Herrmann, K.A., Ficut-Vicas, D., Cigan, P., Heesen, V., Ashley, T., Zhang, H.X.: High-resolution Mass Models of Dwarf Galaxies from LITTLE THINGS. Astron. J. **149**, 180 (2015). doi:10.1088/0004-6256/149/6/180, 1502.01281

Olling, R.P.: NGC 4244: a low mass galaxy with a falling rotation curve and a flaring gas layer. Astron. J. **112**, 457 (1996a). doi:10.1086/118028, astro-ph/9605110

Olling, R.P.: The highly flattened dark matter halo of NGC 4244. Astron. J. **112**, 481 (1996b). doi:10.1086/118029, astro-ph/9605111

Olling, R.P., Merrifield, M.R.: Two measures of the shape of the dark halo of the Milky Way. Mon. Not. R. Astron. Soc. **311**, 361–369 (2000). doi:10.1046/j.1365-8711.2000.03053.x

Oman, K.A., Navarro, J.F., Fattahi, A., Frenk, C.S., Sawala, T., White, S.D.M., Bower, R., Crain, R.A., Furlong, M., Schaller, M., Schaye, J., Theuns, T.: The unexpected diversity of dwarf galaxy rotation curves. Mon. Not. R. Astron. Soc. **452**, 3650–3665 (2015). doi:10.1093/mnras/stv1504, 1504.01437

Oman, K.A., Navarro, J.F., Sales, L.V., Fattahi, A., Frenk, C.S., Sawala, T., Schaller, M., White, S.D.M.: Missing dark matter in dwarf galaxies? Mon. Not. R. Astron. Soc. **460**, 3610–3623 (2016). doi:10.1093/mnras/stw1251, 1601.01026

Oosterloo, T., van Gorkom, J.: A large H I cloud near the centre of the Virgo cluster. Astron. Astrophys. **437**, L19–L22 (2005). doi:10.1051/0004-6361:200500127, astro-ph/0505397

Oosterloo, T., Fraternali, F., Sancisi, R.: The cold gaseous halo of NGC 891. Astron. J. **134**, 1019 (2007). doi:10.1086/520332, 0705.4034

Ostriker, J.P., Peebles, P.J.E.: A numerical study of the stability of flattened galaxies: or, can cold galaxies survive? Astrophys. J. **186**, 467–480 (1973). doi:10.1086/152513

Ostriker, J.P., Peebles, P.J.E., Yahil, A.: The size and mass of galaxies, and the mass of the universe. Astrophys. J. Lett. **193**, L1–L4 (1974). doi:10.1086/181617

Ott, J., Walter, F., Brinks, E., Van Dyk, S.D., Dirsch, B., Klein, U.: Evidence for BlowOut in the low-mass dwarf galaxy Holmberg I. Astron. J. **122**, 3070–3091 (2001). doi:10.1086/324101, astro-ph/0110154

Ott, J., Stilp, A.M., Warren, S.R., Skillman, E.D., Dalcanton, J.J., Walter, F., de Blok, W.J.G., Koribalski, B., West, A.A.: VLA-ANGST: a high-resolution H I survey of nearby dwarf galaxies. Astron. J. **144**, 123 (2012). doi:10.1088/0004-6256/144/4/123, 1208.3737

Peters, S.P.C., van der Kruit, P.C., Allen, R.J., Freeman, K.C.: The shape of dark matter haloes, I. HI observations of edge-on galaxies (2013). ArXiv e-prints 1303.2463
Peters, S.P.C., de Geyter, G., van der Kruit, P.C., Freeman, K.C.: The shape of dark matter haloes IV. The structure of stellar discs in edge-on galaxies (2016a). ArXiv e-prints 1608.05563
Peters, S.P.C., van der Kruit, P.C., Allen, R.J., Freeman, K.C.: The shape of dark matter haloes II. The Galactus HI modelling fitting tool (2016b). ArXiv e-prints 1608.05557
Peters, S.P.C., van der Kruit, P.C., Allen, R.J., Freeman, K.C.: The shape of dark matter haloes III. Kinematics and structure of the HI disc (2016c). ArXiv e-prints 1608.05559
Peters, S.P.C., van der Kruit, P.C., Allen, R.J., Freeman, K.C.: The shape of dark matter haloes, V. Analysis of observations of edge-on galaxies (2016d). ArXiv e-prints 1605.08209
Ponomareva, A.A., Verheijen, M.A.W., Bosma, A.: Detailed HI kinematics of Tully-Fisher calibrator galaxies. Mon. Not. R. Astron. Soc. (2016). doi:10.1093/mnras/stw2213, 1609.00378
Puche, D., Westpfahl, D., Brinks, E., Roy, J.R.: Holmberg II - A laboratory for studying the violent interstellar medium. Astron. J. **103**, 1841–1858 (1992). doi:10.1086/116199
Radburn-Smith, D.J., de Jong, R.S., Streich, D., Bell, E.F., Dalcanton, J.J., Dolphin, A.E., Stilp, A.M., Monachesi, A., Holwerda, B.W., Bailin, J.: Constraining the age of the NGC 4565 H I disk warp: determining the origin of gas warps. Astrophys. J. **780**, 105 (2014). doi:10.1088/0004-637X/780/1/105, 1312.2008
Rand, R.J., Kulkarni, S.R., Hester, J.J.: The distribution of warm ionized gas in NGC 891. Astrophys. J. Lett. **352**, L1–L4 (1990). doi:10.1086/185679
Randriamampandry, T.H., Carignan, C.: Galaxy mass models: MOND versus dark matter haloes. Mon. Not. R. Astron. Soc. **439**, 2132–2145 (2014). doi:10.1093/mnras/stu100, 1401.5619
Read, J.I., Iorio, G., Agertz, O., Fraternali, F.: Understanding the shape and diversity of dwarf galaxy rotation curves in ΛCDM. Mon. Not. R. Astron. Soc. **462**, 3628–3645 (2016). doi:10.1093/mnras/stw1876, 1601.05821
Rich, R.M., Collins, M.L.M., Black, C.M., Longstaff, F.A., Koch, A., Benson, A., Reitzel, D.B.: A tidally distorted dwarf galaxy near NGC 4449. Nature **482**, 192–194 (2012). doi:10.1038/nature10837, 1202.2316
Roberts, M.S.: A high-resolution 21-cm hydrogen-line survey of the Andromeda Nebula. Astrophys. J. **144**, 639 (1966). doi:10.1086/148645
Roberts, M.S.: The gaseous content of galaxies (survey lecture). In: Evans, D.S., Wills, D., Wills, B.J. (eds.) External Galaxies and Quasi-Stellar Objects. IAU Symposium, vol. 44, p. 12 (1972)
Roberts, M.S.: The rotation curve of galaxies. In: Hayli, A. (ed.) Dynamics of Stellar Systems. IAU Symposium, vol. 69, p. 331 (1975)
Roberts, M.S.: How much of the universe do we see? In: Bossi, M., Tucci, P. (eds.) Bicentennial Commemoration of R.G. Boscovich: Milano, September 15–18, Proceedings, pp. 237–247. QB36.B66 B53 (1988)
Roberts, M.S., Rots, A.H.: Comparison of rotation curves of different galaxy types. Astron. Astrophys. **26**, 483–485 (1973)
Roberts, M.S., Whitehurst, R.N.: The rotation curve and geometry of M31 at large galactocentric distances. Astrophys. J. **201**, 327–346 (1975). doi:10.1086/153889
Rogstad, D.H.: Aperture synthesis study of neutral hydrogen in the galaxy M101: II. Discussion. Astron. Astrophys. **13**, 108–115 (1971)
Rogstad, D.H., Shostak, G.S.: Aperture synthesis study of neutral hydrogen in the galaxy M101: I. Observations. Astron. Astrophys. **13**, 99–107 (1971)
Rogstad, D.H., Shostak, G.S.: Gross properties of five Scd galaxies as determined from 21-centimeter observations. Astrophys. J. **176**, 315 (1972). doi:10.1086/151636
Rogstad, D.H., Lockhart, I.A., Wright, M.C.H.: Aperture-synthesis observations of H I in the galaxy M83. Astrophys. J. **193**, 309–319 (1974). doi:10.1086/153164
Rogstad, D.H., Wright, M.C.H., Lockhart, I.A.: Aperture synthesis of neutral hydrogen in the galaxy M33. Astrophys. J. **204**, 703–711 (1976). doi:10.1086/154219
Rogstad, D.H., Chu, K., Crutcher, R.M.: Aperture-synthesis observations of H I in the galaxy NGC 300. Astrophys. J. **229**, 509–513 (1979). doi:10.1086/156983

Roychowdhury, S., Chengalur, J.N., Begum, A., Karachentsev, I.D.: Thick gas discs in faint dwarf galaxies. Mon. Not. R. Astron. Soc. **404**, L60–L63 (2010). doi:10.1111/j.1745-3933.2010.00835.x, 1002.4474

Roychowdhury, S., Chengalur, J.N., Karachentsev, I.D., Kaisina, E.I.: The intrinsic shapes of dwarf irregular galaxies. Mon. Not. R. Astron. Soc. **436**, L104–L108 (2013). doi:10.1093/mnrasl/slt123, 1308.6200

Rubin, V.C., Ford, W.K. Jr.: Rotation of the Andromeda Nebula from a spectroscopic survey of emission regions. Astrophys. J. **159**, 379 (1970). doi:10.1086/150317

Rubin, V.C., Thonnard, N., Ford, W.K. Jr.: Extended rotation curves of high-luminosity spiral galaxies. IV - Systematic dynamical properties, SA through SC. Astrophys. J. Lett. **225**, L107–L111 (1978). doi:10.1086/182804

Rubin, V.C., Ford, W.K. Jr., Thonnard, N.: Rotational properties of 21 Sc galaxies with a large range of luminosities and radii, from NGC 4605 /R = 4kpc/ to UGC 2885 /R = 122 kpc/. Astrophys. J. **238**, 471–487 (1980). doi:10.1086/158003

Rubin, V.C., Ford, W.K. Jr., Thonnard, N., Burstein, D.: Rotational properties of 23 Sb galaxies. Astrophys. J. **261**, 439–456 (1982). doi:10.1086/160355

Rubin, V.C., Burstein, D., Ford, W.K. Jr., Thonnard, N.: Rotation velocities of 16 Sa galaxies and a comparison of Sa, Sb, and Sc rotation properties. Astrophys. J. **289**, 81–98 (1985). doi:10.1086/162866

Rupen, M.P.: Neutral hydrogen observations of NGC 4565 and NGC 891. Astron. J. **102**, 48–106 (1991). doi:10.1086/115858

Sackett, P.D., Morrisoni, H.L., Harding, P., Boroson, T.A.: A faint luminous halo that may trace the dark matter around spiral galaxy NGC5907. Nature **370**, 441–443 (1994a). doi:10.1038/370441a0, astro-ph/9407068

Sackett, P.D., Rix, H.W., Jarvis, B.J., Freeman, K.C.: The flattened dark halo of polar ring galaxy NGC 4650A: a conspiracy of shapes? Astrophys. J. **436**, 629–641 (1994b). doi:10.1086/174938, astro-ph/9406015

Saha, K., de Jong, R., Holwerda, B.: The onset of warps in Spitzer observations of edge-on spiral galaxies. Mon. Not. R. Astron. Soc. **396**, 409–422 (2009). doi:10.1111/j.1365-2966.2009.14696.x, 0902.4436

Salpeter, E.E.: Rotation curves in the outer parts of galaxies from H I observations. In: Berkhuijsen, E.M., Wielebinski, R. (eds.) Structure and Properties of Nearby Galaxies. IAU Symposium, vol. 77, pp. 23–26 (1978)

Sancisi, R.: Warped HI disks in galaxies. Astron. Astrophys. **53**, 159 (1976)

Sancisi, R., Allen, R.J.: Neutral hydrogen observations of the edge-on disk galaxy NGC 891. Astron. Astrophys. **74**, 73–84 (1979)

Sancisi, R., Fraternali, F., Oosterloo, T., van der Hulst, T.: Cold gas accretion in galaxies. Astron. Astrophys. Rev. **15**, 189–223 (2008). doi:10.1007/s00159-008-0010-0, 0803.0109

Sandage, A.: The Hubble Atlas of Galaxies. Carnegie Institution, Washington (1961)

Sandage, A., Freeman, K.C., Stokes, N.R.: The intrinsic flattening of e, so, and spiral galaxies as related to galaxy formation and evolution. Astrophys. J. **160**, 831 (1970). doi:10.1086/150475

Sanders, R.H., McGaugh, S.S.: Modified Newtonian dynamics as an alternative to dark matter. Annu. Rev. Astron. Astrophys. **40**, 263–317 (2002). doi:10.1146/annurev.astro.40.060401.093923, astro-ph/0204521

Schmidt, M.: The distribution of mass in M 31. Bull. Astron. Inst. Neth. **14**, 17 (1957)

Schwarzschild, M.: Mass distribution and mass-luminosity ratio in galaxies. Astron. J. **59**, 273 (1954). doi:10.1086/107013

Schweizer, F., Whitmore, B.C., Rubin, V.C.: Colliding and merging galaxies. II - S0 galaxies with polar rings. Astron. J. **88**, 909–925 (1983). doi:10.1086/113377

Serra, P., Oosterloo, T., Morganti, R., Alatalo, K., Blitz, L., Bois, M., Bournaud, F., Bureau, M., Cappellari, M., Crocker, A.F., Davies, R.L., Davis, T.A., de Zeeuw, P.T., Duc, P.A., Emsellem, E., Khochfar, S., Krajnović, D., Kuntschner, H., Lablanche, P.Y., McDermid, R.M., Naab, T., Sarzi, M., Scott N, Trager, S.C., Weijmans, A.M., Young, L.M.: The ATLAS^{3D} project - XIII.

Mass and morphology of H I in early-type galaxies as a function of environment. Mon. Not. R. Astron. Soc. **422**, 1835–1862 (2012). doi:10.1111/j.1365-2966.2012.20219.x, 1111.4241

Serra, P., Oser, L., Krajnović, D., Naab, T., Oosterloo, T., Morganti, R., Cappellari, M., Emsellem, E., Young, L.M., Blitz, L., Davis, T.A., Duc, P.A., Hirschmann, M., Weijmans, A.M., Alatalo, K., Bayet, E., Bois, M., Bournaud F, Bureau, M., Crocker, A.F., Davies, R.L., de Zeeuw, P.T., Khochfar, S., Kuntschner, H., Lablanche, P.Y., McDermid, R.M., Sarzi, M., Scott, N.: The ATLAS^{3D} project - XXVI. H I discs in real and simulated fast and slow rotators. Mon. Not. R. Astron. Soc. **444**, 3388–3407 (2014). doi:10.1093/mnras/stt2496, 1401.3180

Shobbrook, R.R., Robinson, B.J.: 21 cm observations of NGC 300. Aust. J. Phys. **20**, 131 (1967)

Shostak, G.S., van der Kruit, P.C.: Studies of nearly face-on spiral galaxies. II - H I synthesis observations and optical surface photometry of NGC 628. Astron. Astrophys. **132**, 20–32 (1984)

Simon, J.D., Bolatto, A.D., Leroy, A., Blitz, L.: High-resolution measurements of the dark matter halo of NGC 2976: evidence for a shallow density profile. Astrophys. J. **596**, 957–981 (2003). doi:10.1086/378200, astro-ph/0307154

Simon, J.D., Bolatto, A.D., Leroy, A., Blitz, L., Gates, E.L.: High-resolution measurements of the halos of four dark matter-dominated galaxies: deviations from a universal density profile. Astrophys. J. **621**, 757–776 (2005). doi:10.1086/427684, astro-ph/0412035

Stanonik, K., Platen, E., Aragón-Calvo, M.A., van Gorkom, J.H., van de Weygaert, R., van der Hulst, J.M., Peebles, P.J.E.: Polar disk galaxy found in wall between voids. Astrophys. J. Lett. **696**, L6–L9 (2009). doi:10.1088/0004-637X/696/1/L6, 0903.2325

Stilp, A.M., Dalcanton, J.J., Skillman, E., Warren, S.R., Ott, J., Koribalski, B.: Drivers of H I turbulence in dwarf galaxies. Astrophys. J. **773**, 88 (2013a). doi:10.1088/0004-637X/773/2/88, 1306.2321

Stilp, A.M., Dalcanton, J.J., Warren, S.R., Skillman, E., Ott, J., Koribalski, B.: Global H I kinematics in dwarf galaxies. Astrophys. J. **765**, 136 (2013b). doi:10.1088/0004-637X/765/2/136, 1301.1989

Streich, D., de Jong, R.S., Bailin, J., Bell, E.F., Holwerda, B.W., Minchev, I., Monachesi, A., Radburn-Smith, D.J.: Extragalactic archeology with the GHOSTS Survey. I. Age-resolved disk structure of nearby low-mass galaxies. Astron. Astrophys. **585**, A97 (2016). doi:10.1051/0004-6361/201526013, 1509.06647

Sunyaev, R.A.: The interaction of the metagalactic ultraviolet background radiation with galaxies and the limit on the density of the intergalactic gas. Astrophys. Lett. **3**, 33 (1969)

Swaters, R.A., Sancisi, R., van der Hulst, J.M.: The H I Halo of NGC 891. Astrophys. J. **491**, 140–145 (1997). astro-ph/9707150

Swaters, R.A., van Albada, T.S., van der Hulst, J.M., Sancisi, R.: The Westerbork HI survey of spiral and irregular galaxies. I. HI imaging of late-type dwarf galaxies. Astron. Astrophys. **390**, 829–861 (2002). doi:10.1051/0004-6361: 20011755, astro-ph/0204525

Swaters, R.A., Madore, B.F., van den Bosch, F.C., Balcells, M.: The central mass distribution in dwarf and low surface brightness galaxies. Astrophys. J. **583**, 732–751 (2003). doi:10.1086/345426, astro-ph/0210152

Tamburro, D., Rix, H.W., Leroy, A.K., Mac Low, M.M., Walter, F., Kennicutt, R.C., Brinks, E., de Blok, W.J.G.: What is driving the H I velocity dispersion? Astron. J. **137**, 4424–4435 (2009). doi:10.1088/0004-6256/137/5/4424, 0903.0183

Thilker, D.A., Bianchi, L., Meurer, G., Gil de Paz, A., Boissier, S., Madore, B.F., Boselli, A., Ferguson, A.M.N., Muñoz-Mateos, J.C., Madsen, G.J., Hameed S, Overzier, R.A., Forster, K., Friedman, P.G., Martin, D.C., Morrissey, P., Neff, S.G., Schiminovich, D., Seibert, M., Small, T., Wyder, T.K., Donas, J., Heckman, T.M., Lee, Y.W., Milliard, B., Rich, R.M., Szalay, A.S., Welsh, B.Y., Yi, S.K.: A search for extended ultraviolet disk (xuv-disk) galaxies in the local universe. Astrophys. J. Suppl. Ser. **173**, 538–571 (2007). doi:10.1086/523853, 0712.3555

Tollet, E., Macciò, A.V., Dutton, A.A., Stinson, G.S., Wang, L., Penzo, C., Gutcke, T.A., Buck, T., Kang, X., Brook, C., Di Cintio, A., Keller, B.W., Wadsley, J.: NIHAO - IV: core creation and destruction in dark matter density profiles across cosmic time. Mon. Not. R. Astron. Soc. **456**, 3542–3552 (2016). doi:10.1093/mnras/stv2856, 1507.03590

Toloba, E., Guhathakurta, P., Romanowsky, A.J., Brodie, J.P., Martínez-Delgado, D., Arnold, J.A., Ramachandran, N., Theakanath, K.: New spectroscopic technique based on coaddition of surface brightness fluctuations: NGC 4449 and its stellar tidal stream. Astrophys. J. **824**, 35 (2016). doi:10.3847/0004-637X/824/1/35

Toomre, A.: What amplifies the spirals. In: Fall, S.M., Lynden-Bell, D. (eds.) Structure and Evolution of Normal Galaxies, pp. 111–136. Cambridge University Press, Cambridge (1981)

Trott, C.M., Treu, T., Koopmans, L.V.E., Webster, R.L.: Stars and dark matter in the spiral gravitational lens 2237+0305. Mon. Not. R. Astron. Soc. **401**, 1540–1551 (2010). doi:10.1111/j.1365-2966.2009.15780.x, 0812.0748

Uson, J.M., Matthews, L.D.: H I Imaging observations of superthin galaxies. I. UGC 7321. Astron. J. **125**, 2455–2472 (2003). doi:10.1086/374627, astro-ph/0303275

Van Albada, T.S., Sancisi, R.: Dark matter in spiral galaxies. Philos. Trans. R. Soc. Lond. A **320**, 447–464 (1986). doi:10.1098/rsta.1986.0128

van de Hulst, H.C., Raimond, E., van Woerden, H.: Rotation and density distribution of the Andromeda nebula derived from observations of the 21-cm line. Bull. Astron. Inst. Neth. **14**, 1 (1957)

van der Hulst, T., Sancisi, R.: High-velocity gas in M101. Astron. J. **95**, 1354–1359 (1988). doi:10.1086/114731

van der Kruit, P.C.: Optical surface photometry of eight spiral galaxies studied in Westerbork. Astron. Astrophys. Suppl. Ser. **38**, 15–38 (1979)

van der Kruit, P.C.: The thickness of the hydrogen layer and the three-dimensional mass distribution in NGC 891. Astron. Astrophys. **99**, 298–304 (1981)

van der Kruit, P.C.: Truncations of stellar disks and warps of HI-layers in edge-on spiral galaxies. Astron. Astrophys. **466**, 883–893 (2007). doi:10.1051/0004-6361:20066941, astro-ph/0702486

van der Kruit, P.C., Shostak, G.S.: Studies of nearly face-on spiral galaxies. I - The velocity dispersion of the H I gas in NGC 3938. Astron. Astrophys. **105**, 351–358 (1982)

van der Kruit, P.C., Shostak, G.S.: Studies of nearly face-on spiral galaxies. III - H I synthesis observations of NGC 1058 and the mass distribution in galactic disks. Astron. Astrophys. **134**, 258–267 (1984)

van Woerden, H., Bosma, A., Mebold, U.: Distribution and motions of neutral hydrogen in the giant irregular galaxy NGC 4449. In: Weliachew, L. (ed.) La Dynamique des galaxies spirales, vol. 241, p. 483. Editions du Centre National de la Recherche Scientifique, Paris (1975)

van Zee, L.: Discovery of an extended gas disk around UGC 5288. In: American Astronomical Society Meeting Abstracts. Bulletin of the American Astronomical Society, vol. 36, p. 1495 (2004)

Verdes-Montenegro, L., Yun, M.S., Williams, B.A., Huchtmeier, W.K., Del Olmo, A., Perea, J.: Where is the neutral atomic gas in Hickson groups? Astron. Astrophys. **377**, 812–826 (2001). doi:10.1051/0004-6361: 20011127, astro-ph/0108223

Verdes-Montenegro, L., Bosma, A., Athanassoula, E.: Star formation in the warped outer pseudoring of the spiral galaxy NGC 3642. Astron. Astrophys. **389**, 825–835 (2002). doi:10.1051/0004-6361:20020680, astro-ph/0205088

Verheijen, M.A.W., Sancisi, R.: The Ursa Major cluster of galaxies. IV. HI synthesis observations. Astron. Astrophys. **370**, 765–867 (2001). doi:10.1051/0004-6361:20010090, astro-ph/0101404

Volders, L.M.J.S.: Neutral hydrogen in M 33 and M 101. Bull. Astron. Inst. Neth. **14**, 323 (1959)

Walter, F., Brinks, E., de Blok, W.J.G., Bigiel, F., Kennicutt. R.C. Jr., Thornley, M.D., Leroy, A.: THINGS: the H I nearby galaxy survey. Astron. J. **136**, 2563–2647 (2008). doi:10.1088/0004-6256/136/6/2563, 0810.2125

Wang, J., Kauffmann, G., Józsa, G.I.G., Serra, P., van der Hulst, T., Bigiel, F., Brinchmann, J., Verheijen, M.A.W., Oosterloo, T., Wang, E., Li, C., den Heijer, M., Kerp, J.: The Bluedisks project, a study of unusually H I-rich galaxies - I. H I sizes and morphology. Mon. Not. R. Astron. Soc. **433**, 270–294 (2013). doi:10.1093/mnras/stt722, 1303.3538

Wang, J., Koribalski, B.S., Serra, P., van der Hulst, T., Roychowdhury, S., Kamphuis, P., Chengalur, J.N.: New lessons from the H I size-mass relation of galaxies. Mon. Not. R. Astron. Soc. **460**, 2143–2151 (2016). doi:10.1093/mnras/stw1099, 1605.01489

Weiner, B.J.: The dark matter density problem in massive disk galaxies. In: Ryder, S., Pisano, D., Walker, M., Freeman, K. (eds.) Dark Matter in Galaxies. IAU Symposium, vol. 220, p. 265 (2004). astro-ph/0310666

Weiner, B.J., Sellwood, J.A., Williams, T.B.: The disk and dark halo mass of the barred galaxy NGC 4123. II. Fluid-dynamical models. Astrophys. J. **546**, 931–951 (2001). doi:10.1086/318289, astro-ph/0008205

Westfall, K.B., Bershady, M.A., Verheijen, M.A.W.: The DiskMass survey. III. Stellar kinematics via cross-correlation. Astrophys. J. Suppl. Ser. **193**, 21 (2011). doi:10.1088/0067-0049/193/1/21

Westmeier, T., Braun, R., Koribalski, B.S.: Gas and dark matter in the Sculptor group: NGC 300. Mon. Not. R. Astron. Soc. **410**, 2217–2236 (2011). doi:10.1111/j.1365-2966.2010.17596.x, 1009.0317

Westmeier, T., Koribalski, B.S., Braun, R.: Gas and dark matter in the Sculptor group: NGC 55. Mon. Not. R. Astron. Soc. **434**, 3511–3525 (2013). doi:10.1093/mnras/stt1271, 1307.2962

White, S.D.M., Rees, M.J.: Core condensation in heavy halos - a two-stage theory for galaxy formation and clustering. Mon. Not. R. Astron. Soc. **183**, 341–358 (1978). doi:10.1093/mnras/183.3.341

Whitmore, B.C., McElroy, D.B., Schweizer, F.: The shape of the dark halo in polar-ring galaxies. Astrophys. J. **314**, 439–456 (1987). doi:10.1086/165077

Wilkinson, P.N.: The hydrogen array. In: Cornwell, T.J., Perley, R.A. (eds.) IAU Colloqium 131: Radio Interferometry. Theory, Techniques, and Applications, Astronomical Society of the Pacific Conference Series, vol. 19, pp. 428–432 (1991)

Wolfe, S.A., Pisano, D.J., Lockman, F.J., McGaugh, S.S., Shaya, E.J.: Discrete clouds of neutral gas between the galaxies M31 and M33. Nature **497**, 224–226 (2013). doi:10.1038/nature12082, 1305.1631

Wolfinger, K., Kilborn, V.A., Ryan-Weber, E.V., Koribalski, B.S.: The Ursa major cluster redefined as a 'Supergroup'. Publ. Astron. Soc. Aust. **33**, e038 (2016). doi:10.1017/pasa.2016.31, 1608.01417

Zánmar Sánchez, R., Sellwood, J.A., Weiner, B.J., Williams, T.B.: Modeling the gas flow in the bar of NGC 1365. Astrophys. J. **674**, 797–813 (2008). doi:10.1086/524940, 0710.5286

Zheng, Z., Shang, Z., Su, H., Burstein, D., Chen, J., Deng, Z., Byun, Y.I., Chen, R., Chen, W.P., Deng, L., Fan, X., Fang, L.Z., Hester, J.J., Jiang, Z., Li, Y., Lin, W., Sun, W.H., Tsay, W.S., Windhorst, R.A., Wu, H., Xia, X., Xu W, Xue, S., Yan, H., Zheng, Z., Zhou, X., Zhu, J., Zou, Z., Lu, P.: Deep intermediate-band surface photometry of NGC 5907. Astron. J. **117**, 2757–2780 (1999). doi:10.1086/300866

Chapter 8
Ultra-Deep Imaging: Structure of Disks and Haloes

Johan H. Knapen and Ignacio Trujillo

Abstract Deep imaging is a fundamental tool in the study of the outermost structures of galaxies. We review recent developments in ultra-deep imaging of galaxy disks and haloes, highlighting the technical advances as well as the challenges and summarizing observational results in the context of modern theory and simulations. The deepest modern galaxy imaging comes from three main sources: (1) surveys such as the Sloan Digital Sky Survey's Stripe 82 project; (2) very long exposures on small telescopes, including by amateurs; and (3) long exposures on the largest professional telescopes. The technical challenges faced are common in all these approaches and include the treatment of light scattered by atmosphere and telescope/instrument, correct flat fielding and the subtraction of non-galaxy light in the images. We review scientific results on galaxy disks and haloes obtained with deep imaging, including the detection and characterization of stellar haloes, tidal features and stellar streams, disk truncations and thick disks. The area of ultra-deep imaging is still very much unexplored territory, and future work in this area promises significant advances in our understanding of galaxy formation and evolution.

8.1 Introduction

Our knowledge of the properties of disks of galaxies has been driven by deep imaging for the past decades. Until the mid-1980s, such imaging was done with photographic plates, prepared using specialist chemical techniques and then exposed for long times on large telescopes, often by observers spending long and uncomfortable hours in the prime focus cage of the telescope. The early history of this and the subsequent physical parameters derived for disk galaxies have been

---

J.H. Knapen (✉) • I. Trujillo
Instituto de Astrofísica de Canarias, E-38200 La Laguna, Spain

Departamento de Astrofísica, Universidad de La Laguna, E-38206 La Laguna, Spain
e-mail: jhk@iac.es; trujillo@iac.es

J.H. Knapen et al. (eds.), *Outskirts of Galaxies*, Astrophysics and Space Science Library 434, DOI 10.1007/978-3-319-56570-5_8

summarized by, e.g. van der Kruit and Freeman (2011). Main findings relating to the structure of disks include the description of the surface brightness distribution of disks as exponential by Freeman (1970) and the realization that the vertical profiles of disks seen edge-on can be described by an isothermal sheet (van der Kruit and Searle 1981). Although in the early days they were limited by a small field of view (FOV) and flat fielding issues, imaging with charge-coupled devices (CCDs) quickly took over from photographic plates. Large imaging surveys now provide most data, as reviewed below.

A powerful alternative to deep imaging of integrated light from galaxies is imaging and characterizing individual stars in the outskirts of a galaxy, either by using a camera with a large FOV to observe Local Group galaxies like M31 or by using the *Hubble Space Telescope* (*HST*) to observe dwarf galaxies in the Local Group or to resolve the stars in galaxies outside the Local Group but at distances smaller than 16 Mpc (Zackrisson et al. 2012). Exploration in this sense of M31 started with the Isaac Newton Telescope Wide Field Camera survey (Ibata et al. 2001) and has been recently reviewed by Ferguson and Mackey (2016). Key references for other nearby galaxies include Dalcanton et al. (2009), Radburn-Smith et al. (2011), Gallart et al. (2015) and Monachesi et al. (2016). In the current chapter, we will concentrate on imaging of integrated light and refer the interested reader to the chapter by Crnojević (2017) for a review of results based on imaging individual stars.

Over the past decade, advances in detector technology, observing strategies and data reduction procedures have led to the emergence of new lines of research based on what we call here ultra-deep imaging. This can be obtained in large imaging surveys or obtained for small samples of galaxies or for individual ones, using very long exposure times on small telescopes or with large professional telescopes. In this chapter, we will give examples of the first and third category, whereas impressive examples of results from the second category can be found in the chapter by Abraham et al. (2017).

In this chapter, we will first discuss the various challenges that need to be overcome before ultra-deep imaging can be used to distil scientific advances, in particular those due to light scattered in the atmosphere and telescope, flat fielding and sky subtraction. We will then briefly review the main approaches to obtain deep imaging, namely, from imaging surveys, using small telescopes and using large professional telescopes. In Sect. 8.4, which forms the heart of this review, we consider the progress in our understanding of the outskirts of galaxies that has been achieved, thanks to ultra-deep imaging, paying particular attention to the properties of galaxy disks (including thick disks and disk truncations) and stellar haloes but also touching on properties of tidal streams and satellites. When concluding, we will describe future developments, challenges and expected advances.

8.2 The Challenges of Ultra-Deep Imaging

Obtaining ultra-deep imaging of the sky is plagued with difficulties. For this reason, going beyond the 30 mag arcsec^{-2} frontier (approximately 1500 times fainter than the darkest sky on Earth) has remained rather elusive. In this section, we review the most important challenges that need to be addressed carefully if one desires to obtain ultra-deep imaging.

8.2.1 *Sky Brightness*

Professional astronomical observatories are located on the darkest spots on Earth. Even at these locations (where light pollution caused by human activity is minimal), the night sky brightness is substantial: $\mu_V \sim 22$ mag arcsec^{-2}. This brightness is mainly due to various processes in the upper atmosphere, such as the recombination of atoms which were photoionized by the Sun during the day, a phenomenon known as airglow. With long enough integrations, we are able to reduce the noise in this brightness which results from the intrinsic variability of the night sky and easily obtain images in which we can measure features which are much fainter than the intrinsic sky brightness. However, the sky brightness also contains other components. Beyond our atmosphere, a diffuse light component is caused by the reflection of sunlight on the dust plane of our Solar system. This is the zodiacal light, with a brightness of around $\mu_V \sim 23.5$ mag arcsec^{-2}. This brightness affects particularly those regions of the sky around the ecliptic plane, and it contaminates all observations, including those obtained with space telescopes. The intensity of the zodiacal light is variable and depends on the Solar activity.

8.2.2 *Internal Reflections*

Internal reflections are due to the structure of the telescope and the dome. These reflections can appear at different surface brightness levels but are quite common when one reaches levels of $\mu_V \gtrsim 26$ mag arcsec^{-2}. There are two approaches to minimize these reflections. One way is to use telescopes with simple optics (Abraham and van Dokkum 2014; see also Abraham et al. 2017), and another is to implement clever observing strategies which avoid repetition of similar orientation of the camera on the sky (e.g. Trujillo and Fliri 2016).

8.2.3 Flat Fielding

To obtain reliable ultra-deep imaging, an exquisite flat-field correction of the images needs to be performed. This correction is needed to ensure that a uniform illumination of the CCD leads to a uniform output or uniform counts in the image. In a CCD camera, this is not the case by default as the gain and dark current change across the face of the detector, and distortions due to optics can cause non-uniformities. The process of flat fielding removes all these pixel-to-pixel variations in sensitivity and the effects of distortions in the optical path. Flat-field images are often obtained by exposing on uniformly illuminated surfaces.

Any artefact (gradient, pattern, etc.) left behind during the process of creating the flat-field image used to correct the raw science images will introduce a systematic error in the final image, which in turn will prevent reaching the expected surface brightness limit of the observation. The key to creating a good flat-field image is to have a uniform illumination of the CCD of the camera. For most observing cases, a twilight (or even a dome) flat is good enough for this purpose. However, when the goal is to reach very faint details of the image, a different approach is needed, consisting in creating a flat field using the night sky imaging itself. Such a flat field, using a set of science images, is sometimes referred to as a master flat.

To create a good master flat, the set of science images must be taken at different locations on the sky. In fact, depending on the apparent size of the galaxy, the displacement between one science image and the next should be at least as large as the size of the object. Ideally, not all the science images of a specific galaxy should be located at the same position on the CCD (and, ideally, should also not have the same position angle on the sky). That means preparing an observing scheme that includes both a dithering and a rotation pattern (see the example of this procedure in Trujillo and Fliri 2016). In addition, if the observations are taken over different nights (or observing blocks), the best approach is to create a master flat for each night (or observing block). The use of all the science images in a run rather than those of a single night or observing block is not generally a good idea as slight differences from night to night in the focus and the vignetting correction hinder such an approach.

Once all the science images have been acquired, the process of building a master flat is as follows. First, all the objects in each individual science image are generously masked (for instance, using SExtractor; Bertin and Arnouts 1996). Only the pixels outside the masked areas are used to create the final master flat. Second, to guarantee that all the science images are appropriately weighted during their combination to create the master flat, every individual science image of a given night is normalized. Finally, the normalized and masked individual science images are median-combined into a single master flat.

8.2.3.1 Drift Scanning

An alternative approach which has been used to achieve high-quality flat fielding is referred to as drift scanning or a variant on this called time delay and integration (TDI; McGraw et al. 1980; Wright and Mackay 1981). In drift scanning, the reading of the CCD is done at the same slow rate as the CCD is moved across the sky. In TDI, the CCD does not move, but the readout is timed to coincide with the sidereal rate at which the sky passes by. In both cases, an object is sampled by every pixel in a column, thus averaging out all defects and achieving an extremely efficient flat fielding. As in the case of the master flat described above, the background itself is used for flat fielding, assuring a perfect colour match. Further details, as well as a more complete historical overview, are given by Howell (2006).

In practice, in spite of these significant advantages, draft scanning and TDI have not been used much in the literature. The reasons for this vary but include the difficulty of adjusting the relative movement of sky and CCD with the readout, the loss of efficient observing during the ramp-up and ramp-down phases at the start and end of an exposure, image elongation effects and the fact that the exposure time is fixed by the telescope+CCD setup and often rather short.

In the field of imaging nearby galaxies, the most notable exception to this general dearth of TDI results is the Sloan Digital Sky Survey (SDSS, York et al. 2000). As described by Gunn et al. (1998),[1] the SDSS large-format mosaic CCD camera has been designed to image strips of the sky simultaneously in five colour bands using the TDI technique. The approach chosen by the SDSS team has proven to be very successful in terms of imaging to low surface brightness levels. Even though the exposure time of SDSS images is less than 1 min (53.9 s) and the telescope of modest size (2.5 m), the exquisite flat fielding and sky background allow one to reach very low surface brightness levels indeed, down to 26.5 mag arcsec^{-2} (3σ in an area of 10×10 arcsec; Trujillo and Fliri 2016) or down to 27.5 mag arcsec^{-2} when analysing elliptically averaged surface brightness profiles (see, e.g. Pohlen and Trujillo 2006). As illustrated in Sect. 8.3, co-adding series of SDSS images, as is possible in the Stripe 82 survey area, brings that level down by another 2 mag arcsec^{-2}, allowing ground-breaking science to be performed.

8.2.4 Masking and Background Subtraction

Deep images often reveal coherent structure at low levels, which can be due to, e.g. imperfect flat fielding, spatial variations in the background sky or residual point spread function (PSF) effects. Deep images of galaxies also show foreground stars, as well as background galaxies. All these components must be identified, taken

[1]For a more detailed description of the flat-field procedures used by the SDSS, see http://classic.sdss.org/dr5/algorithms/flatfield.html.

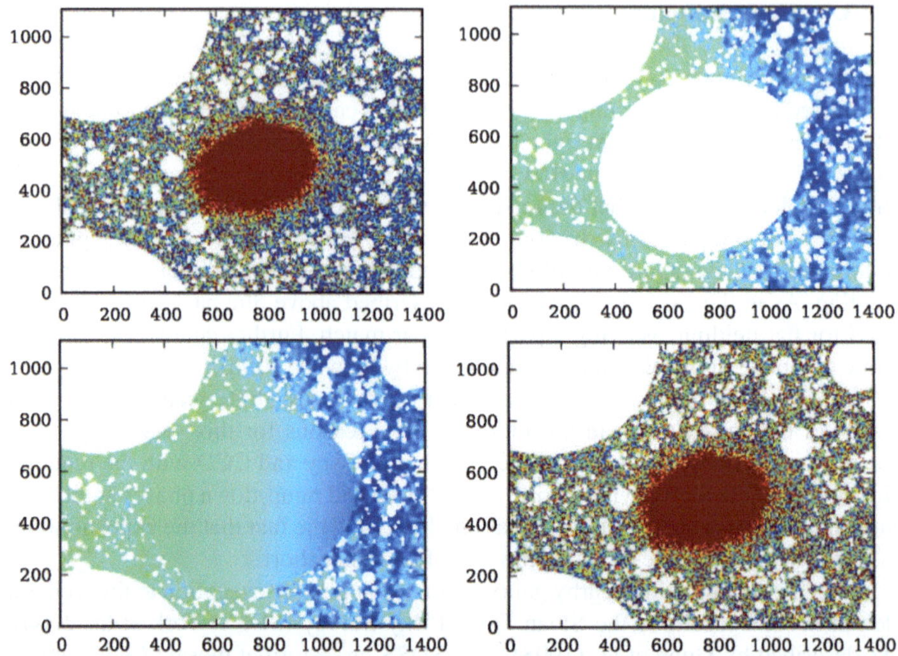

Fig. 8.1 Example of residual background modelling and subtraction, for a Stripe 82 image of the galaxy NGC 941. *Top-left* panel: original image, showing a gradient in the background level. White areas are masked-out images of foreground stars and background galaxies. Image labels are in pixels, with size 0.396 arcsec. North is right, East to the bottom. *Top right*: model for the background, excluding the area of the galaxy. *Lower left*: background model, extrapolated over the area of the galaxy. *Lower right*: final image, with the background model subtracted from the original. Reproduced with permission from Peters et al. (2017)

into account, and/or subtracted before a deep galaxy image can be analysed. As an example of the procedures followed, we show in Fig. 8.1 a SDSS Stripe 82 image of NGC 941 as analysed by Peters et al. (2017). Stars and background galaxies are generously masked, after which a polynomial two-dimensional fit is made to the remaining background pixels. In our example, this is dominated by a left-right gradient in the background level, but the fit also shows smaller-scale structure. The latter could, in principle, be real structure, either related to the galaxy (e.g. tidal streams) or not (e.g. Galactic cirrus; see Sect. 8.2.6). If one looks for structure like tidal streams, a different background modelling technique must be used, for instance, only modelling large-scale fluctuations or gradients. This neatly highlights the difficult nature of this kind of analysis.

In our example, the gradient in the background in all probability is also present at the location of the galaxy, which is why we extrapolate the background model into the galaxy region (lower left panel of Fig. 8.1). This model is then subtracted from the original image, and the result can be used for scientific analysis, in this case studying the shape of the outer regions through analysis of azimuthally averaged radial profiles (see Sect. 8.4.2; this is why the small-scale background structure described in the previous paragraph could safely be subtracted off in this case). The uncertainty limit, down to which these radial profiles can be trusted, is just below 30 mag arcsec^{-2} for the image shown in Fig. 8.1 (data taken from the IAC Stripe82 Legacy Project; Fliri and Trujillo 2016).

8.2.5 Scattered Light

A further and important source of background contamination is produced by all the emitting sources in the image (or even just outside the imaged area). The light of these sources is scattered by the PSF of the instrument across the entire image. Slater et al. (2009) have shown that at $\mu_V \sim 29.5$ mag arcsec^{-2}, an image taken from the ground has all its pixels affected by scattered light from nearby bright sources. Properly removing this contamination is quite challenging and requires an extremely accurate characterization ($<1\%$ at large distances) of the PSF of the camera.

The excess light redistributed by the PSF from both the object of interest and the nearby surrounding sources creates a two-dimensional and highly structured surface that is the main contributor to the background of the image at the faintest surface brightness levels ($\mu_V \gtrsim 29$ mag arcsec^{-2}, Slater et al. 2009). All astronomical images are affected by this scattered light background which is the result of the convolution of the PSF with the light of the sources. The scattered light background of an astronomical image will be more intense if the number of bright sources in the image is large and also if the PSF has significant wings. In this sense, the best option (if feasible) is to select a target within a field devoid of nearby bright surrounding objects. A typical scattered light background is illustrated in Fig. 8.2.

Figure 8.2 illustrates how to deal with the background of scattered light. First, the PSF must be characterized as perfectly as possible by using a reliable extended PSF. An extended PSF with a high signal-to-noise ratio in its outer wings will allow the exploration of the distribution of the light of the nearby brightest sources up to the position of the galaxy under exploration. Second, using the extended PSF, a background scattered light map is created. If the main contributor to the scattered light background is the presence of bright stars, then the scattered light map is created by simply locating the model PSF at the position of the bright stars and scaling the flux of the PSF to these bright sources. Finally, the scattered light map is subtracted from the observed image.

Above, we have discussed the effect of the scattered light created by the surrounding sources around the galaxy of interest. Naturally, the light distribution of the object itself is also convolved with the PSF. In this sense, the scattered light of the targeted galaxy is also creating an artificial excess of light in the outermost region of the galaxy that needs to be addressed in order to explore the properties of the object in its outer regions. We illustrate how to deal with this in Sect. 8.4.4.1 of this chapter.

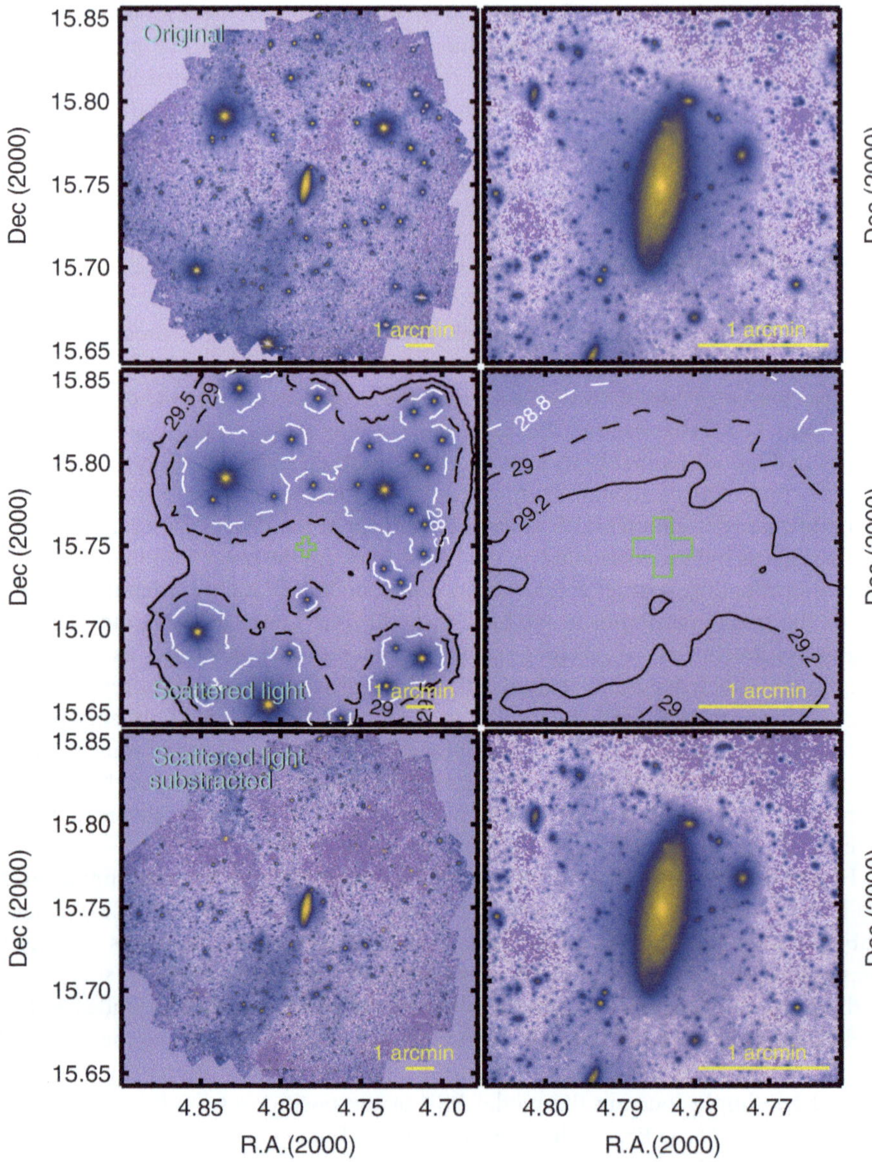

Fig. 8.2 The scattered light around the galaxy UGC 00180 produced by all the stars brighter than $R = 17$ mag in its vicinity. *Top row*: original field (*left*) and a zoom-in of the galaxy. *Middle row*: scattered light. The position of the galaxy is illustrated with a *green cross*. Contours of surface brightness are 28.5, 29 and 29.5 mag arcsec^{-2} (*left*) and 28.8, 29 and 29.2 mag arcsec^{-2} (*right*). *Lower row*: original field after subtraction of the scattered light produced by the brightest sources. Figure taken from Trujillo and Fliri (2016), reproduced with permission of the AAS

8.2.6 *Galactic Cirrus*

Finally, if one is interested in exploring the fainter structures of galaxies, the presence of Galactic cirrus in the images needs to be considered. These filamentary structures are located everywhere on the sky, even at higher Galactic latitudes. For this reason, a careful preselection of the fields to be observed is necessary to minimize the contamination by cirrus. A good illustration of the perils of Galactic cirrus is presented by Davies et al. (2010) for the case of the M81 group. They found that far-infrared emission measured by the *Herschel* satellite correlates very well spatially with narrow-velocity Galactic HI, without any evidence that this far-infrared emission originates in the M81 group. They thus inferred that the optical streams and structures seen in the M81 group are not in fact part of the group, but are rather due to light from our own Galaxy which is back-scattered off Galactic dust.

8.3 Approaches in Ultra-Deep Imaging

The last few years have seen an explosion in the number of works exploring the outermost regions of nearby galaxies using very deep imaging. These works can be grouped into three different flavours: deep ($\sim$1 h) multipurpose surveys with medium-sized (2–4 m) telescopes, extremely long integrations ($\gtrsim$20 h) of particular galaxies with small ($\lesssim$1 m) telescopes or long integrations ($\gtrsim$5 h) with large ($\gtrsim$8 m) telescopes. In what follows, we will summarize some of these efforts.

8.3.1 *Survey Data*

The 3.6 m Canada-France-Hawaii Telescope (CFHT), with its MegaCam wide-field camera comprising 36 CCDs and covering a 1 square degree FOV, has played a significant role in the new generation of deep imaging surveys. This telescope has been used for general-purpose surveys like the Wide Synoptic CFHT Legacy Survey (155 square degrees; Cuillandre et al. 2012) or more specific projects like the Next Generation Virgo Cluster Survey (NGVCS; Ferrarese et al. 2012) or the deep imaging follow-up (Duc et al. 2015) of the ATLAS^{3D} project (Cappellari et al. 2011). These surveys are characterized by having similar depths ($r \sim$ 25–25.5 mag; $S/N = 10$ for point sources) using integration times of 40–60 min, reaching a limiting surface brightness of 28.5–29 mag arcsec^{-2} (3σ in 10×10 arcsec^2).

Among the most important results of the NGVCS is the connection between the distribution of the metal-poor globular clusters and the intra-cluster light (Durrell et al. 2014), suggesting a common origin for these two structures. The deep imaging by Duc et al. (2015) revealed a large number of features surrounding their galaxies.

Both these projects targeted mostly early-type galaxies. Unfortunately, the pipeline used for the reduction of the Wide Synoptic CFHT Legacy Survey removed the low surface brightness features around the brightest extended galaxies in the images, producing obvious "holes" that prevent the use of this reduced dataset for the exploration of the outermost parts of the spiral galaxies. In the later MegaCam surveys, this problem does not occur.

Another telescope that has revolutionized our understanding of galaxies is the Sloan 2.5 m telescope. The Sloan telescope is well known for producing the SDSS (York et al. 2000). Among the different projects that Sloan has covered, of particular interest here is its deep imaging survey in the Southern Galactic cap, popularly known as the "Stripe 82" survey (Jiang et al. 2008; Abazajian et al. 2009). The Stripe 82 survey covers an area of 275 square degrees along the celestial equator ($-50° < \text{R.A.} < 60°$, $-1.25 < \text{Dec.} < 1.25$) and has been observed in all the five SDSS filters: *ugriz*. The typical amount of time on source was $\sim$1.2 h. Being located at the equator, the Stripe 82 area is accessible from most ground-based facilities. A third of all the available SDSS data in the Stripe 82 area were combined by Annis et al. (2014). They reached a depth (50% completeness for point sources) of $r \sim 24.2$ mag. Later on, Jiang et al. (2014) used the entire dataset and reported a gain of 0.3–0.5 mag in depth compared to the previous reduction. None of these reductions were done with the aim of exploring the lowest surface brightness features of the objects. This task was performed by Fliri and Trujillo (2016) and is known as the IAC Stripe 82 Legacy Project. Fliri and Trujillo (2016) reached a depth of $r \sim 24.7$ mag (50% completeness for point sources) and a limiting surface brightness of $\mu_r \sim 28.5\,\text{mag arcsec}^{-2}$ (3σ in $10 \times 10\,\text{arcsec}^2$). The reduced images of Fliri and Trujillo (2016) have been made publicly available through a dedicated webpage (http://www.iac.es/proyecto/stripe82).

8.3.2 *Small Telescopes*

As surface brightness is independent of telescope aperture, in principle one can use small telescopes to reach ultra-faint surface brightness levels. Direct advantages of using small telescopes over larger ones include the larger field of view and the reduced competition for observing time, in particular when private telescopes are used. Using modern CCD technology, this has been done by several workers in the field, aiming to uncover light sources as diverse as intra-cluster light, diffuse galaxies or the outer regions of bright galaxies. The chapter by Abraham et al. (2017) gives significantly more detail on this; we review the basics in this short section.

Mihos and collaborators used the Case Western Reserve University's Burrell Schmidt 0.6 m telescope to obtain ultra-deep imaging of the core region of the Virgo cluster, down to limits of $\mu_V = 28.5\,\text{mag arcsec}^{-2}$ (see Mihos et al. 2016 for a recent update). From these images, they were able to reveal a number of interesting features of the intra-cluster light, including several tidal streamers of more than

100 kpc in length and many smaller tidal tails and bridges between galaxies, and to conclude that cluster assembly appears to be hierarchical in nature rather than the product of smooth accretion around a central galaxy (Mihos et al. 2005; the colours of features in the intra-cluster light were measured by Rudick et al. 2010). Mihos et al. (2015) used the same dataset to find three large and extremely low surface brightness galaxies, only one of which shows the signs of tidal damage that might be expected for such galaxies in a dense cluster environment. The same telescope was used by Mihos et al. (2013) to image the galaxy M101 over an area of 6 deg^2 and to a depth of $\mu_B = 29.5$ mag $arcsec^{-2}$ and by Watkins et al. (2014) to push down to $\mu_B = 30.1$ mag $arcsec^{-2}$ in the M96 galaxy group. The former authors found a number of plumes and spurs but no very extended tidal tails, suggestive of ongoing evolution of the outer disk of the galaxy due to encounters in its group environment, whereas the latter found no optical counterpart to the extended HI surrounding the central elliptical M105 and in general only very subtle interaction signatures in the M96 group.

In a fruitful collaboration with amateur astronomers, Martínez-Delgado et al. (2008, 2010, 2015) have used small private telescopes to obtain deep images of tidal streams around a number of galaxies, most with pre-existing evidence for the presence of some kind of outer structure, and more recently of low surface brightness galaxies in the fields of large nearby galaxies (e.g. Javanmardi et al. 2016). This group uses very long exposures taken at dark sites, imaging through wide filters. Among the most beautiful and well-known results obtained by Martínez-Delgado's group is the discovery of the optical analogues to the morphologies predicted from N-body models of stellar haloes constructed from satellite accretion (e.g. Bullock and Johnston 2005; Johnston et al. 2008). While the resemblance between models and observations is indeed striking and important, it must be kept in mind that most of the structure seen in the models is at lower or much lower surface brightness levels than even the deepest currently available imaging and that most galaxies observed by Martínez-Delgado et al. were targeted specifically to have some previous evidence for tidal structure and are thus not representative of the general galaxy population. In fact, many nearby galaxies show no evidence at all for any tidal or other disturbances in deep images (e.g. Duc et al. 2015, see also Merritt et al. 2016).

A third strand, besides using small existing research telescopes or amateur installations, is to use simple, small, custom-built telescopes optimized for deep imaging through simple optics and a very careful treatment of systematics. The best-known example of this is the Dragonfly Telephoto Array (Abraham and van Dokkum 2014), a set of up to 48 commercial telephoto lenses with excellent coatings coupled to CCD cameras, which minimizes the amount of scattered light produced inside the telescope. In the current configuration, the array is equivalent to a 1 m aperture telescope with a field of view of 6 square degrees (Abraham and van Dokkum 2014; Abraham 2016; see also Abraham et al. 2017).

Among the results obtained with Dragonfly images and profiles (which go down to $\mu_g \sim 28$ mag arcsec^{-2}; 3σ in 12×12 arcsec boxes; Merritt et al. 2016) are the finding that there is a significant spread in the stellar mass fraction surrounding galaxies (van Dokkum et al. 2014; Merritt et al. 2016; see also Sect. 8.4.4) and a study of a so-called ultra-diffuse galaxies in the Coma cluster (van Dokkum et al. 2016 and references therein).

8.3.3 *Large Telescopes*

Large (8–10 m class) telescopes have barely been used so far to obtain very deep imaging of nearby galaxies. To the best of our knowledge, the first successful attempt of going ultra-deep (i.e. surpassing the 30 mag arcsec^{-2} barrier) was conducted by Jablonka et al. (2010), who used the imaging mode of the VIsible MultiObject Spectrograph (VIMOS) on ESO's Very Large Telescope (VLT) to target the edge-on S0 galaxy NGC 3957 reaching surface brightness limits (Vega system) of $\mu_R = 30.6$ mag arcsec^{-2} (1σ; 6 h) and $\mu_V = 31.4$ mag arcsec^{-2} (1σ; 7 h). These authors found that the stellar halo of this galaxy, calculated between 5 and 8 kpc above the disk plane, is consistent with an old and preferentially metal-poor normal stellar population, like that revealed in nearby galaxy haloes from studies of their resolved stellar content. Also worth mentioning is the work by Galaz et al. (2015), who used the more "modest" 6.5 m Magellan Telescope to observe the extremely large galaxy Malin 1. Galaz et al. (2015) used the Megacam camera to image the galaxy for about 4.5 h in *g* and *r*, reaching $\mu_B \sim 28$ mag arcsec^{-2}. Using these images, they obtain an impressive result for the diameter of Malin 1 of 160 kpc, ~50 kpc larger than previous estimates. Their analysis shows that the observed spiral arms reach very low luminosity and mass surface densities, to levels much lower than the corresponding values for the Milky Way.

The currently deepest ever image of a nearby galaxy was recently obtained by Trujillo and Fliri (2016). These authors pointed the Gran Telescopio Canarias (GTC, a 10.4 m telescope) at the galaxy UGC 00180, an object similar to M31 but located at a distance of ~150 Mpc (see Fig. 8.3). Their *r*-band image reached a limiting surface brightness of 31.5 mag arcsec^{-2} (3σ; 10×10 arcsec^2) after 8.1 h on-source integration. This image revealed a stellar halo with significant structure surrounding the galaxy. The stellar halo has a mass fraction of ~3% of the total stellar mass of the galaxy. This value is close to the one found in the Milky Way and M31 using star counting techniques. This is the first time that integrated-light observations of galaxies reach a surface brightness depth close to that achieved using star counting techniques in nearby galaxies. It is a major step forwards as it allows the exploration of the stellar haloes in hundreds of galaxies beyond the Local Group, in particular when viewed in the context of future imaging possibilities with the Large Synoptic Survey Telescope (LSST; see Sect. 8.5).

Fig. 8.3 The galaxy UGC 00180 and its surrounding field observed down to a surface brightness limit of 31.5 mag arcsec^{-2} in the r-band (around ten times deeper than most of the previous deep images obtained from the ground). The image is a combination of SDSS imaging (*colour part*) and 8.1 h of imaging with the GTC; *grey part*). In addition to the stellar halo of the galaxy, the image shows a plethora of details. Among the most remarkable is the filamentary emission from dust of our own Galaxy located in the bottom-left part of the image. There are also distant galaxies which are seen to be merging with other objects and a high-redshift cluster towards the bottom-left corner of the galaxy where the intra-cluster light is visible. Credit: GTC, Gabriel Pérez and Ignacio Trujillo (IAC)

8.4 Disk and Stellar Halo Properties from Ultra-Deep Imaging

8.4.1 Thick Disks

A thick disk is seen as an excess of flux above and below the midplane of edge-on and highly inclined galaxies, typically at a few times the scale height of the thin disk component and with a larger exponential scale height than the thin disk which is dominant in the midplane regions. Thick disks in external galaxies were discovered and defined by Tsikoudi (1979) and Burstein (1979) and have since been found to be very common, in all kinds of galaxies (see, e.g. Dalcanton and Bernstein 2002; Yoachim and Dalcanton 2006; Comerón et al. 2011a,b). Our Milky Way has also been known for a long time to have a thick disk component (e.g. Yoshii 1982; Gilmore and Reid 1983), with stars in the thick disk having significantly lower metallicity than those in the thin component (Gilmore and Wyse 1985).

There is an ongoing debate on the issue of the definition of a thick disk. Early work on the thick disk in the Milky Way, and practically all work on extragalactic thick disks, basically fits stellar density distributions with thin and thick disk components, thus defining them *geometrically*. It is natural to assume that the stars forming part of the thus defined thick component have higher velocity dispersion and possibly older ages and lower metallicity, than those in the thin disk. In our Milky Way, one can observe individual stars, and one can thus define the thick disk *chemically*, for instance, by measuring their α-element abundances relative to iron. Some of the interesting consequences of this are the confirmation by Bensby et al. (2014) that the scale length of the thick disk is much shorter than that of the thin disk or even the statement by Bovy et al. (2012) that the Milky Way does not have a distinct thick disk at all but rather a continuous distribution of disk thicknesses (but see comments by Haywood et al. 2013). This is an ongoing discussion which we note but will not review here. The interested reader may find more discussion and references in a recent conference discussion published by Kawata and Chiappini (2016).

Different formation mechanisms, which can be grouped into three main families, have been proposed for thick disks. All mechanisms directly link the properties of thick disks to the evolution of galaxies, and some indicate clear constraints on the formation of disks. The latter category includes the first broad class of models, in which the high-velocity dispersion of the material forming the disk at high redshift led to the thick disk; the thin disk formed later from lower-dispersion material (e.g. Samland and Gerhard 2003; Brook et al. 2004; Elmegreen and Elmegreen 2006; Bournaud et al. 2009). The second class of models implies a secular origin for thick disks, by stipulating that they are caused by vertical heating and/or the radial migration of stars (e.g. Villumsen 1985; Schönrich and Binney 2009; see also the review by Debattista et al. 2017). In the third class of models, the thick disk arises from interactions with satellite galaxies, through the accretion of stars (e.g. Quinn et al. 1993) or through dynamical heating (e.g. Abadi et al. 2003). No consensus exists as yet in the literature as to which model is the most adequate, and it is likely that all play some role in the formation of thick disks.

Focussing now on the detection and characterization of extragalactic thick disks, there are two main techniques in use: star counts and direct imaging. The former can be done only in the nearest of external galaxies and only with the *HST*, the latter, in principle, across much larger samples (see below). Older work on resolved stars with *HST* includes that by Mould (2005), Tikhonov and Galazutdinova (2005), Seth et al. (2005) and Rejkuba et al. (2009), who use resolved red giant branch (RGB) stars to characterize the thick disk component in one or a handful of galaxies. Recent *HST* work includes that by Streich et al. (2016) who used images from the GHOSTS survey to note the absence of a separate thick disk component for three of their survey galaxies (although their choice of galaxies was not optimal: they are of low mass, so dust lanes are weak or absent and hence the inclination cannot be established well—an important point as thick disks are optimally studied in edge-on galaxies).

Among the most comprehensive studies of samples of external galaxies is that by Yoachim and Dalcanton (2006) who used mainly R-band imaging of 34 late-type edge-on disk galaxies to measure the main structural parameters of the thin and thick disk components by fitting one-dimensional analytic expressions based on the generalized function $\mathrm{sech}^{2/n}$. Although such functions give decent fits to profiles, they are essentially ad hoc and, in the case of superposed fits of both a thin and a thick disk, ignore the gravitational interaction between the two disk components and therefore are not well justified physically.

To remedy this, Comerón et al. (2011a) integrate the equations of equilibrium for a set of gravitationally coupled isothermal stellar and gas disks to derive the parameters of the thin and thick disk components, in particular their relative masses. These authors used deep imaging at 3.6 μm from the *Spitzer* Survey of Stellar Structure in Galaxies (S^4G; Sheth et al. 2010), which have the advantage of being essentially unaffected by dust—an important consideration when studying edge-on galaxies. Fitting one-dimensional luminosity profiles obtained from the images of 46 edge-on and highly inclined galaxies, after a careful masking and background modelling and subtraction effort, Comerón et al. (2011a) found that thick disks are not only ubiquitous but also significantly more massive than previously reported. Typically, thick and thin disks have comparable masses (Fig. 8.4), which favours an in situ origin for the thick disk component, with possibly significant additional amounts of stars from satellites that were accreted after the formation of the galaxy or from secular heating of the thin disk. Because of the different mass-to-light ratios in thin and thick disks, the reported higher mass fractions in the thick component lead to higher overall disk masses and thus a reduced need for dark matter.

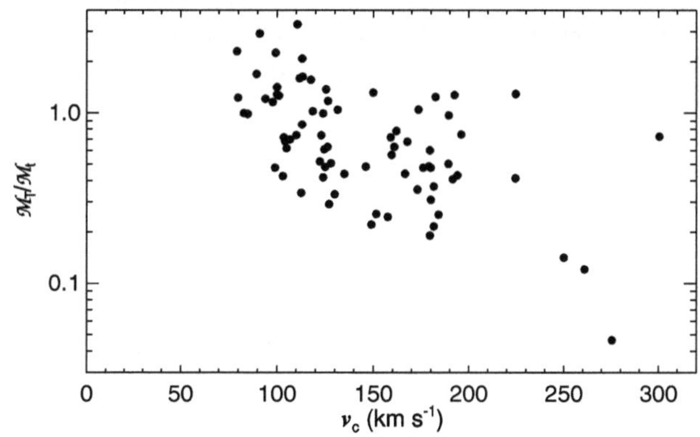

Fig. 8.4 Ratio of the mass of the thick to that of the thin disk, M_T/M_t, as function of circular velocity v_c for a sample of edge-on galaxies expanded from that of Comerón et al. (2012). The new analysis underlying these data incorporates PSF modelling and correction. A value of unity indicates equal thick and thin disk masses. Figure reproduced with permission from S. Comerón et al. (in prep.)

In a later paper, Comerón et al. (2012) confirmed this basic result with a somewhat larger sample of 70 edge-on galaxies, while Comerón et al. (2014) used an analysis of the thin and thick disk components as well as of the central mass concentrations (CMCs; "bulges") to conclude that the ratio of the mass of the dynamically hot components (thick disk and CMC) and that of the cold components (thin and gas disks) is constant and does not depend on the mass of the galaxy. This suggests that both the thick disk and the CMC were formed early on, in a short phase of intense star formation activity and in a turbulent gas disk. The other components were formed later, in a slower phase of lower star formation intensity. Recent work (S. Comerón et al., in prep.) shows that PSF effects in the S^4G imaging are small enough not to influence any of these main conclusions.

Future work will focus on a number of different areas. The first is building larger samples of galaxies with accurately determined thick disk properties, both locally and at higher redshift. The deepest *HST* imaging (e.g. of the *Hubble* Ultra-Deep Fields) can in principle be used for the latter, while the LSST (see Sect. 8.5) should provide the deep imaging of huge samples of nearby galaxies. The second is obtaining detailed information on the kinematics and stellar populations of the thick disks. This is starting to be done now with large telescopes. For instance, Comerón et al. (2015) used observations with the VIMOS instrument on ESO's Very Large Telescope (VLT) to deduce from its kinematics and stellar populations that the thick disk in the galaxy ESO 533-4 formed in a relatively short event, and Comerón et al. (2016) used MUSE on the VLT to conclude that the thick disk in the S0 galaxy ESO 243-49 formed early on in the history of the galaxy and before the star formation in the galaxy was quenched, while Kasparova et al. (2016) and Guérou et al. (2016) provide further evidence that the formation mechanisms of thick disks in galaxies are diverse. Deep spectroscopy with large telescopes, as well as careful colour measurements from LSST imaging, will no doubt bring further progress here and thus shine light on the formation of disk galaxies.

8.4.2 Truncations

The relatively sharp edges of edge-on stellar disks, referred to as truncations, were first noted by van der Kruit (1979). Such truncations, typically occurring at radii of around four or five times the exponential scale length of the inner disk and with scale lengths of less than 1 kpc, appear to be very common, with about three out of four thin disks truncated (Comerón et al. 2012; see the review by van der Kruit and Freeman 2011 for further details).

Truncations are important as they are potential key indicators of the formation and early evolution processes that have shaped disk galaxies, as reviewed in detail by van der Kruit and Freeman (2011), in particular in their Sect. 3.8. For instance, as the truncation corresponds to the maximum in the distribution of specific angular momentum across the disk, it may indicate directly what that distribution was in the protogalaxy, in the case of conservation of angular momentum. Redistribution of material in the disk, as reviewed by Debattista et al. (2017), may have occurred, leading to the current distribution of angular momentum being unrelated to that of the material that formed the disk. In addition, as reviewed by Elmegreen and Hunter (2017), disk breaks and truncations are intimately related to the past and present star formation processes in disks.

While it is thus important to characterize truncations, only in edge-on galaxies are they bright enough to have been observed routinely for the past decades, first on the basis of photographic imaging and later using CCDs. Imaging them in face-on or moderately inclined galaxies has so far proven mostly elusive. Not only will truncations occur at much fainter levels there because of reduced line-of-sight integration through the disk, they can also be masked by the lopsided nature of spiral galaxies (e.g. Zaritsky et al. 2013) or confused by the onset of stellar haloes (Martín-Navarro et al. 2014). In addition, "disk" breaks at relatively high surface brightness and occurring at $\sim 8 \pm 1$ kpc in edge-on galaxies have been confused in the literature with truncations, at $\sim 14 \pm 2$ kpc (Martín-Navarro et al. 2012). For instance, the features labelled as truncations by Pohlen and Trujillo (2006) most probably are disk breaks rather than the face-on counterparts of the truncations observed in edge-on galaxies.

Using some of the deepest imaging available for samples of nearby galaxies, the SDSS Stripe 82 dataset (see Fliri and Trujillo 2016 and Sect. 8.3.1), Peters et al. (2017) have recently attempted to find the elusive truncations in a sample of 22 face-on to moderately inclined galaxies. They used the new data products from Fliri and Trujillo (2016) and added the g', r' and i' images to reach extra depth. They selected their galaxies to be undisturbed, and with well-behaved image backgrounds, and were able to extract surface photometry down to 29–30 r'-mag arcsec^{-2} after performing careful but aggressive masking and modelling of residual background gradients (see Fig. 8.1 for an example). Peters et al. (2017) then used and compared a variety of different analysis and extraction methods to optimize the detection and characterization of truncations in their sample galaxies.

Figure 8.5 illustrates many of the results of Peters et al. (2017). Firstly, disk breaks at relatively high surface brightness were found in the radial profiles of most galaxies (blue dots in the figure), at levels of typically 22–24 r'-mag arcsec^{-2}. Their radius scales with galaxy radius. Secondly, truncations were indeed identified in three of these moderately inclined galaxies, at surface brightness levels of around 28 r'-mag arcsec^{-2}. Thirdly, in most galaxies, a flattening of the radial profile is seen, interpreted as the onset of the stellar halo[2] and starting to dominate the profiles at 28 ± 1 r'-mag arcsec^{-2}. Fourthly, not only do the radii of the onsets of the truncation and the halo components scale with galaxy size, but for some reason, they

[2]Peters et al. (2017) perform a careful PSF modelling to ensure that the observed haloes are not artefacts caused by the PSF.

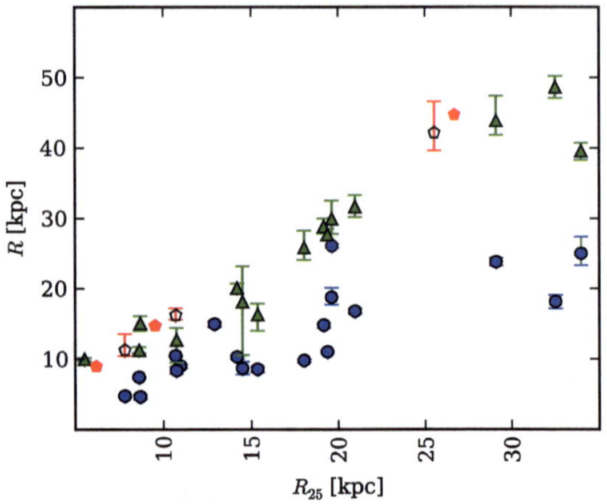

Fig. 8.5 Feature size R as a function of galaxy size, indicated by R_{25}, for the moderately inclined to face-on galaxies in the sample of Peters et al. (2017). *Red pentagons* represent truncations, *green triangles* the onset of the stellar halo, and *blue circles* disk breaks which occur further inside the disk. Reproduced with permission from Peters et al. (2017)

do that in the same manner, so that truncation and halo points (red pentagons and green triangles, respectively, in Fig. 8.5) line up on the same relation. This intriguing relation deserves further attention with other independent surveys, as these features are barely above the surface brightness limit of the Stripe 82.

A further very interesting observation from Peters et al. (2017) is that truncations are only observed in those galaxies from which the halo component is either absent or fainter than usual, although no indications were found from galaxy or feature parameters as to why the halo might be faint or absent. Of the seven galaxies with faint or absent haloes, three were found to host truncations. This is not out of line with the statistical result of Kregel et al. (2002) that at least 20 of their 34 edge-on galaxies were truncated. More detailed study of larger samples of galaxies should shed further light on these important issues, and future developments are promising in this respect (see Sect. 8.5).

8.4.3 Tidal Streams

Tidal streams are the imprints of the ongoing merging activity of galaxies. They are present around a significant fraction of galaxies, showing different shapes, brightness and extensions depending on their progenitor's mass and orbit. The structural (and kinematical) properties of tidal streams can be used to infer information about the dark matter haloes in which the host galaxies are embedded (Bullock and Johnston 2005). These authors conclude that the vast majority of stream features have very low surface brightness ($\mu_V \gtrsim 28.5\,\mathrm{mag\,arcsec^{-2}}$). More precisely, cosmological simulations predict around one detectable stream per galaxy if the observations reach a depth of $\sim 30\,\mathrm{mag\,arcsec^{-2}}$ (e.g. Bullock and Johnston 2005; Johnston et al. 2008; Cooper et al. 2010). This makes the use of very deep surveys mandatory if one wants to have a complete census of past merging activity.

The detailed analysis of our Milky Way and the Andromeda galaxy M31 (using the star count technique) has revealed a plethora of substructure in these objects. Due to the vicinity of these galaxies, the star count technique can reveal features which have surface brightnesses equivalent to $\mu_V \sim 32\,\mathrm{mag\,arcsec^{-2}}$ (e.g. Ibata et al. 2001; Majewski et al. 2003; Belokurov et al. 2006; Bell et al. 2008; McConnachie et al. 2009). These levels are extremely challenging to reach using integrated photometry (the only case published to date is that of the galaxy explored in Trujillo and Fliri 2016). The need, however, for conducting an extensive survey exploring the variety of features around galaxies like the Milky Way is pressing. In fact, according to theory, a large galaxy-to-galaxy variation is expected due to the intrinsic stochasticity of the merger phenomenon. For this reason, to have a proper comparison with theory, it is urgent to have a large survey exploring hundreds of galaxies beyond the Local Group. This is only achievable using the integrated photometry technique. The star count technique becomes unfeasible (with current telescopes) beyond $\sim$16 Mpc (Zackrisson et al. 2012).

The first attempt to systematize the search for streams in galaxies beyond the Local Group has been made by Martínez-Delgado et al. (2010). These authors use modest ($\lesssim$1 m) aperture telescopes using integration times $\sim$10–20 h, allowing them to reach $\mu_g \sim 28.5\,\mathrm{mag/arcsec^2}$. They have found a large variety of tidal stream morphologies in eight nearby galaxies: large arc-like features, giant "umbrellas", shells, plumes, spiral-like patterns, etc. These morphologies are similar, at least qualitatively, to those expected in cosmological simulations. However, the galaxies explored in Martínez-Delgado et al. (2010) are not representative of the general galaxy population as they were preselected to have bright structure surrounding them. In this sense, despite the effort of these authors, the comparison of theory with observations remains currently in its infancy. This is because a large survey, probing hundreds of randomly selected galaxies, is still missing.

8.4.4 Stellar Haloes

Tidal streams and stellar haloes around galaxies are intimately connected. In fact, in observations with high enough resolution and signal-to-noise ratio, stellar haloes should appear as a combination of many different merger events (and consequently, the sum of many tidal streams). Stellar haloes also have a diffuse component which corresponds to those accretion events that happened very early on in a galaxy's evolution. Whereas bright tidal streams inform us about the ongoing or recent merging activity of a galaxy, stellar haloes give us information about the past accretion story of galaxies.

In contrast with numerical simulations, where the characterization of the stellar halo is relatively simple, there is no agreement in the literature on how to quantify the properties of stellar haloes using images of galaxies alone. Simulations (e.g. Cooper et al. 2010; Font et al. 2011; Tissera et al. 2014; Pillepich et al. 2014) show unequivocally that the contribution of accreted stars rises towards the centre of galaxies. In fact, in numerical simulations, the accreted material follows surface mass distributions that can be modelled well with Sérsic (with Sérsic indexes $n \lesssim 3$) or power-law profiles (with logarithmic slope $\sim$ 3.5) for the 3D density profiles. Accreted stars are the main contributors to the stellar mass surface distribution of galaxies in their outer regions; however, towards the inner zones, they contribute much less in comparison with the stars that have formed in situ (a factor of $\sim$100 in projected density; Cooper et al. 2013).

The observational characterization of stellar halo profiles has followed different approaches. Several authors fit the outer (stellar halo-dominated) surface brightness region with some analytical profile, for instance, exponential (e.g. Irwin et al. 2005; Ibata et al. 2007; Trujillo and Fliri 2016) or power law (with logarithmic slopes $\sim$ 2.5; e.g. Tanaka et al. 2010; Courteau et al. 2011; Gilbert et al. 2012). This approach has the disadvantage of assuming a particular shape of the stellar halo towards the inner region of galaxies, motivated by numerical simulations. This shape is derived by extrapolating the fit to the outer region towards the inner area. There is no way of eliminating this assumption observationally with integrated photometry alone, because in the central part of the galaxy, the light is dominated by the bulge and disk. Other authors avoid making any assumption about the inner shape of the stellar halo profile and measure the properties of the stellar halo only in the outer regions where the halo is the dominant contribution. For instance, Buitrago et al. (2016), for elliptical galaxies, use the radial region $10 < R(\mathrm{kpc}) < 50$, whereas Merritt et al. (2016) use $R(\mathrm{kpc}) > 5R_\mathrm{h}$ (with R_h the half-mass radius) for spiral galaxies. This approach has the disadvantage of neglecting the (dominant) contribution of the stellar light of the halo in the innermost region of the galaxy. Any of the two above approaches to measure the relevance of the stellar halo can be followed if a proper comparison (i.e. following the same scheme) is conducted with the numerically simulated galaxies as well. In the literature, it is common to find a comparison of the amount of stellar mass in the stellar halo with the total stellar mass of the galaxy. This relation is then compared with numerical predictions (see, e.g. Fig. 8.6). Observers directly measure neither the light contribution from the stellar haloes nor the stellar mass. To transform from light to stellar mass, it is necessary to estimate a stellar mass-to-light ratio. This can be done using photometric information from several filters (when available).

The fraction of mass in the stellar haloes of disk (Milky Way-like) galaxies is around 1%. However, there are notable exceptions like M101, which does not show evidence for a stellar halo (van Dokkum et al. 2014) even when using deep imaging. Observed stellar haloes are, in general, less massive than expected from theory. Nonetheless, the number of observed galaxies is still too limited to consider this as a serious concern.

Measuring the amount of stellar mass in the stellar halo is a strong test to probe the predictions from the Λ-Cold Dark Matter (ΛCDM) model in the context of galaxy formation. Ultimately, the amount of stars in this component of the galaxy informs us about the merging activity along the full history of the galaxy. Consequently, if the observed stellar haloes are less massive than the theoretical

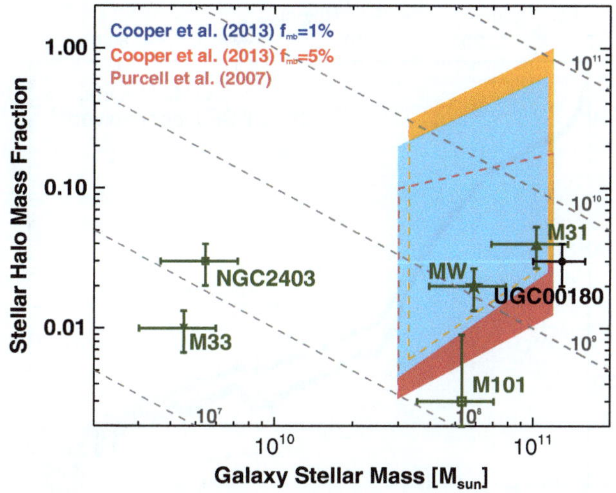

Fig. 8.6 Stellar halo mass fraction versus total galaxy stellar mass, illustrating the location of some nearby galaxies in this plane. In addition, the prediction from several numerical models are overplotted (Purcell et al. 2007; Cooper et al. 2013). The *dashed lines* are the locations of stellar haloes with fixed stellar mass (10^7, 10^8, 10^9, 10^{10} and 10^{11} $M_\odot$). Figure from Trujillo and Fliri (2016), reproduced by permission of the AAS

expectations, this might indicate that the number of accretion events was less than presumed. If this were in fact the case, we could be witnessing a problem for the ΛCDM scheme as relevant as the missing satellite problem (see Sect. 8.4.5).

8.4.4.1 PSF Effects on Imaging the Outermost Regions of Galaxies

As discussed in Sect. 8.2.5, one of the most important sources of light contamination in deep images is the contribution of scattered light from nearby sources surrounding the target galaxy. However, the galaxy itself is usually one of the most important sources of light contamination in the vicinity of the object. This is easy to understand. The surface brightness distribution of the galaxy is convolved with the PSF of the image. That means that a substantial amount of light coming from the central parts of the galaxy is distributed into its outermost region. Sandin (2014, 2015) has nicely illustrated this phenomenon, showing that the scattered light from the object (if not accounted for) can be wrongly interpreted as a bright stellar halo or a thick disk. This was also explored by de Jong (2008). Depending on the shape of the surface brightness distribution of the galaxy, the effect of its own scattered light can be more or less relevant in its outer region. Edge-on disks are normally most severely affected by this effect, whereas face-on galaxies without breaks or truncations are barely affected. As a general rule, the effect will be stronger in those cases where the light distribution is sharper. In Fig. 8.7, we show an example of the effect of the PSF on the light distribution of a spiral galaxy.

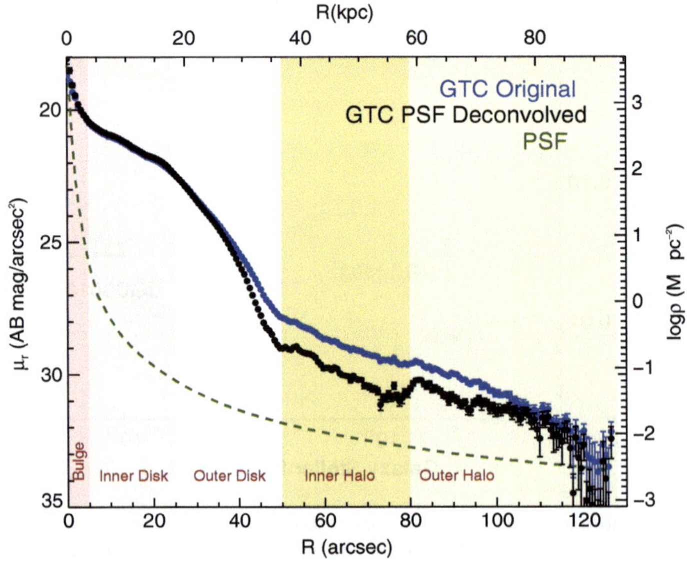

Fig. 8.7 The effect of the PSF on the surface brightness profile of the spiral galaxy UGC 00180. The observed profile (*upper one*) is shown using *blue dots*. The profile corrected for the effect of the PSF (*lower profile*) is illustrated with *black dots*. The *green dashed line* is the surface brightness profile of the PSF of the image. The different parts of the galaxy are indicated with labels. Figure from Trujillo and Fliri (2016), reproduced by permission of the AAS

As Fig. 8.7 illustrates, the effect of the PSF on the outer regions of the galaxy is dramatic. For this particular galaxy, the surface brightness profiles start to deviate at around 26 mag arcsec^{-2}. At $\mu_r \sim 26$ mag arcsec^{-2}, the effect is so important that the surface brightness profiles (affected by the PSF and corrected) are different by about 1 mag. This means that in the outer regions, the scattered light can be as much as three times brighter than the intrinsic light of the stellar halo. For this reason, it is absolutely necessary to model the effect of the PSF on the galaxy itself if one wants to explore the faintest regions of the galaxies.

There are two different approaches to handle the effect of the PSF on the surface brightness distribution of the galaxies. The first one is to apply a deconvolution method directly on the images. This has the advantage that it only requires an exquisitely characterized PSF of the image. However, its main disadvantage is that in the regions where the surface brightness is closest to the noise, the result is quite uncertain. The second approach is to model the intrinsic light distribution of the galaxy and convolve this with the PSF. The convolved model of the galaxy is then fitted to the observed light distribution until a good fit to the data is reached. Once this is achieved, a new image of the object (using the deconvolved model of the galaxy) is created, adding back the residuals of the fit. This is illustrated in Fig. 8.8.

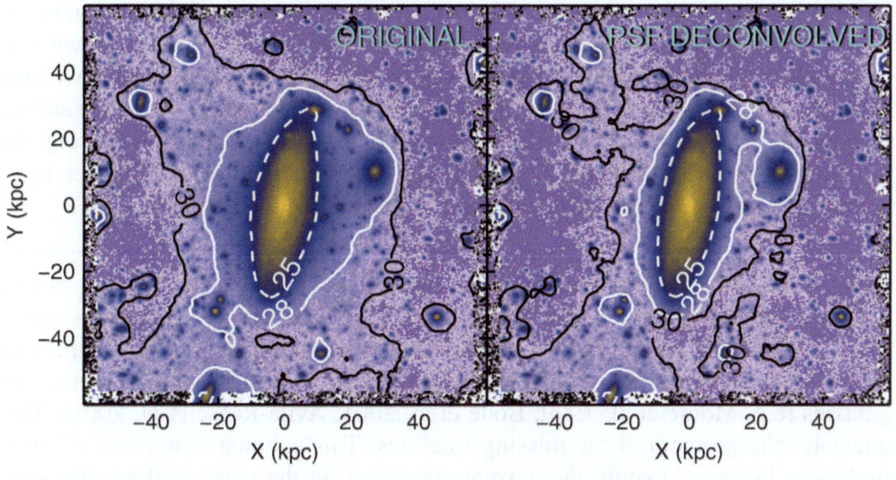

Fig. 8.8 The dramatic effect of the PSF on the surface brightness distribution of the spiral galaxy UGC 00180. Different surface brightness isophotes are indicated with *solid and dashed lines*. After accounting for the effect of the PSF, the amount of light in the outer regions decreases strongly. Figure from Trujillo and Fliri (2016), reproduced by permission of the AAS

Figure 8.8 shows that studies of both thick disks and outer stellar haloes are strongly affected by the PSF. Any deep imaging study of these components needs to account for this effect; otherwise, both the stellar haloes and the thick disks that will be inferred will result significantly brighter than what they really are.

8.4.5 Satellites

Satellites are probably one of the most important pieces to connect the realm of observed galaxies with cosmological modelling. In fact, counting the number of satellites and exploring their properties has become one of the most important tests to explore the predictions of the ΛCDM model on small (i.e. galactic) scales. The analysis of Local Group galaxies has revealed a number of problems related with satellite galaxies. These can be summarized as follows:

- The "missing satellites" problem. Cosmological simulations, in the framework of the ΛCDM model and based only on dark matter particles (i.e. without including the effect of baryons), predict thousands of small dark matter haloes orbiting galaxies like the Milky Way or Andromeda (e.g. Klypin et al. 1999; Moore et al. 1999). However, only a few tens of visible satellites have been observed around these objects (McConnachie 2012).

- The "too big to fail" problem. In the ΛCDM model, the largest satellite galaxies are expected to have a velocity dispersion substantially larger than what is actually measured in any of the dwarf galaxies in the Local Group (Read et al. 2006; Boylan-Kolchin et al. 2011). These big haloes among the satellite population should not have failed to produce dwarf galaxies following the prescription of baryonic physics, so if they exist, dwarf galaxies with such large central velocity dispersion should have been found observationally.

There are a number of solutions to the above cosmological problems at galactic scales. One possibility is that the main constituent of dark matter would not be a heavy (i.e. cold) particle but instead a lighter (i.e. warm) candidate (with a mass of around a keV). This hypothesis has been explored in many papers as it is able to suppress many of the dark matter haloes which should form dwarf galaxies (e.g. Moore et al. 1999; Bode et al. 2001; Avila-Reese et al. 2001). This alleviates the problem of the missing satellites. This solution is not free of other problems, however, mainly the current constraint on the mass of the warm dark matter candidate imposed by the study of the Lyman α forest (i.e. $\gtrsim$4 keV; Baur et al. 2016). This mass is large enough to mimic most of the properties of the ΛCDM scenario (including the production of a large number of dark matter subhaloes). Consequently, warm dark matter particle candidates with such minimum mass would not be of great help. Another alternative is to consider the baryon physics in detail when creating realistic cosmological simulations. It is worth stressing that the above problems arise when comparing simulations of dark matter alone (i.e. devoid of gas and stars) with the frequency and properties of satellites in the Local Group. However, some authors, like Navarro et al. (1996) or Read and Gilmore (2005) have pointed out the interplay between baryons and dark matter. In fact, they have suggested that the dark matter could be heated by impulsive gas mass loss driven by supernova explosions. If so, this could help to reconcile the cosmological simulations with the observations.

The above discrepancies between theory and observations are based on the comparison of the properties of the satellite galaxies of two massive disks, i.e. those of our own Galaxy and Andromeda. Consequently, a natural question arises: is our Local Group anomalous in producing a low number of satellites? To address this question in detail, we need to characterize the satellite population of many galaxies like ours beyond the Local Group. This idea has been pursued by several authors (see, e.g. Liu et al. 2011; Ruiz et al. 2015). Unfortunately, their analyses have only explored the population of the most massive satellites (i.e. down to masses $\sim 10^8$–$10^9\,M_\odot$). To test the discrepancies between the cosmological models and the observations, it is necessary to go at least two orders of magnitudes fainter (i.e. down to $10^6\,M_\odot$). This is where new ultra-deep imaging surveys can play a definitive role (see, e.g. Javanmardi et al. 2016). To explore satellites with masses around $10^6\,M_\odot$ within a sphere of radius $\sim$ 100 Mpc, we need to be able to reach a depth like the one achieved by Stripe 82. If we want to go further and explore a population of satellites with masses of a few times $10^4\,M_\odot$ (which is currently the mass limit of detected satellites in Andromeda), we would need a survey with the depth reached by the current deepest dataset (i.e. the one of Trujillo and Fliri 2016). A survey with such characteristics could be obtained at the end of the LSST programme (Ivezic et al. 2008; see Sect. 8.5).

8.5 Conclusions and Future Developments

In this chapter, we have reviewed how deep imaging is a fundamental tool in the study of the outermost structure of galaxies. Three main sources of imaging are currently used to detect and characterize the outskirts of galaxies: (1) surveys such as the Sloan Digital Sky Survey's Stripe 82 project; (2) very long exposures on small telescopes, including by amateurs; and (3) long exposures on the largest professional telescopes. The technical challenges in overcoming systematic effects are significant and range from the treatment of light scattered by the atmosphere and the telescope and instrument, via flat fielding, to the accurate subtraction of non-galaxy light in the images. We have reviewed recent results on galaxy disks and haloes obtained with deep imaging, including the detection and characterization of thick disks, truncations of stellar disks, tidal streams, stellar haloes and satellites. We have shown how each of these interrelated aspects of the faintest detectable structure in galaxies can shed light on the formation and subsequent evolution of the galaxies and our Universe.

The future is promising in terms of discovering the "low surface brightness Universe" through deep imaging. Current techniques using small and large ground-based telescopes, as reviewed in Sect. 8.3, will continue to be exploited while new facilities will become available. The most promising of these new facilities are poised to be the ground-based LSST and the new space telescopes *JWST* and *Euclid*. LSST (Ivezic et al. 2008) will repeatedly image the whole of the southern sky in six passbands (*ugrizy*), with a planned start date for surveys of around 2021. While each 30 s exposure with the 8.2 m telescope and optimized camera will yield a depth of 24.7 *r*-mag (5-sigma point source depth) over the 9.6 square degree field of view, combining all imaging obtained over the approximately 10 year survey duration could yield a depth of 27.5 *r*-mag (5-sigma point source depth). This roughly means as deep as the current Stripe 82 in one exposure and close to three magnitudes deeper over the whole survey duration. The scientific possibilities offered by this depth of imaging, over an area of sky of over 20,000 $\deg^2$, are tremendous. All the science discussed in this paper could essentially be done with one or a few exposures or easily with the data of say 1 h of observation. This obviously depends critically on whether systematic effects, including but not limited to those discussed in Sect. 8.2, can be properly controlled, modelled and corrected for. This is not trivial and may hinder the full exploitation of the data to their theoretical limits.

The *JWST*, to be launched in 2018, will allow deep imaging but, when compared to, e.g. the *HST*, won't be quite as revolutionary as LSST. The field of view will be limited to just over $2 \times 4\,\mathrm{arcmin}^2$ in the near-IR which will all but exclude deep imaging of the nearest galaxies. Where significant progress can be expected is in the deep imaging of galaxies at redshifts of, say, 0.2 and higher. In particular, near-IR imaging will allow the observation of galaxies at redshifts beyond 1 in rest-frame red passbands, necessary to reduce the effects of both young stellar populations and dust extinction.

Euclid will provide, among other data products, imaging in a very wide optical band ($R+I+Z$) over an area of $15{,}000\,\mathrm{deg}^2$ to a depth of 24.5 mag (10σ for extended source). Compared to LSST, advantages of *Euclid* imaging will be its higher spatial resolution (of $\sim$0.2 arcsec, due to the relatively small telescope aperture of 1.2 m) and better-behaved PSF, thanks to the absence of the Earth atmosphere. A disadvantage is that the visual imaging is done through a very wide filter which excludes the use of colour information. The depth of imaging will be comparable with LSST, though, for *Euclid*'s Deep Survey (over an area of around 40 square degree), which will allow the comparison of high-resolution *Euclid* imaging with the colour information obtained from the LSST images.

The area of ultra-deep imaging is still very much unexplored territory, and future work in this area will be stimulated by the availability of revolutionary new datasets and the continued understanding of the systematics affecting them. It will be a huge technical challenge to properly treat and analyse the upcoming deep imaging, in particular those from LSST and *Euclid*. As we enter previously unexplored territory with ultra-deep imaging, the systematic effects we know about will be challenging to characterize and correct for, and additional difficulties will almost certainly present themselves. But overcoming these issues will definitely pay off in terms of increased understanding of the formation and evolution processes which have led to the Universe and the galaxies as we observe them now. The future of imaging ultra-faint structures is very bright.

Acknowledgements JHK thanks Sébastien Comerón and Carme Gallart for comments on sections of the manuscript. IT has benefitted from multiple conversations with and the hard work of many members of his team. In particular, he thanks Jürgen Fliri, María Cebrián and Javier Román. JHK and IT acknowledge financial support from the Spanish Ministry of Economy and Competitiveness (MINECO) under grants number AYA2013-41243-P and AYA2013-48226-C3-1-P, respectively. JHK acknowledges financial support to the DAGAL network from the People Programme (Marie Curie Actions) of the European Union's Seventh Framework Programme FP7/2007-2013/ under REA grant agreement number PITN-GA-2011-289313.

References

Abadi, M.G., Navarro, J.F., Steinmetz, M., Eke, V.R.: Simulations of galaxy formation in a Λ cold dark matter universe. II. The fine structure of simulated galactic disks. Astrophys. J. **597**, 21–34 (2003). doi:10.1086/378316, astro-ph/0212282

Abazajian, K.N., Adelman-McCarthy, J.K., Agüeros, M.A., Allam, S.S., Allende Prieto, C., An, D., Anderson, K.S.J., Anderson, S.F., Annis, J., Bahcall, N.A., et al.: The seventh data release of the Sloan Digital Sky Survey. Astrophys. J. Suppl. Ser. **182**, 543–558 (2009). doi:10.1088/0067-0049/182/2/543, 0812.0649

Abraham, R., Merritt, A., Zhang, J., van Dokkum, P., Conroy, C., Danieli, S., Mowla, L.: Probing galactic outskirts with dragonfly. In: Gil de Paz, A., Knapen, J.H., Lee, J.C. (eds.) Formation and Evolution of Galaxy Outskirts. IAU Symposium, vol. 321, pp. 137–146 (2017). doi:10.1017/S1743921316012291. http://adsabs.harvard.edu/abs/2017IAUS..321..137A

Abraham, R.G., van Dokkum, P.G.: Ultra-low surface brightness imaging with the dragonfly telephoto array. Publ. Astron. Soc. Pac. **126**, 55–69 (2014). doi:10.1086/674875, 1401.5473

Abraham, R., van Dokkum, P., Conroy, C., Merritt, A., Zhang, J., Lokhorst, D., Danieli, S., Mowla, L.: Future prospects: deep imaging of galaxy outskirts using telescopes large and small. In: Knapen, J.H., Lee, J.C., Gil de Paz, A. (eds.) Outskirts of Galaxies, vol. 434. Springer, Cham (2017). doi: 10.1007/978-3-319-56570-5

Annis, J., Soares-Santos, M., Strauss, M.A., Becker, A.C., Dodelson, S., Fan, X., Gunn, J.E., Hao, J., Ivezić, Ž., Jester, S., Jiang, L., Johnston, D.E., Kubo, J.M., Lampeitl, H., Lin, H., Lupton, R.H., Miknaitis, G., Seo, H.J., Simet, M., Yanny, B.: The Sloan Digital Sky Survey coadd: 275 deg^2 of deep Sloan Digital Sky Survey imaging on stripe 82. Astrophys. J. **794**, 120 (2014). doi:10.1088/0004-637X/794/2/120, 1111.6619

Avila-Reese, V., Colín, P., Valenzuela, O., D'Onghia, E., Firmani, C.: Formation and structure of halos in a warm dark matter cosmology. Astrophys. J. **559**, 516–530 (2001). doi:10.1086/322411, astro-ph/0010525

Baur, J., Palanque-Delabrouille, N., Yèche, C., Magneville, C., Viel, M.: Lyman-alpha forests cool warm dark matter. J. Cosmol. Astropart. Phys. **8**, 012 (2016). doi:10.1088/1475-7516/2016/08/012, 1512.01981

Bell, E.F., Zucker, D.B., Belokurov, V., Sharma, S., Johnston, K.V., Bullock, J.S., Hogg, D.W., Jahnke, K., de Jong, J.T.A., Beers, T.C., Evans, N.W., Grebel, E.K., Ivezić, Ž., Koposov, S.E., Rix, H.W., Schneider, D.P., Steinmetz, M., Zolotov, A.: The accretion origin of the Milky Way's stellar halo. Astrophys. J. **680**, 295–311 (2008). doi:10.1086/588032, 0706.0004

Belokurov, V., Zucker, D.B., Evans, N.W., Gilmore, G., Vidrih, S., Bramich, D.M., Newberg, H.J., Wyse, R.F.G., Irwin, M.J., Fellhauer, M., Hewett, P.C., Walton, N.A., Wilkinson, M.I., Cole, N., Yanny, B., Rockosi, C.M., Beers, T.C., Bell, E.F., Brinkmann, J., Ivezić, Ž., Lupton, R.: The field of streams: sagittarius and its siblings. Astrophys. J. **642**, L137–L140 (2006). doi:10.1086/504797, astro-ph/0605025

Bensby, T., Feltzing, S., Oey, M.S.: Exploring the Milky Way stellar disk. A detailed elemental abundance study of 714 F and G dwarf stars in the solar neighbourhood. Astron. Astrophys. **562**, A71 (2014). doi:10.1051/0004-6361/201322631, 1309.2631

Bertin, E., Arnouts, S.: SExtractor: software for source extraction. Astron. Astrophys. **117**, 393–404 (1996). doi:10.1051/aas:1996164

Bode, P., Ostriker, J.P., Turok, N.: Halo formation in warm dark matter models. Astrophys. J. **556**, 93–107 (2001). doi:10.1086/321541, astro-ph/0010389

Bournaud, F., Elmegreen, B.G., Martig, M.: The thick disks of spiral galaxies as relics from gas-rich, turbulent, clumpy disks at high redshift. Astrophys. J. Lett. **707**, L1–L5 (2009). doi:10.1088/0004-637X/707/1/L1, 0910.3677

Bovy, J., Rix, H.W., Hogg, D.W.: The Milky Way has no distinct thick disk. Astrophys. J. **751**, 131 (2012). doi:10.1088/0004-637X/751/2/131, 1111.6585

Boylan-Kolchin, M., Bullock, J.S., Kaplinghat, M.: Too big to fail? The puzzling darkness of massive Milky Way subhaloes. Mon. Not. R. Astron. Soc. **415**, L40–L44 (2011). doi:10.1111/j.1745-3933.2011.01074.x, 1103.0007

Brook, C.B., Kawata, D., Gibson, B.K., Freeman, K.C.: The emergence of the thick disk in a cold dark matter universe. Astrophys. J. **612**, 894–899 (2004). doi:10.1086/422709, astro-ph/0405306

Buitrago, F., Trujillo, I., Curtis-Lake, E., Cooper, A.P., Bruce, V.A., Montes, M., Perez-Gonzalez, P.G., Cirasuolo, M.: The cosmic assembly of stellar haloes in massive early-type galaxies (2016). ArXiv e-prints 1602.01846

Bullock, J.S., Johnston, K.V.: Tracing galaxy formation with stellar halos. I. Methods. Astrophys. J. **635**, 931–949 (2005). doi:10.1086/497422, astro-ph/0506467

Burstein, D.: Structure and origin of S0 galaxies. III – the luminosity distribution perpendicular to the plane of the disks in S0's. Astrophys. J. **234**, 829–836 (1979). doi:10.1086/157563

Cappellari, M., Emsellem, E., Krajnović, D., McDermid, R.M., Scott, N., Verdoes Kleijn, G.A., Young, L.M., Alatalo, K., Bacon, R., Blitz, L., Bois, M., Bournaud, F., Bureau, M., Davies, R.L., Davis, T.A., de Zeeuw, P.T., Duc, P.A., Khochfar, S., Kuntschner, H., Lablanche, P.Y., Morganti, R., Naab, T., Oosterloo, T., Sarzi, M., Serra, P., Weijmans, A.M.: The ATLAS^{3D} project – I. A volume-limited sample of 260 nearby early-type galaxies: science goals and selection criteria. Mon. Not. R. Astron. Soc. **413**, 813–836 (2011). doi:10.1111/j.1365-2966.2010.18174.x, 1012.1551

Comerón, S., Elmegreen, B.G., Knapen, J.H., Salo, H., Laurikainen, E., Laine, J., Athanassoula, E., Bosma, A., Sheth, K., Regan, M.W., Hinz, J.L., Gil de Paz, A., Menéndez-Delmestre, K., Mizusawa, T., Muñoz-Mateos, J.C., Seibert, M., Kim, T., Elmegreen, D.M., Gadotti, D.A., Ho, L.C., Holwerda, B.W., Lappalainen, J., Schinnerer, E., Skibba, R.: Thick disks of edge-on galaxies seen through the Spitzer survey of stellar structure in galaxies (S^{4}G): lair of missing baryons? Astrophys. J. **741**, 28 (2011a). doi:10.1088/0004-637X/741/1/28, 1108.0037

Comerón, S., Elmegreen, B.G., Knapen, J.H., Sheth, K., Hinz, J.L., Regan, M.W., Gil de Paz, A., Muñoz-Mateos, J.C., Menéndez-Delmestre, K., Seibert, M., Kim, T., Mizusawa, T., Laurikainen, E., Salo, H., Laine, J., Athanassoula, E., Bosma, A., Buta, R.J., Gadotti, D.A., Ho, L.C., Holwerda, B., Schinnerer, E., Zaritsky, D.: The unusual vertical mass distribution of NGC 4013 seen through the Spitzer survey of stellar structure in galaxies (S^{4}G). Astrophys. J. Lett. **738**, L17 (2011b). doi:10.1088/2041-8205/738/2/L17, 1107.0529

Comerón, S., Elmegreen, B.G., Salo, H., Laurikainen, E., Athanassoula, E., Bosma, A., Knapen, J.H., Gadotti, D.A., Sheth, K., Hinz, J.L., Regan, M.W., Gil de Paz, A., Muñoz-Mateos, J.C., Menéndez-Delmestre, K., Seibert, M., Kim, T., Mizusawa, T., Laine, J., Ho, L.C., Holwerda, B.: Breaks in thin and thick disks of edge-on galaxies imaged in the Spitzer survey stellar structure in galaxies (S^{4}G). Astrophys. J. **759**, 98 (2012). doi:10.1088/hnX/759/2/98, 1209.1513

Comerón, S., Elmegreen, B.G., Salo, H., Laurikainen, E., Holwerda, B.W., Knapen, J.H.: Evidence for the concurrent growth of thick discs and central mass concentrations from S^{4}G imaging. Astron. Astrophys. **571**, A58 (2014). doi:10.1051/0004-6361/201424412, 1409.0466

Comerón, S., Salo, H., Janz, J., Laurikainen, E., Yoachim, P.: Galactic archaeology of a thick disc: excavating ESO 533-4 with VIMOS. Astron. Astrophys. **584**, A34 (2015). doi:10.1051/0004-6361/201526815, 1509.03841

Comerón, S., Salo, H., Peletier, R.F., Mentz, J.: A monolithic collapse origin for the thin and thick disc structure of the S0 galaxy ESO 243-49. Astron. Astrophys. **593**, L6 (2016). doi:10.1051/0004-6361/201629292, 1608.04238

Cooper, A.P., Cole, S., Frenk, C.S., White, S.D.M., Helly, J., Benson, A.J., De Lucia, G., Helmi, A., Jenkins, A., Navarro, J.F., Springel, V., Wang, J.: Galactic stellar haloes in the CDM model. Mon. Not. R. Astron. Soc. **406**, 744–766 (2010). doi:10.1111/j.1365-2966.2010.16740.x, 0910.3211

Cooper, A.P., D'Souza, R., Kauffmann, G., Wang, J., Boylan-Kolchin, M., Guo, Q., Frenk, C.S., White, S.D.M.: Galactic accretion and the outer structure of galaxies in the CDM model. Mon. Not. R. Astron. Soc. **434**, 3348–3367 (2013). doi:10.1093/mnras/stt1245, 1303.6283

Courteau, S., Widrow, L.M., McDonald, M., Guhathakurta, P., Gilbert, K.M., Zhu, Y., Beaton, R.L., Majewski, S.R.: The luminosity profile and structural parameters of the andromeda galaxy. Astrophys. J. **739**, 20 (2011). doi:10.1088/0004-637X/739/1/20, 1106.3564

Crnojević, D.: Resolved stellar populations as tracers of outskirts. In: Knapen, J.H., Lee, J.C., Gil de Paz, A. (eds.) Outskirts of Galaxies, vol. 434. Springer, Cham (2017). doi: 10.1007/978-3-319-56570-5

Cuillandre, J.C.J., Withington, K., Hudelot, P., Goranova, Y., McCracken, H., Magnard, F., Mellier, Y., Regnault, N., Bétoule, M., Aussel, H., Kavelaars, J.J., Fernique, P., Bonnarel, F., Ochsenbein, F., Ilbert, O.: Introduction to the CFHT legacy survey final release (CFHTLS T0007). In: Proceedings of SPIE Observatory Operations: Strategies, Processes, and Systems IV, vol. 8448, p. 84480M (2012). doi:10.1117/12.925584

Dalcanton, J.J., Bernstein, R.A.: A structural and dynamical study of late-type, edge-on galaxies. II. Vertical color gradients and the detection of ubiquitous thick disks. Astron. J. **124**, 1328–1359 (2002). doi:10.1086/342286, astro-ph/0207221

Dalcanton, J.J., Williams, B.F., Seth, A.C., Dolphin, A., Holtzman, J., Rosema, K., Skillman, E.D., Cole, A., Girardi, L., Gogarten, S.M., Karachentsev, I.D., Olsen, K., Weisz, D., Christensen, C., Freeman, K., Gilbert, K., Gallart, C., Harris, J., Hodge, P., de Jong, R.S., Karachentseva, V., Mateo, M., Stetson, P.B., Tavarez, M., Zaritsky, D., Governato, F., Quinn, T.: The ACS nearby galaxy survey treasury. Astrophys. J. Suppl. Ser. **183**, 67–108 (2009), doi:10.1088/0067-0049/183/1/67, 0905.3737

Davies, J.I., Wilson, C.D., Auld, R., Baes, M., Barlow, M.J., Bendo, G.J., Bock, J.J., Boselli, A., Bradford, M., Buat, V., Castro-Rodriguez, N., Chanial, P., Charlot, S., Ciesla, L., Clements, D.L., Cooray, A., Cormier, D., Cortese, L., Dwek, E., Eales, S.A., Elbaz, D., Galametz, M., Galliano, F., Gear, W.K., Glenn, J., Gomez, H.L., Griffin, M., Hony, S., Isaak, K.G., Levenson, L.R., Lu, N., Madden, S., O'Halloran, B., Okumura, K., Oliver, S., Page, M.J., Panuzzo, P., Papageorgiou, A., Parkin, T.J., Perez-Fournon, I., Pohlen, M., Rangwala, N., Rigby, E.E., Roussel, H., Rykala, A., Sacchi, N., Sauvage, M., Schulz, B., Schirm, M.R.P., Smith, M.W.L., Spinoglio, L., Stevens, J.A., Srinivasan, S., Symeonidis, M., Trichas, M., Vaccari, M., Vigroux, L., Wozniak, H., Wright, G.S., Zeilinger, W.W.: On the origin of M81 group extended dust emission. Mon. Not. R. Astron. Soc. **409**, 102–108 (2010). doi:10.1111/j.1365-2966.2010.17774.x, 1010.4770

de Jong, R.S.: Point spread function tails and the measurements of diffuse stellar halo light around edge-on disc galaxies. Mon. Not. R. Astron. Soc. **388**, 1521–1527 (2008). doi:10.1111/j.1365-2966.2008.13505.x, 0807.0229

Debattista, V.P., Roškar, R., Loebman, S.R.: The impact of stellar migration on disk outskirts. In: Knapen, J.H., Lee, J.C., Gil de Paz, A. (eds.) Outskirts of Galaxies, vol. 434. Springer, Cham (2017). doi: 10.1007/978-3-319-56570-5

Duc, P.A., Cuillandre, J.C., Karabal, E., Cappellari, M., Alatalo, K., Blitz, L., Bournaud, F., Bureau, M., Crocker, A.F., Davies, R.L., Davis, T.A., de Zeeuw, P.T., Emsellem, E., Khochfar, S., Krajnović, D., Kuntschner, H., McDermid, R.M., Michel-Dansac, L., Morganti, R., Naab, T., Oosterloo, T., Paudel, S., Sarzi, M., Scott, N., Serra, P., Weijmans, A.M., Young, L.M.: The ATLAS^{3D} project – XXIX. The new look of early-type galaxies and surrounding fields disclosed by extremely deep optical images. Mon. Not. R. Astron. Soc. **446**, 120–143 (2015). doi:10.1093/mnras/stu2019, 1410.0981

Durrell, P.R., Côté, P., Peng, E.W., Blakeslee, J.P., Ferrarese, L., Mihos, J.C., Puzia, T.H., Lançon, A., Liu, C., Zhang, H., Cuillandre, J.C., McConnachie, A., Jordán, A., Accetta, K., Boissier, S., Boselli, A., Courteau, S., Duc, P.A., Emsellem, E., Gwyn, S., Mei, S., Taylor, J.E.: The next generation virgo cluster survey. VIII. The spatial distribution of globular clusters in the virgo cluster. Astrophys. J. **794**, 103 (2014). doi:10.1088/0004-637X/794/2/103, 1408.2821

Elmegreen, B.G., Elmegreen, D.M.: Observations of thick disks in the hubble space telescope ultra deep field. Astrophys. J. **650**, 644–660 (2006). doi:10.1086/507578, astro-ph/0607540

Elmegreen, B.G., Hunter, D.A.: Outskirts of nearby disk galaxies: star formation and stellar populations. In: Knapen, J.H., Lee, J.C., Gil de Paz, A. (eds.) Outskirts of Galaxies, vol. 434. Springer, Cham (2017). doi: 10.1007/978-3-319-56570-5

Ferguson, A.M.N., Mackey, A.D.: Substructure and tidal streams in the andromeda galaxy and its satellites. In: Newberg, H.J., Carlin, J.L. (eds.) Astrophysics and Space Science Library, Astrophysics and Space Science Library, vol. 420, p. 191 (2016). doi:10.1007/978-3-319-19336-6_8, 1603.01993

Ferrarese, L., Côté, P., Cuillandre, J.C., Gwyn, S.D.J., Peng, E.W., MacArthur, L.A., Duc, P.A., Boselli, A., Mei, S., Erben, T., McConnachie, A.W., Durrell, P.R., Mihos, J.C., Jordán, A., Lançon, A., Puzia, T.H., Emsellem, E., Balogh, M.L., Blakeslee, J.P., van Waerbeke, L., Gavazzi, R., Vollmer, B., Kavelaars, J.J., Woods, D., Ball, N.M., Boissier, S., Courteau, S., Ferriere, E., Gavazzi, G., Hildebrandt, H., Hudelot, P., Huertas-Company, M., Liu, C., McLaughlin, D., Mellier, Y., Milkeraitis, M., Schade, D., Balkowski, C., Bournaud, F., Carlberg, R.G., Chapman, S.C., Hoekstra, H., Peng, C., Sawicki, M., Simard, L., Taylor, J.E., Tully, R.B., van Driel, W., Wilson, C.D., Burdullis, T., Mahoney, B., Manset, N.: The next generation virgo cluster survey (NGVS). I. Introduction to the survey. Astrophys. J. Suppl. Ser. **200**, 4 (2012). doi:10.1088/0067-0049/200/1/4

Fliri, J., Trujillo, I.: The IAC stripe 82 legacy project: a wide-area survey for faint surface brightness astronomy. Mon. Not. R. Astron. Soc. **456**, 1359–1373 (2016). doi:10.1093/mnras/stv2686, 1603.04474

Font, A.S., McCarthy, I.G., Crain, R.A., Theuns, T., Schaye, J., Wiersma, R.P.C., Dalla Vecchia, C.: Cosmological simulations of the formation of the stellar haloes around disc galaxies. Mon. Not. R. Astron. Soc. **416**, 2802–2820 (2011). doi:10.1111/j.1365-2966.2011.19227.x, 1102.2526

Freeman, K.C.: On the disks of spiral and S0 galaxies. Astrophys. J. **160**, 811 (1970). doi:10.1086/150474

Galaz, G., Milovic, C., Suc, V., Busta, L., Lizana, G., Infante, L., Royo, S.: Deep optical images of Malin 1 reveal new features. Astrophys. J. Lett. **815**, L29 (2015). doi:10.1088/2041-8205/815/2/L29, 1512.01095

Gallart, C., Monelli, M., Mayer, L., Aparicio, A., Battaglia, G., Bernard, E.J., Cassisi, S., Cole, A.A., Dolphin, A.E., Drozdovsky, I., Hidalgo, S.L., Navarro, J.F., Salvadori, S., Skillman, E.D., Stetson, P.B., Weisz, D.R.: The ACS LCID project: on the origin of dwarf galaxy types–a manifestation of the halo assembly bias? Astrophys. J. Lett. **811**, L18 (2015). doi:10.1088/2041-8205/811/2/L18, 1507.08350

Gilbert, K.M., Guhathakurta, P., Beaton, R.L., Bullock, J., Geha, M.C., Kalirai, J.S., Kirby, E.N., Majewski, S.R., Ostheimer, J.C., Patterson, R.J., Tollerud, E.J., Tanaka, M., Chiba, M.: Global properties of M31's stellar halo from the SPLASH survey. I. Surface brightness profile. Astrophys. J. **760**, 76 (2012). doi:10.1088/0004-637X/760/1/76, 1210.3362

Gilmore, G., Reid, N.: New light on faint stars. III – galactic structure towards the south pole and the galactic thick disc. Mon. Not. R. Astron. Soc. **202**, 1025–1047 (1983). doi:10.1093/mnras/202.4.1025

Gilmore, G., Wyse, R.F.G.: The abundance distribution in the inner spheroid. Astron. J. **90**, 2015–2026 (1985). doi:10.1086/113907

Guérou, A., Emsellem, E., Krajnović, D., McDermid, R.M., Contini, T., Weilbacher, P.M.: Exploring the mass assembly of the early-type disc galaxy NGC 3115 with MUSE. Astron. Astrophys. **591**, A143 (2016). doi:10.1051/0004-6361/201628743, 1605.07667

Gunn, J.E., Carr, M., Rockosi, C., Sekiguchi, M., Berry, K., Elms, B., de Haas, E., Ivezić, Ž., Knapp, G., Lupton, R., Pauls, G., Simcoe, R., Hirsch, R., Sanford, D., Wang, S., York, D., Harris, F., Annis, J., Bartozek, L., Boroski, W., Bakken, J., Haldeman, M., Kent, S., Holm, S., Holmgren, D., Petravick, D., Prosapio, A., Rechenmacher, R., Doi, M., Fukugita, M., Shimasaku, K., Okada, N., Hull, C., Siegmund, W., Mannery, E., Blouke, M., Heidtman, D., Schneider, D., Lucinio, R., Brinkman, J.: The sloan digital sky survey photometric camera. Astron. J. **116**, 3040–3081 (1998). doi:10.1086/300645, astro-ph/9809085

Haywood, M., Di Matteo, P., Lehnert, M.D., Katz, D., Gómez, A.: The age structure of stellar populations in the solar vicinity. Clues of a two-phase formation history of the Milky Way disk. Astron. Astrophys. **560**, A109 (2013). doi:10.1051/0004-6361/201321397, 1305.4663

Howell, S.B.: Handbook of CCD Astronomy. Cambridge Observing Handbooks for Research Astronomers, vol. 5, 2nd edn. Cambridge University Press, Cambridge (2006). 2006 ISBN 0521852153

Ibata, R., Irwin, M., Lewis, G., Ferguson, A.M.N., Tanvir, N.: A giant stream of metal-rich stars in the halo of the galaxy M31. Nature **412**, 49–52 (2001). astro-ph/0107090

Ibata, R., Martin, N.F., Irwin, M., Chapman, S., Ferguson, A.M.N., Lewis, G.F., McConnachie, A.W.: The haunted halos of andromeda and triangulum: a panorama of galaxy formation in action. Astrophys. J. **671**, 1591–1623 (2007). doi:10.1086/522574, 0704.1318

Irwin, M.J., Ferguson, A.M.N., Ibata, R.A., Lewis, G.F., Tanvir, N.R.: A minor-axis surface brightness profile for M31. Astrophys. J. Lett. **628**, L105–L108 (2005). doi:10.1086/432718, astro-ph/0505077

Ivezic, Z., Axelrod, T., Brandt, W.N., Burke, D.L., Claver, C.F., Connolly, A., Cook, K.H., Gee, P., Gilmore, D.K., Jacoby, S.H., Jones, R.L., Kahn, S.M., Kantor, J.P., Krabbendam, V.V., Lupton, R.H., Monet, D.G., Pinto, P.A., Saha, A., Schalk, T.L., Schneider, D.P., Strauss, M.A., Stubbs, C.W., Sweeney, D., Szalay, A., Thaler, J.J., Tyson, J.A., LSST Collaboration: Large synoptic survey telescope: from science drivers to reference design. Serbian Astron. J. **176**, 1–13 (2008). doi:10.2298/SAJ0876001I

Jablonka, P., Tafelmeyer, M., Courbin, F., Ferguson, A.M.N.: Direct detection of galaxy stellar halos: NGC 3957 as a test case. Astron. Astrophys. **513**, A78 (2010). doi:10.1051/0004-6361/200913320, 1001.3067

Javanmardi, B., Martinez-Delgado, D., Kroupa, P., Henkel, C., Crawford, K., Teuwen, K., Gabany, R.J., Hanson, M., Chonis, T.S., Neyer, F.: DGSAT: Dwarf galaxy survey with amateur telescopes. I. Discovery of low surface brightness systems around nearby spiral galaxies. Astron. Astrophys. **588**, A89 (2016). doi:10.1051/0004-6361/201527745, 1511.04446

Jiang, L., Fan, X., Annis, J., Becker, R.H., White, R.L., Chiu, K., Lin, H., Lupton, R.H., Richards, G.T., Strauss, M.A., Jester, S., Schneider, D.P.: A survey of $z \sim 6$ quasars in the Sloan digital sky survey deep stripe. I. A flux-limited sample at $z_{AB} < 21$. Astron. J. **135**, 1057–1066 (2008). doi:10.1088/0004-6256/135/3/1057, 0708.2578

Jiang, L., Fan, X., Bian, F., McGreer, I.D., Strauss, M.A., Annis, J., Buck, Z., Green, R., Hodge, J.A., Myers, A.D., Rafiee, A., Richards, G.: The Sloan digital sky survey stripe 82 imaging data: depth-optimized co-adds over 300 deg^2 in five filters. Astrophys. J. Suppl. Ser. **213**, 12 (2014). doi:10.1088/0067-0049/213/1/12, 1405.7382

Johnston, K.V., Bullock, J.S., Sharma, S., Font, A., Robertson, B.E., Leitner, S.N.: Tracing galaxy formation with stellar halos. II. Relating substructure in phase and abundance space to accretion histories. Astrophys. J. **689**, 936–957 (2008). doi:10.1086/592228, 0807.3911

Kasparova, A.V., Katkov, I.Y., Chilingarian, I.V., Silchenko, O.K., Moiseev, A.V., Borisov, S.B.: The diversity of thick galactic discs. Mon. Not. R. Astron. Soc. **460**, L89–L93 (2016). doi:10.1093/mnrasl/slw083, 1604.07624

Kawata, D., Chiappini, C.: Milky Way's thick and thin disk: is there a distinct thick disk? Astron. Nachr. **337**, 976 (2016). doi:10.1002/asna.201612421, 1608.01698

Klypin, A., Kravtsov, A.V., Valenzuela, O., Prada, F.: Where are the missing galactic satellites? Astrophys. J. **522**, 82–92 (1999). doi:10.1086/307643, astro-ph/9901240

Kregel, M., van der Kruit, P.C., de Grijs, R.: Flattening and truncation of stellar discs in edge-on spiral galaxies. Mon. Not. R. Astron. Soc. **334**, 646–668 (2002). doi:10.1046/j.1365-8711.2002.05556.x, astro-ph/0204154

Liu, L., Gerke, B.F., Wechsler, R.H., Behroozi, P.S., Busha, M.T.: How common are the magellanic clouds? Astrophys. J. **733**, 62 (2011). doi:10.1088/0004-637X/733/1/62, 1011.2255

Majewski, S.R., Skrutskie, M.F., Weinberg, M.D., Ostheimer, J.C.: A two micron all sky survey view of the sagittarius dwarf galaxy. I. Morphology of the sagittarius core and tidal arms. Astrophys. J. **599**, 1082–1115 (2003). doi:10.1086/379504, astro-ph/0304198

Martín-Navarro, I., Bakos, J., Trujillo, I., Knapen, J.H., Athanassoula, E., Bosma, A., Comerón, S., Elmegreen, B.G., Erroz-Ferrer, S., Gadotti, D.A., Gil de Paz, A., Hinz, J.L., Ho, L.C., Holwerda, B.W., Kim, T., Laine, J., Laurikainen, E., Menéndez-Delmestre, K., Mizusawa, T., Muñoz-Mateos, J.C., Regan, M.W., Salo, H., Seibert, M., Sheth, K.: A unified picture of breaks and truncations in spiral galaxies from SDSS and S^4G imaging. Mon. Not. R. Astron. Soc. **427**, 1102–1134 (2012). doi:10.1111/j.1365–2966.2012.21929.x, 1208.2893

Martín-Navarro, I., Trujillo, I., Knapen, J.H., Bakos, J., Fliri, J.: Stellar haloes outshine disc truncations in low-inclined spirals. Mon. Not. R. Astron. Soc. **441**, 2809–2814 (2014). doi:10.1093/mnras/stu767, 1401.3749

Martínez-Delgado, D., Peñarrubia, J., Gabany, R.J., Trujillo, I., Majewski, S.R., Pohlen, M.: The ghost of a dwarf galaxy: fossils of the hierarchical formation of the nearby spiral galaxy NGC 5907. Astrophys. J. **689**, 184–193 (2008). doi:10.1086/592555, 0805.1137

Martínez-Delgado, D., Gabany, R.J., Crawford, K., Zibetti, S., Majewski, S.R., Rix, H.W., Fliri, J., Carballo-Bello, J.A., Bardalez-Gagliuffi, D.C., Peñarrubia, J., Chonis, T.S., Madore, B., Trujillo, I., Schirmer, M., McDavid, D.A.: Stellar tidal streams in spiral galaxies of the local volume: a pilot survey with modest aperture telescopes. Astron. J. **140**, 962–967 (2010). doi:10.1088/0004-6256/140/4/962, 1003.4860

Martínez-Delgado, D., D'Onghia, E., Chonis, T.S., Beaton, R.L., Teuwen, K., GaBany, R.J., Grebel, E.K., Morales, G.: A stellar tidal stream around the whale galaxy, NGC 4631. Astron. J. **150**, 116 (2015). doi:10.1088/0004-6256/150/4/116, 1410.6368

McConnachie, A.W.: The observed properties of dwarf galaxies in and around the local group. Astron. J. **144**, 4 (2012). doi:10.1088/0004-6256/144/1/4, 1204.1562

McConnachie, A.W., Irwin, M.J., Ibata, R.A., Dubinski, J., Widrow, L.M., Martin, N.F., Côté, P., Dotter, A.L., Navarro, J.F., Ferguson, A.M.N., Puzia, T.H., Lewis, G.F., Babul, A., Barmby, P., Bienaymé, O., Chapman, S.C., Cockcroft, R., Collins, M.L.M., Fardal, M.A., Harris, W.E., Huxor, A., Mackey, A.D., Peñarrubia, J., Rich, R.M., Richer, H.B., Siebert, A., Tanvir, N., Valls-Gabaud, D., Venn, K.A.: The remnants of galaxy formation from a panoramic survey of the region around M31. Nature **461**, 66–69 (2009). doi:10.1038/nature08327, 0909.0398

McGraw, J.T., Angel, J.R.P., Sargent, T.A.: A charge-coupled device/CCD/transit-telescope survey for galactic and extragalactic variability and polarization. In: Elliott, D.A. (ed.) Proceedings of SPIE Conference on Applications of Digital Image Processing to Astronomy, vol. 264, pp. 20–28 (1980)

Merritt, A., van Dokkum, P., Abraham, R., Zhang, J.: The dragonfly nearby Galaxies survey. I. Substantial variation in the diffuse stellar halos around spiral Galaxies. Astrophys. J. **830**, 62 (2016). doi:10.3847/0004-637X/830/2/62, 1606.08847

Mihos, J.C., Harding, P., Feldmeier, J., Morrison, H.: Diffuse light in the virgo cluster. Astrophys. J. Lett. **631**, L41–L44 (2005). doi:10.1086/497030, astro-ph/0508217

Mihos, J.C., Harding, P., Spengler, C.E., Rudick, C.S., Feldmeier, J.J.: The extended optical disk of M101. Astrophys. J. **762**, 82 (2013). doi:10.1088/0004-637X/762/2/82, 1211.3095

Mihos, J.C., Durrell, P.R., Ferrarese, L., Feldmeier, J.J., Côté, P., Peng, E.W., Harding, P., Liu, C., Gwyn, S., Cuillandre, J.C.: Galaxies at the extremes: ultra-diffuse Galaxies in the virgo cluster. Astrophys. J. Lett. **809**, L21 (2015). doi:10.1088/2041-8205/809/2/L21, 1507.02270

Mihos, J.C., Harding, P., Feldmeier, J.J., Rudick, C., Janowiecki, S., Morrison, H., Slater, C., Watkins, A.: The Burrell schmidt deep virgo survey: tidal debris, galaxy halos, and diffuse intracluster light in the virgo cluster (2016). ArXiv e-prints 1611.04435

Monachesi, A., Bell, E.F., Radburn-Smith, D.J., Bailin, J., de Jong, R.S., Holwerda, B., Streich, D., Silverstein, G.: The GHOSTS survey – II. The diversity of halo colour and metallicity profiles of massive disc galaxies. Mon. Not. R. Astron. Soc. **457**, 1419–1446 (2016). doi:10.1093/mnras/stv2987, 1507.06657

Moore, B., Ghigna, S., Governato, F., Lake, G., Quinn, T., Stadel, J., Tozzi, P.: Dark matter substructure within galactic halos. Astrophys. J. Lett. **524**, L19–L22 (1999). doi:10.1086/312287, astro-ph/9907411

Mould, J.: Red thick disks of nearby galaxies. Astron. J. **129**, 698–711 (2005). doi:10.1086/427248, astro-ph/0411231

Navarro, J.F., Eke, V.R., Frenk, C.S.: The cores of dwarf galaxy haloes. Mon. Not. R. Astron. Soc. **283**, L72–L78 (1996). doi:10.1093/mnras/283.3.L72, astro-ph/9610187

Peters, S.P.C., van der Kruit, P.C., Knapen, J.H., Trujillo, I., Fliri, J., Cisternas, M., Kelvin, L.S.: Stellar disc truncations and extended haloes in face-on spiral galaxies. Mon. Not. R. Astron. Soc. (2017, in press). arXiv:1705.03555

Pillepich, A., Vogelsberger, M., Deason, A., Rodriguez-Gomez, V., Genel, S., Nelson, D., Torrey, P., Sales, L.V., Marinacci, F., Springel, V., Sijacki, D., Hernquist, L.: Halo mass and assembly history exposed in the faint outskirts: the stellar and dark matter haloes of Illustris galaxies. Mon. Not. R. Astron. Soc. **444**, 237–249 (2014). doi:10.1093/mnras/stu1408, 1406.1174

Pohlen, M., Trujillo, I.: The structure of galactic disks. Studying late-type spiral galaxies using SDSS. Astron. Astrophys. **454**, 759–772 (2006). doi:10.1051/0004-6361:20064883, astro-ph/0603682

Purcell, C.W., Bullock, J.S., Zentner, A.R.: Shredded Galaxies as the source of diffuse intrahalo light on varying scales. Astrophys. J. **666**, 20–33 (2007). doi:10.1086/519787, astro-ph/0703004

Quinn, P.J., Hernquist, L., Fullagar, D.P.: Heating of galactic disks by mergers. Astrophys. J. **403**, 74–93 (1993). doi:10.1086/172184

Radburn-Smith, D.J., de Jong, R.S., Seth, A.C., Bailin, J., Bell, E.F., Brown, T.M., Bullock, J.S., Courteau, S., Dalcanton, J.J., Ferguson, H.C., Goudfrooij, P., Holfeltz, S., Holwerda, B.W., Purcell, C., Sick, J., Streich, D., Vlajic, M., Zucker, D.B.: The GHOSTS survey. I. Hubble space telescope advanced camera for surveys data. Astrophys. J. Suppl. Ser. **195**, 18 (2011). doi:10.1088/0067-0049/195/2/18

Read, J.I., Gilmore, G.: Mass loss from dwarf spheroidal galaxies: the origins of shallow dark matter cores and exponential surface brightness profiles. Mon. Not. R. Astron. Soc. **356**, 107–124 (2005). doi:10.1111/j.1365-2966.2004.08424.x, astro-ph/0409565

Read, J.I., Wilkinson, M.I., Evans, N.W., Gilmore, G., Kleyna, J.T.: The importance of tides for the local group dwarf spheroidals. Mon. Not. R. Astron. Soc. **367**, 387–399 (2006). doi:10.1111/j.1365-2966.2005.09959.x, astro-ph/0511759

Rejkuba, M., Mouhcine, M., Ibata, R.: The stellar population content of the thick disc and halo of the Milky Way analogue NGC 891. Mon. Not. R. Astron. Soc. **396**, 1231–1246 (2009). doi:10.1111/j.1365-2966.2009.14821.x, 0903.4211

Rudick, C.S., Mihos, J.C., Harding, P., Feldmeier, J.J., Janowiecki, S., Morrison, H.L.: Optical colors of intracluster light in the virgo cluster core. Astrophys. J. **720**, 569–580 (2010). doi:10.1088/0004-637X/720/1/569, 1003.4500

Ruiz, P., Trujillo, I., Mármol-Queraltó, E.: The abundance of satellites depends strongly on the morphology of the host galaxy. Mon. Not. R. Astron. Soc. **454**, 1605–1619 (2015). doi:10.1093/mnras/stv2030, 1504.02777

Samland, M., Gerhard, O.E.: The formation of a disk galaxy within a growing dark halo. Astron. Astrophys. **399**, 961–982 (2003). doi:10.1051/0004-6361:20021842, astro-ph/0301499

Sandin, C.: The influence of diffuse scattered light. I. The PSF and its role in observations of the edge-on galaxy NGC 5907. Astron. Astrophys. **567**, A97 (2014). doi:10.1051/0004-6361/201423429, 1406.5508

Sandin, C.: The influence of diffuse scattered light. II. Observations of galaxy haloes and thick discs and hosts of blue compact galaxies. Astron. Astrophys. **577**, A106 (2015). doi:10.1051/0004-6361/201425168, 1502.07244

Schönrich, R., Binney, J.: Chemical evolution with radial mixing. Mon. Not. R. Astron. Soc. **396**, 203–222 (2009). doi:10.1111/j.1365-2966.2009.14750.x, 0809.3006

Seth, A.C., Dalcanton, J.J., de Jong, R.S.: A study of edge-on galaxies with the hubble space telescope advanced camera for surveys. II. Vertical distribution of the resolved stellar population. Astron. J. **130**, 1574–1592 (2005). doi:10.1086/444620, astro-ph/0506117

Sheth, K., Regan, M., Hinz, J.L., Gil de Paz, A., Menéndez-Delmestre, K., Muñoz-Mateos, J.C., Seibert, M., Kim, T., Laurikainen, E., Salo, H., Gadotti, D.A., Laine, J., Mizusawa, T., Armus, L., Athanassoula, E., Bosma, A., Buta, R.J., Capak, P., Jarrett, T.H., Elmegreen, D.M., Elmegreen, B.G., Knapen, J.H., Koda, J., Helou, G., Ho, L.C., Madore, B.F., Masters, K.L.,

Mobasher, B., Ogle, P., Peng, C.Y., Schinnerer, E., Surace, J.A., Zaritsky, D., Comerón, S., de Swardt, B., Meidt, S.E., Kasliwal, M., Aravena, M.: The Spitzer survey of stellar structure in Galaxies (S^4G). Publ. Astron. Soc. Pac. **122**, 1397–1414 (2010). doi:10.1086/657638, 1010.1592

Slater, C.T., Harding, P., Mihos, J.C.: Removing internal reflections from deep imaging data sets. Publ. Astron. Soc. Pac. **121**, 1267–1278 (2009). doi:10.1086/648457, 0909.3320

Streich, D., de Jong, R.S., Bailin, J., Bell, E.F., Holwerda, B.W., Minchev, I., Monachesi, A., Radburn-Smith, D.J.: Extragalactic archeology with the GHOSTS Survey. I. Age-resolved disk structure of nearby low-mass galaxies. Astron. Astrophys. **585**, A97 (2016). doi:10.1051/0004-6361/201526013, 1509.06647

Tanaka, M., Chiba, M., Komiyama, Y., Guhathakurta, P., Kalirai, J.S., Iye, M.: Structure and population of the andromeda stellar halo from a subaru/suprime-cam survey. Astrophys. J. **708**, 1168–1203 (2010). doi:10.1088/0004-637X/708/2/1168, 0908.0245

Tikhonov, N.A., Galazutdinova, O.A.: Stellar disks and halos of edge-on spiral galaxies: NGC 891, NGC 4144, and NGC 4244. Astrophysics **48**, 221–236 (2005). doi:10.1007/s10511-005-0021-8, astro-ph/0503235

Tissera, P.B., Beers, T.C., Carollo, D., Scannapieco, C.: Stellar haloes in Milky Way mass galaxies: from the inner to the outer haloes. Mon. Not. R. Astron. Soc. **439**, 3128–3138 (2014). doi:10.1093/mnras/stu181, 1309.3609

Trujillo, I., Fliri, J.: Beyond 31 mag arcsec^2: The frontier of low surface brightness imaging with the largest optical telescopes. Astrophys. J. **823**, 123 (2016). doi:10.3847/0004-637X/823/2/123, 1510.04696

Tsikoudi, V.: Photometry and structure of lenticular galaxies. I – NGC 3115. Astrophys. J. **234**, 842–853 (1979). doi:10.1086/157565

van der Kruit, P.C.: Optical surface photometry of eight spiral galaxies studied in Westerbork. Astr. Astrophys. Suppl. **38**, 15–38 (1979)

van der Kruit, P.C., Freeman, K.C.: Galaxy disks. Ann. Rev. Astron. Astrophys. **49**, 301–371 (2011). doi:10.1146/annurev-astro-083109-153241, 1101.1771

van der Kruit, P.C., Searle, L.: Surface photometry of edge-on spiral galaxies. I – a model for the three-dimensional distribution of light in galactic disks. Astron. Astrophys. **95**, 105–115 (1981)

van Dokkum, P.G., Abraham, R., Merritt, A.: First results from the dragonfly telephoto array: the apparent lack of a stellar halo in the massive spiral galaxy M101. Astrophys. J. Lett. **782**, L24 (2014). doi:10.1088/2041-8205/782/2/L24, 1401.5467

van Dokkum, P., Abraham, R., Brodie, J., Conroy, C., Danieli, S., Merritt, A., Mowla, L., Romanowsky, A., Zhang, J.: A high stellar velocity dispersion and $\sim$100 globular clusters for the ultra-diffuse Galaxy dragonfly 44. Astrophys. J. Lett. **828**, L6 (2016). doi:10.3847/2041-8205/828/1/L6, 1606.06291

Villumsen, J.V.: Evolution of the velocity distribution in galactic disks. Astrophys. J. **290**, 75–85 (1985). doi:10.1086/162960

Watkins, A.E., Mihos, J.C., Harding, P., Feldmeier, J.J.: Searching for diffuse light in the M96 Galaxy group. Astrophys. J. **791**, 38 (2014). doi:10.1088/0004-637X/791/1/38, 1406.6982

Wright, J.F., Mackay, C.D.: The Cambridge charge-coupled device/CCD/system. In: Proceedings of SPIE Society of Photo-Optical Instrumentation Engineers (SPIE) Conference Series, vol. 290, p. 160 (1981)

Yoachim, P., Dalcanton, J.J.: Structural parameters of thin and thick disks in edge-on disk Galaxies. Astron. J. **131**, 226–249 (2006). doi:10.1086/497970, astro-ph/0508460

York, D.G., Adelman, J., Anderson, J.E. Jr., Anderson, S.F., Annis, J., Bahcall, N.A., Bakken, J.A., Barkhouser, R., Bastian, S., Berman, E., Boroski, W.N., Bracker, S., Briegel, C., Briggs, J.W., Brinkmann, J., Brunner, R., Burles, S., Carey, L., Carr, M.A., Castander, F.J., Chen, B., Colestock, P.L., Connolly, A.J., Crocker, J.H., Csabai, I., Czarapata, P.C., Davis, J.E., Doi, M., Dombeck, T., Eisenstein, D., Ellman, N., Elms, B.R., Evans, M.L., Fan, X., Federwitz, G.R., Fiscelli, L., Friedman, S., Frieman, J.A., Fukugita, M., Gillespie, B., Gunn, J.E., Gurbani, V.K., de Haas, E., Haldeman, M., Harris, F.H., Hayes, J., Heckman, T.M., Hennessy, G.S., Hindsley, R.B., Holm, S., Holmgren, D.J., Huang, C.H., Hull, C., Husby, D., Ichikawa, S.I., Ichikawa,

T., Ivezić, Ž., Kent, S., Kim, R.S.J., Kinney, E., Klaene, M., Kleinman, A.N., Kleinman, S., Knapp, G.R., Korienek, J., Kron, R.G., Kunszt, P.Z., Lamb, D.Q., Lee, B., Leger, R.F., Limmongkol, S., Lindenmeyer, C., Long, D.C., Loomis, C., Loveday, J., Lucinio, R., Lupton, R.H., MacKinnon, B., Mannery, E.J., Mantsch, P.M., Margon, B., McGehee, P., McKay, T.A., Meiksin, A., Merelli, A., Monet, D.G., Munn, J.A., Narayanan, V.K., Nash, T., Neilsen, E., Neswold, R., Newberg, H.J., Nichol, R.C., Nicinski, T., Nonino, M., Okada, N., Okamura, S., Ostriker, J.P., Owen, R., Pauls, A.G., Peoples, J., Peterson, R.L., Petravick, D., Pier, J.R., Pope, A., Pordes, R., Prosapio, A., Rechenmacher, R., Quinn, T.R., Richards, G.T., Richmond, M.W., Rivetta, C.H., Rockosi, C.M., Ruthmansdorfer, K., Sandford, D., Schlegel, D.J., Schneider, D.P., Sekiguchi, M., Sergey, G., Shimasaku, K., Siegmund, W.A., Smee, S., Smith, J.A., Snedden, S., Stone, R., Stoughton, C., Strauss, M.A., Stubbs, C., SubbaRao, M., Szalay, A.S., Szapudi, I., Szokoly, G.P., Thakar, A.R., Tremonti, C., Tucker, D.L., Uomoto, A., Vanden Berk, D., Vogeley, M.S., Waddell, P., Wang, S.i., Watanabe, M., Weinberg, D.H., Yanny, B., Yasuda, N., SDSS Collaboration: The Sloan Digital Sky Survey: technical summary. Astron. J. **120**, 1579–1587 (2000). doi:10.1086/301513, astro-ph/0006396

Yoshii, Y.: Density distribution of faint stars in the direction of the north galactic pole. Publ. Astron. Soc. Jpn. **34**, 365 (1982)

Zackrisson, E., de Jong, R.S., Micheva, G.: Unlocking the secrets of stellar haloes using combined star counts and surface photometry. Mon. Not. R. Astron. Soc. **421**, 190–201 (2012). doi:10.1111/j.1365-2966.2011.20290.x, 1112.1696

Zaritsky, D., Salo, H., Laurikainen, E., Elmegreen, D., Athanassoula, E., Bosma, A., Comerón, S., Erroz-Ferrer, S., Elmegreen, B., Gadotti, D.A., Gil de Paz, A., Hinz, J.L., Ho, L.C., Holwerda, B.W., Kim, T., Knapen, J.H., Laine, J., Laine, S., Madore, B.F., Meidt, S., Menendez-Delmestre, K., Mizusawa, T., Muñoz-Mateos, J.C., Regan, M.W., Seibert, M., Sheth, K.: On the origin of lopsidedness in galaxies as determined from the Spitzer survey of stellar structure in Galaxies (S^4G). Astrophys. J. **772**, 135 (2013). doi:10.1088/0004-637X/772/2/135, 1305.2940

Chapter 9
Outskirts of Distant Galaxies in Absorption

Hsiao-Wen Chen

Abstract QSO absorption spectroscopy provides a sensitive probe of both the neutral medium and diffuse ionized gas in the distant Universe. It extends 21 cm maps of gaseous structures around low-redshift galaxies both to lower gas column densities and to higher redshifts. Combining galaxy surveys with absorption-line observations of gas around galaxies enables comprehensive studies of baryon cycles in galaxy outskirts over cosmic time. This chapter presents a review of the empirical understanding of the cosmic neutral gas reservoir from studies of damped Lyα absorbers (DLAs). It describes the constraints on the star formation relation and chemical enrichment history in the outskirts of distant galaxies from DLA studies. A brief discussion of available constraints on the ionized circumgalactic gas from studies of lower column density Lyα absorbers and associated ionic absorption transitions is presented at the end.

9.1 Introduction

Absorption-line spectroscopy complements emission surveys and provides a powerful tool for studying the diffuse, large-scale baryonic structures in the distant Universe (e.g. Rauch 1998; Wolfe et al. 2005; Prochaska and Tumlinson 2009). Depending on the physical conditions of the gas (including gas density, temperature, ionization state and metallicity), a high-density region in the foreground is expected to imprint various absorption transitions of different line strengths in the spectrum of a background QSO. Observing the absorption features imprinted in QSO spectra enables a uniform survey of diffuse gas in and around galaxies, as well as detailed studies of the physical conditions of the gas at redshifts as high as the background sources can be observed.

Figure 9.1 displays an example of optical and near-infrared spectra of a high-redshift QSO. The QSO is at redshift $z_{\rm QSO} = 4.13$, and the spectra are retrieved from the XQ-100 archive (Lopez et al. 2016). At the QSO redshift, multiple broad

H.-W. Chen (✉)
Department of Astronomy & Astrophysics, and Kavli Institute for Cosmological Physics, The University of Chicago, Chicago, IL 60637, USA
e-mail: hchen@oddjob.uchicago.edu

J.H. Knapen et al. (eds.), *Outskirts of Galaxies*, Astrophysics and Space Science Library 434, DOI 10.1007/978-3-319-56570-5_9

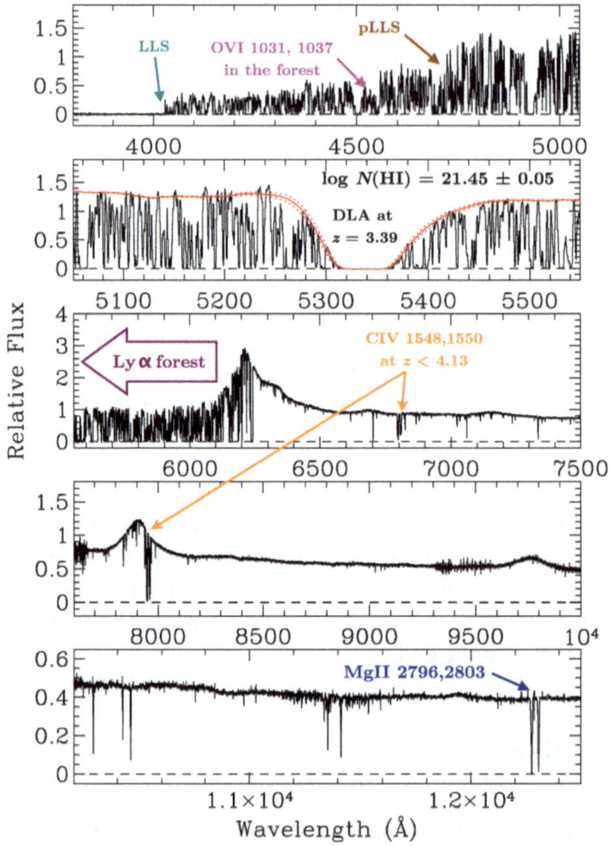

Fig. 9.1 Example of the wealth of information for intervening gas revealed in the optical and near-infrared spectrum of a QSO at $z = 4.13$. In addition to broad emission lines intrinsic to the QSO, such as Lyα/N V at $\approx$6200 Å, a forest of Lyα λ 1215 absorption lines is observed bluewards of 6200 Å. These Lyα forest lines arise in relatively high gas density regions at $z_{\rm abs} \lesssim z_{\rm QSO}$ along the line of sight. The Lyα absorbers span over ten decades in neutral hydrogen column densities (N(H I)) and include (1) neutral damped Lyα absorbers (DLAs), (2) optically thick Lyman limit systems (LLS), (3) partial LLS (pLLS) and (4) highly ionized Lyα absorbers (see text for a quantitative definition of these different classes). The DLAs are characterized by pronounced damping wings (*second panel from the top*), while LLS and pLLS are identified based on the apparent flux discontinuities in QSO spectra (*top panel*). Many of these strong Lyα absorbers are accompanied with metal absorption transitions such as the O VI $\lambda\lambda$ 1031, 1037 doublet transitions which occur in the Lyα forest, and the C IV $\lambda\lambda$ 1548, 1550 and Mg II $\lambda\lambda$ 2796, 2803 doublets. Together, these metal lines constrain the ionization state and chemical enrichment of the gas

emission lines are observed, including the Lyα/N V emission at $\approx$6200 Å, C IV emission at $\approx$7900 Å and C III] emission at $\approx$ 9800 Å. Bluewards of the Lyα emission line are a forest of Lyα λ 1215 absorption lines produced by intervening overdense regions at $z_{\rm abs} \lesssim z_{\rm QSO}$ along the QSO sightline. These overdense regions span a wide range in H I column density (N(H I)), from neutral interstellar gas of

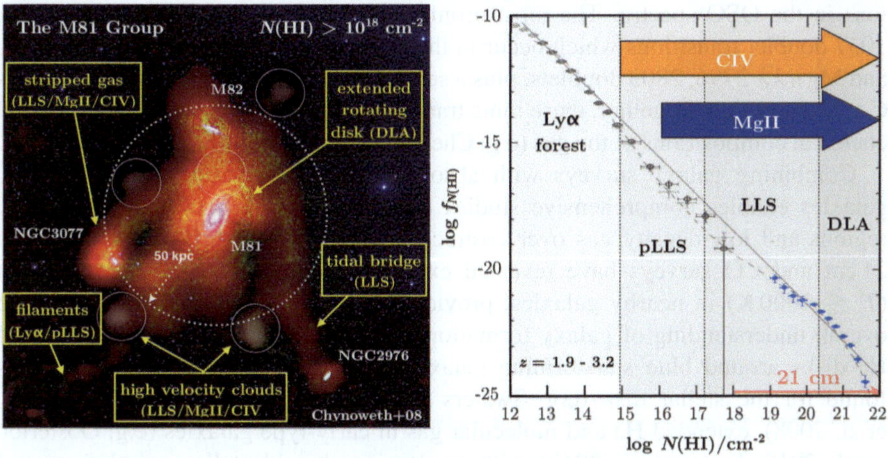

Fig. 9.2 Mapping galaxy outskirts in 21 cm and in QSO absorption-line systems. *Left*: deep 21 cm image of the M81 group, revealing a complex interface between stars and gas in the group. The observed neutral hydrogen column densities range from $N(\text{H I}) \sim 10^{18}\,\text{cm}^{-2}$ in the filamentary structures to $N(\text{H I}) > 10^{21}\,\text{cm}^{-2}$ in the star-forming disks of group members (Yun et al. 1994; Chynoweth et al. 2008). The 21 cm image reveals a diverse array of gaseous structures in this galaxy group, but these observations become extremely challenging beyond redshift $z \approx 0.2$ (e.g. Verheijen et al. 2007; Fernández et al. 2013). *Right*: The H I column density distribution function of Lyα absorbers, $f_{N(\text{H I})}$, uncovered at $z = 1.9$–3.2 along sightlines towards random background QSOs (adapted from Kim et al. 2013). Quasar absorbers in different categories are mapped onto different H I structures both seen and missed in the 21 cm image in the *left panel*. Specifically, DLAs probe the star-forming ISM and extended rotating disks, LLS probe the gaseous streams connecting different group members as well as stripped gas and high-velocity clouds around galaxies and pLLS and strong Lyα absorbers trace ionized gas that is not observed in 21 cm signals. Among the quasar absorbers, C IV absorption transitions are commonly observed in strong Lyα absorbers of $N(\text{H I}) \gtrsim 10^{15}\,\text{cm}^{-2}$ (see, e.g. Kim et al. 2013; D'Odorico et al. 2016), and Mg II absorption transitions are seen in most high-N(H I) absorbers of $N(\text{H I}) \gtrsim 10^{16}\,\text{cm}^{-2}$ (see, e.g. Rigby et al. 2002). These metal-line absorbers trace chemically enriched gas in and around galaxies

$N(\text{H I}) \geq 10^{20.3}\,\text{cm}^{-2}$ to optically opaque Lyman limit systems (LLS) of $N(\text{H I}) > 10^{17.2}\,\text{cm}^{-2}$, to optically thin partial LLS (pLLS) with $N(\text{H I}) = 10^{15-17.2}\,\text{cm}^{-2}$ and to highly ionized Lyα forest lines with $N(\text{H I}) = 10^{12-15}\,\text{cm}^{-2}$ (right panel of Fig. 9.2).

The large N(H I) in the neutral medium produces pronounced damping wings in the QSO spectrum. These absorbers are commonly referred to as damped Lyα absorbers (DLAs). An example is shown in the second panel from the top in Fig. 9.1. In this particular case, a simultaneous fit to the QSO continuum and the damping wings (red curve in the second panel from the top) yields a best-fit $\log N(\text{H I}) = 21.45 \pm 0.05$ for the DLA. At intermediate N(H I), LLS and pLLS are identified based on the apparent flux discontinuities in QSO spectra (top panel). A significant fraction of these strong Lyα absorbers have been enriched with heavy elements which produce additional absorption features due to heavy ions in the QSO spectra. The most prominent features include the O VI $\lambda\lambda$ 1031, 1037 doublet transitions which occur in the Lyα forest and the C IV $\lambda\lambda$ 1548, 1550 and Mg II $\lambda\lambda$ 2796, 2803 doublets, plus a series of low-ionization transitions such as C II, Si II and Fe II. Together, these ionic transitions constrain the ionization state and chemical compositions of the gas (e.g. Chen and Prochaska 2000; Werk et al. 2014).

Combining galaxy surveys with absorption-line observations of gas around galaxies enables comprehensive studies of baryon cycles between star-forming regions and low-density gas over cosmic time. At low redshifts, $z \lesssim 0.2$, deep 21 cm and CO surveys have revealed exquisite details of the cold gas content ($T \lesssim 1000$ K) in nearby galaxies, providing both new clues and puzzles in the overall understanding of galaxy formation and evolution. These include extended H I disks around blue star-forming galaxies with the H I extent $\approx 2\times$ what is found for the stellar disk (e.g. Swaters et al. 2002; Walter et al. 2008; Leroy et al. 2008), extended H I and molecular gas in early-type galaxies (e.g. Oosterloo et al. 2010; Serra et al. 2012) with predominantly old stellar populations and little or no ongoing star formation (Salim and Rich 2010) and widespread H I streams connecting regular-looking galaxies in group environments (e.g. Verdes-Montenegro et al. 2001; Chynoweth et al. 2008).

Figure 9.2 (left panel) showcases an example of a deep 21 cm image of the M81 group, a poor group of dynamical mass $M_{\rm dyn} \sim 10^{12}\,M_\odot$ (Karachentsev and Kashibadze 2006). Prominent group members include the grand design spiral galaxy M81 at the centre, the proto-starburst galaxy M82 and several other lower-mass satellites (Burbidge and Burbidge 1961). The 21 cm image displays a diverse array of gaseous structures in the M81 group, from extended rotating disks, warps, high-velocity clouds (HVCs), tidal tails and filaments to bridges connecting what appear to be optically isolated galaxies. High column density gaseous streams of $N(\rm H\,I) \gtrsim 10^{18}\,cm^{-2}$ are seen extending beyond 50 kpc in projected distance from M81, despite the isolated appearances of M81 and other group members in optical images. These spatially resolved imaging observations of different gaseous components serve as important tests for theoretical models of galaxy formation and evolution (e.g. Agertz et al. 2009; Marasco et al. 2016). However, 21 cm imaging observations are insensitive to warm ionized gas of $T \sim 10^4$ K and become extremely challenging for galaxies beyond redshift $z = 0.2$ (e.g. Verheijen et al. 2007; Fernández et al. 2013).

QSO absorption spectroscopy extends 21 cm maps of gaseous structures around galaxies to both lower gas column density and higher redshifts. Based on the characteristic $N(\rm H\,I)$, direct analogues can be drawn between different types of QSO absorbers and different gaseous components seen in deep 21 cm images of nearby galaxies. For example, DLAs probe the neutral gas in the interstellar medium (ISM) and extended rotating disks, LLS probe optically thick gaseous streams and high-velocity clouds in galaxy haloes and pLLS and strong Lyα absorbers of $N(\rm H\,I) \approx 10^{14-17}\,cm^{-2}$ trace ionized halo gas and starburst outflows (e.g. supergalactic winds in M 82 Lehnert et al. 1999) that cannot be reached with 21 cm observations.

The right panel of Fig. 9.2 displays the H I column density distribution function, $f_{N(\rm H\,I)}$, for all Lyα absorbers uncovered at $z = 1.9$–3.2 along random QSO sightlines

(Kim et al. 2013). $f_{N(\mathrm{H\,I})}$, defined as the number of Lyα absorbers per unit absorption path length per unit H I column density interval, is a key statistical measure of the Lyα absorber population. It represents a cross-section weighted surface density profile of hydrogen gas in a cosmological volume. With sufficiently high spectral resolution and high signal-to-noise ratio, $S/N \gtrsim 30$, QSO absorption spectra probe tenuous gas with N(H I) as low as $N(\mathrm{H\,I}) \sim 10^{12}\,\mathrm{cm}^{-2}$. The steeply declining $f_{N(\mathrm{H\,I})}$ with increasing N(H I) shows that the occurrence (or areal coverage) of pLLS and strong Lyα absorbers of $N(\mathrm{H\,I}) \approx 10^{14-17}\,\mathrm{cm}^{-2}$ is $\approx$10 times higher than that of optically thick LLS along a random sightline and $\approx$100 times higher than the incidence of DLAs. Such a differential frequency distribution is qualitatively consistent with the spatial distribution of H I gas recorded in local 21 cm surveys (e.g. Fig. 9.2, left panel), where gaseous disks with N(H I) comparable to DLAs cover a much smaller area on the sky than streams and HVCs with N(H I) comparable to LLS. If a substantial fraction of optically thin absorbers originate in galaxy haloes, then their higher incidence implies a gaseous halo of size at least three times what is seen in deep 21 cm images.

In addition, many of these strong Lyα absorbers exhibit associated transitions due to heavy ions. In particular, C IV absorption transitions are commonly observed in strong Lyα absorbers of $N(\mathrm{H\,I}) \gtrsim 10^{15}\,\mathrm{cm}^{-2}$ (see, e.g. Kim et al. 2013; D'Odorico et al. 2016), and Mg II absorption transitions are seen in most high-N(H I) absorbers of $N(\mathrm{H\,I}) \gtrsim 10^{16}\,\mathrm{cm}^{-2}$ (e.g. Rigby et al. 2002). While Mg II absorbers are understood to originate in photoionized gas of temperature $T \sim 10^4$ K (e.g. Bergeron and Stasińska 1986), C IV absorbers are more commonly seen in complex, multiphase media (e.g. Rauch et al. 1996; Boksenberg and Sargent 2015). These metal-line absorbers therefore offer additional probes of chemically enriched gas in and around galaxies.

This chapter presents a brief review of the current state of knowledge on the outskirts of distant galaxies from absorption-line studies. The review will first focus on the properties of the neutral gas reservoir probed by DLAs and then outline the insights into star formation and chemical enrichment in the outskirts of distant galaxies from searches of DLA galaxies. A comprehensive review of DLAs is already available in Wolfe et al. (2005). Therefore, the emphasis here focusses on new findings over the past decade. Finally, a brief discussion will be presented on the empirical properties and physical understandings of the ionized circumgalactic gas as probed by strong Lyα and various metal-line absorbers.

9.2 Tracking the Neutral Gas Reservoir over Cosmic Time

DLAs are historically defined as Lyα absorbers with neutral hydrogen column densities exceeding $N(\mathrm{H\,I}) = 2 \times 10^{20}\,\mathrm{cm}^{-2}$ (Wolfe et al. 2005), corresponding to a surface mass density limit of $\Sigma_{\mathrm{atomic}} \approx 2\,M_{\odot}\,\mathrm{pc}^{-2}$ for atomic gas (including helium). The large gas surface mass densities revealed in high-redshift DLAs are comparable to what is seen in 21 cm observations of nearby star-forming galaxies (e.g. Walter et al. 2008; Leroy et al. 2008), making DLAs a promising signpost of young galaxies in the distant Universe (Wolfe et al. 1986). In addition, the $N(\text{H I})$ threshold ensures that the gas is neutral under the metagalactic ionizing radiation field (e.g. Viegas 1995; Prochaska and Wolfe 1996; Prochaska et al. 2002). Neutral gas provides the seeds necessary for sustaining star formation. Therefore, observations of DLAs not only help establish a census of the cosmic evolution of the neutral gas reservoir (e.g. Neeleman et al. 2016a) but also offer a unique window into star formation physics in distant galaxies (e.g. Lanzetta et al. 2002; Wolfe and Chen 2006).

While the utility of DLAs for probing the young Universe is clear, these objects are relatively rare (see the right panel of Fig. 9.2), and establishing a statistically representative sample of these rare systems requires a large sample of QSO spectra. Over the last decade, significant progress has been made in characterizing the DLA population at $z \gtrsim 2$, owing to the rapidly growing spectroscopic sample of high-redshift QSOs from the Sloan Digital Sky Survey (SDSS; York et al. 2000). The blue points in the right panel of Fig. 9.2 are based on ~1000 DLAs and ~500 strong LLS identified at $z \approx 2$–5 in an initial SDSS DLA sample (Noterdaeme et al. 2009). The sample of known DLAs at $z \gtrsim 2$ has continued to grow, reaching ~10,000 DLAs found in the SDSS spectroscopic QSO sample (e.g. Noterdaeme et al. 2012).

The large number of known DLAs has led to an accurate characterization of the neutral gas reservoir at high redshifts. Figure 9.3a displays the observed $N(\text{H I})$ distribution function, f_{DLA}, based on ~7000 DLAs identified at $z \approx 2$–5 (Noterdaeme et al. 2012). The plot shows that f_{DLA} is well represented by a Schechter function (Schechter 1976) at $\log N(\text{H I}) \lesssim 22$ following

$$f_{\text{DLA}} \equiv f_{N(\text{H I})}(\log N(\text{H I}) \geq 20.3) \propto \left[\frac{N(\text{H I})}{N_*(\text{H I})}\right]^{\alpha} \exp[-N(\text{H I})/N_*(\text{H I})], \quad (9.1)$$

with a shallow power-law index of $\alpha \approx -1.3$ below the characteristic H I column density $\log N_*(\text{H I}) \approx 21.3$ and a steep exponential decline at larger $N(\text{H I})$ (Noterdaeme et al. 2009, 2012). At $\log N(\text{H I}) > 22$, the observations clearly deviate from the best-fit Schechter function. However, DLAs are also exceedingly rare in this high-$N(\text{H I})$ regime. Only eight such strong DLAs have been found in this large DLA sample (Noterdaeme et al. 2012), making measurements of f_{DLA} in the two highest-$N(\text{H I})$ bins very uncertain. In comparison to $f_{N(\text{H I})}$ established from 21 cm maps of nearby galaxies (Zwaan et al. 2005), the amplitude of f_{DLA} at $z \gtrsim 2$ is $\approx 2\times$ higher than $f_{N(\text{H I})}$ at $z \approx 0$, but the overall shapes are remarkably similar at both low- and high-$N(\text{H I})$ regimes (Fig. 9.3a; see also Sánchez-Ramírez et al. 2016; Rafelski et al. 2016).

At $\log N(\text{H I}) > 21$, numerical simulations have shown that the predicted shape in f_{DLA} is sensitive to the detailed ISM physics, including the formation of molecules (H_2) and different feedback processes (e.g. Altay et al. 2011, 2013; Bird et al. 2014). Comparison of the observed and predicted f_{DLA} therefore provides an independent and critical test for the prescriptions of these physical processes in cosmological simulations. However, the constant exponentially declining trend at

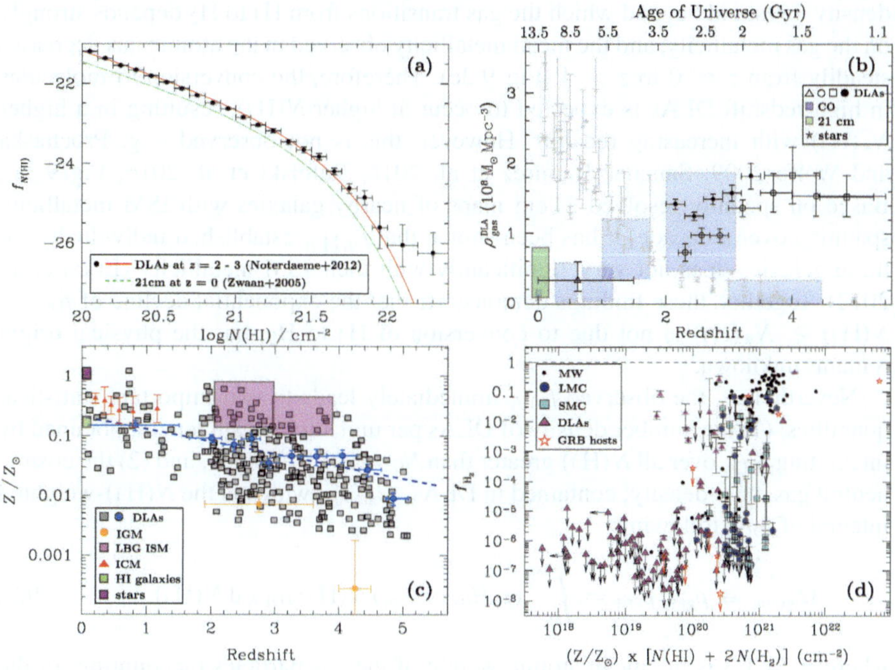

Fig. 9.3 Summary of known DLA properties: (**a**) evolving neutral hydrogen column density distribution functions, $f_{N(\mathrm{H\,I})}$, from DLAs at $z = 2$–3 (Noterdaeme et al. 2012) to H I galaxies at $z \approx 0$ (Zwaan et al. 2005); (**b**) declining cosmic neutral gas mass density with increasing Universe age (or decreasing redshift) from observations of DLAs (*solid points* from Noterdaeme et al. 2012, *open circles* from Prochaska and Wolfe 2009, *open squares* from Crighton et al. 2015 and *open triangle* from Neeleman et al. 2016a) following Eq. (9.2), local H I galaxies (*green-shaded box*, a compilation from Neeleman et al. 2016a) and molecular gas (*blue-shaded boxes*, Decarli et al. 2016), in comparison to increasing cosmic stellar mass density in galaxies with increasing Universe age (*grey asterisks*, a compilation from Madau and Dickinson 2014); (**c**) gas-phase metallicity (Z) relative to Solar ($Z_\odot$) as a function of redshift in DLAs (*grey squares* for individual absorbers and *blue points* for N(H I)-weighted mean from Marc Rafelski, Rafelski et al. 2012, 2014), IGM at $z \gtrsim 2$ (*orange circles*, Aguirre et al. 2008; Simcoe 2011), ISM of starburst galaxies at $z \approx 2$–4 (*light magenta boxes*, Pettini et al. 2001; Pettini 2004; Erb et al. 2006a; Maiolino et al. 2008; Mannucci et al. 2009), intra-cluster medium in X-ray luminous galaxy clusters at $z \lesssim 1$ (*red triangles*, Balestra et al. 2007), H I-selected galaxies (*green box*, Zwaan et al. 2005) and stars at $z = 0$ (*dark purple box*, Gallazzi et al. 2008); and (**d**) molecular gas fraction, f_{H_2}, versus total surface density of neutral gas scaled by gas metallicity for high-redshift DLAs in *triangles* (Noterdaeme et al. 2008, 2016), γ-ray burst host ISM in *star symbols* (e.g. Noterdaeme et al. 2015) and local ISM in the Milky Way (Wolfire et al. 2008) and Large and Small Magellanic Clouds (Tumlinson et al. 2002) in *dots*, *blue circles* and *cyan squares*, respectively

$N(\mathrm{H\,I}) \gtrsim 2 \times 10^{21}\,\mathrm{cm}^{-2}$ between low-redshift H I galaxies and high-redshift DLAs presents a puzzle.

At $z = 0$, the rapidly declining $f_{N(\mathrm{H\,I})}$ at $N(\mathrm{H\,I}) \gtrsim N_*(\mathrm{H\,I})$ has been interpreted as due to the conversion of atomic gas to molecular gas (Zwaan and Prochaska 2006; Braun 2012). As illustrated at the end of this section and in Fig. 9.3d, the column density threshold beyond which the gas transitions from H I to H_2 depends strongly on the gas metallicity, and the mean metallicity observed in the atomic gas decreases steadily from $z \approx 0$ to $z > 4$ (Fig. 9.3c). Therefore, the conversion to molecules in high-redshift DLAs is expected to occur at higher $N(\mathrm{H\,I})$, resulting in a higher $N_*(\mathrm{H\,I})$ with increasing redshift. However, this is not observed (e.g. Prochaska and Wolfe 2009; Sánchez-Ramírez et al. 2016; Rafelski et al. 2016; Fig. 9.3a). Based on spatially resolved 21 cm maps of nearby galaxies with ISM metallicity spanning over a decade, it has been shown that $f_{N(\mathrm{H\,I})}$ established individually for these galaxies does not vary significantly with their ISM metallicity (Erkal et al. 2012). Together, these findings demonstrate that the exponential decline of f_{DLA} at $N(\mathrm{H\,I}) \gtrsim N_*(\mathrm{H\,I})$ is not due to conversion of H I to H_2, but the physical origin remains unknown.

Nevertheless, the observed f_{DLA} immediately leads to two important statistical quantities: (1) the number density of DLAs per unit survey path length, obtained by integrating f_{DLA} over all $N(\mathrm{H\,I})$ greater than $N_0 = 2\times10^{20}\,\mathrm{cm}^{-2}$, and (2) the cosmic neutral gas mass density, contained in DLAs, Ω_{atomic}, which is the $N(\mathrm{H\,I})$-weighted integral of f_{DLA} following

$$\Omega_{\mathrm{atomic}} \equiv \rho_{\mathrm{gas}}/\rho_{\mathrm{crit}} = \int_{N_0}^{\infty} (\mu\, H_0/c/\rho_{\mathrm{crit}})\, N(\mathrm{H\,I}) f_{\mathrm{DLA}}\, \mathrm{d}N(\mathrm{H\,I}), \tag{9.2}$$

where $\mu = 1.3$ is the mean atomic weight of the gas particles (accounting for the presence of helium), H_0 is the Hubble constant, c is the speed of light and ρ_{crit} is the critical density of the Universe (e.g. Lanzetta et al. 1991; Wolfe et al. 1995). The shallow power-law index α in the best-fit f_{DLA}, together with a steep exponential decline at high $N(\mathrm{H\,I})$ from the Schechter function in Eq. (9.1), indicates that while DLAs of $N(\mathrm{H\,I}) < N_*(\mathrm{H\,I})$ dominate the neutral gas cross-section (and therefore the number density), strong DLAs of $N(\mathrm{H\,I}) \sim N_*(\mathrm{H\,I})$ contribute predominantly to the neutral mass density in the Universe (e.g. Zwaan et al. 2005). A detailed examination of the differential Ω_{atomic} distribution as a function of $N(\mathrm{H\,I})$ indeed confirms that the bulk of neutral gas is contained in DLAs of $N(\mathrm{H\,I}) \approx 2\times10^{21}\,\mathrm{cm}^{-2}$ (e.g. Noterdaeme et al. 2012).

The cosmic evolution of ρ_{gas} observed in DLAs, from Eq. (9.2), is shown in black points in Fig. 9.3b. Only measurements based on blind DLA surveys are presented in the plot.[1] These include an early sample of $\approx$700 DLAs at $z = 2.5$–5 in the SDSS Data Release (DR) 5 (open circles; Prochaska and Wolfe 2009), an expanded

[1]At $z \lesssim 1.6$, DLA surveys require QSO spectroscopy carried out in space and have been limited to the number of UV-bright QSOs available for absorption-line searches. Consequently, the number of known DLAs from blind surveys is small, $\approx$15 (see Neeleman et al. 2016a for a compilation). To increase substantially the sample of known DLAs at low redshifts, Rao & Turnshek (Rao et al. 2006) devised a clever space programme to search for new DLAs in known Mg II absorbers. Their strategy yielded a substantial gain, tripling the total sample size of $z \lesssim 1.6$ DLAs. However, the Mg II-selected DLA sample also includes a survey bias that is not well understood. It has been shown that excluding Mg II-selected DLAs reduces the inferred Ω_{atomic} by more than a factor of 4 (e.g. Neeleman et al. 2016a). For consistency, only measurements of Ω_{atomic} based on blind DLA surveys are included in the plot.

sample of ≈7000 DLAs in the SDSS DR12 (solid points; Noterdaeme et al. 2012), an expanded high-redshift sample of DLAs at $z = 4$–5 (open squares; Crighton et al. 2015) and a sample of ≈14 DLAs at $z \lesssim 1.6$ from an exhaustive search in the *Hubble Space Telescope* (*HST*) UV spectroscopic archive (open triangle; Neeleman et al. 2016a).

A range of mean H I mass density at $z \approx 0$ has been reported from different 21 cm surveys (see Neeleman et al. 2016a for a recent compilation). These measurements are included in the green box in Fig. 9.3b. Despite a relatively large scatter between different 21 cm surveys and between DLA surveys, a steady decline in $\Omega_{\rm atomic}$ is observed from $z \approx 4$ to $z \approx 0$. For comparison, the cosmic evolution of the molecular gas mass density obtained from a recent blind CO survey (Decarli et al. 2016) is also included as blue-shaded boxes in Fig. 9.3b, along with the cosmic evolution of stellar mass density measured in different galaxy surveys, shown in grey asterisks (data from Madau and Dickinson 2014). Figure 9.3b shows that the decline in the neutral gas mass density with decreasing redshift is coupled with an increase in the mean stellar mass density in galaxies, which is qualitatively consistent with the expectation that neutral gas is being consumed to form stars. However, it is also clear that atomic gas alone is insufficient to explain the observed order of magnitude gain in the total stellar mass density from $z \approx 3$ to $z \approx 0$, which implies the need for replenishing the neutral gas reservoir with accretion from the intergalactic medium (IGM) (e.g. Kereš et al. 2009; Prochaska and Wolfe 2009). At the same time, new blind CO surveys have shown that molecular gas contributes roughly an equal amount of neutral gas mass density as atomic gas observed in DLAs at $z \lesssim 3$ (e.g. Walter et al. 2014; Decarli et al. 2016), although the uncertainties are still very large. Together with the knowledge of an extremely low molecular gas fraction in DLAs (see the discussion on the next page and Fig. 9.3d), these new CO surveys indicate that previous estimates of the total neutral gas mass density based on DLAs alone have been underestimated by as much as a factor of 2. An expanded blind CO survey over a cosmological volume is needed to reduce the uncertainties in the observed molecular gas mass densities at different redshifts, which will cast new insights into the connections between the star formation, the neutral gas reservoir and the ionized IGM over cosmic time.

Observations of the chemical compositions of DLAs provide additional clues to the connection between the neutral gas probed by DLAs and star formation (e.g. Pettini 2004). In particular, because the gas is predominantly neutral, the dominant ionization for most heavy elements (such as Mg, Si, S, Fe, Zn, etc.) is in the singly ionized state, and therefore the observed abundances of these low-ionization species place direct and accurate constraints on the elemental abundances of the gas (e.g. Viegas 1995; Prochaska and Wolfe 1996; Vladilo et al. 2001; Prochaska et al. 2002). Additional constraints on the dust content and on the sources that drive the chemical enrichment history in DLAs can be obtained by comparing the relative abundances of different elements. Specifically, comparing the relative abundances between refractory (such as Cr and Fe) and non-refractory elements (such as S and Zn) indicates the presence of dust in the neutral gas, the amount of which increases with metallicity (e.g. Meyer et al. 1989; Pettini et al. 1990; Savage and Sembach 1996; Wolfe et al. 2005). The relative abundances of α- to Fe-peak elements determine whether core-collapse supernovae (SNe) or SNe Ia dominate the chemical enrichment history, and DLAs typically exhibit an α-element-enhanced abundance pattern (e.g. Lu et al. 1996; Pettini et al. 1999; Prochaska and Wolfe 1999).

Figure 9.3c presents a summary of gas metallity (Z) relative to Solar ($Z_\odot$) measured for >250 DLAs at $z \lesssim 5$ (grey squares from Rafelski et al. 2012, 2014). The cosmic mean gas metallicity in DLAs as a function of redshift can be determined based on a N(H I)-weighted average over an ensemble of DLAs in each redshift bin (blue points), which is found to increase steadily with decreasing redshift following a best-fit mean relation of $\langle Z/Z_\odot \rangle = [-0.20 \pm 0.03]\, z - [0.68 \pm 0.09]$ (dashed blue line, Rafelski et al. 2014). For comparison, the figure also includes measurements for stars (dark purple box, Gallazzi et al. 2008) and H I-selected galaxies (green box, Zwaan et al. 2005) at $z = 0$, iron abundances in the intra-cluster medium in X-ray luminous galaxy clusters at $z \lesssim 1$ (red triangles, Balestra et al. 2007), ISM of starburst galaxies (light magenta boxes) at $z \approx 2$–3 (Pettini et al. 2001; Pettini 2004; Erb et al. 2006a) and at $z = 3$–4 (Maiolino et al. 2008; Mannucci et al. 2009) and IGM at $z \gtrsim 2$ (orange circles, Aguirre et al. 2008; Simcoe 2011).

It is immediately clear from Fig. 9.3c that there exists a large scatter in the observed metallicity in DLAs at all redshifts. In addition, while the cosmic mean metallicity in DLAs is significantly higher than what is observed in the low-density IGM, it remains lower than what is observed in the star-forming ISM at $z = 2$–4 and a factor of $\approx$5 below the mean values observed in stars at $z = 0$. The chemical enrichment level in DLAs is also lower than the iron abundances seen in the intra-cluster medium at intermediate redshifts. The observed low metallicity relative to the measurements in and around known luminous galaxies raised the question of whether or not the DLAs probe preferentially low-metallicity, gas-rich galaxies and are not representative of more luminous, metal-rich galaxies found in large-scale surveys (e.g. Pettini 2004).

The large scatter in the observed metallicity in DLAs is found to be explained by a combination of two factors (Chen et al. 2005): (1) the mass-metallicity (or luminosity-metallicity) relation in which more massive galaxies on average exhibit higher global ISM metallicities (e.g. Tremonti et al. 2004; Erb et al. 2006a; Neeleman et al. 2013; Christensen et al. 2014) and (2) metallicity gradients commonly seen in star-forming disks with lower metallicities at larger distances (e.g. Zaritsky et al. 1994; van Zee et al. 1998; Sánchez et al. 2014; Wuyts et al. 2016). If DLAs sample a representative galaxy population including both low-mass and massive galaxies and probe both inner and outer disks of these galaxies, then a large metallicity spread is expected.

The observed low metallicity in DLAs, relative to star-forming ISM, is also understood as due to a combination of DLAs being a gas cross-section selected sample and the presence of metallicity gradients in disk galaxies (Chen et al. 2005). A cross-section selected sample contains a higher fraction of absorbers originating in galaxy outskirts than in the inner regions, and the presence of metallicity gradients indicates that galaxy outskirts have lower metallicities than what is observed in inner disks (see Sect. 9.3 and Fig. 9.4 below for more details). Indeed, including both factors, a gas cross-section weighting scheme and a metallicity gradient, for local H I galaxies resulted in a mean metallicity comparable to what is observed in DLAs (green box in Fig. 9.3c; Zwaan et al. 2005).

While DLAs exhibit a moderate level of chemical enrichment, searches for molecular gas in DLAs have yielded only a few detections (e.g. Noterdaeme et al. 2008, 2016; Jorgenson et al. 2014). Figure 9.3d displays the observed molecular gas fraction, which is defined as $f_{H_2} \equiv 2N(H_2)/[N(\text{H I}) + 2N(H_2)]$, versus metallicity-scaled total hydrogen column density for $\approx$100 DLAs at $z \approx 2$–4 (triangles). The DLAs span roughly two decades in $N(\text{H I})$ from $N(\text{H I}) \approx 2 \times 10^{20}\,\text{cm}^{-2}$ to

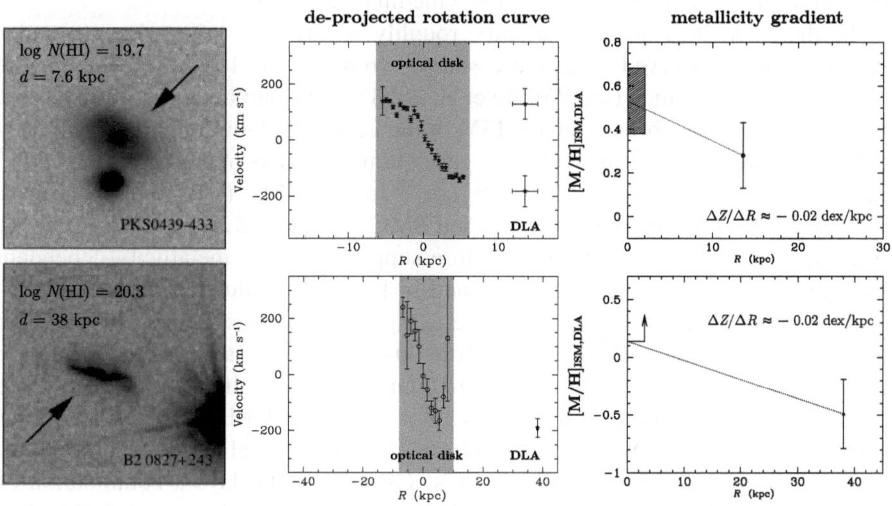

Fig. 9.4 Neutral gas kinematics and metallicity revealed by the presence of a DLA in the outskirts of two L_* galaxies (adapted from Chen et al. 2005). The *top row* presents a DLA found at $d = 7.6$ kpc from a disk galaxy at $z = 0.101$, which also exhibits widespread CO emission in the disk (Neeleman et al. 2016b). The *bottom row* presents a DLA at $d = 38$ kpc from an edge-on disk at $z = 0.525$. Deep r-band images of the galaxies are presented in the *left panels*, which display spatially resolved disk morphologies and enable accurate measurements of the inclination and orientation of the optical disk. The *middle panels* present the optical rotation curves deprojected along the disk plane (points in shaded area) based on the inclination angle determined from the optical image of each galaxy (Eqs. (9.3) and (9.4)). If the DLAs occur in extended disks, the corresponding galactocentric distances of the two galaxies from Eq. (9.3) are $R = 13.6$ kpc (*top*) and $R = 38$ kpc (*bottom*). The DLA in the *top panel* is resolved into two components of comparable ionic column densities (Som et al. 2015) but an order of magnitude difference in $N(H_2)$ (Muzahid et al. 2015). The component with a lower $N(H_2)$ appears to be corotating with the optical disk (lower DLA data point), while the component with stronger $N(H_2)$ appears to be counterrotating, possibly due to a satellite (upper DLA data point). The DLA in the *bottom panel* displays simpler gas kinematics consistent with an extended rotating disk out to $\approx$40 kpc. The *right panels* present the metallicity gradient observed in the gaseous disks based on comparisons of ISM gas-phase metallicity and metallicity of the DLA beyond the optical disks. In both cases, the gas metallicity declines with increasing radius according to $\Delta Z/\Delta R = -0.02\,\text{dex kpc}^{-1}$

$N(\text{H I}) \approx 2.5 \times 10^{22}\,\text{cm}^{-2}$. Strong limits have been placed for f_{H_2} for the majority of DLAs at $f_{H_2} \lesssim 10^{-5}$ with only $\approx$10% displaying the presence of H_2 and two having $f_{H_2} > 0.1$. In contrast, the ISM of the Milky Way (MW), at comparable $N(\text{H I})$, displays a much higher f_{H_2} than the DLAs at high redshifts.

The formation of molecules is understood to depend on two competing factors: (1) the ISM radiation field which photodissociates molecules and (2) dust which facilitates molecule formation (e.g. Elmegreen 1993; Cazaux and Spaans 2004). Dust is considered a more dominant factor because of its dual roles in both forming molecules and shielding them from the ISM radiation field. In star-forming galaxies, the dust-to-gas mass ratio is observed to correlate strongly with ISM gas-phase metallicity (e.g. Leroy et al. 2011; Rémy-Ruyer et al. 2014). It is therefore expected that the observed molecular gas fraction should correlate with gas metallicity (e.g. Elmegreen 1989; Krumholz et al. 2009; Gnedin et al. 2009).

In the MW ISM with metallicity roughly Solar, $Z \approx Z_{\odot}$, the molecular gas fraction is observed to increase sharply from $f_{H_2} < 10^{-4}$ to $f_{H_2} \gtrsim 0.1$ at $N(\text{H I}) \approx 2 \times 10^{20}\,\text{cm}^{-2}$ (see Wolfire et al. 2008). The sharp transition from atomic to molecular is also observed in the ISM of the Large and Small Magellanic Clouds (LMC and SMC), but occurs at higher gas column densities of $N(\text{H I}) \approx 10^{21}\,\text{cm}^{-2}$ for the LMC and $N(\text{H I}) \approx 3 \times 10^{21}\,\text{cm}^{-2}$ for the SMC (see Tumlinson et al. 2002). The ISM metallicities of LMC and SMC are $Z \approx 0.5\,Z_{\odot}$ and $Z \approx 0.15\,Z_{\odot}$, respectively. These observations therefore support a simple metallicity-dependent transitional gas column density illustrated in Fig. 9.3d. Following the metallicity-scaling relation, it is clear that despite a high $N(\text{H I})$, most DLAs do not have sufficiently high metallicity (and therefore dust content) to facilitate the formation of molecules (Gnedin and Kravtsov 2010; Gnedin and Draine 2014; Noterdaeme et al. 2015). This finding also applies to γ-ray burst (GRB) host galaxies (star symbols in Fig. 9.3d). With few exceptions (Prochaska et al. 2009; Krühler et al. 2013; Friis et al. 2015), the ISM in most GRB hosts displays a combination of very high $N(\text{H I})$ and low f_{H_2} (e.g. Tumlinson et al. 2007; Ledoux et al. 2009). The observed absence of H_2 in DLAs, together with a large molecular mass density revealed in blind CO surveys (e.g. Walter et al. 2014; Decarli et al. 2016), shows that a complete census for the cosmic evolution of the neutral gas reservoir requires complementary surveys of molecular gas over a broad redshift range. In addition, as described in Sect. 9.4 below, the observed low molecular gas content also has important implications for star formation properties in metal-deficient, high neutral gas surface density environments.

9.3 Probing the Neutral Gas Phase in Galaxy Outskirts

Considerable details have been learned about the physical properties and chemical enrichment in neutral atomic gas from DLA studies. To apply the knowledge of DLAs for a better understanding of distant galaxies, it is necessary to first identify

DLA galaxies and compare them with the general galaxy population. Searches for DLA galaxies are challenging, because distant galaxies are faint and because the relatively small extent of high-N(H I) gas around galaxies places the absorbing galaxies at small angular distances from the bright background QSOs. Based on a well-defined H I size-mass relation observed in local H I galaxies (e.g. Broeils and Rhee 1997; Verheijen and Sancisi 2001; Swaters et al. 2002), the characteristic projected separation (accounting for weighting by cross-section) between a DLA and an L_* absorbing galaxy is $\approx$16 kpc and smaller for lower-mass galaxies. At $z = 1$–2, a projected distance of 16 kpc corresponds to an angular separation of $\lesssim 2''$ and greater at lower and higher redshifts.

While fewer DLAs are known at $z \lesssim 1$ (see Sect. 9.2), a large number ($\approx$40) of these low-redshift DLAs have their galaxy counterparts (or candidates) found based on a combination of photometric and spectroscopic techniques (e.g. Chen and Lanzetta 2003; Rao et al. 2003, 2011; Péroux et al. 2016). It has been shown based on this low-redshift DLA galaxy sample that DLAs probe a representative galaxy population in luminosity and colour. DLA galaxies are consistent with an H I cross-section selected sample with a large fraction of DLAs found at projected distance $d \gtrsim 10$ kpc from the absorbing galaxies (e.g. Chen and Lanzetta 2003; Rao et al. 2011). In addition to regular disk galaxies, two DLAs have been found in a group environment (e.g. Bergeron and Boissé 1991; Chen and Lanzetta 2003; Kacprzak et al. 2010; Péroux et al. 2011), suggesting that stripped gas from galaxy interactions could also contribute to the incidence of DLAs. The low-redshift DLA sample is expected to continue to grow dramatically with new discoveries from the SDSS (e.g. Straka et al. 2015). In contrast, the search for DLA galaxies at $z > 2$ has been less successful despite extensive efforts (e.g. Warren et al. 2001; Møller et al. 2002; Péroux et al. 2012; Fumagalli et al. 2015). To date, only $\approx$12 DLA galaxies have been found at $z > 2$ (Krogager et al. 2012; Fumagalli et al. 2015).

In addition to a general characterization of the DLA galaxy population, individual DLA and galaxy pairs provide a unique opportunity to probe neutral gas in the outskirts of distant galaxies. Figure 9.4 shows two examples of constraining the kinematics and chemical enrichment in the outskirts of neutral disks from combining resolved optical imaging and spectroscopy of the galaxy with an absorption-line analysis of the DLA. In the first example (top row), a DLA of $\log N(\mathrm{H\,I}) = 19.7$ is found at $d = 7.6$ kpc from an L_* galaxy at $z = 0.101$, which also exhibits widespread CO emission in the disk (Neeleman et al. 2016b). The galaxy disk is resolved in the ground-based r-band image (upper-left panel), which enables accurate measurements of the disk inclination and orientation (Chen et al. 2005). While the observed N(H I) falls below the nominal threshold of a DLA, the gas is found to be largely neutral (e.g. Chen et al. 2005; Som et al. 2015). In addition, abundant H_2 is detected in the absorbing gas (Muzahid et al. 2015). Optical spectra of the galaxy clearly indicate a strong velocity shear along the disk, suggesting an organized rotation motion (Chen et al. 2005) which is confirmed by recent CO observations (Neeleman et al. 2016b). At the same time, the DLA is resolved into two components of comparable ionic column densities (Som et al. 2015) but an order of magnitude difference in $N(H_2)$ (Muzahid et al. 2015). A rotation curve of the gaseous disk extending beyond 10 kpc (top-centre panel) can be established based on the observed velocity shear ($v_{\rm obs}$) and deprojection onto the disk plane following

$$\frac{R}{d} = \sqrt{1 + \sin^2(\phi)\tan^2(i)} \tag{9.3}$$

and

$$v = \frac{v_{\rm obs}}{\cos(\phi)\sin(i)} \sqrt{1 + \sin^2(\phi)\tan^2(i)}, \tag{9.4}$$

where R is the galactocentric radius along the disk, v is the deprojected rotation velocity, i is the inclination angle of the disk and ϕ is the azimuthal angle from the major axis of the disk where the DLA is detected (Chen et al. 2005; see also Steidel et al. 2002 for an alternative formalism). For the two absorbing components in this DLA, it is found that the component with a lower $N(H_2)$ appears to be corotating with the optical disk (lower DLA data point), while the component with stronger $N(H_2)$ appears to be counterrotating, possibly due to a satellite (upper DLA data point). Comparing the ISM gas-phase metallicity and the metallicity of the DLA shows a possible gas metallicity gradient of $\Delta Z/\Delta R = -0.02$ dex kpc^{-1} out to $R \approx 14$ kpc.

The bottom row of Fig. 9.4 presents a DLA at $d = 38$ kpc from an edge-on disk at $z = 0.525$. A strong velocity shear is also seen along the disk of this L_* galaxy. Because the QSO sightline occurs along the extended edge-on disk, Eqs. (9.3) and (9.4) directly lead to $R \approx d$ and $v \approx v_{\rm obs}$ for this system. This DLA galaxy presents a second example for galaxies with an extended rotating disk out to $\approx$40 kpc. At the same time, the deep r-band image (lower-left panel) from *HST* suggests that the disk is warped near the QSO sightline, which is also reflected by the presence of a disturbed rotation velocity at $R > 5$ kpc (bottom-centre panel). The metallicity measured in the gas phase (bottom-right panel) displays a similar gradient of $\Delta Z/\Delta R = -0.02$ dex kpc^{-1} to the galaxy at the top, which is also comparable to what is seen in the ISM of nearby disk galaxies (e.g. Zaritsky et al. 1994; van Zee et al. 1998; Sánchez et al. 2014). A declining gas-phase metallicity from the inner ISM to neutral gas at larger distances appears to hold for most DLA galaxies at $z \lesssim 1$, and the declining trend continues into ionized halo gas traced by strong LLS of $N(\mathrm{H\,I}) = 10^{19-20}$ cm^{-2} (e.g. Péroux et al. 2016).

At $z > 2$, spatially resolved observations of ISM gas kinematics become significantly more challenging, because the effective radii of L_* galaxies are typically $r_e = 1$–3 kpc (e.g. Law et al. 2012), corresponding to $\lesssim 0.3''$, and smaller for fainter or lower-mass objects. Star-forming regions in these distant galaxies are barely resolved in ground-based, seeing-limited observations (e.g. Law et al. 2007; Förster Schreiber et al. 2009; Wright et al. 2009). Beam smearing can result in significant bias in interpreting the observed velocity shear and distributions of heavy elements (e.g. Davies et al. 2011; Wuyts et al. 2016). However, accurate measurements can be obtained to differentiate ISM metallicities of DLA galaxies from metallicities of neutral gas beyond the star-forming regions. Using the small sample of known DLA galaxies at $z \gtrsim 2$, a metallicity gradient of $\Delta Z/\Delta R = -0.02\,\mathrm{dex\,kpc^{-1}}$ is also found in these distant star-forming galaxies (Christensen et al. 2014; Jorgenson and Wolfe 2014).

9.4 The Star Formation Relation in the Early Universe

While direct identifications of galaxies giving rise to $z > 2$ DLAs have proven extremely challenging, critical insights into the star formation relation in the early Universe can still be gained from comparing the incidence of DLAs with the spatial distribution of star formation rate (SFR) per unit area uncovered in deep imaging data (Lanzetta et al. 2002; Wolfe and Chen 2006). Specifically, the SFR per unit area ($\Sigma_{\rm SFR}$) is correlated with the surface mass density of neutral gas ($\Sigma_{\rm gas}$), following a Schmidt-Kennicutt relation in nearby galaxies (e.g. Schmidt 1959; Kennicutt 1998). The global star formation relation, $\Sigma_{\rm SFR} = 2.5 \times 10^{-4}\,(\Sigma_{\rm gas}/1\,M_\odot\,\mathrm{pc}^{-2})^{1.4}\,M_\odot\,\mathrm{yr}^{-1}\,\mathrm{kpc}^{-2}$ (dashed line in Fig. 9.5), is established using a sample of local spiral galaxies and nuclear starbursts (solid grey points in Fig. 9.5) over a broad range of $\Sigma_{\rm gas}$, from $\Sigma_{\rm gas} \approx 10\,M_\odot\,\mathrm{pc}^{-2}$ to $\Sigma_{\rm gas} \approx 10^4\,M_\odot\,\mathrm{pc}^{-2}$.

Empirical constraints for a Schmidt-Kennicutt relation at high redshifts require observations of the neutral gas content in star-forming galaxies. Although observations of individual galaxies in H I emission remain out of reach, the sample of $z = 1$–3 galaxies with resolved CO maps is rapidly growing (e.g. Baker et al. 2004; Genzel et al. 2010; Tacconi et al. 2013). The observed $\Sigma_{\rm SFR}$ versus $\Sigma_{\rm molecular}$ for the high-redshift CO-detected sample is shown in open squares in Fig. 9.5, which occur at high surface densities of $\Sigma_{\rm molecular} \gtrsim 100\,M_\odot\,\mathrm{pc}^{-2}$. Considering only $\Sigma_{\rm molecular}$ is appropriate for these galaxies, because locally it has been shown that at this high surface density regime, molecular gas dominates (e.g. Martin and Kennicutt 2001; Wong and Blitz 2002; Bigiel et al. 2008). In contrast, DLAs probe neutral gas with $N(\mathrm{H\,I})$ ranging from $N(\mathrm{H\,I}) = 2 \times 10^{20}\,\mathrm{cm}^{-2}$ to $N(\mathrm{H\,I}) \approx 5 \times 10^{22}\,\mathrm{cm}^{-2}$. The range in $N(\mathrm{H\,I})$ corresponds to a range in surface mass density of atomic gas from $\Sigma_{\rm atomic} \approx 2\,M_\odot\,\mathrm{pc}^{-2}$ to $\Sigma_{\rm atomic} \gtrsim 200\,M_\odot\,\mathrm{pc}^{-2}$, which is comparable to the global average of total neutral gas surface mass density in local disk galaxies (e.g. Fig. 9.5). Therefore, DLAs offer an important laboratory for investigating the star formation relation in the distant Universe, and direct constraints can be obtained from searches of in situ star formation in DLAs.

In principle, the Schmidt-Kennicutt relation can be rewritten in terms of $N(\mathrm{H\,I})$ for pure atomic gas following

$$\Sigma_{\rm SFR} = K \times [N(\mathrm{H\,I})/N_0]^{\beta} \quad M_\odot\,\mathrm{yr}^{-1}\,\mathrm{kpc}^{-2}, \tag{9.5}$$

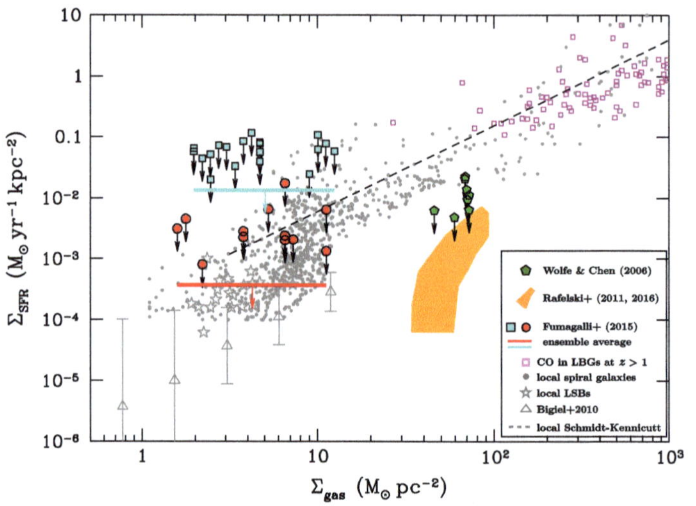

Fig. 9.5 The global star formation relation observed in nearby galaxies and at high redshifts. The correlation between the SFR per unit area ($\Sigma_{\rm SFR}$) and the total surface gas mass density ($\Sigma_{\rm gas}$), combining both atomic (H I) and molecular (H_2) for nearby spiral and starburst galaxies, is shown in *small filled circles* (Kennicutt 1998; Graciá-Carpio et al. 2008; Leroy et al. 2008), together with the best-fit Schmidt-Kennicutt relation shown by the *dashed line* (Kennicutt 1998). A reduced star formation efficiency is observed both in low surface brightness galaxies and in the outskirts of normal spirals, which are shown in *grey star symbols* and *open triangles*, respectively (Wyder et al. 2009; Bigiel et al. 2010). CO molecules have been detected in many massive starburst galaxies ($M_{\rm star} > 2.5\times10^{10}\,M_\odot$) at $z = 1$–3 (e.g. Baker et al. 2004; Genzel et al. 2010; Tacconi et al. 2013), which occur at the high surface density regime of the global star formation relation (*open squares*). In contrast, searching for in situ star formation in DLAs has revealed a reduced star formation efficiency in this metal-deficient gas. Specifically, *green points* and *orange-shaded area* represent the constraints obtained from comparing the sky coverage of low surface brightness emission with the incidence of DLAs (Wolfe and Chen 2006; Rafelski et al. 2011, 2016). *Cyan squares and red circles* represent the limits inferred from imaging searches of galaxies associated with individual DLAs, and the *cyan and red bars* represent the limiting $\Sigma_{\rm SFR}$ based on ensemble averages of the two samples (Fumagalli et al. 2015). The level of star formation observed in high-N(H I) DLAs (*green pentagons and orange-shaded* area) is comparable to what is seen in nearby low surface brightness galaxies and in the outskirts of normal spirals. See the main text for a detailed discussion

which is justified for regions probed by DLAs with a low molecular gas content (see Sect. 9.2 and Fig. 9.3d). For reference, the local Schmidt-Kennicutt relation has $K = 2.5\times10^{-4}\,M_\odot\,{\rm yr}^{-1}\,{\rm kpc}^{-2}$, $\beta = 1.4$ and $N_0 = 1.25\times10^{20}\,{\rm cm}^{-2}$ for a pure atomic hydrogen gas. Following Eq. (9.5), the N(H I) distribution function, $f_{N({\rm H\,I})}$ (e.g. Fig. 9.3a), can then be expressed in terms of the $\Sigma_{\rm SFR}$ distribution function, $h(\Sigma_{\rm SFR})$, which is the projected proper area per $d\Sigma_{\rm SFR}$ interval per comoving volume (Lanzetta et al. 2002). The $\Sigma_{\rm SFR}$ distribution function $h(\Sigma_{\rm SFR})$ is related to $f_{N({\rm H\,I})}$ according to $h(\Sigma_{\rm SFR})\,{\rm d}\Sigma_{\rm SFR} = (H_0/c)\,f_{N({\rm H\,I})}\,{\rm d}N({\rm H\,I})$.

This exercise immediately leads to two important observable quantities. First, the sky covering fraction (C_A) of star-forming regions in the redshift range, $[z_1, z_2]$, with an observed SFR per unit area in the interval of $\Sigma_{\rm SFR}$ and $\Sigma_{\rm SFR} + {\rm d}\Sigma_{\rm SFR}$ is determined following

$$C_{\rm A}[\Sigma_{\rm SFR}|N({\rm H\,I})] = \int_{z_1}^{z_2} \frac{c\,(1+z)^2}{H(z)} h(\Sigma_{\rm SFR})\,{\rm d}\Sigma_{\rm SFR}\,{\rm d}z, \tag{9.6}$$

where c is the speed of light and $H(z)$ is the Hubble expansion rate. Equation (9.6) is equivalent to $f_{N({\rm H\,I})}{\rm d}N({\rm H\,I}){\rm d}X$, where ${\rm d}X \equiv (1+z)^2\,H_0/H(z)\,{\rm d}z$ is the comoving absorption path length. In addition, the first moment of $h(\Sigma_{\rm SFR})$ leads to the comoving SFR density (Lanzetta et al. 2002; Hopkins et al. 2005),

$$\dot{\rho}_*(>\Sigma_{\rm SFR}^{\rm min}) = \int_{\Sigma_{\rm SFR}^{\rm min}}^{\Sigma_{\rm SFR}^{\rm max}} \Sigma_{\rm SFR} h(\Sigma_{\rm SFR})\,{\rm d}\Sigma_{\rm SFR}. \tag{9.7}$$

Constraints on the star formation relation at high redshift, namely, K and β in Eq. (9.5), can then be obtained by comparing $f_{N({\rm H\,I})}$-inferred $C_{\rm A}$ and $\dot{\rho}_*$ with results from searches of low surface brightness emission in deep galaxy survey data. Furthermore, estimates of missing light in low surface brightness regions can also be obtained using Eq. (9.7) (e.g. Lanzetta et al. 2002; Rafelski et al. 2011).

In practice, Eq. (9.5) is a correct representation only if disks are not well formed and a spherical symmetry applies to the DLAs. For randomly oriented disks, corrections for projection effects are necessary, and detailed formalisms are presented in Wolfe and Chen (2006) and Rafelski et al. (2011). In addition, the inferred surface brightness of in situ star formation in the DLA gas is extremely low after accounting for the cosmological surface brightness dimming. At $z = 2$–3, only DLAs at the highest-$N({\rm H\,I})$ end of $f_{N({\rm H\,I})}$ are expected to be visible in ultra-deep imaging data (cf. Lanzetta et al. 2002; Wolfe and Chen 2006). For example, DLAs of $N({\rm H\,I}) > 1.6 \times 10^{21}\,{\rm cm}^{-2}$ at $z \approx 3$ are expected to have V-band (corresponding roughly to rest-frame 1500 Å at $z = 3$) surface brightness $\mu_V \lesssim 28.4\,{\rm mag\,arcsec}^{-2}$, assuming the local Schmidt-Kennicutt relation. The expected low surface brightness of UV photons from young stars in high-redshift DLAs dictates the galaxy survey depth necessary to uncover star formation associated with the DLA gas. At $N({\rm H\,I}) > 1.6 \times 10^{21}\,{\rm cm}^{-2}$, roughly 3% of the sky ($C_{\rm A} \approx 0.03$) is expected to be covered by extended low surface brightness emission of $\mu_V \lesssim 28.4\,{\rm mag\,arcsec}^{-2}$. For comparison, the sky covering fraction of luminous starburst galaxies at $z = 2$–3 is less than 0.1%.

Available constraints for the star formation efficiency at $z = 1$–3 are shown in colour symbols in Fig. 9.5. Specifically, the Hubble Ultra-Deep Field (HUDF; Beckwith et al. 2006) V-band image offers sufficient depth for detecting objects of $\mu_V \approx 28.4\,{\rm mag\,arcsec}^{-2}$. Under the assumption that DLAs originate in regions distinct from known star-forming galaxies, an exhaustive search for extended low surface brightness emission in the HUDF has uncovered only a small number of these faint objects, far below the expectation from applying the local Schmidt-Kennicutt relation for DLAs of $N(\mathrm{H\,I}) > 1.6 \times 10^{21}\,\mathrm{cm}^{-2}$ following Eq. (9.6). Consequently, matching the observed limit on $\dot{\rho_*}$ from these faint objects with expectations from Eq. (9.7) has led to the conclusion that the star formation efficiency in metal-deficient atomic gas is more than 10× lower than expectations from the local Schmidt-Kennicutt relation (Wolfe and Chen 2006; green pentagons in Fig. 9.5).

On the other hand, independent observations of DLA galaxies at $z = 2$–3 have suggested that these absorbers are associated with typical star-forming galaxies at high redshifts. These include a comparable clustering amplitude of DLAs and these galaxies (e.g. Cooke et al. 2006), the findings of a few DLA galaxies with mass and SFR comparable to luminous star-forming galaxies found in deep surveys (e.g. Møller et al. 2002, 2004; Christensen et al. 2007) and detections of a DLA feature in the ISM of star-forming galaxies (e.g. Pettini et al. 2002; Chen et al. 2009; Dessauges-Zavadsky et al. 2010). If DLAs originate in neutral gas around known star-forming galaxies, then these luminous star-forming galaxies should be more spatially extended than has been realized. Searches for low surface brightness emission in the outskirts of these galaxies based on stacked images have indeed uncovered extended low surface brightness emission out to more than twice the optical extent of a single image. However, repeating the exercise of computing the cumulative $\dot{\rho}_*$ from Eq. (9.7) has led to a similar conclusion that the star formation efficiency is more than 10× lower in metal-deficient atomic gas at $z = 1$–3 than expectations from the local Schmidt-Kennicutt relation (Rafelski et al. 2011, 2016). The results are shown as the orange-shaded area in Fig. 9.5. In addition, the amount of missing light in the outskirts of these luminous star-forming galaxies is found to be ≈10% of what is observed in the core (Rafelski et al. 2011).

At the same time, imaging searches of individual DLA galaxies have been conducted for ≈30 DLAs identified along QSO sightlines that have high-redshift LLS serving as a natural coronagraph to block the background QSO glare, improving the imaging depth in areas immediate to the QSO sightline (Fumagalli et al. 2015). These searches have yielded only null results, leading to upper limits on the underlying surface brightness of the DLA galaxies (cyan squares and red circles in Fig. 9.5). While the survey depth is not sufficient for detecting associated star-forming regions in most DLAs in the survey sample of Fumagalli et al. (2015) based on the local Schmidt-Kennicutt relation, the ensemble average is beginning to place interesting limits (cyan and red arrows).

The lack of in situ star formation in DLAs may not be surprising given the low molecular gas content. In the local Universe, it is understood that the Schmidt-Kennicutt relation is driven primarily by molecular gas mass ($\Sigma_{\mathrm{molecular}}$), while the surface density of atomic gas (Σ_{atomic}) "saturates" at $\sim 10\,M_\odot\,\mathrm{pc}^{-2}$ beyond which the gas transitions into the molecular phase (e.g. Martin and Kennicutt 2001; Wong and Blitz 2002; Bigiel et al. 2008). As described in Sect. 9.2 and Fig. 9.3d, the transitional surface density from atomic to molecular is metallicity dependent. Therefore, the low star formation efficiency observed in DLA gas can be understood as a metallicity-dependent Schmidt-Kennicutt relation. This is qualitatively consistent with the observed low $\Sigma_{\rm SFR}$ in nearby low surface brightness galaxies (e.g. Wyder et al. 2009; star symbols in Fig. 9.5) and in the outskirts of normal spirals (e.g. Bigiel et al. 2010; open triangles in Fig. 9.5), where the ISM is found to be metal-poor (e.g. McGaugh 1994; Zaritsky et al. 1994; Bresolin et al. 2012). Numerical simulations incorporating a metallicity dependence in the H_2 production rate have also confirmed that the observed low star formation efficiency in DLAs can be reproduced in metal-poor gas (e.g. Gnedin and Kravtsov 2010).

A metallicity-dependent Schmidt-Kennicutt relation has wide-ranging implications in extragalactic research, from the physical origin of DLAs at high redshifts to star formation and chemical enrichment histories in different environments and to detailed properties of distant galaxies such as morphologies, sizes and cold gas content. It is clear from Fig. 9.5 that there exists a significant gap in the gas surface densities, between $\Sigma_{\rm gas} \approx 10\,M_\odot\,{\rm pc}^{-2}$ probed by these direct DLA galaxy searches and $\Sigma_{\rm gas} \approx 100\,M_\odot\,{\rm pc}^{-2}$ probed by CO observations of high-redshift starburst systems (open squares in Fig. 9.5). Continuing efforts targeting high-N(H I) DLAs (and therefore high $\Sigma_{\rm gas}$) at sufficient imaging depths are expected to place critical constraints on the star formation relation in low-metallicity environments at high redshifts. Similarly, spatially resolved maps of star formation and neutral gas at $z > 1$ to mean surface densities of $\Sigma_{\rm SFR} < 0.1\,M_\odot\,{\rm yr}^{-1}\,{\rm kpc}^{-2}$ and $\Sigma_{\rm atomic,\ molecular} \approx 10-100\,M_\odot\,{\rm pc}^{-2}$ will bridge the gap of existing observations and offer invaluable insights into the star formation relation in different environments.

9.5 From Neutral ISM to the Ionized Circumgalactic Medium

Beyond the neutral ISM, strong Lyα absorbers of N(H I) $\approx 10^{14-20}\,{\rm cm}^{-2}$ and associated metal-line absorbers offer a sensitive probe of the diffuse circumgalactic medium (CGM) to projected distances $d \approx$ 100–500 kpc (e.g. Fig. 9.2). But because the circumgalactic gas is significantly more ionized in the LLS and lower-N(H I) regime, measurements of its ionization state and metallicity bear considerable uncertainties and should be interpreted with caution.

Several studies have attempted to constrain the ionization state and metallicity of the CGM by considering the relative abundances of different ions at low- and high-ionization states (e.g. Savage et al. 2002; Stocke et al. 2006). For example, attributing observed O VI absorbers to cool ($T \sim 10^4$ K), photoionized gas irradiated by the metagalactic ionizing radiation field, the observed column density ratios between O VI and low-ionization transitions (such as C III and C IV) require extremely low gas densities of $n_{\rm H} \sim 10^{-5}\,{\rm cm}^{-3}$. Combining the inferred low gas density with observed N(O VI), which are typically $\gtrsim 10^{14.5}$ cm^{-2} in galactic haloes (e.g. Tumlinson et al. 2011), leads to a moderate gas metallicity of $\gtrsim 1/10$ Solar and unphysically large cloud sizes of $l_c \sim 1$ Mpc (e.g. Tripp et al. 2001; Savage et al. 2002; Stocke et al. 2006).[2] Excluding O VI due to possible origins in shocks or turbulent mixing layers (e.g. Heckman et al. 2002) and considering only relative abundances of low-ionization species increase estimated gas densities to $n_{\rm H} \sim 10^{-4}$–10^{-3} cm^{-3}. The inferred cloud sizes remain large with $l_c \sim 10$–100 kpc, in tension with what is observed locally for the HVCs. The implied thermal pressures in the cool gas phase are still two orders of magnitude lower than what is expected from pressure equilibrium with a hot ($T \approx 10^6$ K) medium (e.g. Stocke et al. 2013; Werk et al. 2014), indicating that these clouds would be crushed quickly. Considering non-equilibrium conditions (e.g. Gnat and Sternberg 2007; Oppenheimer and Schaye 2013) and the presence of local ionizing sources may help alleviate these problems (e.g. Cantalupo 2010), but the systematic uncertainties are difficult to quantify.

Nevertheless, exquisite details concerning extended halo gas have been learned over the past decade based on various samples of close galaxy and background QSO pairs. Because luminous QSOs are rare, roughly one QSO of $g \lesssim 18$ mag per square degree (e.g. Richards et al. 2006), absorption-line studies of the CGM against background QSO light have been largely limited to one probe per galaxy. Only in a few cases are multiple QSOs found at $d \lesssim 300$ kpc from a foreground galaxy (e.g. Norman et al. 1996; Keeney et al. 2013; Davis et al. 2015; Lehner et al. 2015; Bowen et al. 2016) for measuring coherence in spatial distribution and kinematics of extended gas around the galaxy. All of these cases are in the local Universe, because the relatively large angular extent of these galaxies on the sky increases the probability of finding more than one background QSO. This local sample has now been complemented with new studies, utilizing multiply lensed QSOs and close projected QSO pairs, which provide spatially resolved CGM absorption properties for a growing sample of galaxies at intermediate redshifts (e.g. Chen et al. 2014; Rubin et al. 2015; Zahedy et al. 2016).

With one QSO probe per halo, a two-dimensional map of CGM absorption properties can be established based on an ensemble average of a large sample of QSO-galaxy pairs ($N_{\rm pair} \sim 100$–1000). Figure 9.6 summarizes some of the observable quantities of the CGM. First, panels (a) and (b) at the top display the radial profiles of rest-frame absorption equivalent width ($W_{\rm r}$) for different absorption transitions, including hydrogen Lyα; low-ionization C II and Si II; intermediate-ionization Si III, Si IV and C IV; and high-ionization O VI absorption transitions, colour-coded in black, red, orange, green, blue, magenta and dark purple, respectively. For transitions that are not detected, a 2-σ upper limit is shown as a downward arrow. Because of the large number of data points, the upper limits are shown in pale colours for

[2] For comparison, the sizes of extended HVC complexes at $d \sim 10$ kpc from the MW disk are a few to 15 kpc across (e.g. Putman et al. 2012). HVCs at larger distances are found to be more compact, $\lesssim 2$ kpc (e.g. Westmeier et al. 2008; Lockman et al. 2012; Giovanelli et al. 2013).

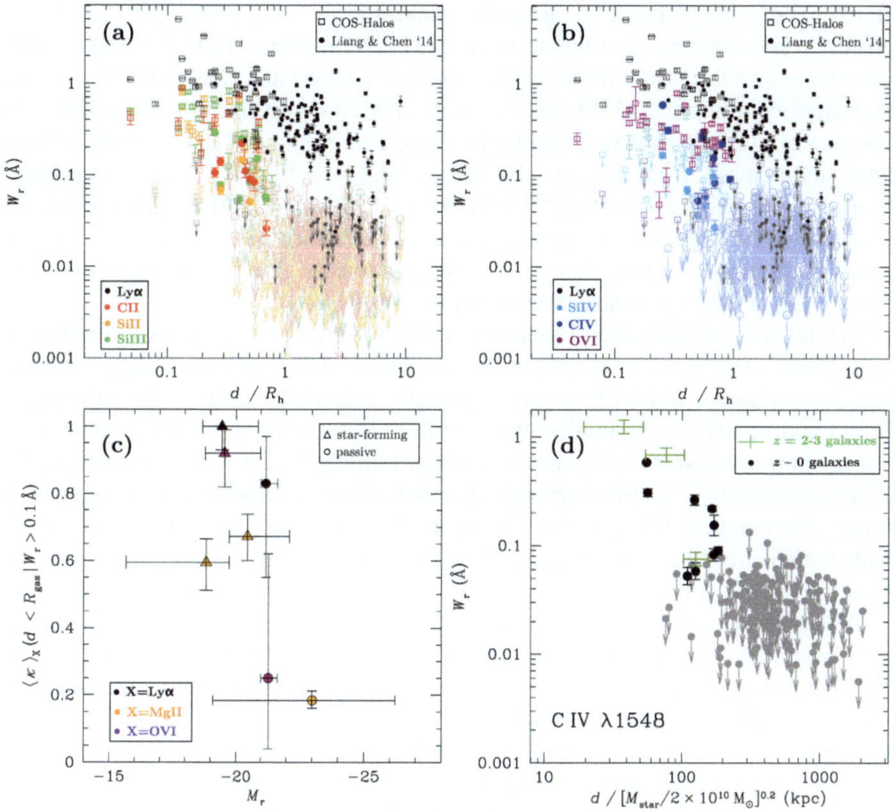

Fig. 9.6 Observed absorption properties of halo gas around galaxies. The *top panels* display the radial profiles of rest-frame absorption equivalent width (W_r) versus halo-radius R_h-normalized projected distance for different absorption transitions. Low-ionization transitions are presented in panel (**a**) and high-ionization transitions in panel (**b**). Lyα data points are presented in both panels for cross comparison. The galaxy sample includes 44 galaxies at $z \approx 0.25$ from the COS-Halos project (*open squares*; Tumlinson et al. 2011, 2013; Werk et al. 2013) and ~200 galaxies at $z \approx 0.04$ from public archives (*circles*; Liang and Chen 2014), for which high-quality, ultraviolet QSO spectra are available for constraining the presence or absence of multiple ions in individual haloes. Different transitions are colour-coded to highlight the differences in their spatial distributions. For transitions that are not detected, a 2-σ upper limit is shown by a *downward arrow*. No heavy ions are found beyond $d = R_h$, while Lyα continues to be seen to larger distances. Panel (**c**) displays the ensemble average of gas covering fraction ($\langle\kappa\rangle$) as a function of absolute r-band magnitude (M_r) for Lyα (*black symbols*), Mg II (*orange*) and O VI (*purple*). Star-forming galaxies (*triangles*) on average are fainter and exhibit higher covering fractions of hydrogen and chemically enriched gas probed by both low- and high-ionization species than passive galaxies (*circles*). Measurements of Lyα and O VI absorbing gas are based on COS-Halos galaxies for $R_{gas} = R_h$. Measurements of Mg II absorbing gas are based on ≈260 star-forming galaxies at $z \approx 0.25$ (Chen et al. 2010a) and ~38,000 passive luminous red galaxies at $z \approx 0.5$ (Huang et al. 2016) for $R_{gas} = R_h/3$. Panel (**d**) illustrates the apparent constant nature of mass-normalized radial profiles of CGM absorption since $z \approx 3$ (e.g. Chen 2012; Liang and Chen 2014). The high-redshift observations are based on mean C IV absorption in stacked spectra of ~500 starburst galaxies with a mean stellar mass and dispersion of $\langle \log M_{star} \rangle = 9.9 \pm 0.5$ (Steidel et al. 2010), and the low-redshift observations are for ~200 individual galaxies with $\langle \log M_{star} \rangle = 9.7 \pm 1.1$ and modest SFR (Liang and Chen 2014)

clarity. The galaxy sample includes 44 galaxies at $z \approx 0.25$ from the COS-Halos project (open squares; Tumlinson et al. 2011, 2013; Werk et al. 2013) and $\sim$200 galaxies at $z \approx 0.04$ from public archives (circles; Liang and Chen 2014), for which high-quality, ultraviolet QSO spectra are available for constraining the presence or absence of multiple ions in individual haloes. These galaxies span four decades in total stellar mass, from $M_{\rm star} \approx 10^7\,M_\odot$ to $M_{\rm star} \approx 10^{11}\,M_\odot$, and a wide range in SFR, from SFR $< 0.1\,M_\odot\,{\rm yr}^{-1}$ to SFR $> 10\,M_\odot\,{\rm yr}^{-1}$. Diffuse gas is observed beyond $d = 50$ kpc around distant galaxies, extending the detection limit of H I gas in inner galactic haloes from 21 cm observations (e.g. Fig. 9.2) to lower column density gas at larger distances and higher redshifts.

While W_r is typically found to decline steadily with increasing d for all transitions (e.g. Chen 2012; Werk et al. 2014), the scatters are large. Including the possibility that more massive haloes have more spatially extended halo gas, the halo radius R_h-normalized W_r-d distribution indeed displays substantially reduced scatters in the radial profiles shown in panels (a) and (b) of Fig. 9.6. A reduced scatter in the R_h-normalized W_r-d distribution indicates that galaxy mass plays a dominant role in driving the extent of halo gas. In addition, it also confirms that accurate associations between absorbers and absorbing galaxies have been found for the majority of the systems.

A particularly interesting feature in Fig. 9.6 is a complete absence of heavy ions beyond $d = R_h$, while detections of Lyα continue to larger distances. The absence of heavy ions at $d > R_h$, which is observed for a wide range of ionization states, strongly indicates that chemical enrichment is confined within individual galaxy haloes. This finding applies to both low-mass dwarfs and massive galaxies. However, it should also be noted that heavy ions are observed beyond R_h for galaxies with close neighbours (e.g. Borthakur et al. 2013; Johnson et al. 2015), suggesting that environmental effects play a role in distributing heavy elements beyond the enriched gaseous haloes of individual galaxies. Comparing panels (a) and (b) of Fig. 9.6 also shows that within individual galaxy haloes, a global ionization gradient is seen with more highly ionized gas detected at larger distances. For instance, the observed W_r declines to $<$0.1 Å at $d \approx 0.5\,R_h$ for C II and Si II, while C IV and O VI absorbers of $W_r > 0.1$ Å continue to be found beyond $0.5\,R_h$.

The observed W_r versus d (or d/R_h) based on a blind survey of absorption features in the vicinities of known galaxies also enables measurements of gas covering fraction.[3] The mean gas covering fraction ($\langle\kappa\rangle$) can be measured by a simple accounting of the fraction of galaxies in an annular area displaying associated

[3]A blind survey of absorption features around known galaxies differs fundamentally from a blind survey of galaxies around known absorbers (e.g. Kacprzak et al. 2008). By design, a blind galaxy survey around known absorbers excludes transparent sightlines and does not provide the sample necessary for measuring the incidence and covering fraction of absorbing species. In addition, because of limited survey depths, a blind galaxy survey is more likely to find more luminous members at larger d that are correlated with the true absorbing galaxies which are fainter and closer to the QSO sightline, resulting in a significantly larger scatter in the W_r versus d distribution (e.g. Kacprzak et al. 2008; Nielsen et al. 2013).

absorbers with W_r exceeding some detection threshold W_0, and uncertainties can be estimated based on a binomial distribution function. Dividing the sample into different projected distance bins, it is clear from Fig. 9.6a, b that the fraction of non-detections increases with increasing projected distance, resulting in a declining $\langle\kappa\rangle$ with increasing *d* for all transitions observed (see also Chen et al. 2010a; Werk et al. 2014; Huang et al. 2016).

It is also interesting to examine how $\langle\kappa\rangle$ depends on galaxy properties. Figure 9.6c displays $\langle\kappa\rangle$ observed within a fiducial gaseous radius R_{gas} for star-forming (triangles) and passive (circles) galaxies. The measurements are made for Lyα (black symbols), Mg II (orange) and O VI (purple) with a threshold of $W_0 = 0.1$ Å and shown in relation to the absolute *r*-band magnitude (M_r). Error bars represent the 68% confidence interval. The absolute *r*-band magnitude is a direct observable of a galaxy and serves as a proxy for its underlying total stellar mass. Measurements of Lyα and O VI absorbing gas are based on COS-Halos galaxies for $R_{gas} = R_h$ (see also Johnson et al. (2015) for a sample compiled from the literature). Measurements of Mg II absorbing gas are based on $\approx$260 star-forming galaxies at $z \approx 0.25$ (Chen et al. 2010a) and $\sim$38,000 passive luminous red galaxies at $z \approx 0.5$ (Huang et al. 2016) for $R_{gas} = R_h/3$ (e.g. Chen and Tinker 2008). The larger sample sizes led to better constrained $\langle\kappa\rangle$ for Mg II absorbing gas in galactic haloes. In general, star-forming galaxies on average are fainter and less massive and exhibit a higher covering fraction of chemically enriched gas than passive galaxies (see also Johnson et al. 2015). At the same time, the covering fraction of chemically enriched gas is definitely non-zero around massive quiescent galaxies.

Comparing the radial profiles of CGM absorption at different redshifts offers additional insights into the evolution history of the CGM, which in turn helps distinguish between different models for chemical enrichment in galaxy haloes. The radial profiles of the CGM have been found to evolve little since $z \sim 3$ (e.g. Chen 2012), even though the star-forming properties in galaxies have evolved significantly. Figure 9.6d illustrates the apparent constant nature of mass-normalized radial profiles of C IV absorption in galactic haloes (Liang and Chen 2014). The high-redshift observations are based on stacked spectra of $\sim$500 starburst galaxies with a mean stellar mass and dispersion of $\langle \log M_{star} \rangle = 9.9 \pm 0.5$ (Steidel et al. 2010) and a mean SFR of $\langle \mathrm{SFR} \rangle \approx 30$–$60\,M_\odot\,\mathrm{yr}^{-1}$ (e.g. Erb et al. 2006b; Reddy et al. 2012). The low-redshift galaxy sample contains individual measurements of $\sim$200 galaxies with $\langle \log M_{star} \rangle = 9.7 \pm 1.1$ and more quiescent star-forming activities of $\langle \mathrm{SFR} \rangle \sim 1\,M_\odot\,\mathrm{yr}^{-1}$ (Chen 2012; Liang and Chen 2014). The constant mass-normalized CGM radial profiles between galaxies of very different SFR indicate that mass (rather than SFR) is a dominant factor that determines the CGM properties over a cosmic time interval. This is consistent with previous findings that CGM absorption properties depend strongly on galaxy mass but only weakly on SFR (e.g. Chen et al. 2010b), but at odds with popular models that attribute metal-line absorbers to starburst-driven outflows (e.g. Steidel et al. 2010; Ménard et al. 2011).

A discriminating characteristic of starburst-driven outflows is their distinctly nonspherical distribution in galactic haloes in the presence of a well-formed star-forming disk. Specifically, galactic-scale outflows are expected to travel preferentially along the polar axis where the gas experiences the least resistance (e.g.

Heckman et al. 1990). In contrast, accretion of the IGM is expected to proceed along the disk plane with $\lesssim$10% covering fraction on the sky (e.g. Faucher-Giguère and Kereš 2011; Fumagalli et al. 2011). Such azimuthal dependence of the spatial distribution of infalling and outflowing gas is fully realized in state-of-the-art cosmological zoom-in simulations (e.g. Shen et al. 2013; Agertz and Kravtsov 2015). Observations of $z \approx 0.7$ galaxies have shown that at $d < 50$ kpc, the mean Mg II absorption equivalent width within 45° of the minor axis is twice of the mean value found within 45° of the major axis, although such azimuthal dependence is not observed at $d > 50$ kpc (Bordoloi et al. 2011). The observed azimuthal dependence of the mean Mg II absorption strength is qualitatively consistent with the expectation that these heavy ions originate in starburst-driven outflows, and the lack of such azimuthal dependence implies that starburst outflows are confined to the inner halo of $d \lesssim 50$ kpc.

Many subsequent studies have generalized this observed azimuthal dependence at $d < 50$ kpc to larger distances and attributed absorbers detected near the minor axis to starburst-driven outflows and those found near the major axis to accretion (e.g. Bouché et al. 2012; Kacprzak et al. 2015). However, a causal connection between the observed absorbing gas and either outflows or accretion remains to be established. While gas metallicity may serve as a discriminator with the expectation of starburst outflows being more metal enriched relative to the low-density IGM, uncertainties arise due to poorly understood chemical mixing and metal transport (e.g. Tumlinson 2006). Incidentally, a relatively strong Mg II absorber has been found at $d \approx 60$ kpc along the minor axis of a starburst galaxy, but the metallicity of the absorbing gas is ten times lower than what is observed in the ISM (Kacprzak et al. 2014), highlighting the caveat of applying gas metallicity as the sole parameter for distinguishing between accretion and outflows.

Figure 9.7 presents visual comparisons of the geometric alignment of galaxy major axis relative to the QSO sightline and the observed CGM absorption strength. The figure at the top displays the observed O VI column density, N(O VI), versus d for COS-Halos galaxies at $z \approx 0.2$ (Tumlinson et al. 2013). The bottom figure displays comparisons of N(O VI) and N(Mg II) for these galaxies. The absorption-line measurements are adopted from Werk et al. (2013). When spatially resolved images are available, the data points are replaced with an image panel of the absorbing galaxy. Each panel is 25 proper kpc on a side and is oriented such that the QSO sightline falls on the y-axis at the corresponding N(O VI) of the galaxy. The relative alignment between galaxy major axis and the background QSO sightline cannot be determined, if the galaxies are face-on with a minor-to-major axis ratio > 0.7 or if the galaxies display irregular/asymmetric morphologies. These galaxies are labelled "F" and "A", respectively. For galaxies that clearly display a smooth and elongated morphology, the orientation of the major axis can be accurately measured. Galaxies with the QSO located within 30° of the minor axis are labelled "m", while galaxies with the QSO located within 30° of the major axis are labelled "M". Galaxies with the QSO sightline occuring intermediate (30–60°) between the minor and major axis are labelled "45". Star-forming galaxies are colour-coded in blue,

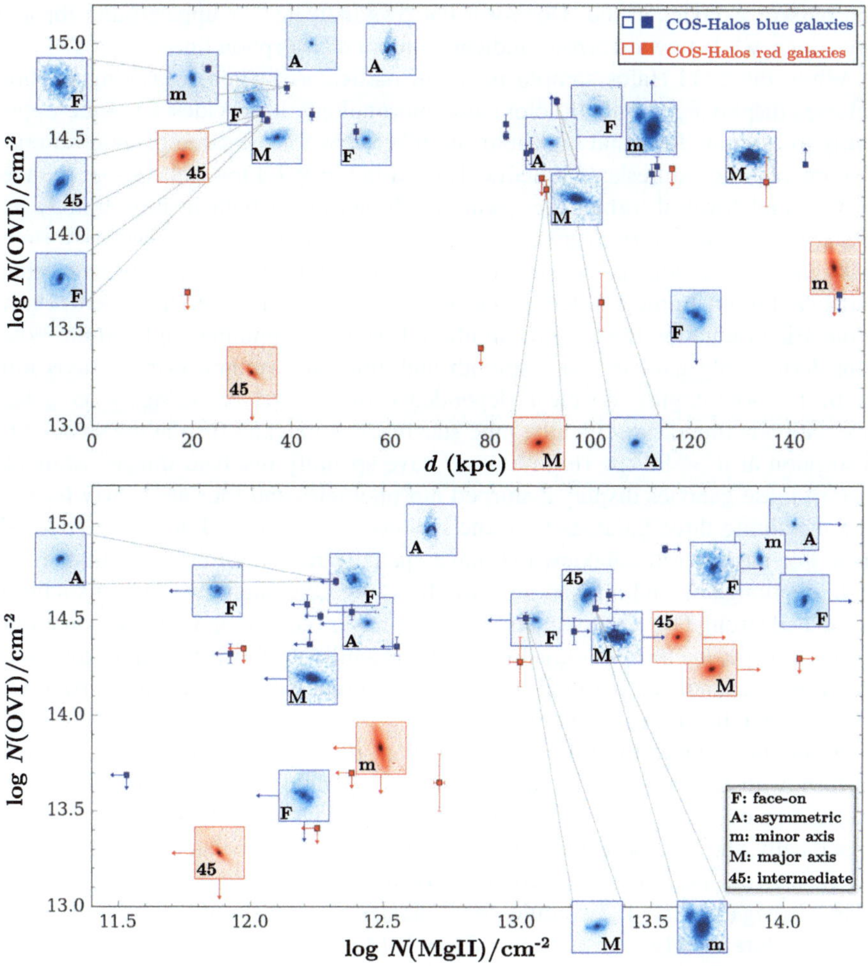

Fig. 9.7 Visual comparisons of the geometric alignment of galaxy major axis relative to the QSO sightline and the observed CGM absorption strength (by Rebecca Pierce). *Top*: observed O VI column density, N(O VI), versus d for COS-Halos star-forming (in *blue*) and passive (in *red*) galaxies (Tumlinson et al. 2011). *Bottom*: comparisons of N(O VI) and N(Mg II) for the COS-Halos galaxies from Werk et al. (2013). When spatially resolved images are available, the data points are replaced with an image panel of the absorbing galaxy. Each panel is 25 proper kpc on a side and is oriented such that the QSO sightline occurs on the y-axis at the corresponding O VI column density of the galaxy. Disk alignments cannot be determined for face-on galaxies (minor-to-major axis ratio > 0.7) and galaxies displaying irregular/asymmetric morphologies, which are labeled "F" and "A", respectively. Galaxies with the QSO located within 30° of the minor axis are labeled "m" in the lower-left corner, while galaxies with the QSO located within 30° of the major axis are labelled "M". Galaxies with the QSO sightline occurring intermediate (30–60°) between the minor and major axis are labelled "45". Downward arrows indicate 2-σ upper limits for non-detections, while upward arrows indicate saturated absorption lines. The COS-Halos galaxy sample provides a unique opportunity to examine low- and high-ionization halo gas for the same galaxies at once. Galaxies surrounded by O VI and Mg II absorbing gas clearly exhibit a broad range both in morphology and in disk orientation. In addition, the observed N(Mg II) displays a significantly larger scatter than N(O VI)

and passive galaxies in red. Downward arrows indicate 2-σ upper limits for non-detections, while upward arrows indicate saturated absorption lines.

While the COS-Halos sample is small, particularly when restricting to those galaxies displaying a smooth, elongated morphology, it provides a unique opportunity to examine low- and high-ionization halo gas for the same galaxies at once. Two interesting features are immediately clear in Fig. 9.7. First, galaxies surrounded by O VI and Mg II absorbing gas exhibit a broad range both in morphology and in star formation history, from compact quiescent galaxies to regular star-forming disks and to interacting pairs. The diverse morphologies in O VI and Mg II absorbing galaxies illuminate the challenge and uncertainties in characterizing their relative geometric orientation to the QSO sightline based on azimuthal angle alone. When considering only galaxies with smooth and elongated (minor-to-major axis ratio < 0.7) morphologies, no clear dependence of N(O VI) or N(Mg II) on galaxy orientation is found. Specifically, nine star-forming galaxies displaying strong O VI absorption at $d < 80$ kpc (log N(O VI)) have spatially resolved images available. Two of these galaxies display disturbed morphologies and four are nearly face-on. The remaining three galaxies have the inclined disks oriented at 0°, 45° and 90° each. For passive red galaxies, two have spatially resolved images available, and both are elongated and aligned at ≈45° from the QSO sightline. One displays an associated strong O VI absorber, and the other has no corresponding O VI detections. At $d > 80$ kpc, the morphology distribution is similar to those at smaller distances. No strong dependence is found between the presence or absence of a strong O VI absorber and the galaxy orientation.

In addition, while the observed N(O VI) at $d < 100$ kpc appears to be more uniformly distributed with a mean and scatter of log $N(\text{O VI}) = 14.5 \pm 0.3$ (Tumlinson et al. 2011), the observed N(Mg II) displays a significantly larger scatter. Specifically, the face-on galaxy at $d \approx 32$ kpc with an associated O VI absorber of log $N(\text{O VI}) \approx 14.7$ does not have an associated Mg II absorber detected to a limit of log $N(\text{Mg II}) \approx 12.4$. Two quiescent galaxies at $z \approx 20$ and 90 kpc (red panels) exhibit saturated Mg II absorption of log $N(\text{Mg II}) > 13.5$ and similarly strong O VI of log $N(\text{O VI}) \approx 14.3$. A small scatter implies a more uniformly distributed medium, while a large scatter implies a more clumpy nature of the absorbing gas or a larger variation between different galaxy haloes. Such distinct spatial distributions between low- and high-ionization gas further highlight the complex nature of the chemically enriched CGM, which depends on more than the geometric alignment of the galaxies. A three-dimensional model of gas kinematics that takes full advantage of the detailed morphologies and star formation history of the galaxies is expected to offer a deeper understanding of the physical origin of chemically enriched gas in galaxy haloes (e.g. Gauthier and Chen 2012; Chen et al. 2014; Diamond-Stanic et al. 2016).

9.6 Summary

QSO absorption spectroscopy provides a sensitive probe of both neutral medium and diffuse ionized gas in the distant Universe. It extends 21 cm maps of gaseous structures around low-redshift galaxies both to lower gas column densities and to higher redshifts. Specifically, DLAs of $N(\text{H I}) \gtrsim 2 \times 10^{20}\,\text{cm}^{-2}$ probe neutral gas in the ISM of distant star-forming galaxies, LLS of $N(\text{H I}) > 10^{17}\,\text{cm}^{-2}$ probe optically thick HVCs and gaseous streams in and around galaxies and strong Lyα absorbers of $N(\text{H I}) \approx 10^{14-17}\,\text{cm}^{-2}$ and associated metal-line absorption transitions, such as Mg II, C IV and O VI, trace chemically enriched, ionized gas and starburst outflows. Over the last decade, an unprecedentedly large number of $\sim$10,000 DLAs have been identified along random QSO sightlines to provide robust statistical characterizations of the incidence and mass density of neutral atomic gas at $z \lesssim 5$. Extensive follow-up studies have yielded accurate measurements of chemical compositions and molecular gas content for this neutral gas cross-section selected sample from $z \approx 5$ to $z \approx 0$ (Sect. 9.2). Combining galaxy surveys with absorption-line observations of gas around galaxies has enabled comprehensive studies of baryon cycles between star-forming regions and low-density gas over cosmic time. DLAs, while being rare as a result of a small cross-section of neutral medium in the Universe, have offered a unique window into gas dynamics and chemical enrichment in the outskirts of star-forming disks (Sect. 9.3), as well as star formation physics at high redshifts (Sect. 9.4). Observations of strong Lyα absorbers and associated ionic transitions around galaxies have also demonstrated that galaxy mass is a dominant factor in driving the extent of chemically enriched halo gas and that chemical enrichment is well confined within galactic haloes for both low-mass dwarfs and massive galaxies (Sect. 9.5).

With new observations carried out using new, multiplex instruments, continuing progress is expected in further advancing our understanding of baryonic cycles in the outskirts of galaxies over the next few years. These include, but are not limited to, (1) direct constraints for the star formation relation in different environments (e.g. Gnedin and Kravtsov 2010), particularly for star-forming galaxies at $z \gtrsim 2$ in low surface density regimes of $\Sigma_{\text{SFR}} < 0.1\,M_\odot\,\text{yr}^{-1}\,\text{kpc}^{-2}$ and $\Sigma_{\text{gas}} \approx 10\text{–}100\,M_\odot\,\text{pc}^{-2}$; (2) an empirical understanding of galaxy environmental effects in distributing heavy elements to large distances based on deep galaxy surveys carried out in a large number of QSO fields (e.g. Johnson et al. 2015); and (3) a three-dimensional map of gas flows in the circumgalactic space that combines absorption-line kinematics along multiple sightlines with optical morphologies of the absorbing galaxies and emission morphologies of extended gas around the galaxies (e.g. Rubin et al. 2011; Chen et al. 2014; Zahedy et al. 2016). Wide-field IFUs on existing large ground-based telescopes substantially increase the efficiency in faint galaxy surveys (e.g. Bacon et al. 2015) and in revealing extended low surface brightness emission features around high-redshift galaxies (e.g. Cantalupo et al. 2014; Borisova et al. 2016). The *James Webb Space Telescope* (*JWST*), which is scheduled to be launched in October 2018, will expand the sensitivity of detecting faint star-forming galaxies in the early Universe. Combining deep infrared images from *JWST* and CO (or dust continuum) maps from ALMA will lead to critical constraints for the star formation relation in low surface density regimes.

Acknowledgements The author wishes to dedicate this review to the memory of Arthur M. Wolfe for his pioneering and seminal work on the subject of damped Lyα absorbers and for inspiring generations of scientists to pursue original and fundamental research. The author thanks Nick Gnedin, Sean Johnson, Rebecca Pierce, Marc Rafelski and Fakhri Zahedy for providing helpful input and comments. In preparing this review, the author has made use of NASA's Astrophysics Data System Bibliographic Services.

References

Agertz, O., Kravtsov, A.V.: On the interplay between star formation and feedback in galaxy formation simulations. Astrophys. J. **804**, 18 (2015). doi:10.1088/0004-637X/804/1/18, 1404.2613

Agertz, O., Teyssier, R., Moore, B.: Disc formation and the origin of clumpy galaxies at high redshift. Mon. Not. R. Astron. Soc. **397**, L64–L68 (2009). doi:10.1111/j.1745-3933.2009.00685.x, 0901.2536

Aguirre, A., Dow-Hygelund, C., Schaye, J., Theuns, T.: Metallicity of the intergalactic medium using pixel statistics. IV. Oxygen. Astrophys. J. **689**, 851–864 (2008). doi:10.1086/592554, 0712.1239

Altay, G., Theuns, T., Schaye, J., Crighton, N.H.M., Dalla Vecchia, C.: Through thick and thin–H I absorption in cosmological simulations. Astrophys. J. Lett. **737**, L37 (2011). doi:10.1088/2041-8205/737/2/L37, 1012.4014

Altay, G., Theuns, T., Schaye, J., Booth, C.M., Dalla Vecchia, C.: The impact of different physical processes on the statistics of Lyman-limit and damped Lyman α absorbers. Mon. Not. R. Astron. Soc. **436**, 2689–2707 (2013). doi:10.1093/mnras/stt1765, 1307.6879

Bacon, R., Brinchmann, J., Richard, J., Contini, T., Drake, A., Franx, M., Tacchella, S., Vernet, J., Wisotzki, L., Blaizot, J., Bouché, N., Bouwens, R., Cantalupo, S., Carollo, C.M., Carton, D., Caruana, J., Clément, B., Dreizler, S., Epinat, B., Guiderdoni, B., Herenz, C., Husser, T.O., Kamann, S., Kerutt, J., Kollatschny, W., Krajnovic, D., Lilly, S., Martinsson, T., Michel-Dansac, L., Patricio, V., Schaye, J., Shirazi, M., Soto, K., Soucail, G., Steinmetz, M., Urrutia, T., Weilbacher, P., de Zeeuw, T.: The MUSE 3D view of the Hubble Deep Field South. Astron. Astrophys. **575**, A75 (2015). doi:10.1051/0004-6361/201425419, 1411.7667

Baker, A.J., Tacconi, L.J., Genzel, R., Lehnert, M.D., Lutz, D.: Molecular gas in the lensed Lyman break galaxy cB58. Astrophys. J. **604**, 125–140 (2004). doi:10.1086/381798, astro-ph/0312099

Balestra, I., Tozzi, P., Ettori, S., Rosati, P., Borgani, S., Mainieri, V., Norman, C., Viola, M.: Tracing the evolution in the iron content of the intra-cluster medium. Astron. Astrophys. **462**, 429–442 (2007). doi:10.1051/0004-6361:20065568, astro-ph/0609664

Beckwith, S.V.W., Stiavelli, M., Koekemoer, A.M., Caldwell, J.A.R., Ferguson, H.C., Hook, R., Lucas, R.A., Bergeron, L.E., Corbin, M., Jogee, S., Panagia, N., Robberto, M., Royle, P., Somerville, R.S., Sosey, M.: The Hubble ultra deep field. Astron. J. **132**, 1729–1755 (2006). doi:10.1086/507302, astro-ph/0607632

Bergeron, J., Boissé, P.: A sample of galaxies giving rise to Mg II quasar absorption systems. Astron. Astrophys. **243**, 344–366 (1991)

Bergeron, J., Stasińska, G.: Absorption line systems in QSO spectra - properties derived from observations and from photoionization models. Astron. Astrophys. **169**, 1–13 (1986)

Bigiel, F., Leroy, A., Walter, F., Brinks, E., de Blok, W.J.G., Madore, B., Thornley, M.D.: The star formation law in nearby galaxies on sub-Kpc scales. Astron. J. **136**, 2846–2871 (2008). doi:10.1088/0004-6256/136/6/2846, 0810.2541

Bigiel, F., Leroy, A., Walter, F., Blitz, L., Brinks, E., de Blok, W.J.G., Madore, B.: Extremely inefficient star formation in the outer disks of nearby galaxies. Astron. J. **140**, 1194–1213 (2010). doi:10.1088/0004-6256/140/5/1194, 1007.3498

Bird, S., Vogelsberger, M., Haehnelt, M., Sijacki, D., Genel, S., Torrey, P., Springel, V., Hernquist, L.: Damped Lyman α absorbers as a probe of stellar feedback. Mon. Not. R. Astron. Soc. **445**, 2313–2324 (2014). doi:10.1093/mnras/stu1923, 1405.3994

Boksenberg, A., Sargent, W.L.W.: Properties of QSO metal-line absorption systems at high redshifts: nature and evolution of the absorbers and new evidence on escape of ionizing radiation from galaxies. Astrophys. J. Suppl. Ser. **218**, 7 (2015). doi:10.1088/0067-0049/218/1/7, 1410.3784

Bordoloi, R., Lilly, S.J., Knobel, C., Bolzonella, M., Kampczyk, P., Carollo, C.M., Iovino, A., Zucca, E., Contini, T., Kneib, J.P., Le Fevre, O., Mainieri, V., Renzini, A., Scodeggio, M., Zamorani, G., Balestra, I., Bardelli, S., Bongiorno, A., Caputi, K., Cucciati, O., de la Torre, S., de Ravel, L., Garilli, B., Kovač, K., Lamareille, F., Le Borgne, J.F., Le Brun, V., Maier, C., Mignoli, M., Pello, R., Peng, Y., Perez Montero, E., Presotto, V., Scarlata, C., Silverman, J., Tanaka, M., Tasca, L., Tresse, L., Vergani, D., Barnes, L., Cappi, A., Cimatti, A., Coppa, G., Diener, C., Franzetti, P., Koekemoer, A., López-Sanjuan, C., McCracken, H.J., Moresco, M., Nair, P., Oesch, P., Pozzetti, L., Welikala, N.: The radial and azimuthal profiles of Mg II absorption around $0.5 < z < 0.9$ zCOSMOS Galaxies of different colors, masses, and environments. Astrophys. J. **743**, 10 (2011). doi:10.1088/0004-637X/743/1/10, 1106.0616

Borisova, E., Cantalupo, S., Lilly, S.J., Marino, R.A., Gallego, S.G., Bacon, R., Blaizot, J., Bouché, N., Brinchmann, J., Carollo, C.M., Caruana, J., Finley, H., Herenz, E.C., Richard, J., Schaye, J., Straka, L.A., Turner, M.L., Urrutia, T., Verhamme, A., Wisotzki, L.: Ubiquitous giant Ly α nebulae around the brightest quasars at $z \sim 3.5$ revealed with MUSE (2016). ArXiv e-prints 1605.01422

Borthakur, S., Heckman, T., Strickland, D., Wild, V., Schiminovich, D.: The impact of starbursts on the circumgalactic medium. Astrophys. J. **768**, 18 (2013). doi:10.1088/0004-637X/768/1/18, 1303.1183

Bouché, N., Hohensee, W., Vargas, R., Kacprzak, G.G., Martin, C.L., Cooke, J., Churchill, C.W. Physical properties of galactic winds using background quasars. Mon. Not. R. Astron. Soc. **426**, 801–815 (2012) doi:10.1111/j.1365-2966.2012.21114.x, 1110.5877

Bowen, D.V., Chelouche, D., Jenkins, E.B., Tripp, T.M., Pettini, M., York, D.G., Frye, B.L. The structure of the circumgalactic medium of galaxies: cool accretion inflow around NGC 1097. Astrophys. J. **826**, 50 (2016). doi:10.3847/0004-637X/826/1/50, 1605.04907

Braun, R.: Cosmological evolution of atomic gas and implications for 21 cm H I absorption. Astrophys. J. **749**, 87 (2012). doi:10.1088/0004-637X/749/1/87, 1202.1840

Bresolin, F., Kennicutt, R.C., Ryan-Weber, E.: Gas metallicities in the extended disks of NGC 1512 and NGC 3621. Chemical signatures of metal mixing or enriched gas accretion? Astrophys. J. **750**, 122 (2012). doi:10.1088/0004-637X/750/2/122, 1203.0956

Broeils, A.H., Rhee, M.H.: Short 21-cm WSRT observations of spiral and irregular galaxies. HI properties. Astron. Astrophys. **324**, 877–887 (1997)

Burbidge, E.M., Burbidge, G.R. Recent investigations of groups and clusters of galaxies. Astron. J. **66**, 541–550 (1961). doi:10.1086/108461

Cantalupo, S.: Stars quenching stars: how photoionization by local sources regulates gas cooling and galaxy formation. Mon. Not. R. Astron. Soc. **403**, L16–L20 (2010). doi:10.1111/j.1745-3933.2010.00806.x, 0912.4149

Cantalupo, S., Arrigoni-Battaia, F., Prochaska, J.X., Hennawi, J.F., Madau, P.: A cosmic web filament revealed in Lyman-α emission around a luminous high-redshift quasar. Nature **506**, 63–66 (2014). doi:10.1038/nature12898, 1401.4469

Cazaux, S., Spaans, M.: Molecular hydrogen formation on dust grains in the high-redshift universe. Astrophys. J. **611**, 40–51 (2004). doi:10.1086/422087, astro-ph/0404340

Chen, H.W.: The unchanging circumgalactic medium over the past 11 billion years. Mon. Not. R. Astron. Soc. **427**, 1238–1244 (2012). doi:10.1111/j.1365-2966.2012.22053.x, 1209.1094

Chen, H.W., Lanzetta, K.M.: The nature of damped Lyα absorbing galaxies at z <= 1: a photometric redshift survey of damped Lyα absorbers. Astrophys. J. **597**, 706–729 (2003). doi:10.1086/378635, astro-ph/0308190
Chen, H.W., Prochaska, J.X.: The origin of a chemically enriched Lyα absorption system at Z = 0.167. Astrophys. J. Lett. **543**, L9–L13 (2000). doi:10.1086/318179, astro-ph/0009001
Chen, H.W., Tinker, J.L.: The Baryon content of dark matter halos: empirical constraints from Mg II absorbers. Astrophys. J. **687**, 745–756 (2008). doi:10.1086/591927, 0801.2169
Chen, H.W., Kennicutt R.C. Jr., Rauch, M.: Abundance profiles and kinematics of damped Lyα absorbing galaxies at z < 0.651. Astrophys. J. **620**, 703–722 (2005). doi:10.1086/427088, astro-ph/0411006
Chen, H.W., Perley, D.A., Pollack, L.K., Prochaska, J.X., Bloom, J.S., Dessauges-Zavadsky, M., Pettini, M., Lopez, S., Dall'aglio, A., Becker, G.D.: High-redshift starbursting dwarf galaxies revealed by γ-ray burst afterglows. Astrophys. J. **691**, 152–174 (2009). doi:10.1088/0004-637X/691/1/152, 0809.2608
Chen, H.W., Helsby, J.E., Gauthier, J.R., Shectman, S.A., Thompson, I.B., Tinker, J.L.: An empirical characterization of extended cool gas around Galaxies using Mg II absorption features. Astrophys. J. **714**, 1521–1541 (2010a). doi:10.1088/0004-637X/714/2/1521, 1004.0705
Chen, H.W., Wild, V., Tinker, J.L., Gauthier, J.R., Helsby, J.E., Shectman, S.A., Thompson, I.B.: What determines the incidence and extent of Mg II absorbing gas around galaxies? Astrophys. J. Lett. **724**, L176–L182 (2010b). doi:10.1088/2041-8205/724/2/L176, 1011.0735
Chen, H.W., Gauthier, J.R., Sharon, K., Johnson, S.D., Nair, P., Liang, C.J.: Spatially resolved velocity maps of halo gas around two intermediate-redshift galaxies. Mon. Not. R. Astron. Soc. **438**, 1435–1450 (2014). doi:10.1093/mnras/stt2288, 1312.0016
Christensen, L., Wisotzki, L., Roth, M.M., Sánchez, S.F., Kelz, A., Jahnke, K.: An integral field spectroscopic survey for high redshift damped Lyman-α galaxies. Astron. Astrophys. **468**, 587–601 (2007). doi:10.1051/0004-6361:20066410, 0704.0654
Christensen, L., Møller, P., Fynbo, J.P.U., Zafar, T.: Verifying the mass-metallicity relation in damped Lyman α selected galaxies at 0.1 < z < 3.2. Mon. Not. R. Astron. Soc. **445**, 225–238 (2014). doi:10.1093/mnras/stu1726, 1404.6529
Chynoweth, K.M., Langston, G.I., Yun, M.S., Lockman, F.J., Rubin, K.H.R., Scoles, S.A.: Neutral hydrogen clouds in the M81/M82 group. Astron. J. **135**, 1983–1992 (2008). doi:10.1088/0004-6256/135/6/1983, 0803.3631
Cooke, J., Wolfe, A.M., Gawiser, E., Prochaska, J.X.: Survey for Galaxies associated with z$\sim$3 damped Lyα systems. II. Galaxy-absorber correlation functions. Astrophys. J. **652**, 994–1010 (2006). doi:10.1086/507476, astro-ph/0607149
Crighton, N.H.M., Murphy, M.T., Prochaska, J.X., Worseck, G., Rafelski, M., Becker, G.D., Ellison, S.L., Fumagalli, M., Lopez, S., Meiksin, A., O'Meara, J.M.: The neutral hydrogen cosmological mass density at z = 5. Mon. Not. R. Astron. Soc. **452**, 217–234 (2015). doi:10.1093/mnras/stv1182, 1506.02037
Davies, R., Förster Schreiber, N.M., Cresci, G., Genzel, R., Bouché, N., Burkert, A., Buschkamp, P., Genel, S., Hicks, E., Kurk, J., Lutz, D., Newman, S., Shapiro, K., Sternberg, A., Tacconi, L.J., Wuyts, S.: How well can we measure the intrinsic velocity dispersion of distant disk galaxies? Astrophys. J. **741**, 69 (2011). doi:10.1088/0004-637X/741/2/69, 1108.0285
Davis, J.D., Keeney, B.A., Danforth, C.W., Stocke, J.T.: On the size and mass of photoionized clouds in extended spiral galaxy halos. Astrophys. J. **810**, 92 (2015). doi:10.1088/0004-637X/810/2/92, 1506.04095
Decarli, R., Walter, F., Aravena, M., Carilli, C., Bouwens, R., da Cunha, E., Daddi, E., Ivison, R.J., Popping, G., Riechers, D., Smail, I., Swinbank, M., Weiss, A., Anguita, T., Assef, R., Bauer, F., Bell, E.F., Bertoldi, F., Chapman, S., Colina, L., Cortes, P.C., Cox, P., Dickinson, M., Elbaz, D., Gónzalez-López, J., Ibar, E., Infante, L., Hodge, J., Karim, A., Le Fevre, O., Magnelli, B., Neri, R., Oesch, P., Ota, K., Rix, H.W., Sargent, M., Sheth, K., van der Wel, A., van der Werf, P., Wagg, J.: ALMA spectroscopic survey in the Hubble Ultra Deep Field: CO luminosity functions and the evolution of the cosmic density of molecular gas (2016). arXiv e-prints 1607.06770

Dessauges-Zavadsky, M., D'Odorico, S., Schaerer, D., Modigliani, A., Tapken, C., Vernet, J.: Rest-frame ultraviolet spectrum of the gravitationally lensed galaxy "the 8 o'clock arc": stellar and interstellar medium properties. Astron. Astrophys. **510**, A26 (2010). doi:10.1051/0004-6361/200913337, 0912.4384

Diamond-Stanic, A.M., Coil, A.L., Moustakas, J., Tremonti, C.A., Sell, P.H., Mendez, A.J., Hickox, R.C., Rudnick, G.H.: Galaxies probing Galaxies at high resolution: co-rotating gas associated with a milky way analog at z = 0.4. Astrophys. J. **824**, 24 (2016). doi:10.3847/0004-637X/824/1/24, 1507.01945

D'Odorico, V., Cristiani, S., Pomante, E., Carswell, R.F., Viel, M., Barai, P., Becker, G.D., Calura, F., Cupani, G., Fontanot, F., Haehnelt, M.G., Kim, T.S., Miralda-Escudé, J., Rorai, A., Tescari, E., Vanzella, E.: Metals in the $z{\sim}3$ intergalactic medium: results from an ultra-high signal-to-noise ratio UVES quasar spectrum. Mon. Not. R. Astron. Soc. (2016) doi:10.1093/mnras/stw2161, 1608.06116

Elmegreen, B.G.: A pressure and metallicity dependence for molecular cloud correlations and the calibration of mass. Astrophys. J. **338**, 178–196 (1989). doi:10.1086/167192

Elmegreen, B.G.: The H to H2 transition in galaxies - totally molecular galaxies. Astrophys. J. **411**, 170–177 (1993). doi:10.1086/172816

Erb, D.K., Shapley, A.E., Pettini, M., Steidel, C.C., Reddy, N.A., Adelberger, K.L.: The mass-metallicity relation at $z > \sim 2$. Astrophys. J. **644**, 813–828 (2006a). doi:10.1086/503623, astro-ph/0602473

Erb, D.K., Steidel, C.C., Shapley, A.E., Pettini, M., Reddy, N.A., Adelberger, K.L.: Hα Observations of a large sample of galaxies at $z \sim 2$: implications for star formation in high-redshift galaxies. Astrophys. J. **647**, 128–139 (2006b). doi:10.1086/505341, astro-ph/0604388

Erkal, D., Gnedin, N.Y., Kravtsov, A.V.: On the origin of the high column density turnover in the H I column density distribution. Astrophys. J. **761**, 54 (2012). doi:10.1088/0004-637X/761/1/54, 1201.3653

Faucher-Giguère, C.A., Kereš, D.: The small covering factor of cold accretion streams. Mon. Not. R. Astron. Soc. **412**, L118–L122 (2011). doi:10.1111/j.1745-3933.2011.01018.x, 1011.1693

Fernández, X., van Gorkom, J.H., Hess, K.M., Pisano, D.J., Kreckel, K., Momjian, E., Popping, A., Oosterloo, T., Chomiuk, L., Verheijen, M.A.W., Henning, P.A., Schiminovich, D., Bershady, M.A., Wilcots, E.M., Scoville, N.: A pilot for a very large array H I deep field. Astrophys. J. Lett. **770**, L29 (2013). doi:10.1088/2041-8205/770/2/L29, 1303.2659

Förster Schreiber, N.M., Genzel, R., Bouché, N., Cresci, G., Davies, R., Buschkamp, P., Shapiro, K., Tacconi, L.J., Hicks, E.K.S., Genel, S., Shapley, A.E., Erb, D.K., Steidel, C.C., Lutz, D., Eisenhauer, F., Gillessen, S., Sternberg, A., Renzini, A., Cimatti, A., Daddi, E., Kurk, J., Lilly, S., Kong, X., Lehnert, M.D., Nesvadba, N., Verma, A., McCracken, H., Arimoto, N., Mignoli, M., Onodera, M.: The SINS survey: SINFONI integral field spectroscopy of $z \sim 2$ star-forming galaxies. Astrophys. J. **706**, 1364–1428 (2009). doi:10.1088/0004-637X/706/2/1364, 0903.1872

Friis, M., De Cia, A., Krühler, T., Fynbo, J.P.U., Ledoux, C., Vreeswijk, P.M., Watson, D.J., Malesani, D., Gorosabel, J., Starling, R.L.C., Jakobsson, P., Varela, K., Wiersema, K., Drachmann, A.P., Trotter, A., Thöne, C.C., de Ugarte Postigo, A., D'Elia, V., Elliott, J., Maturi, M., Goldoni, P., Greiner, J., Haislip, J., Kaper, L., Knust, F., LaCluyze, A., Milvang-Jensen, B., Reichart, D., Schulze, S., Sudilovsky, V., Tanvir, N., Vergani, S.D.: The warm, the excited, and the molecular gas: GRB 121024A shining through its star-forming galaxy. Mon. Not. R. Astron. Soc. **451**, 167–183 (2015). doi:10.1093/mnras/stv960, 1409.6315

Fumagalli, M., Prochaska, J.X., Kasen, D., Dekel, A., Ceverino, D., Primack, J.R.: Absorption-line systems in simulated galaxies fed by cold streams. Mon. Not. R. Astron. Soc. **418**, 1796–1821 (2011). doi:10.1111/j.1365-2966.2011.19599.x, 1103.2130

Fumagalli, M., O'Meara, J.M., Prochaska, J.X., Rafelski, M., Kanekar, N.: Directly imaging damped Ly α galaxies at $z > 2$ - III. The star formation rates of neutral gas reservoirs at $z{\sim}2.7$. Mon. Not. R. Astron. Soc. **446**, 3178–3198 (2015). doi:10.1093/mnras/stu2325, 1409.2880

Gallazzi, A., Brinchmann, J., Charlot, S., White, S.D.M.: A census of metals and baryons in stars in the local Universe. Mon. Not. R. Astron. Soc. **383**, 1439–1458 (2008). doi:10.1111/j.1365-2966.2007.12632.x, 0708.0533

Gauthier, J.R., Chen, H.W.: Empirical constraints of supergalactic winds at $z \gtrsim 0.5$. Mon. Not. R. Astron. Soc. **424**, 1952–1962 (2012). doi:10.1111/j.1365-2966.2012.21327.x, 1205.4037

Genzel, R., Tacconi, L.J., Gracia-Carpio, J., Sternberg, A., Cooper, M.C., Shapiro, K., Bolatto, A., Bouché, N., Bournaud, F., Burkert, A., Combes, F., Comerford, J., Cox, P., Davis, M., Schreiber, N.M.F., Garcia-Burillo, S., Lutz, D., Naab, T., Neri, R., Omont, A., Shapley, A., Weiner, B.: A study of the gas-star formation relation over cosmic time. Mon. Not. R. Astron. Soc. **407**, 2091–2108 (2010). doi:10.1111/j.1365-2966.2010.16969.x, 1003.5180

Giovanelli, R., Haynes, M.P., Adams, E.A.K., Cannon, J.M., Rhode, K.L., Salzer, J.J., Skillman, E.D., Bernstein-Cooper, E.Z., McQuinn, K.B.W.: ALFALFA discovery of the nearby gas-rich dwarf galaxy Leo P. I. H I observations. Astron. J. **146**, 15 (2013). doi:10.1088/0004-6256/146/1/15, 1305.0272

Gnat, O., Sternberg, A.: Time-dependent ionization in radiatively cooling gas. Astrophys. J. Suppl. Ser. **168**, 213–230 (2007). doi:10.1086/509786, astro-ph/0608181

Gnedin, N.Y., Draine, B.T.: Line overlap and self-shielding of molecular hydrogen in Galaxies. Astrophys. J. **795**, 37 (2014). doi:10.1088/0004-637X/795/1/37, 1406.4129

Gnedin, N.Y., Kravtsov, A.V.: On the Kennicutt-Schmidt relation of low-metallicity high-redshift Galaxies. Astrophys. J. **714**, 287–295 (2010). doi:10.1088/0004-637X/714/1/287, 0912.3005

Gnedin, N.Y., Tassis, K., Kravtsov, A.V.: Modeling molecular hydrogen and star formation in cosmological simulations. Astrophys. J. **697**, 55–67 (2009). doi:10.1088/0004-637X/697/1/55, 0810.4148

Graciá-Carpio, J., García-Burillo, S., Planesas, P., Fuente, A., Usero, A.: Evidence of enhanced star formation efficiency in luminous and ultraluminous infrared galaxies. Astron. Astrophys. **479**, 703–717 (2008). doi:10.1051/0004-6361:20078223, 0712.0582

Heckman, T.M., Armus, L., Miley, G.K.: On the nature and implications of starburst-driven galactic superwinds. Astrophys. J. Suppl. Ser. **74**, 833–868 (1990). doi:10.1086/191522

Heckman, T.M., Norman, C.A., Strickland, D.K., Sembach, K.R.: On the physical origin of O VI absorption-line systems. Astrophys. J. **577**, 691–700 (2002). doi:10.1086/342232, astro-ph/0205556

Hopkins, A.M., Rao, S.M., Turnshek, D.A.: The star formation history of damped Lyα absorbers. Astrophys. J. **630**, 108–114 (2005). doi:10.1086/432046, astro-ph/0505418

Huang, Y.H., Chen, H.W., Johnson, S.D., Weiner, B.J.: Characterizing the chemically enriched circumgalactic medium of $\sim$ 38000 luminous red galaxies in SDSS DR12. Mon. Not. R. Astron. Soc. **455**, 1713–1727 (2016) . doi:10.1093/mnras/stv2327, 1510.01336

Johnson, S.D., Chen, H.W., Mulchaey, J.S.: On the possible environmental effect in distributing heavy elements beyond individual gaseous haloes. Mon. Not. R. Astron. Soc. **449**, 3263–3273 (2015). doi:10.1093/mnras/stv553, 1503.04199

Jorgenson, R.A., Wolfe, A.M.: Spatially resolved emission of a high-redshift DLA Galaxy with the Keck/OSIRIS IFU. Astrophys. J. **785**, 16 (2014). doi:10.1088/0004-637X/785/1/16, 1311.0045

Jorgenson, R.A., Murphy, M.T., Thompson, R., Carswell, R.F.: The Magellan uniform survey of damped Lyman α systems - II. Paucity of strong molecular hydrogen absorption. Mon. Not. R. Astron. Soc. **443**, 2783–2800 (2014). doi:10.1093/mnras/stu1314, 1407.1111

Kacprzak, G.G., Churchill, C.W., Steidel, C.C., Murphy, M.T.: Halo gas cross sections and covering fractions of Mg II absorption selected galaxies. Astron. J. **135**, 922–927 (2008). doi:10.1088/0004-6256/135/3/922, 0710.5765

Kacprzak, G.G., Murphy, M.T., Churchill, C.W.: Galaxy group at z = 0.3 associated with the damped Lyman α system towards quasar Q1127-145. Mon. Not. R. Astron. Soc. **406**, 445–459 (2010). doi:10.1111/j.1365-2966.2010.16667.x, 1003.3669

Kacprzak, G.G., Martin, C.L., Bouché, N., Churchill, C.W., Cooke, J., LeReun, A., Schroetter, I., Ho, S.H., Klimek, E.: New perspective on Galaxy outflows from the first detection of both intrinsic and traverse metal-line absorption. Astrophys. J. Lett. **792**, L12 (2014). doi:10.1088/2041-8205/792/1/L12, 1407.4126

Kacprzak, G.G., Muzahid, S., Churchill, C.W., Nielsen, N.M., Charlton, J.C.: The azimuthal dependence of outflows and accretion detected using O VI absorption. Astrophys. J. **815**, 22 (2015). doi:10.1088/0004-637X/815/1/22, 1511.03275

Karachentsev, I.D., Kashibadze, O.G.: Masses of the local group and of the M81 group estimated from distortions in the local velocity field. Astrophysics **49**, 3–18 (2006). doi:10.1007/s10511-006-0002-6

Keeney, B.A., Stocke, J.T., Rosenberg, J.L., Danforth, C.W., Ryan-Weber, E.V., Shull, J.M., Savage, B.D., Green, J.C.: HST/COS spectra of three QSOs that probe the circumgalactic medium of a single spiral galaxy: evidence for gas recycling and outflow. Astrophys. J. **765**, 27 (2013). doi:10.1088/0004-637X/765/1/27, 1301.4242

Kennicutt, R.C. Jr.: The global Schmidt law in star-forming Galaxies. Astrophys. J. **498**, 541–552 (1998). doi:10.1086/305588, astro-ph/9712213

Kereš, D., Katz, N., Fardal, M., Davé, R., Weinberg, D.H.: Galaxies in a simulated ΛCDM Universe - I. Cold mode and hot cores. Mon. Not. R. Astron. Soc. **395**, 160–179 (2009). doi:10.1111/j.1365-2966.2009.14541.x, 0809.1430

Kim, T.S., Partl, A.M., Carswell, R.F., Müller, V.: The evolution of H I and C IV quasar absorption line systems at $1.9 < z < 3.2$. Astron. Astrophys. **552**, A77 (2013). doi:10.1051/0004-6361/201220042, 1302.6622

Krogager, J.K., Fynbo, J.P.U., Møller, P., Ledoux, C., Noterdaeme, P., Christensen, L., Milvang-Jensen, B., Sparre, M.: On the sizes of $z\gtrsim2$ damped Lyα absorbing galaxies. Mon. Not. R. Astron. Soc. **424**, L1–L5 (2012). doi:10.1111/j.1745-3933.2012.01272.x, 1204.2833

Krühler, T., Ledoux, C., Fynbo, J.P.U., Vreeswijk, P.M., Schmidl, S., Malesani, D., Christensen, L., De Cia, A., Hjorth, J., Jakobsson, P., Kann, D.A., Kaper, L., Vergani, S.D., Afonso, P.M.J., Covino, S., de Ugarte Postigo, A., D'Elia, V., Filgas, R., Goldoni, P., Greiner, J., Hartoog, O.E., Milvang-Jensen, B., Nardini, M., Piranomonte, S., Rossi, A., Sánchez-Ramírez, R., Schady, P., Schulze, S., Sudilovsky, V., Tanvir, N.R., Tagliaferri, G., Watson, D.J., Wiersema, K., Wijers, R.A.M.J., Xu, D.: Molecular hydrogen in the damped Lyman α system towards GRB 120815A at $z = 2.36$. Astron. Astrophys. **557**, A18 (2013). doi:10.1051/0004-6361/201321772, 1304.7003

Krumholz, M.R., McKee, C.F., Tumlinson, J.: The atomic-to-molecular transition in Galaxies. II: H I and H_2 column densities. Astrophys. J. **693**, 216–235 (2009). doi:10.1088/0004-637X/693/1/216, 0811.0004

Lanzetta, K.M., Wolfe, A.M., Turnshek, D.A., Lu, L., McMahon, R.G., Hazard, C.: A new spectroscopic survey for damped Ly-alpha absorption lines from high-redshift galaxies. Astrophys. J. Suppl. Ser. **77**, 1–57 (1991). doi:10.1086/191596

Lanzetta, K.M., Yahata, N., Pascarelle, S., Chen, H.W., Fernández-Soto, A.: The star formation rate intensity distribution function: implications for the cosmic star formation rate history of the universe. Astrophys. J. **570**, 492–501 (2002). doi:10.1086/339774, astro-ph/0111129

Law, D.R., Steidel, C.C., Erb, D.K., Larkin, J.E., Pettini, M., Shapley, A.E., Wright, S.A.: Integral field spectroscopy of high-redshift star-forming Galaxies with laser-guided adaptive optics: evidence for dispersion-dominated kinematics. Astrophys. J. **669**, 929–946 (2007). doi:10.1086/521786, 0707.3634

Law, D.R., Steidel, C.C., Shapley, A.E., Nagy, S.R., Reddy, N.A., Erb, D.K.: An HST/WFC3-IR morphological survey of Galaxies at z = 1.5-3.6. I. Survey description and morphological properties of star-forming Galaxies. Astrophys. J. **745**, 85 (2012). doi:10.1088/0004-637X/745/1/85, 1107.3137

Ledoux, C., Vreeswijk, P.M., Smette, A., Fox, A.J., Petitjean, P., Ellison, S.L., Fynbo, J.P.U., Savaglio, S.: Physical conditions in high-redshift GRB-DLA absorbers observed with VLT/UVES: implications for molecular hydrogen searches. Astron. Astrophys. **506**, 661–675 (2009). doi:10.1051/0004-6361/200811572, 0907.1057

Lehner, N., Howk, J.C., Wakker, B.P.: Evidence for a massive, extended circumgalactic medium around the andromeda Galaxy. Astrophys. J. **804**, 79 (2015). doi:10.1088/0004-637X/804/2/79, 1404.6540

Lehnert, M.D., Heckman, T.M., Weaver, K.A.: Very extended X-ray and Hα emission in M82: implications for the superwind phenomenon. Astrophys. J. **523**, 575–584 (1999). doi:10.1086/307762, astro-ph/9904227

Leroy, A.K., Walter, F., Brinks, E., Bigiel, F., de Blok, W.J.G., Madore, B., Thornley, M.D.: The star formation efficiency in nearby galaxies: measuring where gas forms stars effectively. Astron. J. **136**, 2782–2845 (2008). doi:10.1088/0004-6256/136/6/2782, 0810.2556

Leroy, A.K., Bolatto, A., Gordon, K., Sandstrom, K., Gratier, P., Rosolowsky, E., Engelbracht, C.W., Mizuno, N., Corbelli, E., Fukui, Y., Kawamura, A.: The CO-to-H_2 conversion factor from infrared dust emission across the local group. Astrophys. J. **737**, 12 (2011). doi:10.1088/0004-637X/737/1/12, 1102.4618

Liang, C.J., Chen, H.W.: Mining circumgalactic baryons in the low-redshift universe. Mon. Not. R. Astron. Soc. **445**, 2061–2081 (2014). doi:10.1093/mnras/stu1901, 1402.3602

Lockman, F.J., Free, N.L., Shields, J.C.: The neutral hydrogen bridge between M31 and M33. Astron. J. **144**, 52 (2012). doi:10.1088/0004-6256/144/2/52, 1205.5235

Lopez, S., D'Odorico, V., Ellison, S.L., Becker, G.D., Christensen, L., Cupani, G., Denney, K.D., Paris, I., Worseck, G., Berg, T.A.M., Cristiani, S., Dessauges-Zavadsky, M., Haehnelt, M., Hamann, F., Hennawi, J., Irsic, V., Kim, T.S., Lopez, P., Saust, R.L., Menard, B., Perrotta, S., Prochaska, J.X., Sanchez-Ramirez, R., Vestergaard, M., Viel, M., Wisotzki, L.: XQ-100: a legacy survey of one hundred 3.5 < z < 4.5 quasars observed with VLT/XSHOOTER (2016) . ArXiv e-prints 1607.08776

Lu, L., Sargent, W.L.W., Barlow, T.A., Churchill, C.W., Vogt, S.S.: Abundances at high redshifts: the chemical enrichment history of damped Ly alpha Galaxies. Astrophys. J. Suppl. Ser. **107**, 475 (1996). doi:10.1086/192373, astro-ph/9606044

Madau, P., Dickinson, M.: Cosmic star-formation history. Ann. Rev. Astron. Astrophys. **52**, 415–486 (2014). doi:10.1146/annurev-astro-081811-125615, 1403.0007

Maiolino, R., Nagao, T., Grazian, A., Cocchia, F., Marconi, A., Mannucci, F., Cimatti, A., Pipino, A., Ballero, S., Calura, F., Chiappini, C., Fontana, A., Granato, G.L., Matteucci, F., Pastorini, G., Pentericci, L., Risaliti, G., Salvati, M., Silva, L.: AMAZE. I. The evolution of the mass-metallicity relation at z > 3. Astron. Astrophys. **488**, 463–479 (2008). doi:10.1051/0004-6361:200809678, 0806.2410

Mannucci, F., Cresci, G., Maiolino, R., Marconi, A., Pastorini, G., Pozzetti, L., Gnerucci, A., Risaliti, G., Schneider, R., Lehnert, M., Salvati, M.: LSD: Lyman-break galaxies Stellar populations and Dynamics - I. Mass, metallicity and gas at z ∼ 3.1. Mon. Not. R. Astron. Soc. **398**, 1915–1931 (2009). doi:10.1111/j.1365-2966.2009.15185.x, 0902.2398

Marasco, A., Crain, R.A., Schaye, J., Bahé, Y.M., van der Hulst, T., Theuns, T., Bower, R.G.: The environmental dependence of H I in galaxies in the EAGLE simulations. Mon. Not. R. Astron. Soc. **461**, 2630–2649 (2016). doi:10.1093/mnras/stw1498, 1606.06288

Martin, C.L., Kennicutt, R.C. Jr.: Star formation thresholds in galactic disks. Astrophys. J. **555**, 301–321 (2001). doi:10.1086/321452, astro-ph/0103181

McGaugh, S.S.: Oxygen abundances in low surface brightness disk galaxies. Astrophys. J. **426**, 135–149 (1994). doi:10.1086/174049, astro-ph/9311064

Ménard, B., Wild, V., Nestor, D., Quider, A., Zibetti, S., Rao, S., Turnshek, D.: Probing star formation across cosmic time with absorption-line systems. Mon. Not. R. Astron. Soc. **417**, 801–811 (2011). doi:10.1111/j.1365-2966.2011.18227.x, 0912.3263

Meyer, D.M., Welty, D.E., York, D.G.: Element abundances at high redshift. Astrophys. J. Lett. **343**, L37–L40 (1989). doi:10.1086/185505

Møller, P., Warren, S.J., Fall, S.M., Fynbo, J.U., Jakobsen, P.: Are high-redshift damped Lyα galaxies Lyman break galaxies? Astrophys. J. **574**, 51–58 (2002). doi:10.1086/340934, astro-ph/0203361

Møller, P., Fynbo, J.P.U., Fall, S.M.: Detection of Lyman-α emission from a DLA galaxy: possible implications for a luminosity-metallicity relation at z = 2-3. Astron. Astrophys. **422**, L33–L37 (2004). doi:10.1051/0004-6361:20040194, astro-ph/0406307

Muzahid, S., Srianand, R., Charlton, J.: An HST/COS survey of molecular hydrogen in DLAs & sub-DLAs at z < 1: molecular fraction and excitation temperature. Mon. Not. R. Astron. Soc. **448**, 2840–2853 (2015). doi:10.1093/mnras/stv133, 1410.3828

Neeleman, M., Wolfe, A.M., Prochaska, J.X., Rafelski, M.: The fundamental plane of damped Lyα systems. Astrophys. J. **769**, 54 (2013). doi:10.1088/0004-637X/769/1/54, 1303.7239

Neeleman, M., Prochaska, J.X., Ribaudo, J., Lehner, N., Howk, J.C., Rafelski, M., Kanekar, N.: The H I content of the universe over the past 10 Gyrs. Astrophys. J. **818**, 113 (2016a). doi:10.3847/0004-637X/818/2/113, 1601.01691

Neeleman, M., Prochaska, J.X., Zwaan, M.A., Kanekar, N., Christensen, L., Dessauges-Zavadsky, M., Fynbo, J.P.U., van Kampen, E., Møller, P., Zafar, T.: First connection between cold gas in emission and absorption: CO emission from a Galaxy-Quasar pair. Astrophys. J. Lett. **820**, L39 (2016b). doi:10.3847/2041-8205/820/2/L39, 1604.05720

Nielsen, N.M., Churchill, C.W., Kacprzak, G.G.: MAGIICAT II. general characteristics of the Mg II absorbing circumgalactic medium. Astrophys. J. **776**, 115 (2013). doi:10.1088/0004-637X/776/2/115, 1211.1380

Norman, C.A., Bowen, D.V., Heckman, T., Blades, C., Danly, L.: Hubble space telescope observations of QSO absorption lines associated with starburst galaxy outflows. Astrophys. J. **472**, 73 (1996). doi:10.1086/178042

Noterdaeme, P., Ledoux, C., Petitjean, P., Srianand, R.: Molecular hydrogen in high-redshift damped Lyman-α systems: the VLT/UVES database. Astron. Astrophys. **481**, 327–336 (2008). doi:10.1051/0004-6361:20078780, 0801.3682

Noterdaeme, P., Petitjean, P., Ledoux, C., Srianand, R.: Evolution of the cosmological mass density of neutral gas from Sloan Digital Sky Survey II - Data Release 7. Astron. Astrophys. **505**, 1087–1098 (2009). doi:10.1051/0004-6361/200912768, 0908.1574

Noterdaeme, P., Petitjean, P., Carithers, W.C., Pâris, I., Font-Ribera, A., Bailey, S., Aubourg, E., Bizyaev, D., Ebelke, G., Finley, H., Ge, J., Malanushenko, E., Malanushenko, V., Miralda-Escudé, J., Myers, A.D., Oravetz, D., Pan, K., Pieri, M.M., Ross, N.P., Schneider, D.P., Simmons, A., York, D.G.: Column density distribution and cosmological mass density of neutral gas: Sloan Digital Sky Survey-III Data Release 9. Astron. Astrophys. **547**, L1 (2012). doi:10.1051/0004-6361/201220259, 1210.1213

Noterdaeme, P., Petitjean, P., Srianand, R.: The elusive H$\text{I}\rightarrow$H$_2$ transition in high-z damped Lyman-α systems. Astron. Astrophys. **578**, L5 (2015). doi:10.1051/0004-6361/201526018, 1505.04997

Noterdaeme, P., Krogager, J.K., Balashev, S., Ge, J., Gupta, N., Krühler, T., Ledoux, C., Murphy, M.T., Pâris, I., Petitjean, P., Rahmani, H., Srianand, R., Ubachs, W.: Discovery of a Perseus-like cloud in the early Universe: HI-to-H2 transition, carbon monoxide and small dust grains at zabs=2.53 towards the quasar J0000+0048 (2016) . ArXiv e-prints 1609.01422

Oosterloo, T., Morganti, R., Crocker, A., Jütte, E., Cappellari, M., de Zeeuw, T., Krajnović, D., McDermid, R., Kuntschner, H., Sarzi, M., Weijmans, A.M.: Early-type galaxies in different environments: an HI view. Mon. Not. R. Astron. Soc. **409**, 500–514 (2010). doi:10.1111/j.1365-2966.2010.17351.x, 1007.2059

Oppenheimer, B.D., Schaye, J.: Non-equilibirum ionization and cooling of metal-enriched gas in the presence of a photoionization background. Mon. Not. R. Astron. Soc. **434**, 1043–1062 (2013). doi:10.1093/mnras/stt1043, 1302.5710

Péroux, C., Bouché, N., Kulkarni, V.P., York, D.G., Vladilo, G.: A SINFONI integral field spectroscopy survey for galaxy counterparts to damped Lyman α systems - I. New detections and limits for intervening and associated absorbers. Mon. Not. R. Astron. Soc. **410**, 2237–2250 (2011). doi:10.1111/j.1365-2966.2010.17598.x, 1009.0025

Péroux, C., Bouché, N., Kulkarni, V.P., York, D.G., Vladilo, G.: A SINFONI integral field spectroscopy survey for galaxy counterparts to damped Lyman α systems - III. Three additional detections. Mon. Not. R. Astron. Soc. **419**, 3060–3073 (2012). doi:10.1111/j.1365-2966.2011.19947.x, 1110.0666

Péroux, C., Quiret, S., Rahmani, H., Kulkarni, V.P., Epinat, B., Milliard, B., Straka, L.A., York, D.G., Rahmati, A., Contini, T.: A SINFONI integral field spectroscopy survey for galaxy counterparts to damped Lyman α systems - VI. Metallicity and geometry as gas flow probes. Mon. Not. R. Astron. Soc. **457**, 903–916 (2016). doi:10.1093/mnras/stw016, 1601.02796

Pettini, M.: Element abundances through the cosmic ages. In: Esteban, C., García López, R., Herrero, A., Sánchez, F. (eds.) Cosmochemistry. The Melting Pot of the Elements, pp. 257–298 (2004). astro-ph/0303272

Pettini, M., Boksenberg, A., Hunstead, R.W.: Metal enrichment, dust, and star formation in galaxies at high redshifts. I - The Z = 2.3091 absorber toward PHL 957. Astrophys. J. **348**, 48–56 (1990). doi:10.1086/168212

Pettini, M., Ellison, S.L., Steidel, C.C., Bowen, D.V.: Metal abundances at z<1.5: fresh clues to the chemical enrichment history of damped Lyα systems. Astrophys. J. **510**, 576–589 (1999). doi:10.1086/306635, astro-ph/9808017

Pettini, M., Shapley, A.E., Steidel, C.C., Cuby, J.G., Dickinson, M., Moorwood, A.F.M., Adelberger, K.L., Giavalisco, M.: The rest-frame optical spectra of Lyman break galaxies: star formation, extinction, abundances, and kinematics. Astrophys. J. **554**, 981–1000 (2001). doi:10.1086/321403, astro-ph/0102456

Pettini, M., Rix, S.A., Steidel, C.C., Adelberger, K.L., Hunt, M.P., Shapley, A.E.: New observations of the interstellar medium in the Lyman break galaxy MS 1512-cB58. Astrophys. J. **569**, 742–757 (2002). doi:10.1086/339355, astro-ph/0110637

Prochaska, J.X., Tumlinson, J.: Baryons: what, when and where? Astrophys. Space Sci. Proc. **10**, 419 (2009). doi:10.1007/978-1-4020-9457-6_16, 0805.4635

Prochaska, J.X., Wolfe, A.M.: A Keck HIRES investigation of the metal abundances and kinematics of the Z = 2.46 damped LY alpha system toward Q0201+365. Astrophys. J. **470**, 403 (1996). doi:10.1086/177875, astro-ph/9604042

Prochaska, J.X., Wolfe, A.M.: Chemical abundances of the damped Lyα systems at z>1.5. Astrophys. J. Suppl. Ser. **121**, 369–415 (1999). doi:10.1086/313200, astro-ph/9810381

Prochaska, J.X., Wolfe, A.M.: On the (Non)evolution of H I gas in galaxies over cosmic time. Astrophys. J. **696**, 1543–1547 (2009). doi:10.1088/0004-637X/696/2/1543, 0811.2003

Prochaska, J.X., Howk, J.C., O'Meara, J.M., Tytler, D., Wolfe, A.M., Kirkman, D., Lubin, D., Suzuki, N.: The UCSD HIRES/Keck I damped Lyα abundance database. III. An empirical study of photoionization in the damped Lyα system toward GB 1759+7539. Astrophys. J. **571**, 693–711 (2002). doi:10.1086/340066, astro-ph/0202140

Prochaska, J.X., Sheffer, Y., Perley, D.A., Bloom, J.S., Lopez, L.A., Dessauges-Zavadsky, M., Chen, H.W., Filippenko, A.V., Ganeshalingam, M., Li, W., Miller, A.A., Starr, D.: The first positive detection of molecular gas in a GRB host galaxy. Astrophys. J. Lett. **691**, L27–L32 (2009). doi:10.1088/0004-637X/691/1/L27, 0901.0556

Putman, M.E., Peek, J.E.G., Joung, M.R.: Gaseous galaxy halos. Ann. Rev. Astron. Astrophys. **50**, 491–529 (2012). doi:10.1146/annurev-astro-081811-125612, 1207.4837

Rafelski, M., Wolfe, A.M., Chen, H.W.: Star formation from DLA gas in the outskirts of Lyman break galaxies at z $\sim$ 3. Astrophys. J. **736**, 48 (2011). doi:10.1088/0004-637X/736/1/48, 1011.6390

Rafelski, M., Wolfe, A.M., Prochaska, J.X., Neeleman, M., Mendez, A.J.: Metallicity evolution of damped Lyα systems out to z $\sim$5. Astrophys. J. **755**, 89 (2012). doi:10.1088/0004-637X/755/2/89, 1205.5047

Rafelski, M., Neeleman, M., Fumagalli, M., Wolfe, A.M., Prochaska, J.X.: The rapid decline in metallicity of damped Lyα systems at z$\sim$5. Astrophys. J. Lett. **782**, L29 (2014). doi:10.1088/2041-8205/782/2/L29, 1310.6042

Rafelski, M., Gardner, J.P., Fumagalli, M., Neeleman, M., Teplitz, H.I., Grogin, N., Koekemoer, A.M., Scarlata, C.: The star formation rate efficiency of neutral atomic-dominated hydrogen gas in the outskirts of star-forming galaxies from z $\sim$ 1 to z$\sim$3. Astrophys. J. **825**, 87 (2016). doi:10.3847/0004-637X/825/2/87, 1604.08597

Rao, S.M., Nestor, D.B., Turnshek, D.A., Lane, W.M., Monier, E.M., Bergeron, J.: Low-redshift damped Lyα galaxies toward the Quasars B2 0827+243, PKS 0952+179, PKS 1127-145, and PKS 1629+120. Astrophys. J. **595**, 94–108 (2003). doi:10.1086/377331, astro-ph/0211297

Rao, S.M., Turnshek, D.A., Nestor, D.B.: Damped Lyα systems at z<1.65: the expanded Sloan Digital Sky Survey hubble space telescope sample. Astrophys. J. **636**, 610–630 (2006). doi:10.1086/498132, astro-ph/0509469

Rao, S.M., Belfort-Mihalyi, M., Turnshek, D.A., Monier, E.M., Nestor, D.B., Quider, A.: A ground-based imaging study of galaxies causing damped Lyman α (DLA), sub-DLA and Lyman limit system absorption in quasar spectra. Mon. Not. R. Astron. Soc. **416**, 1215–1249 (2011). doi:10.1111/j.1365-2966.2011.19119.x, 1103.4047

Rauch, M.: The Lyman alpha forest in the spectra of QSOs. Ann. Rev. Astron. Astrophys. **36**, 267–316 (1998). doi:10.1146/annurev.astro.36.1.267, astro-ph/9806286

Rauch, M., Sargent, W.L.W., Womble, D.S., Barlow, T.A.: Temperature and kinematics of C IV absorption systems. Astrophys. J. Lett. **467**, L5 (1996). doi:10.1086/310187, astro-ph/9606041

Reddy, N.A., Pettini, M., Steidel, C.C., Shapley, A.E., Erb, D.K., Law, D.R.: The characteristic star formation histories of galaxies at redshifts z$\sim$2-7. Astrophys. J. **754**, 25 (2012). doi:10.1088/0004-637X/754/1/25, 1205.0555

Rémy-Ruyer, A., Madden, S.C., Galliano, F., Galametz, M., Takeuchi, T.T., Asano, R.S., Zhukovska, S., Lebouteiller, V., Cormier, D., Jones, A., Bocchio, M., Baes, M., Bendo, G.J., Boquien, M., Boselli, A., DeLooze, I., Doublier-Pritchard, V., Hughes, T., Karczewski, O.Ł., Spinoglio, L.: Gas-to-dust mass ratios in local galaxies over a 2 dex metallicity range. Astron. Astrophys. **563**, A31 (2014). doi:10.1051/0004-6361/201322803, 1312.3442

Richards, G.T., Strauss, M.A., Fan, X., Hall, P.B., Jester, S., Schneider, D.P., Vanden Berk, D.E., Stoughton, C., Anderson, S.F., Brunner, R.J., Gray, J., Gunn, J.E., Ivezić, Ž., Kirkland, M.K., Knapp, G.R., Loveday, J., Meiksin, A., Pope, A., Szalay, A.S., Thakar, A.R., Yanny, B., York, D.G., Barentine, J.C., Brewington, H.J., Brinkmann, J., Fukugita, M., Harvanek, M., Kent, S.M., Kleinman, S.J., Krzesiński, J., Long, D.C., Lupton, R.H., Nash T., Neilsen, E.H. Jr., Nitta, A., Schlegel, D.J., Snedden, S.A.: The sloan digital sky survey Quasar survey: Quasar luminosity function from data release 3. Astron. J. **131**, 2766–2787 (2006). doi:10.1086/503559, astro-ph/0601434

Rigby, J.R., Charlton, J.C., Churchill, C.W.: The population of weak Mg II absorbers. II. The properties of single-cloud systems. Astrophys. J. **565**, 743–761 (2002). doi:10.1086/324723, astro-ph/0110191

Rubin, K.H.R., Prochaska, J.X., Ménard, B., Murray, N., Kasen, D., Koo, D.C., Phillips, A.C.: Low-ionization line emission from a starburst galaxy: a new probe of a galactic-scale outflow. Astrophys. J. **728**, 55 (2011). doi:10.1088/0004-637X/728/1/55, 1008.3397

Rubin, K.H.R., Hennawi, J.F., Prochaska, J.X., Simcoe, R.A., Myers, A., Lau, M.W.: (2015). Dissecting the gaseous halos of $z$$\sim$2 damped Ly$\alpha$ systems with close Quasar pairs. Astrophys. J. **808**, 38. doi:10.1088/0004-637X/808/1/38, 1411.6016

Salim, S., Rich, R.M.: Star formation signatures in optically quiescent early-type galaxies. Astrophys. J. Lett. **714**, L290–L294 (2010). doi:10.1088/2041-8205/714/2/L290, 1004.2041

Sánchez, S.F., Rosales-Ortega, F.F., Iglesias-Páramo, J., Mollá, M., Barrera-Ballesteros, J., Marino, R.A., Pérez, E., Sánchez-Blazquez, P., González Delgado, R., Cid Fernandes, R., de Lorenzo-Cáceres, A., Mendez-Abreu, J., Galbany, L., Falcon-Barroso, J., Miralles-Caballero, D., Husemann, B., García-Benito, R., Mast, D., Walcher, C.J., Gil de Paz, A., García-Lorenzo, B., Jungwiert, B., Vílchez, J.M., Jílková, L., Lyubenova, M., Cortijo-Ferrero, C., Díaz, A.I., Wisotzki, L., Márquez, I., Bland-Hawthorn, J., Ellis, S., van de Ven, G., Jahnke, K., Papaderos, P., Gomes, J.M., Mendoza, M.A., López-Sánchez, Á.R.: A characteristic oxygen abundance gradient in galaxy disks unveiled with CALIFA. Astron. Astrophys. **563**, A49 (2014). doi:10.1051/0004-6361/201322343, 1311.7052

Sánchez-Ramírez, R., Ellison, S.L., Prochaska, J.X., Berg, T.A.M., López, S., D'Odorico, V., Becker, G.D., Christensen, L., Cupani, G., Denney, K.D., Pâris, I., Worseck, G., Gorosabel, J.: The evolution of neutral gas in damped Lyman α systems from the XQ-100 survey. Mon. Not. R. Astron. Soc. **456**, 4488–4505 (2016). doi:10.1093/mnras/stv2732, 1511.05003

Savage, B.D., Sembach, K.R.: Interstellar abundances from absorption-line observations with the hubble space telescope. Ann. Rev. Astron. Astrophys. **34**, 279–330 (1996). doi:10.1146/annurev.astro.34.1.279

Savage, B.D., Sembach, K.R., Tripp, T.M., Richter, P.: Far ultraviolet spectroscopic explorer and space telescope imaging spectrograph observations of intervening O VI absorption line systems in the spectrum of PG 0953+415. Astrophys. J. **564**, 631–649 (2002). doi:10.1086/324288

Schechter, P.: An analytic expression for the luminosity function for galaxies. Astrophys. J. **203**, 297–306 (1976). doi:10.1086/154079

Schmidt, M.: The rate of star formation. Astrophys. J. **129**, 243 (1959). doi:10.1086/146614

Serra, P., Oosterloo, T., Morganti, R., Alatalo, K., Blitz, L., Bois, M., Bournaud, F., Bureau, M., Cappellari, M., Crocker, A.F., Davies, R.L., Davis, T.A., de Zeeuw, P.T., Duc, P.A., Emsellem, E., Khochfar, S., Krajnović, D., Kuntschner, H., Lablanche, P.Y., McDermid, R.M., Naab, T., Sarzi, M., Scott, N., Trager, S.C., Weijmans, A.M., Young, L.M.: The ATLAS^{3D} project - XIII. Mass and morphology of H I in early-type galaxies as a function of environment. Mon. Not. R. Astron. Soc. **422**, 1835–1862 (2012). doi:10.1111/j.1365-2966.2012.20219.x, 1111.4241

Shen, S., Madau, P., Guedes, J., Mayer, L., Prochaska, J.X., Wadsley, J.: The circumgalactic medium of massive galaxies at $z \sim 3$: a test for stellar feedback, galactic outflows, and cold streams. Astrophys. J. **765**, 89 (2013). DOI 10.1088/0004-637X/765/2/89, 1205.0270

Simcoe, R.A.: The carbon content of intergalactic gas at $z = 4.25$ and its evolution toward $z = 2.4$. Astrophys. J. **738**, 159 (2011). doi:10.1088/0004-637X/738/2/159, 1106.2810

Som, D., Kulkarni, V.P., Meiring, J., York, D.G., Péroux, C., Lauroesch, J.T., Aller, M.C., Khare, P.: Hubble space telescope observations of sub-damped Lyα absorbers at $z < 0.5$, and implications for galaxy chemical evolution. Astrophys. J. **806**, 25 (2015). doi:10.1088/0004-637X/806/1/25, 1502.01989

Steidel, C.C., Kollmeier, J.A., Shapley, A.E., Churchill, C.W., Dickinson, M., Pettini, M.: The kinematic connection between absorbing gas toward QSOs and galaxies at intermediate redshift. Astrophys. J. **570**, 526–542 (2002). doi:10.1086/339792, astro-ph/0201353

Steidel, C.C., Erb, D.K., Shapley, A.E., Pettini, M., Reddy, N., Bogosavljević, M., Rudie, G.C., Rakic, O.: The structure and kinematics of the circumgalactic medium from far-ultraviolet spectra of $z \sim = 2$-3 galaxies. Astrophys. J. **717**, 289–322 (2010). doi:10.1088/0004-637X/717/1/289, 1003.0679

Stocke, J.T., Penton, S.V., Danforth, C.W., Shull, J.M., Tumlinson, J., McLin, K.M.: The galaxy environment of O VI absorption systems. Astrophys. J. **641**, 217–228 (2006). doi:10.1086/500386, astro-ph/0509822

Stocke, J.T., Keeney, B.A., Danforth, C.W., Shull, J.M., Froning, C.S., Green, J.C., Penton, S.V., Savage, B.D.: Characterizing the circumgalactic medium of nearby galaxies with HST/COS and HST/STIS absorption-line spectroscopy. Astrophys. J. **763**, 148 (2013). doi:10.1088/0004-637X/763/2/148, 1212.5658

Straka, L.A., Noterdaeme, P., Srianand, R., Nutalaya, S., Kulkarni, V.P., Khare, P., Bowen, D., Bishof, M., York, D.G.: Galactic nebular lines in the fibre spectra of background QSOs: reaching a hundred QSO-galaxy pairs with spectroscopic and photometric measurements. Mon. Not. R. Astron. Soc. **447**, 3856–3872 (2015). doi:10.1093/mnras/stu2739, 1412.6661

Swaters, R.A., van Albada, T.S., van der Hulst, J.M., Sancisi, R.: The Westerbork HI survey of spiral and irregular galaxies. I. HI imaging of late-type dwarf galaxies. Astron. Astrophys. **390**, 829–861 (2002). doi:10.1051/0004-6361:20011755, astro-ph/0204525

Tacconi, L.J., Neri, R., Genzel, R., Combes, F., Bolatto, A., Cooper, M.C., Wuyts, S., Bournaud, F., Burkert, A., Comerford, J., Cox, P., Davis, M., Förster Schreiber, N.M., García-Burillo, S., Gracia-Carpio, J., Lutz, D., Naab, T., Newman, S., Omont, A., Saintonge, A., Shapiro Griffin, K., Shapley, A., Sternberg, A., Weiner, B.: PHIBSS: molecular gas content and scaling relations in $z \sim 1$-3 massive, main-sequence star-forming galaxies. Astrophys. J. **768**, 74 (2013). doi:10.1088/0004-637X/768/1/74, 1211.5743

Tremonti, C.A., Heckman, T.M., Kauffmann, G., Brinchmann, J., Charlot, S., White, S.D.M., Seibert, M., Peng, E.W., Schlegel, D.J., Uomoto, A., Fukugita, M., Brinkmann, J.: The origin of the mass-metallicity relation: insights from 53,000 star-forming galaxies in the Sloan Digital Sky Survey. Astrophys. J. **613**, 898–913 (2004). doi:10.1086/423264, astro-ph/0405537

Tripp, T.M., Giroux, M.L., Stocke, J.T., Tumlinson, J., Oegerle, W.R.: The ionization and metallicity of the intervening O VI absorber at z=0.1212 in the spectrum of H1821+643. Astrophys. J. **563**, 724–735 (2001). doi:10.1086/323965, astro-ph/0108371

Tumlinson, J.: Chemical evolution in hierarchical models of cosmic structure. I. Constraints on the early Stellar initial mass function. Astrophys. J. **641**, 1–20 (2006). doi:10.1086/500383, astro-ph/0507442

Tumlinson, J., Shull, J.M., Rachford, B.L., Browning, M.K., Snow, T.P., Fullerton, A.W., Jenkins, E.B., Savage, B.D., Crowther, P.A., Moos, H.W., Sembach, K.R., Sonneborn, G., York, D.G.: A far ultraviolet spectroscopic explorer survey of interstellar molecular hydrogen in the small and large magellanic clouds. Astrophys. J. **566**, 857–879 (2002). doi:10.1086/338112, astro-ph/0110262

Tumlinson, J., Prochaska, J.X., Chen, H.W., Dessauges-Zavadsky, M., Bloom, J.S.: Missing molecular hydrogen and the physical conditions of GRB host galaxies. Astrophys. J. **668**, 667–673 (2007). doi:10.1086/521294, astro-ph/0703666

Tumlinson, J., Thom, C., Werk, J.K., Prochaska, J.X., Tripp, T.M., Weinberg, D.H., Peeples, M.S., O'Meara, J.M., Oppenheimer, B.D., Meiring, J.D., Katz, N.S., Davé, R., Ford, A.B., Sembach, K.R.: The large, oxygen-rich halos of star-forming galaxies are a major reservoir of galactic metals. Science **334**, 948 (2011). doi:10.1126/science.1209840, 1111.3980

Tumlinson, J., Thom, C., Werk, J.K., Prochaska, J.X., Tripp, T.M., Katz, N., Davé, R., Oppenheimer, B.D., Meiring, J.D., Ford, A.B., O'Meara, J.M., Peeples, M.S., Sembach, K.R., Weinberg, D.H.: The COS-Halos survey: rationale, design, and a census of circumgalactic neutral hydrogen. Astrophys. J. **777**, 59 (2013). doi:10.1088/0004-637X/777/1/59, 1309.6317

van Zee, L., Salzer, J.J., Haynes, M.P., O'Donoghue, A.A., Balonek, T.J.: Spectroscopy of outlying H II regions in spiral galaxies: abundances and radial gradients. Astron. J. **116**, 2805–2833 (1998). doi:10.1086/300647, astro-ph/9808315

Verdes-Montenegro, L., Yun, M.S., Williams, B.A., Huchtmeier, W.K., Del Olmo, A., Perea, J.: Where is the neutral atomic gas in Hickson groups? Astron. Astrophys. **377**, 812–826 (2001). doi:10.1051/0004-6361:20011127, astro-ph/0108223

Verheijen, M., van Gorkom, J.H., Szomoru, A., Dwarakanath, K.S., Poggianti, B.M., Schiminovich, D.: WSRT ultradeep neutral hydrogen imaging of galaxy clusters at z $\sim$0.2: a pilot survey of Abell 963 and Abell 2192. Astrophys. J. Lett. **668**, L9–L13 (2007). doi:10.1086/522621, 0708.3853

Verheijen, M.A.W., Sancisi, R.: The Ursa Major cluster of galaxies. IV. HI synthesis observations. Astron. Astrophys. **370**, 765–867 (2001). doi:10.1051/0004-6361:20010090, astro-ph/0101404

Viegas, S.M.: Abundances at high redshift: ionization correction factors. Mon. Not. R. Astron. Soc. **276**, 268–272 (1995). doi:10.1093/mnras/276.1.268

Vladilo, G., Centurión, M., Bonifacio, P., Howk, J.C.: Ionization properties and elemental abundances in damped Lyα systems. Astrophys. J. **557**, 1007–1020 (2001). doi:10.1086/321650, astro-ph/0104298

Walter, F., Brinks, E., de Blok, W.J.G., Bigiel, F., Kennicutt, R.C. Jr., Thornley, M.D., Leroy, A.: THINGS: the H I nearby galaxy survey. AJ **136**, 2563–2647 (2008). doi:10.1088/0004-6256/136/6/2563, 0810.2125

Walter, F., Decarli, R., Sargent, M., Carilli, C., Dickinson, M., Riechers, D., Ellis, R., Stark, D., Weiner, B., Aravena, M., Bell, E., Bertoldi, F., Cox, P., Da Cunha, E., Daddi, E., Downes, D., Lentati, L., Maiolino, R., Menten, K.M., Neri, R., Rix, H.W., Weiss, A.: A molecular line scan in the hubble deep field north: constraints on the CO luminosity function and the cosmic H_2 density. Astrophys. J. **782**, 79 (2014). doi:10.1088/0004-637X/782/2/79, 1312.6365

Warren, S.J., Møller, P., Fall, S.M., Jakobsen, P.: NICMOS imaging search for high-redshift damped Lyα galaxies. Mon. Not. R. Astron. Soc. **326**, 759–773 (2001). doi:10.1046/j.1365-8711.2001.04629.x, astro-ph/0105032

Werk, J.K., Prochaska, J.X., Thom, C., Tumlinson, J., Tripp, T.M., O'Meara, J.M., Peeples, M.S.: The COS-Halos survey: an empirical description of metal-line absorption in the low-redshift circumgalactic medium. Astrophys. J. Suppl. Ser. **204**, 17 (2013). doi:10.1088/0067-0049/204/2/17, 1212.0558

Werk, J.K., Prochaska, J.X., Tumlinson, J., Peeples, M.S., Tripp, T.M., Fox, A.J., Lehner, N., Thom, C., O'Meara, J.M., Ford, A.B., Bordoloi, R., Katz, N., Tejos, N., Oppenheimer, B.D., Davé, R., Weinberg, D.H.: The COS-Halos survey: physical conditions and baryonic mass in the low-redshift circumgalactic medium. Astrophys. J. **792**, 8 (2014). doi:10.1088/0004-637X/792/1/8, 1403.0947

Westmeier, T., Brüns, C., Kerp, J.: Relics of structure formation: extra-planar gas and high-velocity clouds around the Andromeda Galaxy. Mon. Not. R. Astron. Soc. **390**, 1691–1709 (2008). doi:10.1111/j.1365-2966.2008.13858.x, 0808.3611

Wolfe, A.M., Chen, H.W.: Searching for low surface brightness galaxies in the hubble ultra deep field: implications for the star formation efficiency in neutral gas at $z \sim 3$. Astrophys. J. **652**, 981–993 (2006). doi:10.1086/507574, astro-ph/0608040

Wolfe, A.M., Turnshek, D.A., Smith, H.E., Cohen, R.D.: Damped Lyman-alpha absorption by disk galaxies with large redshifts. I - The Lick survey. Astrophys. J. Suppl. Ser. **61**, 249–304 (1986). doi:10.1086/191114

Wolfe, A.M., Lanzetta, K.M., Foltz, C.B., Chaffee, F.H.: The large bright QSO survey for damped LY alpha absorption systems. Astrophys. J. **454**, 698 (1995). doi:10.1086/176523

Wolfe, A.M., Gawiser, E., Prochaska, J.X.: Damped Ly α systems. Ann. Rev. Astron. Astrophys. **43**, 861–918 (2005). doi:10.1146/annurev.astro.42.053102.133950, astro-ph/0509481

Wolfire, M.G., Tielens, A.G.G.M., Hollenbach, D., Kaufman, M.J.: Chemical rates on small grains and PAHs: C^+ recombination and H_2 formation. Astrophys. J. **680**, 384–397 (2008). doi:10.1086/587688, 0803.0138

Wong, T., Blitz, L.: The relationship between gas content and star formation in molecule-rich spiral galaxies. Astrophys. J. **569**, 157–183 (2002). doi:10.1086/339287, astro-ph/0112204

Wright, S.A., Larkin, J.E., Law, D.R., Steidel, C.C., Shapley, A.E., Erb, D.K.: Dynamics of galactic disks and mergers at $z \sim 1.6$: spatially resolved spectroscopy with Keck laser guide star adaptive optics. Astrophys. J. **699**, 421–440 (2009). doi:10.1088/0004-637X/699/1/421, 0810.5599

Wuyts, E., Wisnioski, E., Fossati, M., Förster Schreiber, N.M., Genzel, R., Davies, R., Mendel, J.T., Naab, T., Röttgers, B., Wilman, D.J., Wuyts, S., Bandara, K., Beifiori, A., Belli, S., Bender, R., Brammer, G.B., Burkert, A., Chan, J., Galametz, A., Kulkarni, S.K., Lang, P., Lutz, D., Momcheva, I.G., Nelson, E.J., Rosario, D., Saglia, R.P., Seitz, S., Tacconi, L.J., Tadaki, K.I., Übler, H., van Dokkum, P.: The evolution of metallicity and metallicity gradients from $z = 2.7$ to 0.6 with KMOS^{3D}. Astrophys. J. **827**, 74 (2016). doi:10.3847/0004-637X/827/1/74, 1603.01139

Wyder, T.K., Martin, D.C., Barlow, T.A., Foster, K., Friedman, P.G., Morrissey, P., Neff, S.G., Neill, J.D., Schiminovich, D., Seibert, M., Bianchi, L., Donas, J., Heckman, T.M., Lee, Y.W., Madore, B.F., Milliard, B., Rich, R.M., Szalay, A.S., Yi, S.K.: The star formation law at low surface density. Astrophys. J. **696**, 1834–1853 (2009). doi:10.1088/0004-637X/696/2/1834, 0903.3015

York, D.G., Adelman, J., Anderson, J.E. Jr., Anderson, S.F., Annis, J., Bahcall, N.A., Bakken, J.A., Barkhouser, R., Bastian, S., Berman, E., Boroski, W.N., Bracker, S., Briegel, C., Briggs, J.W., Brinkmann, J., Brunner, R., Burles, S., Carey, L., Carr, M.A., Castander, F.J., Chen, B., Colestock, P.L., Connolly, A.J., Crocker, J.H., Csabai, I., Czarapata, P.C., Davis, J.E., Doi, M., Dombeck, T., Eisenstein, D., Ellman, N., Elms, B.R., Evans, M.L., Fan, X., Federwitz, G.R., Fiscelli, L., Friedman, S., Frieman, J.A., Fukugita, M., Gillespie, B., Gunn, J.E., Gurbani, V.K., de Haas, E., Haldeman, M., Harris, F.H., Hayes, J., Heckman, T.M., Hennessy, G.S., Hindsley, R.B., Holm, S., Holmgren, D.J., Huang, Ch., Hull, C., Husby, D., Ichikawa, S.I., Ichikawa, T., Ivezić, Ž., Kent, S., Kim, R.S.J., Kinney, E., Klaene, M., Kleinman, A.N., Kleinman, S., Knapp, G.R., Korienek, J., Kron, R.G., Kunszt, P.Z., Lamb, D.Q., Lee, B., Leger R.F., Limmongkol, S., Lindenmeyer, C., Long, D.C., Loomis, C., Loveday, J., Lucinio, R., Lupton, R.H., MacKinnon, B., Mannery, E.J., Mantsch, P.M., Margon, B., McGehee, P., McKay, T.A., Meiksin, A., Merelli, A., Monet, D.G., Munn, J.A., Narayanan, V.K., Nash, T., Neilsen, E., Neswold, R., Newberg, H.J., Nichol, R.C., Nicinski, T., Nonino, M., Okada, N., Okamura, S., Ostriker, J.P., Owen, R., Pauls, A.G., Peoples, J., Peterson, R.L., Petravick, D., Pier, J.R., Pope,

A., Pordes, R., Prosapio, A., Rechenmacher, R., Quinn, T.R., Richards, G.T., Richmond, M.W., Rivetta, C.H., Rockosi, C.M., Ruthmansdorfer, K., Sandford, D., Schlegel, D.J., Schneider, D.P., Sekiguchi, M., Sergey, G., Shimasaku, K., Siegmund, W.A., Smee, S., Smith, J.A., Snedden, S., Stone, R., Stoughton, C., Strauss, M.A., Stubbs, C., SubbaRao, M., Szalay, A.S., Szapudi, I., Szokoly, G.P., Thakar, A.R., Tremonti, C., Tucker, D.L., Uomoto, A., Vanden Berk, D., Vogeley, M.S., Waddell, P., Wang, S.I., Watanabe, M., Weinberg, D.H., Yanny, B., Yasuda, N., SDSS Collaboration: The Sloan Digital Sky Survey: technical summary. Astron. J. **120**, 1579–1587 (2000). doi:10.1086/301513, astro-ph/0006396

Yun, M.S., Ho, P.T.P., Lo, K.Y.: A high-resolution image of atomic hydrogen in the M81 group of galaxies. Nature **372**, 530–532 (1994). doi:10.1038/372530a0

Zahedy, F.S., Chen, H.W., Rauch, M., Wilson, M.L., Zabludoff, A.: Probing the cool interstellar and circumgalactic gas of three massive lensing galaxies at z = 0.4–0.7. Mon. Not. R. Astron. Soc. **458**, 2423–2442 (2016). doi:10.1093/mnras/stw484, 1510.04307

Zaritsky, D., Kennicutt, R.C. Jr., Huchra, J.P.: H II regions and the abundance properties of spiral galaxies. Astrophys. J. **420**, 87–109 (1994). doi:10.1086/173544

Zwaan, M.A.: Prochaska JX (2006) Where is the molecular hydrogen in damped Lyα absorbers? Astrophys. J. **643**, 675–679 doi:10.1086/503191, astro-ph/0601655

Zwaan, M.A., van der Hulst, J.M., Briggs, F.H., Verheijen, M.A.W., Ryan-Weber, E.V.: Reconciling the local galaxy population with damped Lyman α cross-sections and metal abundances. Mon. Not. R. Astron. Soc. **364**, 1467–1487 (2005). doi:10.1111/j.1365-2966.2005.09698.x, astro-ph/0510127

Chapter 10
Future Prospects: Deep Imaging of Galaxy Outskirts Using Telescopes Large and Small

Roberto Abraham, Pieter van Dokkum, Charlie Conroy, Allison Merritt, Jielai Zhang, Deborah Lokhorst, Shany Danieli, and Lamiya Mowla

Abstract The Universe is almost totally unexplored at low surface brightness levels. In spite of great progress in the construction of large telescopes and improvements in the sensitivity of detectors, the limiting surface brightness of imaging observations has remained static for about 40 years. Recent technical advances have at last begun to erode the barriers preventing progress. In this

R. Abraham (✉) • D. Lokhorst
Department of Astronomy and Astrophysics, University of Toronto, 50 St. George Street, Toronto, ON, Canada M5S 3H4

Dunlap Institute for Astronomy and Astrophysics, University of Toronto, 50 St. George Street, Toronto, ON, Canada M5S 3H4
e-mail: abraham@astro.utoronto.ca

P. van Dokkum • A. Merritt • L. Mowla
Department of Astronomy, Yale University, Steinbach Hall, 52 Hillhouse Avenue, New Haven, CT 06511, USA

C. Conroy
Harvard-Smithsonian Center for Astrophysics, 60 Garden St, Cambridge, MA 02138, USA

J. Zhang
Department of Astronomy and Astrophysics, University of Toronto, 50 St. George Street, Toronto, ON, Canada M5S 3H4

Dunlap Institute for Astronomy and Astrophysics, University of Toronto, 50 St. George Street, Toronto, ON, Canada M5S 3H4

Canadian Institute for Theoretical Astrophysics, 60 St. George Street, Toronto, ON, Canada M5S 3H8

S. Danieli
Department of Astronomy, Yale University, Steinbach Hall, 52 Hillhouse Avenue, New Haven, CT 06511, USA

Harvard-Smithsonian Center for Astrophysics, 60 Garden St, Cambridge, MA 02138, USA

Department of Physics, Yale University, 266 Whitney Avenue, New Haven, CT 06511, USA

Yale Center for Astronomy and Astrophysics, 52 Hillhouse Avenue, New Haven, CT 06511, USA

J.H. Knapen et al. (eds.), *Outskirts of Galaxies*, Astrophysics and Space Science Library 434, DOI 10.1007/978-3-319-56570-5_10

chapter, we describe the technical challenges to low surface brightness imaging, describe some solutions and highlight some relevant observations that have been undertaken recently with both large and small telescopes. Our main focus will be on discoveries made with the Dragonfly Telephoto Array (Dragonfly), which is a new telescope concept designed to probe the Universe down to hitherto unprecedented low surface brightness levels. We conclude by arguing that these discoveries are probably only scratching the surface of interesting phenomena that are observable when the Universe is explored at low surface brightness levels.

10.1 Motivation

A fundamental prediction of hierarchical galaxy formation models in a dark energy-dominated cold dark matter (ΛCDM) cosmology is that all galaxies are surrounded by a vast and complex network of ultra-low surface brightness filaments and streams, the relics of past merger events (Purcell et al. 2007; Johnston et al. 2008; Cooper et al. 2013; Pillepich et al. 2014). Resolved (star count-based) analyses have revealed the existence of streams and filaments around the Milky Way (e.g. Majewski et al. 2003; Belokurov et al. 2007; Carollo et al. 2007; Bell et al. 2008) and M31 (e.g. Ibata et al. 2001, 2007; Ferguson et al. 2002; Richardson et al. 2008; McConnachie et al. 2009; Gilbert et al. 2012). These impressive results provide strong evidence that some massive spiral galaxies formed, at least in part, hierarchically. In the ΛCDM picture, many thousands of such streams and filaments combine over time to define galactic extended stellar haloes, with the bulk of the material distributed at surface brightnesses well below 30 mag/arcsec^2 (Johnston et al. 2008). In contrast to the star count-based results on stellar streams, the detection of these haloes has proven elusive. The *Hubble Space Telescope* (*HST*) has undertaken some very successful deep pencil-beam star count surveys of a number of galaxies outside the Local Group (e.g. the GHOSTS survey; c.f. Radburn-Smith et al. 2011 and references therein), but since these stellar haloes are very extended (tens of arcminutes for nearby objects), nearby haloes may not be well sampled by pencil beams, and the seemingly much simpler strategy of trying to image them directly using ground-based telescopes is quite attractive. Many such studies have claimed detections of the extended low surface brightness stellar haloes of galaxies on the basis of direct imaging, but these claims are now controversial, with recent investigations dismissing these "haloes" as simply being scattered light. This point was first made by de Jong in 2008, and the putative detection of low surface brightness stellar haloes from unresolved imaging has recently been the subject of two exhaustive investigations by Sandin (2014, 2015), who concludes that most claimed detections are spurious.

10.2 Why Is Low Surface Brightness Imaging Hard?

The faintest galaxies detected at present are about seven magnitudes (over a factor of 600) fainter than the faintest galaxies studied during the "photographic era" of astronomy prior to the mid-1980s. It is therefore quite remarkable that over this same period of time, there has been essentially no improvement in the limiting surface brightness of deep imaging observations of galaxies (e.g. the low surface brightness limits presented in Kormendy and Bahcall 1974 are quite impressive even by modern standards). As we will describe below, this is because low surface brightness imaging is not usually limited by photon statistics or by spatial resolution. Instead, it is limited by imperfect control of systematic errors. The implication is that considerable work can be done by relatively small telescopes, and indeed a number of interesting investigations of tidal structures around nearby galaxies have appeared recently, based on amateur-professional collaborations using small telescopes (Martínez-Delgado et al. 2009, 2010, 2012, 2016).

A major benefit of the work done to date using small telescopes is that the imaging systems tend to be relatively simple. The absence of complicated re-imaging optics[1] tends to produce images from small telescopes that are "cleaner" than those of large telescopes, with fewer internal reflections. (See, e.g. the very striking examples in Duc et al. 2015, one of which is reproduced in Fig. 10.1.) As will be described below, this clean focal plane is an important benefit, because internal reflections and scattering are important systematics that have played the major role in defining the limiting surface brightness of all visible-wavelength imaging for the past few decades (Slater et al. 2009). For certain classes of observation (such as the detection of intra-cluster light or of stellar haloes), the impact of ghosting and scattering on the wide-angle point spread function (PSF) can have a huge impact on the credibility of the results.

On the other hand, provided one is content with probing the Universe at surface brightness levels brighter than $V \sim 28.5\,\mathrm{mag\,arcsec^{-2}}$ (which has effectively been the limit for the last 40 years), ghosting and internal reflections are generally an inconvenience rather than a showstopper. The well-motivated astronomer can take steps to limit the impact of ghosting on large telescopes by using optimal dithering strategies (Tal and van Dokkum 2011), and in extreme cases one can attempt to model internal reflections and subtract them off (Slater et al. 2009; Duc 2016; Roman and Trujillo 2016). Even when such steps are not taken, in general, it is fairly clear when a compact structure on an image is due to scattered light or an internal reflection, as opposed to something intrinsic to a galaxy, such as a tidal feature. All things considered, the authors of this chapter would certainly rather base

[1]Re-imaging optics are an essential component of cameras on large telescopes for many reasons, e.g. to reduce their long native focal lengths in order to provide reasonable image scales on relatively small sensors. Most small telescopes utilize small and simple flat-field correctors or even image straight onto their sensors, since some small telescope optical designs have intrinsically well-corrected flat focal planes that are a good match to CCD pixel sizes.

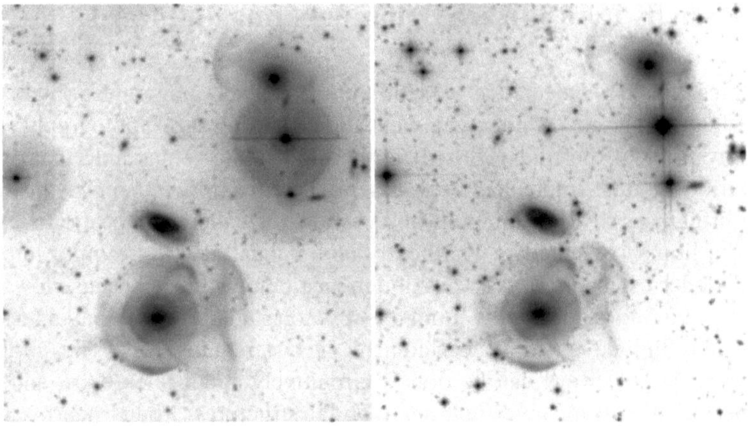

Fig. 10.1 Images of the same field around NGC 474 (to the east) and NGC 467 (to the west) obtained with two different cameras and telescopes: MegaCam on the 3.6 m CFHT (*left*) and a 12 in. amateur telescope (*right*). The integration time is 30× longer on the small telescope than on the CFHT, but the limiting surface brightness is the same, and note the absence of large stellar haloes in the image obtained by the amateur telescope which confuse the CFHT data at low surface brightness levels (Figure reproduced with permission from Duc et al. 2015)

conclusions about the prevalence of most galactic features (with the very notable exception of stellar haloes) on large and fair samples of low-intermediate-redshift galaxies that can be obtained efficiently by large telescopes such as the Canada-France-Hawaii Telescope (CFHT; e.g. Atkinson et al. 2013), rather than base conclusions on "cleaner" images obtained of a relatively few very nearby galaxies using amateur setups.[2] However, "picking" between small and large telescopes in this way, where both are limited to similar surface brightnesses, is not nearly as interesting as trying to figure out *why* both are constrained to the same limiting surface brightness. And it would be far more interesting still to use this information to find a way to break through the limit, in order to study structures that neither can easily detect.

We now return to consideration of the systematic errors that make low surface brightness imaging very hard. Most of the obvious systematics are instrumental, and these find their origin in the optical train (e.g. in scattering and internal reflections, which Slater et al. (2009) show quite clearly is the most severe limitation in the case of the Burrell-Schmidt, the most optimized low surface brightness imager on the planet prior to the system described in Sect. 10.4) or in the detector (e.g. in

[2]Most investigations to date using amateur telescopes have been biased towards imaging "pretty" galaxies with known peculiarities, using imaging setups without standard filter sets or even without any filters at all. In many cases this is because the equipment being used is optimized for aesthetic or artistic imaging. Actually, judged as works of art, we think the absolutely spectacular images generated by many of these amateurs far out-distance the images produced by most professional astronomers.

imperfect flat-fielding and dark current subtraction). Once these have been mastered, nature provides a host of complications external to the imaging system. Some of these complications are well known, such as variability in the sky background introduced by airglow lines in the upper atmosphere, and some are not so well known, such as the non-negligible structure of the telescopic PSF on spatial scales of tens of arcminutes (King 1971; Racine 1996; Sandin 2014). One must also reckon with very significant sources of low surface brightness contamination that have an extraterrestrial origin, such as Galactic cirrus (likely to be the dominant systematic in many cases) and the unresolved sources making up the extragalactic background light (Bernstein 2007).

The reader will probably agree that this is a distressingly long list of systematics that one needs to worry about. In most cases, such as in trying to detect galactic stellar haloes, imperfect corrections made for the systematics just noted will result in *false-positive* detections. For example, it is not hard to undertake deep imaging observations and "find" galactic stellar haloes that in reality may or may not exist (Sandin 2014, 2015). *The true challenge is to hit the required depth with the precision needed to find low surface brightness structures when they really do exist.* But in spite of (or perhaps because of) these issues, the low surface brightness Universe is a treasure trove of almost totally unexplored astrophysical phenomena: galactic stellar haloes are only the beginning. If one could "crack" the systematics preventing us from getting down to really low surface brightness levels, new avenues would be opened up in a large range of subjects. A host of phenomena are known to exist at very low surface brightness, ranging from dust rings around planets in our Solar system (Hamilton et al. 2015) to supernova light echoes (Rest et al. 2012), to high-velocity clouds raining onto the disk of the Milky Way (Simon et al. 2006), to intra-group and intra-cluster light (Montes and Trujillo 2014) and even to new emission mechanisms in rich clusters of galaxies (Yamazaki and Loeb 2015). The challenge is to come up with a telescope design that allows the systematic errors just noted to be addressed.

10.3 Small Telescope Arrays as Better Imaging Mousetraps

General-purpose telescopes have to make certain compromises, some of which place limitations on their effectiveness for low surface brightness imaging. In this section, we are going to examine some of these compromises at a high level and also describe an alternative telescope concept that is in some ways better optimized for wide-field imaging than conventional telescope designs. This discussion will continue at a lower level in the next section by presenting the Dragonfly Telephoto Array, which is an existence proof for the concept. These two sections may be considered something of a digression, and readers who are more interested in science results than in telescopes may wish to skip ahead to Sect. 10.5.

Telescopes are designed on the basis of complicated trades between aperture, field of view, resolution, wavelength coverage and many other factors. In general, the right telescope for the job depends on the job. For a seeing-limited survey of point sources, an appropriate figure of merit Φ for a telescope is:

$$\Phi \propto \frac{D^2 \Omega\, \eta}{\mathrm{d}\Omega}, \quad (10.1)$$

where D is the aperture, Ω is the field of view, η is the throughput and $\mathrm{d}\Omega$ is the resolution. Optimizing for this figure of merit drives one towards a telescope design with a fairly large aperture and a large field of view, such as the innovative optical design chosen for the Large Synoptic Survey Telescope (LSST). On the other hand, if one is more interested in investigating individual targets over a small area of the sky at diffraction-limited resolution, a more appropriate figure of merit is:

$$\Phi \propto \frac{D^4 \Omega\, \eta}{\lambda}, \quad (10.2)$$

where λ is the wavelength. This figure of merit drives the design of telescopes such as the European Extremely Large Telescope (E-ELT) and the Thirty Meter Telescope (TMT), in which both adaptive optics and large aperture play important roles.

Telescopes with designs optimized on the basis of Eqs. (10.1) and (10.2) are on the horizon, and these will provide the next generation of astronomers with a truly formidable set of capabilities. Nevertheless, there are some interesting edge cases that benefit from being looked at in a somewhat different way. For example, consider the situation where one is only interested in imaging very faint structures that are much larger than the limiting resolution of the instrument (or the atmosphere).[3] The figure of merit simply scales with the number of detected photons, so for a very extended source, this will be given by:

$$\Phi \propto \mu\, a\, \eta f^{-2}, \quad (10.3)$$

where μ is the surface brightness, a is the pixel area and f is the focal ratio (also known as the f-ratio). This equation is interesting because the aperture of the telescope does not enter into the equation directly, but the focal ratio does. Therefore, for imaging of very extended sources, one doesn't necessarily want a big telescope. Instead, one wants an *optically fast* (low f-ratio) telescope.

Another perspective on Eqs. (10.1) and (10.2) also depends on considerations of spatial resolution. The D^4 scaling in Eq. (10.2) refers to cases of diffraction-limited

[3]Obviously, for most problems in astronomy, one *does* care about resolution, and on the rare occasions when one doesn't, with digital detectors, one can always rebin data. That said, we recently scratched our heads at the wisdom of having to do a 15×15 rebinning of beautiful 0.5 arcsec seeing CFHT and Gemini imaging data in order to obtain signal-to-noise levels comparable to those obtained using telephoto lenses for a number of very diffuse low surface brightness structures.

imaging, where the resolution scales as λ/D. The D^2 scaling of Eq. (10.1) refers to cases where the telescope is limited by atmospheric resolution, which is around 0.5 arcsec at the best sites. This resolution corresponds to the diffraction limit of a 25 cm (10 in.) telescope working at visible wavelengths. Therefore, all ground-based telescopes with images that are not sharpened by adaptive optics are operating with the resolution of (at best) a 25 cm telescope. In that case, there is no *fundamental* difference between obtaining an image with a large-aperture telescope and stacking the images (obtained at the same time) from an array of smaller aperture telescopes. Whether or not there is a *practical* difference depends on a myriad number of factors, such as the read noise and dark current in the detectors relative to the Poisson noise from the sky background. But, at least in principle, the stacked image is equivalent to that obtained from a ground-based telescope with aperture $D_{\rm eff}$ and focal ratio $f_{\rm eff}$:

$$D_{\rm eff} = \sqrt{N} \times D \quad (10.4)$$

$$f_{\rm eff} = f/\sqrt{N}, \quad (10.5)$$

where N is the number of small telescopes, each of which has aperture D.

Arrays of small telescopes have a number of advantages and disadvantages. One disadvantage is that the need for replication of N instruments to go on the N telescopes, so instruments that are simple to mass-produce (such as cameras) are probably the most suitable ones to deploy on the array. Another disadvantage is computational complexity, though the last few decades have seen innovations in computation far outstripping innovations in classical optics, so relying on Moore's law and the inherent parallelism of image stacking means this disadvantage is no longer very serious. Perhaps the most serious disadvantage is mechanical complexity, since N telescopes mean there are N more things that can go wrong, presenting a management challenge. The best defence against this disadvantage is more parallelism, because if N is large and all telescopes are completely independent, then it does not much matter if a few telescopes are non-operational on a given night. On the other hand, the advantages of this design can be quite remarkable.

- The performance of the system can be excellent, because individual telescopes in the array can be relatively simple all-refractive designs that offer superb performance in terms of scattering and ghosting (see next section). And by making N large enough, systems with $f_{\rm eff} < 0.5$ are possible, which is not possible using monolithic telescopes (since $f = 0.5$ is the thermodynamic limit).
- Individual telescopes and arrays are inexpensive because they can take advantage of the commercial availability of the needed components. For example, conventional large telescopes have long focal lengths and thus require sensors with large pixels (typically $\gtrsim$15 μm) for which there is essentially no commercial demand. Small telescopes have short focal lengths that are excellent matches to cell phone sensors (whose pixels are typically $\sim$1.5 μm in size). Exponential growth in the

demand for such sensors has driven their development to the point where a few dollars can easily buy an off-the-shelf 20 megapixel sensor whose performance is quite comparable to that of a CCD sensor for a large telescope, but at a cost that is about four orders of magnitude lower.

- Arrays can start small and grow large relatively inexpensively. For telescope arrays, the cost scales with aperture as $D_{\rm eff}^2$ for both the telescope and the enclosure, which is a much more favourable scaling relation than is the case for conventional large telescopes. The cost of conventional large telescope designs scales with aperture as $D^{2.85}$ (for pre-1980 designs; Meinel 1978) and $D^{2.5}$ (for post-1980 designs; Oschmann 2004). Furthermore, the cost of an enclosure for a conventional large telescope scales as D^3 (Oschmann 2004).

10.4 The Dragonfly Telephoto Array

The Dragonfly Telephoto Array (Abraham and van Dokkum 2014) was designed to use the ideas in Sect. 10.3 in order to break through most of the systematic errors noted in Sect. 10.2, so we could explore the low surface brightness Universe with high precision. The most intractable problem in low surface brightness imaging is scattering in the optical train, typically from faint stars in the field, though sometimes from bright stars outside the field (see Slater et al. (2009) for a beautiful investigation of the sources of scatter). Because this scattering originates in several of the basic design trades that make large telescopes possible (e.g. an obstructed pupil and reflective surfaces that have high-frequency micro-roughness that backscatters into the optical path and pollutes the focal plane), Dragonfly has an unobstructed pupil and no reflective surfaces at all. The array builds up its effective aperture by multiplexing the latest generation of high-end commercial telephoto lenses that use nano-fabricated coatings with sub-wavelength structures to yield a factor of ten improvement in wide-angle scattered light relative to other astronomical telescopes (see Sandin (2015) for a comparison). The array is designed to increase in aperture with time, and over the last 2 years, a 48-lens Dragonfly array has been assembled gradually in New Mexico as a collaboration between the Universities of Toronto, Yale and Harvard. In its current configuration (half of which is shown in Fig. 10.2), Dragonfly is equivalent to a 1 m aperture $f/0.39$ refracting telescope with a six-square-degree field of view and optical scattering an order of magnitude lower than conventional telescopes. The current detectors have 5.4 μm pixels resulting in an angular resolution of 2.85 arcsec/pixel, so the images are under sampled. This is fine for our science goals, and the array has superb performance when imaging low surface brightness structures on scales larger than about 5 arcsec. However, we can envision a number of programmes that would benefit from improved point source sensitivity, and we are planning a detector switch in the future to improve both resolution and point source limiting depth. We are also planning to add a narrow-band imaging mode, the motivation for which will be described in Sect. 10.5.3.

Fig. 10.2 One of the two 24-lens arrays comprising the 48-lens Dragonfly Telephoto Array. The lenses are co-aligned, and the full array is equivalent to a 1 m aperture $f/0.39$ refractor with a 2×3 deg field of view. The arrays are housed in domes at the New Mexico Skies Observatories and operate as a single telescope slaved to the same robotic control system. The lenses are commercial 400 mm Canon USM IS II lenses that have superb (essentially diffraction-limited) optical quality. This particular lens has very low scatter on account of proprietary nanostructure coatings on key optical surfaces (see Abraham and van Dokkum 2014 for details). Each lens is affixed to a separate CCD camera, and both are controlled by a miniature computer attached to the back of each camera that runs bespoke camera and lens control software that we have made publicly available on a GitHub repository. In the latest incarnation of Dragonfly, each lens is self-configuring and is controlled by its own node.js JavaScript server in an "Internet of Things" configuration that provides a RESTful interface. Growing the array is done by simply bolting a new lens onto the array and plugging in network and power cables

Several of the sources of systematic error in ultra-deep imaging find their origins in the atmosphere, rather than in the instrument. To combat these, the operational model for using Dragonfly is in some ways as innovative as the hardware. When investigating galaxy haloes, the array points only at locations predetermined (on the basis of IRAS imaging) to have low Galactic cirrus contamination, and the array operates in a fully autonomous robotic mode that tracks atmospheric systematics in real time. The latter point is important because once one has greatly reduced the wide-angle scatter inherent to the instrument, the tall pole becomes the atmosphere. The wide-angle telescopic PSF (the "stellar aureole") is not well understood, and at least some of its origin is instrumental (e.g. Bernstein 2007). But a very significant fraction is due to scattering by icy aerosols in the upper atmosphere (a fact well known to atmospheric physicists; see DeVore et al. (2013) for an interesting application of stellar aureole measurements to global warming). This component of the wide-angle scatter is variable (a fact noted by Sandin 2014), and our Dragonfly data shows quite clearly that this variability extends down to a timescale of minutes. Dealing with these sorts of "second-order" non-instrumental scattering issues clearly matters (see Fig. 10.3 for an illustrative example). We note that some

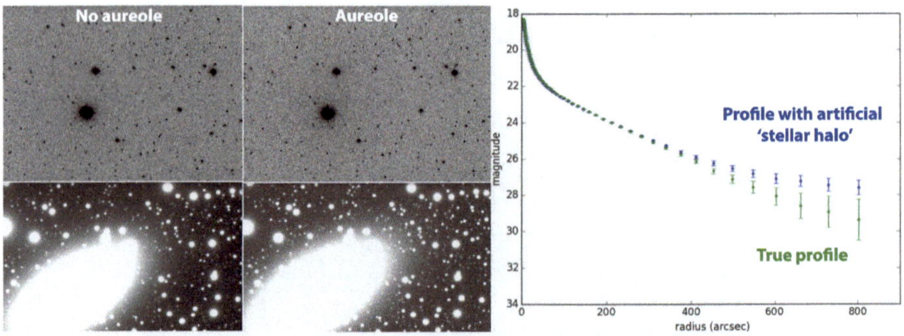

Fig. 10.3 A "spot the difference" exercise for the reader. This figure shows the results from a simulation intended to explore the impact of variability on the atmospheric component of the wide-angle stellar PSF (the "stellar aureole"). The *top panels* in the first two columns show 40 × 60 arcmin regions from single 600 s Dragonfly frames with photometric zero points differing by 0.1 mag. The intensity of the stellar aureole has been found to correlate with the zero point, and investigation of fields with bright stars shows that small variations in the photometric zero points of individual frames at this level correspond to fields with PSFs that have significantly different structure in their wide-angle wings. The *bottom half of the left two columns* show cutouts from simulated observations of a typical target galaxy. The simulation models the results of a ~2 h exposure with Dragonfly, made by stacking 600 frames obtained from 48 lenses on a galaxy with the structural properties of NGC 2841. The images adopt different forms for the wide-angle PSF, with shapes that result in a 0.1 mag difference in the recovered zero points. In the *left image*, we have assumed a standard PSF model out to ~1 arcmin with negligible contribution beyond this scale. The *right image* shows the corresponding simulation with a stellar aureole included. The prescription for the aureole is taken from Racine (1996). Note that the impact of the aureole is nearly invisible to the eye, except for the merest hint of additional low-level scatter. Nevertheless, the aureole has a profound impact on the shape of the surface brightness profile at large radius. The *right-hand panel* presents the azimuthally averaged profile of the target galaxy out to a radius of nearly 0.25°. Because the impact of short-timescale wide-angle PSF variability is significant, in the Dragonfly pipeline, individual frames are automatically calibrated and assessed for image quality as they arrive. In practice, around 20% of data frames that appear fine to the eye are dropped from the final stack because of small variations in their photometric zero points (See Zhang et al. 2017 for details.) (Reproduced with permission)

authors correct for wide-angle PSF contamination by post facto measurement and subsequent modelling (e.g. Trujillo and Fliri 2016), and this is certainly a good thing. However, until the sources of the variability in the atmospheric component of the wide-angle PSF are better understood, we think that real-time monitoring of the large-angle PSF is likely to be important for reliable measurement of stellar haloes.

10.5 The Universe Below 30 mag/arcsec²

10.5.1 Galactic Outskirts

Dragonfly has recently completed a campaign of ultra-deep imaging of 18 nearby galaxies (the Dragonfly Nearby Galaxies Survey; Merritt et al. 2016). All results from this campaign were obtained with Dragonfly in its prototype eight- and ten-lens configurations, where its performance was equivalent to that of a 0.4 m $f/1.0$ refractor and a 0.45 m $f/0.9$ refractor, respectively. An analysis of the (essentially absent) stellar halo of M101 using Dragonfly data was presented in van Dokkum et al. (2014), and an analysis of the satellite content of M101 appeared in Merritt et al. (2014). van Dokkum et al. (2015a,b, 2016) show Dragonfly results on ultra-diffuse galaxies in the Coma cluster (described more in the next section). Our main focus in this section will be to highlight results that have recently appeared in Merritt et al. (2016), which presents data on the stellar halo mass fractions of the first eight galaxies in the Dragonfly Nearby Galaxies Survey. We will also highlight preliminary Dragonfly results on the outer disks of galaxies that will be appearing in Zhang et al. (2017) and Lokhorst et al. (2017, Imaging the cosmic web, private communication).

The data shown here are taken from the Dragonfly Nearby Galaxies Survey (Merritt et al. 2016), which is a sparse-selected ultra-deep imaging study of 18 galaxies selected on the basis of four criteria: (1) galaxies must have $M_B < -19$ mag, (2) galaxies must be further than 3 Mpc away, (3) galaxies must be located at high Galactic latitude within "holes" of low Galactic cirrus determined from IRAS 100 μm imaging, and (4) galaxies must be visible for extended periods of time with low air mass as seen from New Mexico. No other selection criteria were imposed. The third criterion is interesting, because it turns out that most regions of the high Galactic latitude sky are unsuitable for ultra-deep imaging without preselection to avoid cirrus contamination.

Deep images of the first eight galaxies (all spirals) in the Dragonfly Nearby Galaxies Survey are shown in Fig. 10.4, together with a "reconstructed" (by combining star-count data in the outskirts and Dragonfly data in the interior) image of M31 as seen at a distance of 7 Mpc using the same instrument. Each image is 30 arcmin on a side. The familiar high surface brightness appearance of the galaxies is shown in colour; the low surface brightness outskirts are shown using a greyscale. The galaxies show a remarkable diversity in their low surface brightness

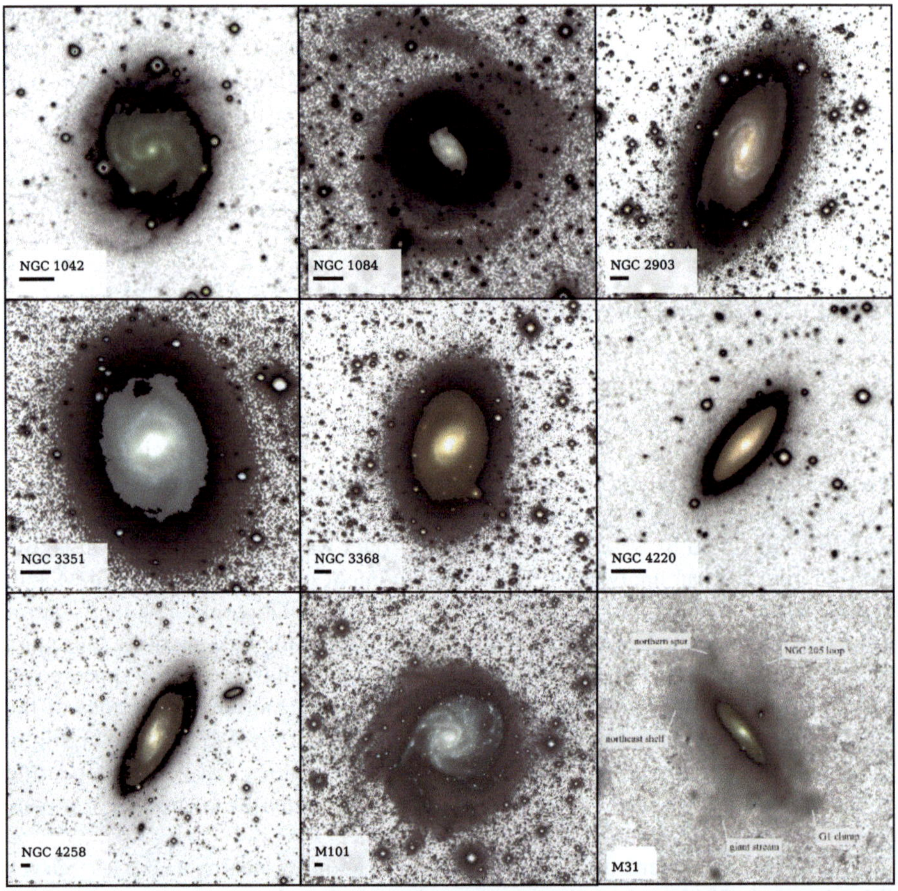

Fig. 10.4 Images of each of the eight galaxies in the sample of Merritt et al. (2016). The pseudo-colour images were created from *g*- and *r*-band images for the high surface brightness regions, and the *greyscale* shows the lower surface brightness outskirts. The *bottom-right panel* shows M31, created from a combination of Dragonfly and PAndAS data (McConnachie et al. 2009; Carlberg 2009) and redshifted to a distance of 7 Mpc (van Dokkum et al. 2014). *Black lines* beneath each galaxy name indicate scales of 1 arcmin (Figure taken from Merritt et al. 2016, reproduced with permission)

structures. Nevertheless, it seems that M31-like haloes, characterized by significant substructure, do not appear to be the norm.

Surface brightness profiles corresponding to the galaxies shown in Fig. 10.4 are shown in Fig. 10.5. All profiles extend down to at least 32 mag/arcsec^2, with some stretching down to 34 mag/arcsec^2. We have restricted these profiles to regions with high signal-to-noise ratio and have incorporated known systematics into the uncertainty estimates. Since all of the galaxies are well studied, Merritt et al. (2016) present careful comparisons between profiles obtained using Dragonfly data and

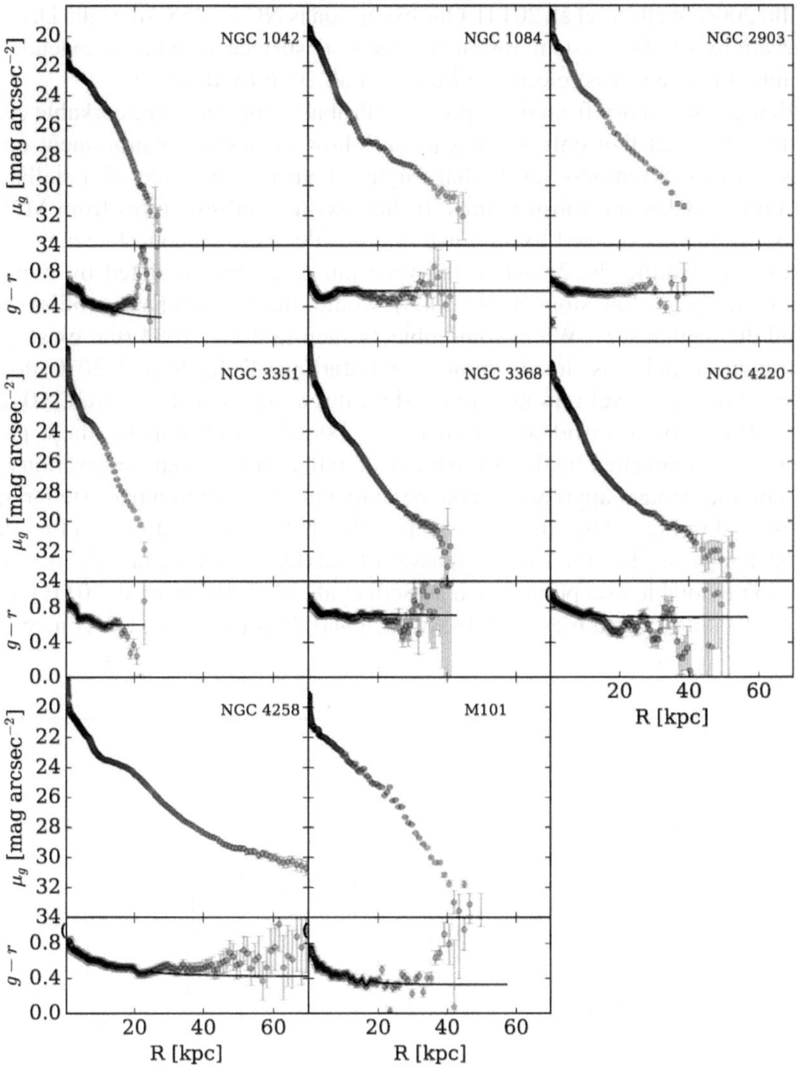

Fig. 10.5 Surface brightness and *g*–*r* colour profiles (with 1σ error bars) for the first eight galaxies explored in the Dragonfly Nearby Galaxies Survey (Figure taken from Merritt et al. 2016, reproduced with permission)

previously published surface brightness profiles for these galaxies. There is an excellent agreement between these profiles in brighter regions where many surveys overlap. Results remain quite consistent until around 29 mag/arcsec^2, at which point relatively little comparison data exists. However, starting at these low surface brightness levels, Dragonfly profiles generally show less evidence for deviations from exponential disks than do profiles from other investigations (e.g. Pohlen and

Trujillo 2006; Watkins et al. 2014). One exception is NGC 4258, since the Dragonfly observations of this system reveal a very low surface brightness extended red structure. (The reader is referred to Merritt et al. 2016 for details.)

What can one learn from these profiles? Probably their most remarkable characteristic is the fact that only a few galaxies show evidence for prominent upturns at low surface brightness levels that might signal the presence of a stellar halo built up by coalescing substructures. If they existed, substructures from M31-like haloes would be expected to dominate the profiles below about 27.5 mag/arcsec^2 (Bakos and Trujillo 2012), while the very faint streams predicted by numerical simulations (e.g. Johnston et al. 2008) would likely dominate profiles below around 30 mag/arcsec^2. While some objects such as NGC 1084 (the existence of substructure which was already known from Martínez-Delgado et al. 2010) do show substructures at a level reminiscent of M31, three objects (M101, NGC 1042 and NGC 3351) show no evidence at all for an extended stellar halo—these profiles appear to be dominated by flux from the disk to the limits of our observations.

A complementary approach to understanding the contribution from stellar haloes is presented in Fig. 10.6, which shows the stellar halo mass fraction as a function of total stellar mass. The total stellar masses of nearby haloes are largely unexplored (with some notable exceptions, such as Seth et al. 2007; Bailin et al. 2011; Greggio et al. 2014; van Dokkum et al. 2014; Streich and de Jong 2015). For concreteness,

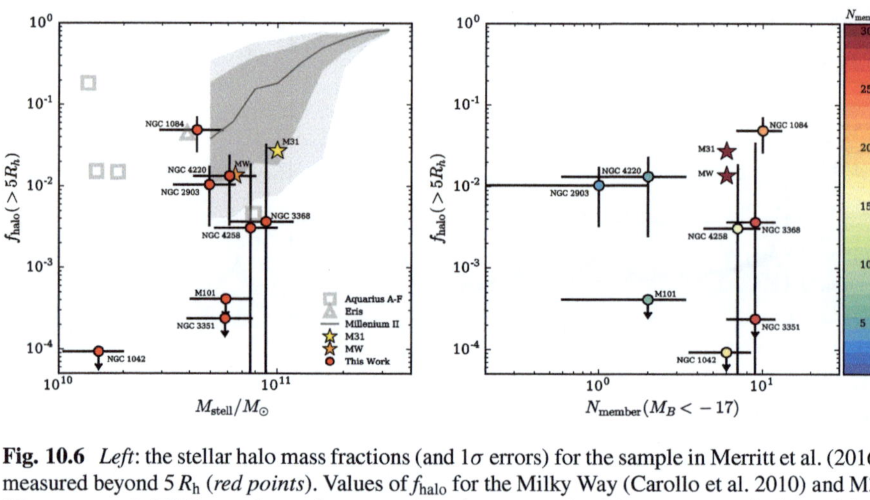

Fig. 10.6 *Left*: the stellar halo mass fractions (and 1σ errors) for the sample in Merritt et al. (2016), measured beyond $5\,R_h$ (*red points*). Values of f_{halo} for the Milky Way (Carollo et al. 2010) and M31 (Courteau et al. 2011) are shown for comparison (*orange and gold stars*, respectively) and have been scaled to the halo mass fraction outside of $5\,R_h$ assuming the structure of the halo of M31 (Irwin et al. 2005; Courteau et al. 2011). Predictions of f_{halo}, measured over $3 \leq r \leq 280$ kpc from the Aquarius simulations (Cooper et al. 2010), over $r \geq 20$ kpc from the Eris simulation (Pillepich et al. 2015) and over $r \geq 3$ kpc from the Millennium II simulation (galaxies with $B/T < 0.9$ only; Cooper et al. 2013), are indicated by *grey open squares, triangles and shaded region*, respectively. *Right*: Environmental richness is parametrized by the number of group members (Makarov and Karachentsev 2011) with $M_B < -17$. The *colour of each symbol* corresponds to the *total* number of known group members for that particular galaxy. The stellar halo mass fractions do not appear to be a function of environment (Figure taken from Merritt et al. 2016, reproduced with permission)

we define the halo mass fraction to be the stellar mass in excess of a disk+bulge model outside of five half-light radii R_h, a region where the stellar halo should start to contribute significantly (Abadi et al. 2006; Johnston et al. 2008; Font et al. 2011; Cooper et al. 2015; Pillepich et al. 2015). Given the relatively narrow range in stellar mass explored by these observations ($2–8 \times 10^{10} M_\odot$), the data display a remarkably wide range in stellar halo mass fractions. One of the galaxies in our sample, NGC 1084, has a stellar halo mass fraction of 0.049 ± 0.02 (even larger than that of M31), while, as noted earlier, three others (NGC 1042, NGC 3351 and M101) have stellar haloes that are undetected in our data. We measure an RMS scatter of $1.01^{+0.09}_{-0.26}$ dex and a peak-to-peak span of a factor of >100. This level of stochasticity is high and certainly exceeds the expectations (Amorisco et al. 2015; Cooper et al. 2010, 2013) from numerical simulations (the grey regions shown in the figure), though they may be qualitatively consistent with variations in the structure and stellar populations of nearby stellar haloes observed in both integrated light and star count studies (e.g. Mouhcine et al. 2007; Barker et al. 2012; Monachesi et al. 2016). The former may be contaminated by scattered light, and the latter (based on pencil beams) may not be fairly sampling the haloes, so the Dragonfly observations provide a robust baseline for future characterization of the stellar halo mass fraction in luminous nearby spirals.

One possible criticism of the results presented so far is that they are based on analyses of surface brightness profiles. The obvious benefit of using profiles is the increase in signal-to-noise ratio they bring because of averaging along isophotes. Another very useful benefit of profiles is that they decrease the dimensionality of the problem being investigated. However, a potential weakness of profiles is their strong sensitivity at large radii to the accuracy of sky background estimates. At a more fundamental level, profiles might also be criticized for making strong assumptions about an underlying symmetry in the images. It is therefore interesting to note that many of the most important conclusions obtained so far can be reinforced, at least at a qualitative level, simply by a careful inspection of the images. This is particularly true when the data are viewed in the appropriate panchromatic context. Figure 10.7 (taken from Zhang et al. 2017) shows a comparison of *GALEX* FUV images (top row), radio observations of HI from THINGS (middle row) and ultra-deep *g*-band Dragonfly imaging (bottom row) for two galaxies with extended UV disks. Lightly binned Dragonfly images probe down to around 30 mag/arcsec^2 even without the benefit of radial averaging, and at these surface brightness levels, it is clear that *at visible wavelengths starlight stretches out to the limits of the HI observations.* In the examples shown, the bulk of the starlight at very large radii is contained within extended disks, rather than in a roughly spherical halo. One suspects that the traditional view that at radio wavelengths disks are about twice as big as they are at visible wavelengths is simply the product of the high sensitivity of the radio data and the limited ability of ground-based telescopes to undertake low surface brightness observations. Careful inspection of Fig. 10.7 shows that at large radii the starlight in the Dragonfly images traces the HI data closely. Smooth starlight also fills in the regions between the UV knots in the *GALEX* images, which (like the evolved starlight) stretch out to the limits of the radio observations. The existence of evolved

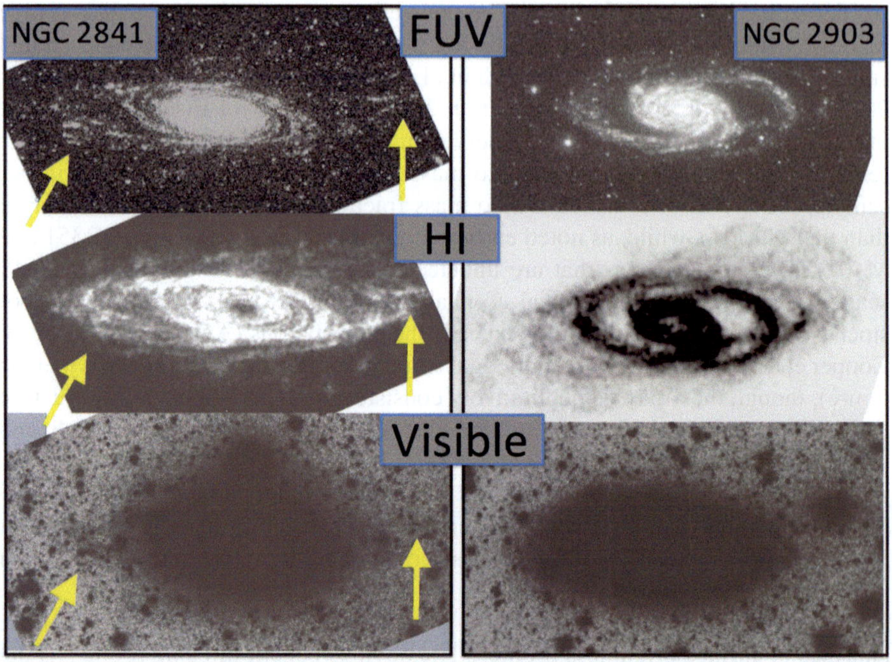

Fig. 10.7 *GALEX* FUV images (*top row*), HI observations from THINGS (*middle row*), and ultra-deep Dragonfly imaging (*bottom row*) for NGC 2841 (*left*) and NGC 2903 (*right*), two systems showing extended UV disks (Thilker et al. 2007) (Taken from Zhang et al. 2017, reproduced with permission)

disk stars at very large radii is another manifestation of the important problem first highlighted by the existence of XUV disks (Thilker et al. 2007): how does one form stars at radii well beyond the bulk of the molecular gas in galaxies and at locations where disks are at least globally Toomre stable? Certainly local regions of instability can emerge from dense pockets of gas compressed by turbulence (Elmegreen and Hunter 2006), but what drives the turbulence, and where does the molecular gas come from? In any case, it is amazing to see in these data how, at least in some cases, giant disks rather than stellar haloes define the faint outskirts of the galaxies.

Returning to profiles, it is interesting to explore the relationship between the gaseous and stellar components of the outskirts of galaxies. A comparison between stars and gas for one well-studied galaxy, NGC 2841, is shown in Fig. 10.8. Between 30 and 50 kpc, the two components appear to track each other rather closely. Beyond 50 kpc, the stellar profile flattens, while the gas profile appears to decline, though the significance of this is presently unclear given the rather large systematic errors at these radii. Perhaps the upturn in the stellar mass density profile signals the onset of a well-defined stellar halo in this system? This seems plausible, but perhaps more ambitious imaging would also reveal a further continuation of the enormous stellar disk already uncovered, stretching out to radii at which the disk-halo interface fuels

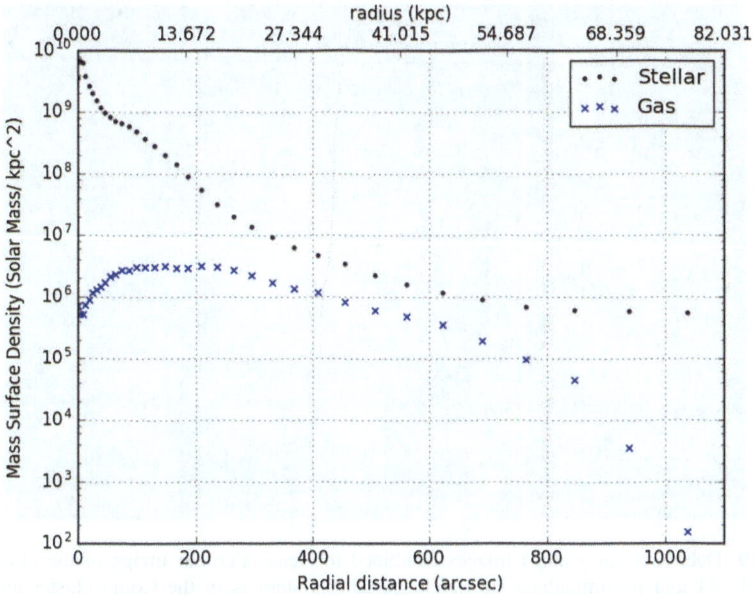

Fig. 10.8 The stellar mass surface density profile (*top*) and neutral gas mass density profile (*bottom*) of NGC 2841 based on the Dragonfly observations shown in the previous figure (Taken from Zhang et al. 2017, reproduced with permission)

the fire of star formation in the galaxies and ultimately connects these systems into the cosmic web of primordial gas (Sect. 10.5.3).

10.5.2 Ultra-Diffuse Galaxies

One of the first science targets imaged with Dragonfly was the Coma cluster of galaxies. This is the nearest rich cluster of galaxies, and it has been studied extensively for nearly a century, yet Dragonfly's first observation of Coma yielded an interesting discovery: the existence of a large number of very faint, large low surface brightness objects that turned out to be a new class of giant spheroidal galaxies (van Dokkum et al. 2015a,b). We named this class of objects *ultra-diffuse galaxies* (UDGs), because they have sizes similar to the Milky Way (half-light radii around 3 kpc) but only 1/100 to 1/1000 the number of stars as the Milky Way. Figure 10.9 shows a now well-known example, Dragonfly 44, along with its surprisingly abundant globular cluster distribution stretching out into the outskirts of the galaxy. The total number of globular clusters can be used to trace Dragonfly 44's dynamical mass via the proportionality between globular cluster numbers and halo mass (Harris et al. 2013, 2015; Hudson et al. 2014), backed up by ultra-deep Keck

Fig. 10.9 Deep Gemini *g* and *i* images combined to create a colour image of the ultra-diffuse Dragonfly 44 and its immediate surroundings. This system is in the Coma cluster and it has a remarkable appearance: it is a low surface brightness, spheroidal system whose outskirts are peppered with faint, compact sources we have identified as globular clusters (Figure taken from van Dokkum et al. 2016, reproduced with permission)

spectroscopy (van Dokkum et al. 2016). This particular system has a dynamical mass similar to that of the Milky Way.

The discovery of UDGs has generated tremendous interest in the community, from observers who are rapidly enlarging the UDG samples (e.g. Koda et al. 2015; Mihos et al. 2015; van der Burg et al. 2016), from simulators who must now try to understand the origin and evolution of these galaxies (e.g. Yozin and Bekki 2015; Amorisco and Loeb 2016) and even from alternative-gravity researchers who claim their existence challenges dark matter models (Milgrom 2015). The existence of so many presumably "delicate" UDGs (Koda et al. put their number at $\sim$800 in Coma) in a rich cluster poses the immediate question of why they are not being ripped apart by the tidal field of the cluster. They may be short-lived and be on their first infall and about to be shredded, but this seems unlikely given their predominantly old and red stellar populations and smooth morphologies. If they have survived for several orbits in the Coma cluster, then simple stability arguments suggest that they must have significantly higher masses than implied by their stellar populations; in fact, in order to survive, their dark matter fractions need to be $>$98% (van Dokkum et al. 2015a) within their half-light radii.

A central goal of future Dragonfly research is to determine the origin of UDGs in rich clusters. It has been suggested that UDGs are low-mass galaxies that are anomalously large ("inflated dwarfs") because they are undergoing tidal disruption (Collins et al. 2014) or have a high spin (Amorisco and Loeb 2016). As noted earlier, an alternative idea is that they are very massive dark matter haloes that are greatly under-abundant in stars because they did not manage to form normal stellar populations (van Dokkum et al. 2015a; Agertz and Kravtsov 2015; Yozin and Bekki 2015; van der Burg et al. 2016). In this picture UDGs are "failed giant" galaxies. Distinguishing between inflated dwarfs and failed giants can be done using dynamical information obtained from absorption line widths, though these are biased towards the interior of the systems. As noted earlier, an alternative approach (that is also being pursued with *HST*) is to probe the dark matter content of UDGs by imaging their globular cluster distributions. Studies with *HST* (and soon with the *James Webb Space Telescope*) will be a marked improvement on ground-based imaging such as that shown in Fig. 10.9. Intriguingly, Peng and Lim (2016) claim that one object with *HST* data (Dragonfly 17) represents the most extreme conceivable case of a "failed giant", namely, a system comprised almost entirely of dark matter in which the visible stars are all the product of the initial formation of a stellar halo that somehow failed to initiate the formation phase of the bulk of the galaxy. In this case, we need not probe the outskirts of the system to explore the halo—the whole galaxy is a halo! We find a similar situation for Dragonfly 44 (van Dokkum et al. 2016), though there are also many systems with lower mass, suggesting UDGs are a mixed set of objects, and a healthy fraction are dwarfs, as noted by Beasley and Trujillo (2016) and Roman and Trujillo (2016).

10.5.3 Imaging the Cosmic Web: The Next Frontier?

The ultimate limits of the Dragonfly concept are not yet known. Our robotic operational model and tight control of systematics (in particular our real-time modelling of sky variability) allow unusually long integration times to be undertaken. With the 10-lens array (as the 48-lens array has only just come on-line), our longest integrations are 50 h in duration (spread over many nights), and over this period of time, we have not yet become limited by any systematics (e.g. scattering, sky variability, flat-field accuracy) that would make longer integrations pointless. It appears that with the current configuration, the ultimate systematic limit on depth will probably be source confusion from unresolved background galaxies, but we can get a factor of 2–3 improvement in resolution with new CMOS detectors using smaller pixels. (The relevant detector technology is driven by demand for better mobile phone camera sensors, so advances happen quickly.) As a result, a $\sim$500-lens Dragonfly array (equivalent to a $\sim$3 m aperture $f/0.13$ refractor) may well be perfectly feasible, and we can already envision ways to grow Dragonfly up to that scale.

What would one do with a 500-lens Dragonfly? Certainly it would be exciting to go much deeper on our targets, and it would be wonderful to be able to do everything in this chapter 10$\times$ faster. It might be even more exciting to embark on new science objectives such as taking the exploration of galactic outskirts to a completely new level by adding a narrow-band capability to Dragonfly. This would allow detailed exploration of feedback at the disk-halo interface of the circumgalactic environment.

In narrow-band imaging mode, Dragonfly could be used to study extremely faint nebular emission lines (primarily Hα and [OIII]) in the circumgalactic medium (CGM) of nearby galaxies to probe the disk-halo interface (Putman et al. 2012). Maps of gas flows into and out of galaxies (as inflowing material "fuels the fire" of star formation while winds drive gas back into haloes) would be interesting and would do much to constrain galactic feedback models.

More speculatively, a large array of small telescopes working in narrow-band imaging mode might allow one to directly image ionized gas in the brighter portions of the filamentary cosmic web. The intergalactic medium (IGM) and its close cousin, the CGM, are arguably the most important (since they contain the most baryons) and least understood (since they're relatively difficult to detect) baryonic components of the Universe. Of course the denser pockets of the IGM/CGM have long been studied in absorption using UV lines (e.g. the Lyα forest) and in 21 cm emission using radio telescopes. However, both approaches are quite limited. Absorption line studies probe limited lines of sight. Furthermore, Lyα must be cosmologically band-shifted in order to be accessible using ground-based telescopes, so more is known about the IGM/CGM at high redshifts than is known locally. At much longer wavelengths, single-dish radio telescopes have the required sensitivity to probe 21 cm emission from HI in haloes in the nearby Universe ($z < 0.1$), but they lack the needed resolution, while radio interferometers have the required resolution, but they lack the necessary dynamic range. Given these limitations, direct observation of local Lyα emission (the primary coolant for the IGM) from structures in the cosmic web would be extraordinarily interesting. The UV is inaccessible from the ground, but direct imaging of Lyα is already one of the core motivations for the proposed French/Chinese *MESSIER* satellite. Could something similar be achieved from the ground?

Detecting Hα emission from the IGM using a ground-based telescope would be a huge challenge, because only $\sim$5% of Lyα photons ultimately wind up being reemitted as Hα photons. Simulations (Lokhorst et al., 2017, Imaging the cosmic web, private communication) suggest that small telescope arrays could target three aspects of the IGM/CGM: (1) the fluorescent "skin" of local "dark" HI clouds, (2) extended haloes analogous to the extended Lyα haloes/blobs recently detected around Lyman break galaxies at high redshifts and (3) emission from filaments of the IGM itself. Detection of the first two of these would be highly interesting and would likely be achievable with the current 48-lens Dragonfly array using long integration times (tens of hours). The third component poses a much greater challenge. Figure 10.10, taken from Lokhorst et al. (2017, Imaging the cosmic web, private communication), presents the predicted surface brightness of Lyα from the diffuse gas and dense gas clumps in the Bertone and Schaye (2012) hydrodynamical simulation of the $z = 2$ IGM. We use these simulations to model the corresponding distribution of Hα signal-to-noise ratio and surface brightness for various regions in the simulation. Assuming qualitatively similar structures and surface brightnesses in the local Universe and assuming the required very long integration times can be undertaken without hitting a "wall" of systematic errors, direct imaging of dense clumps of the IGM is within the capability of the existing 48-lens Dragonfly imaging

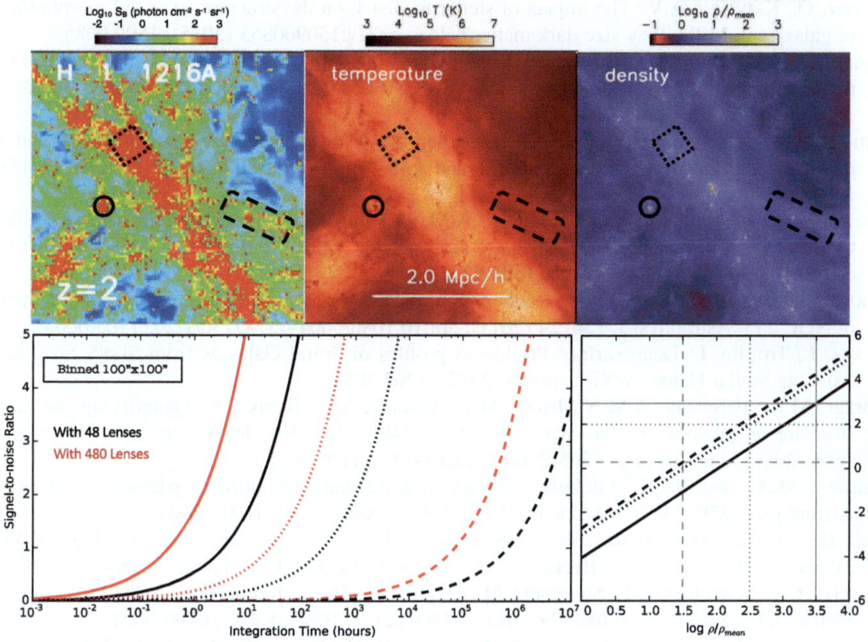

Fig. 10.10 *Top*: the simulated Lyα surface brightness, temperature and density maps from Figs. 2 and 4 of Bertone and Schaye (2012), showing the intersection of IGM filaments. The side of the box is 4 comoving Mpc h^{-1}. Regions corresponding to $\log_{10}(T) \sim 6$ and $\log_{10}(T) \sim 4$–5 are indicated by *dashed boxes and solid circles*, respectively. *Bottom left*: the SNR as a function of exposure time for the surface brightness of Hα corresponding to the *boxed and circled regions* in the *upper panel*. Regions have been binned to 100 × 100 arcsec boxes, and *red and black lines* denote the existing Dragonfly (*red*) and a system with 10× the current number of lenses. *Bottom right*: the Hα surface brightness as a function of the hydrogen density in units of the mean density of the gas (Taken from Lokhorst et al., 2017, Imaging the cosmic web, private communication, reproduced with permission)

in Hα when operating in a highly binned low-resolution mode, but investigation of the dense IGM at high resolution, or of more representative emission from the IGM, would require a 500-lens Dragonfly. With such an instrument, one could take the exploration of galactic outskirts to its logical conclusion, by pushing out beyond galactic haloes and into the cosmic web.

References

Abadi, M.G., Navarro, J.F., Steinmetz, M.: Stars beyond galaxies: the origin of extended luminous haloes around galaxies. Mon. Not. R. Astron. Soc. **365**, 747–758 (2006). doi:10.1111/j.1365-2966.2005.09789.x, arXiv:astro-ph/0506659

Abraham, R.G., van Dokkum, P.G.: Ultra-low surface brightness imaging with the dragonfly telephoto array. Publ. Astron. Soc. Pac. **126**, 55–69 (2014). doi:10.1086/674875, 1401.5473

Agertz, O., Kravtsov, A.V.: The impact of stellar feedback on the structure, size and morphology of galaxies in Milky Way size dark matter haloes. arXiv:150900853 (2015), 1509.00853

Amorisco, N.C., Loeb, A.: Ultradiffuse galaxies: the high-spin tail of the abundant dwarf galaxy population. Mon. Not. R. Astron. Soc. **459**, L51–L55 (2016). doi:10.1093/mnrasl/slw055, 1603.00463

Amorisco, N.C., Martinez-Delgado, D., Schedler, J.: A dwarf galaxy's transformation and a massive galaxy's edge: autopsy of kill and killer in NGC 1097. ArXiv e-prints (2015), 1504.03697

Atkinson, A.M., Abraham, R.G., Ferguson, A.M.N.: Faint tidal features in Galaxies within the Canada-France-Hawaii telescope legacy survey wide fields. Astrophys. J. **765**, 28 (2013). doi:10.1088/0004-637X/765/1/28, 1301.4275

Bailin, J., Bell, E.F., Chappell, S.N., Radburn-Smith, D.J., de Jong, R.S.: The resolved stellar halo of NGC 253. Astrophys. J. **736**, 24 (2011). doi:10.1088/0004-637X/736/1/24, 1105.0005

Bakos, J., Trujillo, I.: Deep surface brightness profiles of Spiral Galaxies from SDSS Stripe82: touching Stellar Halos. ArXiv e-prints (2012), 1204.3082

Barker, M.K., Ferguson, A.M.N., Irwin, M.J., Arimoto, N., Jablonka, P.: Quantifying the faint structure of galaxies: the late-type spiral NGC 2403. Mon. Not. R. Astron. Soc. **419**, 1489–1506 (2012). doi:10.1111/j.1365-2966.2011.19814.x, 1109.2625

Beasley, M.A., Trujillo, I.: Globular clusters indicate that ultra-diffuse galaxies are dwarfs. Astrophys. J. **830**, 23 (2016). doi:10.3847/0004-637X/830/1/23, 1604.08024

Bell, E.F., Zucker, D.B., Belokurov, V., Sharma, S., Johnston, K.V., Bullock, J.S., Hogg, D.W., Jahnke, K., de Jong, J.T.A., Beers, T.C., Evans, N.W., Grebel, E.K., Ivezić, Ž., Koposov, S.E., Rix, H.W., Schneider, D.P., Steinmetz, M., Zolotov, A.: The accretion origin of the Milky Way's Stellar Halo. Astrophys. J. **680**, 295–311 (2008). doi:10.1086/588032, 0706.0004

Belokurov, V., Evans, N.W., Irwin, M.J., Lynden-Bell, D., Yanny, B., Vidrih, S., Gilmore, G., Seabroke, G., Zucker, D.B., Wilkinson, M.I., Hewett, P.C., Bramich, D.M., Fellhauer, M., Newberg, H.J., Wyse, R.F.G., Beers, T.C., Bell, E.F., Barentine, J.C., Brinkmann, J., Cole, N., Pan, K., York, D.G.: An orphan in the "field of streams". Astrophys. J. **658**, 337–344 (2007). doi:10.1086/511302. arXiv:astro-ph/0605705

Bernstein, R.A.: The optical extragalactic background light: revisions and further comments. Astrophys. J. **666**, 663–673 (2007). doi:10.1086/519824

Bertone, S., Schaye, J.: Rest-frame ultraviolet line emission from the intergalactic medium at $2 \leq z \leq 5$. Mon. Not. R. Astron. Soc. **419**, 780–798 (2012). doi:10.1111/j.1365-2966.2011.19742.x, 1008.1791

Carlberg, R.G.: Star stream folding by Dark Galactic Subhalos. Astrophys. J. **705**, L223–L226 (2009). doi:10.1088/0004-637X/705/2/L223, 0908.4345

Carollo, D., Beers, T.C., Lee, Y.S., Chiba, M., Norris, J.E., Wilhelm, R., Sivarani, T., Marsteller, B., Munn, J.A., Bailer-Jones, C.A.L., Fiorentin, P.R., York, D.G.: Two stellar components in the halo of the Milky Way. Nature **450**, 1020–1025 (2007). doi:10.1038/nature06460, 0706.3005

Carollo, D., Beers, T.C., Chiba, M., Norris, J.E., Freeman, K.C., Lee, Y.S., Ivezić, Ž., Rockosi, C.M., Yanny, B.: Structure and kinematics of the Stellar Halos and thick disks of the Milky Way based on calibration stars from Sloan Digital Sky Survey DR7. Astrophys. J. **712**, 692–727 (2010). doi:10.1088/0004-637X/712/1/692, 0909.3019

Collins, M.L.M., Chapman, S.C., Rich, R.M., Ibata, R.A., Martin, N.F., Irwin, M.J., Bate, N.F., Lewis, G.F., Peñarrubia, J., Arimoto, N., Casey, C.M., Ferguson, A.M.N., Koch, A., McConnachie, A.W., Tanvir, N.: The masses of local group dwarf spheroidal galaxies: the death of the universal mass profile. Astrophys. J. **783**, 7 (2014). doi:10.1088/0004-637X/783/1/7, 1309.3053

Cooper, A.P., Cole, S., Frenk, C.S., White, S.D.M., Helly, J., Benson, A.J., De Lucia, G., Helmi, A., Jenkins, A., Navarro, J.F., Springel, V., Wang, J.: Galactic stellar haloes in the CDM model. Mon. Not. R. Astron. Soc. **406**, 744–766 (2010). doi:10.1111/j.1365-2966.2010.16740.x, 0910.3211

Cooper, A.P., D'Souza, R., Kauffmann, G., Wang, J., Boylan-Kolchin, M., Guo, Q., Frenk, C.S., White, S.D.M.: Galactic accretion and the outer structure of galaxies in the CDM model. ArXiv e-prints (2013), 1303.6283

Cooper, A.P., Parry, O.H., Lowing, B., Cole, S., Frenk, C.: Formation of in situ stellar haloes in Milky Way-mass galaxies. Mon. Not. R. Astron. Soc. **454**, 3185–3199 (2015). doi:10.1093/mnras/stv2057, 1501.04630

Courteau, S., Widrow, L.M., McDonald, M., Guhathakurta, P., Gilbert, K.M., Zhu, Y., Beaton, R.L., Majewski, S.R.: The luminosity profile and structural parameters of the andromeda galaxy. Astrophys. J. **739**, 20 (2011). doi:10.1088/0004-637X/739/1/20, 1106.3564

DeVore, J.G., Kristl, J.A., Rappaport, S.A.: Retrieving cirrus microphysical properties from stellar aureoles. J. Geophys. Res. Atmos. **118**(11), 5679–5697 (2013). doi:10.1002/jgrd.50440

Duc, P.A.: Using deep images and simulations to trace collisional debris around massive galaxies. ArXiv e-prints (2016), 1604.08364

Duc, P.A., Cuillandre, J.C., Karabal, E., Cappellari, M., Alatalo, K., Blitz, L., Bournaud, F., Bureau, M., Crocker, A.F., Davies, R.L., Davis, T.A., de Zeeuw, P.T., Emsellem, E., Khochfar, S., Krajnović, D., Kuntschner, H., McDermid, R.M., Michel-Dansac, L., Morganti, R., Naab, T., Oosterloo, T., Paudel, S., Sarzi, M., Scott, N., Serra, P., Weijmans, A.M., Young, L.M.: The ATLAS^{3D} project - XXIX. The new look of early-type galaxies and surrounding fields disclosed by extremely deep optical images. Mon. Not. R. Astron. Soc. **446**, 120–143 (2015). doi:10.1093/mnras/stu2019, 1410.0981

Elmegreen, B.G., Hunter, D.A.: Radial profiles of Star formation in the far outer regions of galaxy disks. Astrophys. J. **636**, 712–720 (2006). doi:10.1086/498082, astro-ph/0509190

Ferguson, A.M.N., Irwin, M.J., Ibata, R.A., Lewis, G.F., Tanvir, N.R.: Evidence for stellar substructure in the halo and outer disk of M31. Astron. J. **124**, 1452–1463 (2002). doi:10.1086/342019, astro-ph/0205530

Font, A.S., McCarthy, I.G., Crain, R.A., Theuns, T., Schaye, J., Wiersma, R.P.C., Dalla Vecchia, C.: Cosmological simulations of the formation of the stellar haloes around disc galaxies. Mon. Not. R. Astron. Soc. **416**, 2802–2820 (2011). doi:10.1111/j.1365-2966.2011.19227.x, 1102.2526

Gilbert, K.M., Guhathakurta, P., Beaton, R.L., Bullock, J., Geha, M.C., Kalirai, J.S., Kirby, E.N., Majewski, S.R., Ostheimer, J.C., Patterson, R.J., Tollerud, E.J., Tanaka, M., Chiba, M.: Global properties of M31's Stellar Halo from the SPLASH survey. I. Surface brightness profile. Astrophys. J. **760**, 76 (2012). doi:10.1088/0004-637X/760/1/76, 1210.3362

Greggio, L., Rejkuba, M., Gonzalez, O.A., Arnaboldi, M., Iodice, E., Irwin, M., Neeser, M.J., Emerson, J.: A panoramic VISTA of the stellar halo of NGC 253. Astron. Astrophys. **562**, A73 (2014). doi:10.1051/0004-6361/201322759, 1401.1665

Hamilton, D.P., Skrutskie, M.F., Verbiscer, A.J., Masci, F.J.: Small particles dominate Saturn's Phoebe ring to surprisingly large distances. Nature **522**, 185–187 (2015). doi:10.1038/nature14476

Harris, W.E., Harris, G.L.H., Alessi, M.: A catalog of globular cluster systems: what determines the size of a Galaxy's globular cluster population? Astrophys. J. **772**, 82 (2013). doi:10.1088/0004-637X/772/2/82, 1306.2247

Harris, W.E., Harris, G.L., Hudson, M.J.: Dark matter halos in galaxies and globular cluster populations. II. Metallicity and morphology. Astrophys. J. **806**, 36 (2015). doi:10.1088/0004-637X/806/1/36, 1504.03199

Hudson, M.J., Harris, G.L., Harris, W.E.: Dark matter halos in galaxies and globular cluster populations. Astrophys. J. **787**, L5 (2014). doi:10.1088/2041-8205/787/1/L5, 1404.1920

Ibata, R., Lewis, G.F., Irwin, M., Totten, E., Quinn, T.: Great circle tidal streams: evidence for a nearly spherical massive dark halo around the Milky Way. Astrophys. J. **551**, 294–311 (2001)

Ibata, R., Martin, N.F., Irwin, M., Chapman, S., Ferguson, A.M.N., Lewis, G.F., McConnachie, A.W.: The haunted halos of Andromeda and Triangulum: a panorama of galaxy formation in action. Astrophys. J. **671**, 1591–1623 (2007). doi:10.1086/522574, 0704.1318

Irwin, M.J., Ferguson, A.M.N., Ibata, R.A., Lewis, G.F., Tanvir, N.R.: A minor-axis surface brightness profile for M31. Astrophys. J. **628**, L105–L108 (2005). doi:10.1086/432718, astro-ph/0505077

Johnston, K.V., Bullock, J.S., Sharma, S., Font, A., Robertson, B.E., Leitner, S.N.: Tracing galaxy formation with stellar halos. II. Relating substructure in phase and abundance space to accretion histories. Astrophys. J. **689**, 936–957 (2008). doi:10.1086/592228, 0807.3911

King, I.R.: The profile of a star image. Publ. Astron. Soc. Pac. **83**, 199 (1971). doi:10.1086/129100

Koda, J., Yagi, M., Yamanoi, H., Komiyama, Y.: Approximately a thousand ultra-diffuse galaxies in the coma cluster. Astrophys. J. **807**, L2 (2015). doi:10.1088/2041-8205/807/1/L2, 1506.01712

Kormendy, J., Bahcall, J.N.: Faint envelopes of galaxies. Astron. J. **79**, 671–677 (1974). doi:10.1086/111595

Majewski, S.R., Skrutskie, M.F., Weinberg, M.D., Ostheimer, J.C.: A two micron all sky survey view of the sagittarius dwarf galaxy. I. Morphology of the sagittarius core and tidal arms. Astrophys. J. **599**, 1082–1115 (2003). doi:10.1086/379504, astro-ph/0304198

Makarov, D., Karachentsev, I.: Galaxy groups and clouds in the local ($z \sim 0.01$) Universe. Mon. Not. R. Astron. Soc. **412**, 2498–2520 (2011). doi:10.1111/j.1365-2966.2010.18071.x, 1011.6277

Martínez-Delgado, D., Pohlen, M., Gabany, R.J., Majewski, S.R., Peñarrubia, J., Palma, C.: Discovery of a giant stellar tidal stream around the disk galaxy NGC 4013. Astrophys. J. **692**, 955–963 (2009). doi:10.1088/0004-637X/692/2/955, 0812.3219

Martínez-Delgado, D., Gabany, R.J., Crawford, K., Zibetti, S., Majewski, S.R., Rix, H.W., Fliri, J., Carballo-Bello, J.A., Bardalez-Gagliuffi, D.C., Peñarrubia, J., Chonis, T.S., Madore, B., Trujillo, I., Schirmer, M., McDavid, D.A.: Stellar tidal streams in spiral galaxies of the local volume: a pilot survey with modest aperture telescopes. Astron. J. **140**, 962–967 (2010). doi:10.1088/0004-6256/140/4/962, 1003.4860

Martínez-Delgado, D., Romanowsky, A.J., Gabany, R.J., Annibali, F., Arnold, J.A., Fliri, J., Zibetti, S., van der Marel, R.P., Rix, H.W., Chonis, T.S., Carballo-Bello, J.A., Aloisi, A., Macciò, A.V., Gallego-Laborda, J., Brodie, J.P., Merrifield, M.R.: Dwarfs gobbling dwarfs: a stellar tidal stream around NGC 4449 and hierarchical galaxy formation on small scales. Astrophys. J. **748**, L24 (2012). doi:10.1088/2041-8205/748/2/L24, 1112.2154

Martínez-Delgado, D., Läsker, R., Sharina, M., Toloba, E., Fliri, J., Beaton, R., Valls-Gabaud, D., Karachentsev, I.D., Chonis, T.S., Grebel, E.K., Forbes, D.A., Romanowsky, A.J., Gallego-Laborda, J., Teuwen, K., Gómez-Flechoso, M.A., Wang, J., Guhathakurta, P., Kaisin, S., Ho, N.: Discovery of an ultra-diffuse galaxy in the Pisces–Perseus supercluster. Astron. J. **151**, 96 (2016). doi:10.3847/0004-6256/151/4/96, 1601.06960

McConnachie, A.W., Irwin, M.J., Ibata, R.A., Dubinski, J., Widrow, L.M., Martin, N.F., Côté, P., Dotter, A.L., Navarro, J.F., Ferguson, A.M.N., Puzia, T.H., Lewis, G.F., Babul, A., Barmby, P., Bienaymé, O., Chapman, S.C., Cockcroft, R., Collins, M.L.M., Fardal, M.A., Harris, W.E., Huxor, A., Mackey, A.D., Peñarrubia, J., Rich, R.M., Richer, H.B., Siebert, A., Tanvir, N., Valls-Gabaud, D., Venn, K.A.: The remnants of galaxy formation from a panoramic survey of the region around M31. Nature **461**, 66–69 (2009). doi:10.1038/nature08327, 0909.0398

Meinel, A.B.: An overview of the technological possibilities of future telescopes. In: Pacini, F., Richter, W., Wilson, R.N. (eds.) Optical Telescopes of the Future, pp. 13–26. Southern European Observatory, Geneva (1978)

Merritt, A., van Dokkum, P., Abraham, R.: The discovery of seven extremely low surface brightness galaxies in the field of the nearby spiral galaxy M101. Astrophys. J. **787**, L37 (2014). doi:10.1088/2041-8205/787/2/L37, 1406.2315

Merritt, A., van Dokkum, P., Abraham, R., Zhang, J.: The dragonfly nearby galaxies survey. I. Substantial variation in the diffuse stellar halos around spiral galaxies. Astrophys. J. **830**, 62 (2016). doi:10.3847/0004-637X/830/2/62, 1606.08847

Mihos, J.C., Durrell, P.R., Ferrarese, L., Feldmeier, J.J., Côté, P., Peng, E.W., Harding, P., Liu, C., Gwyn, S., Cuillandre, J.C.: Galaxies at the extremes: ultra-diffuse galaxies in the virgo cluster. Astrophys. J. **809**, L21 (2015). doi:10.1088/2041-8205/809/2/L21, 1507.02270

Milgrom, M.: Ultra-diffuse cluster galaxies as key to the MOND cluster conundrum. Mon. Not. R. Astron. Soc. **454**, 3810–3815 (2015). doi:10.1093/mnras/stv2202. 1508.04001

Monachesi, A., Bell, E.F., Radburn-Smith, D.J., de Jong, R.S., Bailin, J., Holwerda, B., Streich, D.: Resolving the stellar halos of six massive disk galaxies beyond the Local Group. In:

Bragaglia, A., Arnaboldi, M., Rejkuba, M., Romano, D. (eds.) The General Assembly of Galaxy Halos: Structure, Origin and Evolution. IAU Symposium, vol. 317, pp. 222–227 (2016). doi:10.1017/S1743921315008558, http://adsabs.harvard.edu/abs/2016IAUS..317..222M

Montes, M., Trujillo, I.: Intracluster light at the frontier: A2744. Astrophys. J. **794**, 137 (2014). doi:10.1088/0004-637X/794/2/137, 1405.2070

Mouhcine, M., Rejkuba, M., Ibata, R.: The stellar halo of the edge-on galaxy NGC 891. Mon. Not. R. Astron. Soc. **381**, 873–880 (2007). doi:10.1111/j.1365-2966.2007.12291.x

Oschmann, J.M. (ed.): The scaling relationship between telescope cost and aperture size for very large telescopes. In: SPIE Conference Series, vol. 5489 (2004). doi:10.1117/12.552181

Peng, E.W., Lim, S.: A rich globular cluster system in Dragonfly 17: are ultra-diffuse galaxies pure stellar halos? Astrophys. J. **822**, L31 (2016). doi:10.3847/2041-8205/822/2/L31, 1604.07496

Pillepich, A., Vogelsberger, M., Deason, A., Rodriguez-Gomez, V., Genel, S., Nelson, D., Torrey, P., Sales, L.V., Marinacci, F., Springel, V., Sijacki, D., Hernquist, L.:Halo mass and assembly history exposed in the faint outskirts: the stellar and dark matter haloes of Illustris galaxies. Mon. Not. R. Astron. Soc. **444**, 237–249 (2014). doi:10.1093/mnras/stu1408, 1406.1174

Pillepich, A., Madau, P., Mayer, L.: Building late-type spiral galaxies by in-situ and ex-situ star formation. Astrophys. J. **799**, 184 (2015). doi:10.1088/0004-637X/799/2/184, 1407.7855

Pohlen, M., Trujillo, I.: The structure of galactic disks. Studying late-type spiral galaxies using SDSS. Astron. Astrophys. **454**, 759–772 (2006). doi:10.1051/0004-6361:20064883, astro-ph/0603682

Purcell, C.W., Bullock, J.S., Zentner, A.R.: Shredded galaxies as the source of diffuse intrahalo light on varying scales. Astrophys. J. **666**, 20–33 (2007). doi:10.1086/519787, astro-ph/0703004

Putman, M.E., Peek, J.E.G., Joung, M.R.: Gaseous galaxy halos. Ann. Rev. Astron. Astrophys. **50**, 491–529 (2012). doi:10.1146/annurev-astro-081811-125612, 1207.4837

Racine, R.: The telescope point spread function. Publ. Astron. Soc. Pac. **108**, 699 (1996). doi:10.1086/133788

Radburn-Smith, D.J., de Jong, R.S., Seth, A.C., Bailin, J., Bell, E.F., Brown, T.M., Bullock, J.S., Courteau, S., Dalcanton, J.J., Ferguson, H.C., Goudfrooij, P., Holfeltz, S., Holwerda, B.W., Purcell, C., Sick, J., Streich, D., Vlajic, M., Zucker, D.B.: The GHOSTS survey. I. Hubble space telescope advanced camera for surveys data. Astrophys. J. Suppl. Ser. **195**, 18 (2011). doi:10.1088/0067-0049/195/2/18

Rest, A., Prieto, J.L., Walborn, N.R., Smith, N., Bianco, F.B., Chornock, R., Welch, D.L., Howell, D.A., Huber, M.E., Foley, R.J., Fong, W., Sinnott, B., Bond, H.E., Smith, R.C., Toledo, I., Minniti, D., Mandel, K.: Light echoes reveal an unexpectedly cool η Carinae during its nineteenth-century Great Eruption. Nature **482**, 375–378 (2012). doi:10.1038/nature10775, 1112.2210

Richardson, J.C., Ferguson, A.M.N., Johnson, R.A., Irwin, M.J., Tanvir, N.R., Faria, D.C., Ibata, R.A., Johnston, K.V., Lewis, G.F., McConnachie, A.W., Chapman, S.C.: The nature and origin of substructure in the outskirts of M31. I. Surveying the stellar content with the hubble space telescope advanced camera for surveys. Astron. J. **135**, 1998–2012 (2008). doi:10.1088/0004-6256/135/6/1998, 0803.2614

Roman, J., Trujillo, I.: The spatial distribution of ultra diffuse galaxies within large scale structures. ArXiv e-prints (2016), 1603.03494

Sandin, C.: The influence of diffuse scattered light. I. The PSF and its role in observations of the edge-on galaxy NGC 5907. Astron. Astrophys. **567**, A97 (2014). doi:10.1051/0004-6361/201423429, 1406.5508

Sandin, C.: The influence of diffuse scattered light. II. Observations of galaxy haloes and thick discs and hosts of blue compact galaxies. Astron. Astrophys. **577**, A106 (2015). doi:10.1051/0004-6361/201425168, 1502.07244

Seth, A., de Jong, R., Dalcanton, J., GHOSTS Team: Detection of a stellar halo in NGC 4244. In: Vazdekis, A., Peletier, R. (eds.) Stellar Populations as Building Blocks of Galaxies, IAU Symposium, vol. 241, pp. 523–524 (2007). doi:10.1017/S1743921307009003, astro-ph/0701704

Simon, J.D., Blitz, L., Cole, A.A., Weinberg, M.D., Cohen, M.: The cosmological significance of high-velocity cloud complex H. Astrophys. J. **640**, 270–281 (2006). doi:10.1086/499914, astro-ph/0511542

Slater, C.T., Harding, P., Mihos, J.C.: Removing internal reflections from deep imaging data sets. Publ. Astron. Soc. Pac. **121**, 1267–1278 (2009). doi:10.1086/648457, 0909.3320

Streich, D., de Jong, R.S.: Vertical structure of stellar populations in galaxy disks. Highlights Astron. **16**, 343–343 (2015). doi:10.1017/S1743921314011156

Tal, T., van Dokkum, P.G.: The faint stellar halos of massive red galaxies from stacks of more than 42,000 SDSS LRG images. Astrophys. J. **731**, 89 (2011). doi:10.1088/0004-637X/731/2/89, 1102.4330

Thilker, D.A., Bianchi, L., Meurer, G., Gil de Paz, A., Boissier, S., Madore, B.F., Boselli, A., Ferguson, A.M.N., Muñoz-Mateos, J.C., Madsen, G.J., Hameed, S., Overzier, R.A., Forster, K., Friedman, P.G., Martin, D.C., Morrissey, P., Neff, S.G., Schiminovich, D., Seibert, M., Small, T., Wyder, T.K., Donas, J., Heckman, T.M., Lee, Y.W., Milliard, B., Rich, R.M., Szalay, A.S., Welsh, B.Y., Yi, S.K.: A search for extended ultraviolet disk (XUV-Disk) galaxies in the local universe. Astrophys. J. Suppl. Ser. **173**, 538–571 (2007). doi:10.1086/523853, 0712.3555

Trujillo, I., Fliri, J.: Beyond 31 mag arcsec^2: The frontier of low surface brightness imaging with the largest optical telescopes. Astrophys. J. **823**, 123 (2016). doi:10.3847/0004-637X/823/2/123, 1510.04696

van der Burg, R.F.J., Muzzin, A., Hoekstra, H.: The abundance and spatial distribution of ultra-diffuse galaxies in nearby galaxy clusters. Astron. Astrophys. **590**, A20 (2016). doi:10.1051/0004-6361/201628222, 1602.00002

van Dokkum, P.G., Abraham, R., Merritt, A.: First results from the dragonfly telephoto array: the apparent lack of a stellar halo in the massive spiral galaxy M101. Astrophys. J. **782**, L24 (2014). doi:10.1088/2041-8205/782/2/L24, 1401.5467

van Dokkum, P.G., Abraham, R., Merritt, A., Zhang, J., Geha, M., Conroy, C.: Forty-seven Milky Way-sized, extremely diffuse galaxies in the coma cluster. Astrophys. J. **798**, L45 (2015a). doi:10.1088/2041-8205/798/2/L45, 1410.8141

van Dokkum, P.G., Romanowsky, A.J., Abraham, R., Brodie, J.P., Conroy, C., Geha, M., Merritt, A., Villaume, A., Zhang, J.: Spectroscopic confirmation of the existence of large, diffuse galaxies in the coma cluster. Astrophys. J. **804**, L26 (2015b). doi:10.1088/2041-8205/804/1/L26, 1504.03320

van Dokkum, P., Abraham, R., Brodie, J., Conroy, C., Danieli, S., Merritt, A., Mowla, L., Romanowsky, A., Zhang, J.: A high stellar velocity dispersion and ~100 globular clusters for the ultra-diffuse galaxy dragonfly 44. Astrophys. J. **828**, L6 (2016). doi:10.3847/2041-8205/828/1/L6, 1606.06291

Watkins, A.E., Mihos, J.C., Harding, P., Feldmeier, J.J.: Searching for diffuse light in the M96 galaxy group. Astrophys. J. **791**, 38 (2014). doi:10.1088/0004-637X/791/1/38, 1406.6982

Yamazaki, R., Loeb, A.: Optical inverse-Compton emission from clusters of galaxies. Mon. Not. R. Astron. Soc. **453**, 1990–1998 (2015). doi:10.1093/mnras/stv1757, 1506.07414

Yozin, C., Bekki, K.: The quenching and survival of ultra diffuse galaxies in the Coma cluster. Mon. Not. R. Astron. Soc. **452**, 937–943 (2015). doi:10.1093/mnras/stv1073, 1507.05161

Zhang, J., Abraham, R.G., van Dokkum, P.G., Merritt, A.: A giant stellar disk in NGC 2841. Astrophys. J. (2017, submitted)

Index

J.H. Knapen et al. (eds.), *Outskirts of Galaxies*, Astrophysics and Space Science Library 434, DOI 10.1007/978-3-319-56570-5

MIX
Papier aus verantwortungsvollen Quellen
Paper from responsible sources
FSC® C105338

If you have any concerns about our products,
you can contact us on
ProductSafety@springernature.com

In case Publisher is established outside the EU,
the EU authorized representative is:
Springer Nature Customer Service Center GmbH
Europaplatz 3, 69115 Heidelberg, Germany

Printed by Libri Plureos GmbH
in Hamburg, Germany